Fundamentals of
THE FUNGI

Second Edition

Fundamentals of
THE FUNGI

Elizabeth Moore-Landecker

Glassboro State College

PRENTICE-HALL, INC., *Englewood Cliffs, New Jersey 07632*

Library of Congress Cataloging in Publication Data

MOORE-LANDECKER, ELIZABETH.
  Fundamentals of the fungi.

  Includes bibliographies and indexes.
  1. Mycology. 2. Fungi. I. Title. [DNLM: 1. Fungi.
QK 603 M825f]
QK603.M62 1982   589.2   81-21149
ISBN 0-13-339200-7   AACR2

© 1982, 1972 by Prentice-Hall, Inc., Englewood Cliffs, New Jersey 07632

Printed in the United States of America

10 9 8 7 6 5 4 3 2 1

Editorial-production supervision by Paul Spencer
Cover design by Diane Saxe
Interior design by Paul Spencer
Manufacturing buyer: John Hall

ISBN 0-13-339200-7

Prentice-Hall International, Inc., *London*
Prentice-Hall of Australia Pty. Limited, *Sydney*
Prentice-Hall of Canada, Ltd., *Toronto*
Prentice-Hall of India Private Limited, *New Delhi*
Prentice-Hall of Japan, Inc., *Tokyo*
Prentice-Hall of Southeast Asia Pte. Ltd., *Singapore*
Whitehall Books Limited, *Wellington, New Zealand*

To

RICHARD P. KORF

for his inspiration as a teacher
and for his continued friendship

Contents

Part One

**MORPHOLOGY
AND TAXONOMY**

Preface
to the Second Edition

Like the first edition, this book is intended to serve as a textbook for students who are first being introduced to mycology. The classification system used is an adaptation of that proposed by G. C. Ainsworth (1973, *The Fungi—An Advanced Treatise*, 4 A: 1-7). In this edition, the section ''Morphology and Taxonomy'' has been greatly expanded. Several additional orders, specific fungi as examples, and the Trichomycetes have been included. Increased attention has been given to many aspects of fungal ultrastructure, morphology, development, and evolution. Examples of this expanded coverage include zoosporogenesis and zoospore structure, centrum development, and evolution of the Zygomycetes. The sections ''Physiology and Reproduction'' and ''Ecology and Utilization by Humans'' have been updated and some new material has been added. New material includes sections on chitin synthesis and aflatoxins, as well as recent developments in some topics considered in the first edition (for example, hormonal control and mycorrhizae). The chapter summaries that appeared in the first edition have been deleted to save space. Otherwise, there have not been any substantial deletions or condensations.

Numerous individuals have given invaluable assistance in the preparation of this textbook. Dr. Lekh Batra is thanked for the loan of many of his personal reference materials and for assistance in obtaining photographs. The staff at the library in the Academy of Natural Sciences in Philadelphia were most helpful during my work there. Many individuals contributed photographs or permission to use published illustrations, and are acknowledged in the figure legends accompanying the illustrations. Drs. Lekh Batra, Harold Burdsall, Garry C. Cole, Terry W. Johnson, Jr., and anonymous reviewers read part or all of the manuscript and provided constructive comments. The following people assisted in typing portions of the manuscript: Miss Barbara Ennis, Miss Dawn Hobbs, Miss Mary Ott, and Mrs. Louise Willis. All of these individuals are gratefully thanked for their aid.

Turnersville, New Jersey   E.M.-L.

Preface
to the First Edition

The science of mycology had modest beginnings in the eighteenth century, often as a Sunday hobby for physicians. Now fungi are the concern of the taxonomist, morphologist, geneticist, ecologist, plant pathologist, physician, biochemist, and commercial microbiologist. This list reflects the ubiquity of the fungi, their usefulness as research organisms, and their involvement in many facets of our everyday lives.

This book provides a broad introduction to the field of mycology; it explores the morphology, taxonomy, physiology, ecology, and commercial utilization of the fungi. The level of the book is suitable for the individual who has a basic understanding of biology and chemistry. It is hoped that this book will enrich the background of those who will not study mycology further and that it will serve as a background for those who will study mycology more intensively.

I would like to acknowledge the assistance of many people who made this book possible. Dr. George G. Kent of the Plant Pathology Department at Cornell University very generously made photographs from that department available for use in this book. Mr. Howard Lyon, also of that department, prepared many of the photographs in this book. Dr. William C. Denison and other professional colleagues who read part or all of the manuscript made many helpful suggestions. Many individuals, companies, and publishers provided photographs or permission to use published materials (these are individually acknowledged in the text). LeMoyne College provided financial assistance for part of the preparation of the manuscript in the form of a Faculty Research Grant. Mrs. Ina Taylor, Mrs. Florence Barnard, and Mrs. Joyce Marks typed portions of the manuscript. Finally, my husband, Peter, provided constant help and encouragement. All these individuals and concerns are gratefully thanked.

Part One

MORPHOLOGY AND TAXONOMY

ONE

An Introduction to the Fungi

Laymen are generally familiar with many fungi such as mushrooms in forests or fields, bracket fungi on the side of decaying logs or standing trees, puffballs which send forth clouds of spores when kicked by a playful child, and perhaps colorful cup fungi on logs or soil. Homemakers use yeast to leaven their bread, and brewers use this same organism to form alcohol in beer, wine, or whiskey. Perhaps farmers drown their sorrows with this alcoholic brew after their crop is destroyed by a devastating fungus disease, such as corn smut or black stem rust of wheat. The city dweller who may not be familiar with fungi in their natural habitats may be waging a battle with ringworm or athlete's foot, fungus diseases of humans. Perhaps this same city dweller is involved in manufacture of antibiotics from fungi, of fungus-processed cheeses, or of drugs which are chemically modified and perfected by chemical activities of fungi. Meanwhile, a tribal rite may be taking place in Mexico where hallucinogenic mushrooms are consumed and vivid dreams are experienced. Many fungi rarely come to the attention of the layperson. These include microscopic inhabitants of almost every habitat, including those fungi which trap nematodes (roundworms). Some fungi eject their spores over a precise trajectory and with a precision force, much as a missile is precisely fired into space. They may become partners with algae, insects, or higher plants and enter into complex biological relationships with these other organisms.

What are the fungi, and how do they differ from other organisms?

Although the fungi are a diverse group of organisms, a feature which they have in common with each other is their mode of nutrition. Like animals, fungi are *heterotrophic* organisms which must consume preformed organic matter. They may live as *saprophytes*, which digest and consume dead plants or animals, their parts, or their wastes. Alternatively, fungi may live as *parasites* and assimilate tissues of living plants

3

and animals. In all cases, digestive enzymes are liberated from the fungal cell into the immediate environment, where food molecules are simplified and the nutrients pass into the fungal cell as a watery solution. Nutrition is discussed in greater detail in Chapters 7 and 11.

Other heterotrophic organisms include bacteria and animals. How does a fungus differ from these organisms? Bacteria lack a nucleus, while fungi possess a nucleus similar to that in higher plants and animals. Animals, unlike fungi, ingest large masses of food which are then digested within the organism. In addition, the typical fungus cell is bounded by a rigid wall, unlike animal cells.

CELLS, HYPHAE, AND TISSUES

Fungi may exist as single-celled organisms. If this is the case, then the single cell must accomplish everything from assimilating nutrients to reproducing. Single-celled fungi or fungi which form a thallus of only a few cells may be found among the chytrids (Chapter 2), and the yeasts and their allies (Chapter 4).

The vast majority of the fungi are multicellular organisms. Like other multicellular organisms, they typically have a subdivision of roles for their different structural parts. Part of the thallus is buried within the food source. There it obtains nutrients that are transported to the remainder of the thallus, where reproduction may be occurring. In most fungi, the assimilative portion of the thallus is an extensive *mycelium,* a loosely organized mass of elongated cellular threads that permeate the substratum (such as the soil, a diseased plant, or a dead tree) and is not visible to the naked eye. Each individual thread of the mycelium is a *hypha*. When the mycelial fungi reproduce, a specific portion of the thallus differentiates and functions in either a nonsexual or sexual mode. The reproductive structures do not assimilate nutrients but rather are nourished by the assimilative mycelium. The reproductive structures may vary in size from microscopic to massive and visible to the naked eye.

Macroscopic fungal structures are sometimes formed; these usually achieve their large structure because they are at least partially composed of tissues. These tissues may consist of hyphae that have aggregated but retain their typical elongated appearance, or they may range to types in which the hyphae have formed rounded, more or less isodiametric cells. This latter type of tissue resembles the parenchyma of higher plants and is often called *pseudoparenchyma* (Fig. 1-1). Macroscopic fungal structures may include specialized resistant bodies such as *sclerotia* (discussed further on pages 159), as well as reproductive structures. Unlike the vegetative hyphae, these macroscopic structures are not involved in nutrient assimilation.

An important aspect of the organization of mycelial fungi is that there must be a mechanism for transport of nutrients and materials from one part of the fungus to another. For example, it is important that nutrients be transported from that part of the fungus which is buried in the food source to that part which is reproducing, or from the matured portions of the mycelium to the growing hyphal tips. Cytoplasmic streaming occurs, but unlike cyclical movement in the higher plants it is unidirectional. Immediately behind the advancing hyphal tip, movement of the cytoplasm is generally slow and uniform, with a velocity about 1 to $1\frac{1}{2}$ times the rate of advance of the hyphal apex. Further back in the mycelium, the cytoplasm may appear to be stationary or may have relatively rapid movement (up to 40 millimeters/hour), from an area of mycelium which either is being evacuated or is associated with nuclear migration (Isaac, 1964). The nuclei themselves may move through the hyphae, and in one

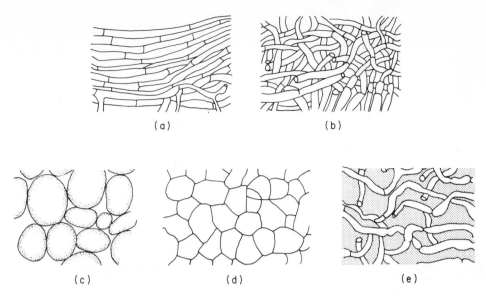

Fig. 1-1. Representative fungal tissues. One type may merge into another, and modifications of the types shown here may occur: (a) Tissue consisting of hyphae that remain distinct and are more or less parallel to each other; (b) Tissue consisting of intertwined hyphae; (c) Tissue consisting of rounded cells with intercellular spaces; (d) Tissue with cells that are angular owing to mutual pressure and lacking intercellular spaces; (e) Tissue consisting of hyphae interspersed through a watery gel (shaded).

fungus their speed of movement was found to be 4 to 5 millimeters/hour, a rate 2 to 3 times that at which the hyphal tip was advancing. This movement occurred independently of the cytoplasmic streaming (Dowding and Buller, 1940).

Important considerations are the cellular organization of the hyphae, the structure of cells, and the mode of nuclear division. These topics follow.

Hyphae and Septa

The cells that comprise the hyphae are elongated and are typically binucleate or multinucleate. One cell in the hyphae is separated from another by a cross wall, or *septum*. There are some interesting differences in septal structure among the fungi. Some members of the Division Mastigomycota (Chapter 2) form *pseudosepta,* septa that are perforated by so many pores that they are sievelike. In many fungi, the septum is perforated only by a single pore or a few pores, which are small but nevertheless allow nuclei and cytoplasm to pass from cell to cell. Such fungi include all or most members of the Trichomycetes and Ascomycotina, many members of the Deuteromycotina, and some members of the Basidiomycotina (in the orders Uredinales and Tremellales). (These particular fungi are explored further in Chapters 3, 4, 6, and 5, respectively.)

In the Ascomycotina, the septum is relatively simple and its wall tapers gradually and evenly towards the pore, forming a wedge in cross section, and a *Woronin body* occurs adjacent to it. A Woronin body is a rounded organelle that is surrounded by a double unit membrane and has granular contents. While the hypha is healthy, the Woronin body remains in its usual position in the cytoplasm adjacent to a pore

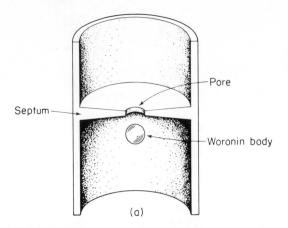

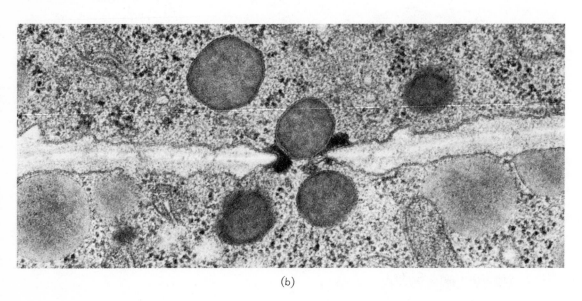

Fig. 1-2. Ascomycete-type septum. (a) Diagrammatic interpretation of a simple septum together with a Woronin body; (b) Woronin bodies in the vicinity of the septal pore, × 55,500. [(a) Adapted from R. T. Moore and J. H. McAlear, 1962. *Am. J. Botany* **49**:86–94. (b) Courtesy T. M. Hammill.]

(Fig. 1-2), but when a hyphal cell ages or becomes damaged, the Woronin body moves into the pore and becomes a plug. This plug effectively separates the cytoplasm of the aged or damaged cell from the cytoplasm of cells that are still healthy.

In most members of the Basidiomycotina and some members of the Deuteromycotina (Chapters 5 and 6), the septum is more complex than those described above. This more complex structure is often called a *dolipore septum*. Near the center of the septum there is an enlarged ring, which surrounds the septal pore and resembles a doughnut in its overall form. The most distinctive feature of the septum is the formation of a thickened, porous, electron-dense *septal pore cap* covering the septal swelling and septal pore. This cap is formed from the endoplasmic reticulum lying close to and parallel to the cross walls on both sides (Fig. 1-3) and may differ slightly in structure from one part of the fungus to another (Wells, 1978). Cytoplasm and small organelles can pass through the septum from cell to cell, but nuclear migration presumably can occur with some modification of the pores. In *Rhizoctonia solani,* the nuclei and other organelles constrict as they pass through the discontinuities in the septal pore cap and subsequently through the septal pore, which enlarges to accommodate their passage (Bracker and Butler, 1964). Alternatively the dolipore septum may break down to leave a simple septum (with no thickenings or caps), through which nuclei subsequently migrate from cell to cell (Giesey and Day, 1965). Blockage of the septum may occur if the central ring expands to close the pore entirely or if some electron-dense material plugs the openings (Wells, 1978).

An important general characteristic of hyphae is that they either are nonseptate or consist of cells which are incompletely separated from each other. The perforated septa allow cytoplasmic continuity from cell to cell, which facilitates cytoplasmic streaming from one cell to another or from one part of the hypha to another. The phenotype of a given cell not only is under the control of its included nuclei but also is affected by the control exerted on the cytoplasm in neighboring cells. Because of the perforated septa, the hypha functions essentially as does a *coenocyte* (a multinucleate, elongated cell) in other organisms regardless of whether the nuclei occur only in uninucleate or binucleate cells. Even those septa that appear to be nonperforated (this is common in those septa that separate reproductive structures from the vegetative mycelium) may be traversed by *plasmodesmata,* thin cytoplasmic strands.

Fig. 1-3. (a) Electronmicrograph of a dolipore septal pore with cap, × 11,500. (b) Diagrammatic interpretation. Key: C, septal pore cap; CD, discontinuity in septal pore cap; ER, endoplasmic reticulum; S, septal swelling; SP, septum. [Courtesy C. E. Bracker and E. E. Butler, 1963, *Mycologia* **55**:35–58.]

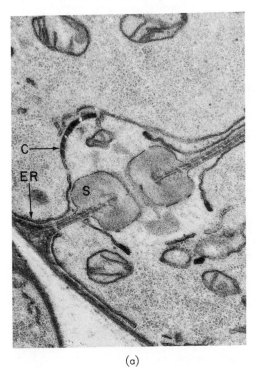

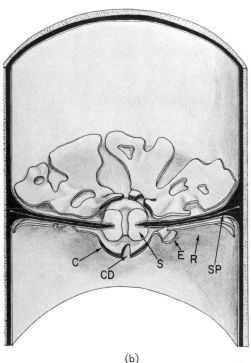

(a)   (b)

Cell Structure

What is a "typical" fungal cell like? Fungal cells are basically similar to those in higher plants and animals, although there are some important differences. The general features of fungal cells will be described below. These characteristics are shared by the vast majority of cells, whether they are isolated single cells, cells within a hypha, or nonspecialized reproductive cells. There are some specialized features, such as the structure of and formation of certain spore types, which will be considered in subsequent chapters.

Cell walls of fungi are typically comparatively thin in vegetative cells, where the cell walls are usually only about 0.2 microns thick, but they may be thicker in some specialized cells or spores. Major constituents of the walls include a variety of polysaccharides, which usually represent about 80% to 90% of the wall. Other important constituents include protein, which is often conjugated with the polysaccharides, and a small amount of lipid. The most important polysaccharides are chitin and cellulose, both polymers of glucose (Fig. 1-4). The majority of the fungi have chitin but not cellulose in their walls, while in a few fungi the reverse is true. The chitin or cellulose occurs in microfibrils, which are intertwined and embedded in an amorphous matrix which cements them together (Fig. 1-5), thereby providing a skeletal framework which gives the walls their morphological form. The matrix material contains protein and other polysaccharides such as glucans or mannans, which are polymers of glucose and mannose, respectively. By using specialized techniques it may be determined that the wall typically consists of two or more layers. The skeletal framework of chitin or cellulose microfibrils typically occurs in a single layer, while the remaining layers appear to be homogenous and consist primarily of amorphous proteins and/or carbohydrates (Aronson, 1965; Bartnicki-Garcia, 1968; Burnett, 1979). In a few cases the glucans may be organized into a coarse reticulated pattern or may occur as a mosaic of short rodlets.

Fig. 1-4. Major components of the cell walls of fungi.

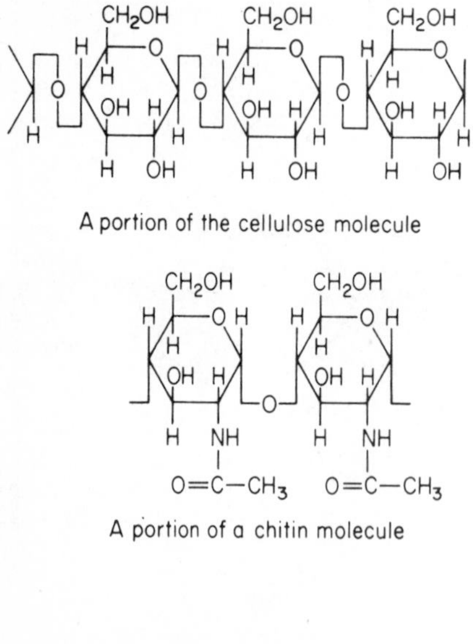

A portion of the cellulose molecule

A portion of a chitin molecule

Fig. 1-5. Microfibrils from the internal surface of a fungal cell wall, × 55,000. [From S. Bartnicki-Garcia and E. Reyes, 1968, *Biochim. Biophys. Acta* **165**:32-42.]

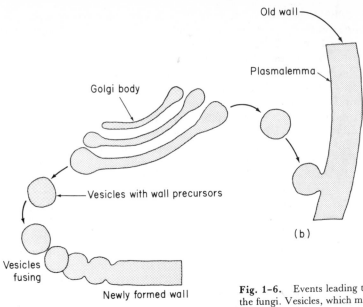

Old wall

Plasmalemma

Golgi body

Vesicles with wall precursors

(b)

Vesicles
fusing

Newly formed wall

(a)

Fig. 1-6. Events leading to wall formation or enlargement in the fungi. Vesicles, which may originate from a Golgi apparatus as diagrammed here, may fuse to form a new wall (sequence *a*), or may empty their contents between the plasmalemma and existing wall by fusing with the plasmalemma (sequence *b*).

During the life of the fungus, walls of vegetative cells such as hyphal tips grow by extension, new septa may form in hyphal cells or in sexual cells, new walls may be initiated when spores are formed endogenously within a cell, or walls may thicken. All these processes have been observed to involve *vesicle* formation in some fungi that have been carefully studied. Vesicles are small bodies in the cytoplasm that appear bubblelike and are surrounded by a single unit membrane. They have been found in unusually high numbers in those parts of the cell where wall growth or deposition is occurring. In electronmicrographs some of these vesicles appear to be emptying their contents between the plasmalemma and cell wall or to be merging (Fig. 1-6). Some vesicles contain enzymes that will soften the existing wall, making it pliable and capable of extension, while other vesicles contain wall precursors. Extremely small vesicles may contain chitin synthetase, an enzyme that assists in the assembly of a microfibril of chitin from its precursors and is active where the wall is being constructed (Bartnicki-Garcia et al., 1979). Vesicles that appear to be involved in wall formation have not been found in all fungi examined for their presence (Heath, 1976).

Inside the cell wall the protoplast is surrounded by a *plasmalemma,* which is generally similar to that in other organisms. A major structural feature of the protoplast is the *endoplasmic reticulum.* The endoplasmic reticulum of the fungi, unlike that of the higher plants and animals, is usually comparatively sparse. It is a continuation of the nuclear envelope, but continuity of the endoplasmic reticulum with the plasmalemma is either rare or absent in the fungi. Most often the unit membranes of the endoplasmic reticulum are sheetlike, but sometimes they may be in the form of tubules. Frequently the endoplasmic reticulum produces isolated vesicles that appear to form by a pinching off of the membranous segments of the endoplasmic reticulum.

These vesicles may contain materials that were previously within the endoplasmic reticulum and may also transport these materials to other parts of the cell. Some of the vesicles may be involved in wall growth. The morphology of the endoplasmic reticulum may vary with the age and physiological condition of the cell as well as with the role of the cell. The endoplasmic reticulum is usually more abundant in young or metabolically active cells than in aged or inactive cells. In cells which may be secreting a material that is emptied outside of the cell, the endoplasmic reticulum may be a branching system of smooth tubules or it may be sheetlike but dilated and swollen so that the paired membranes are far apart from each other. Dilated endoplasmic reticulum may also occur in cells in which wall deposition is occurring. Closely stacked endoplasmic reticulum does not commonly occur but may be present in some metabolically active cells, such as those involved in secretion or those that are forming wall material.

Ribosomes, similar to those found in higher plants and animals, occur in the cytoplasm of fungal cells. The ribosomes are usually distributed throughout the cytoplasm and are not concentrated along the endoplasmic reticulum (Figs. 1–7 and 1–8). Therefore the endoplasmic reticulum is most often the *smooth* type. *Rough* endoplasmic reticulum, which is marked by an accumulation of ribosomes, occurs infrequently, but occurs in some cells which are especially active metabolically. For example, rough endoplasmic reticulum occurs typically in the growing hyphal tip. Also ribosomes may be observed on the surface of dilated endoplasmic reticulum that is apparently involved in secretion (e.g., Bracker, 1967; Hill, 1977).

A Golgi apparatus, similar to that found in higher plants and animals, has been detected in some fungi although it is not of wide occurrence among the fungi (Figs. 1–6 and 1–8). The Golgi dictyosome consists of a stack of adjacent cisternae, which appear to be pinching off vesicles. Usually 2 to 5 cisternae are found in fungal Golgi dictyosomes. However, in some fungi lacking the stacked cisternae, apparently isolated and dilated cisternae of the endoplasmic reticulum may be morphologically similar to a single Golgi cisterna. In higher plants and animals, these vesicles sometimes carry slimy secretory products to the exterior of the cell, but this function has not been generally attributed to the Golgi apparatus in the fungi, as direct evidence is lacking. In some fungi the Golgi apparatus produces vesicles that are involved in wall formation (e.g., Grove et al., 1970; Heath, 1976) (Fig. 1–6).

The fungal cell may also contain *microbodies* (Fig. 1–8), a special class of small vesicles that are similar to those found in higher plants, where they originate from the endoplasmic reticulum (Frederick et al., 1975). The microbodies are surrounded by a single unit membrane, are usually round or oval in shape, and have somewhat dense contents with a granular matrix and possibly a crystalline core. The microbodies all contain catalase as well as a variety of other enzymes. Some contain enzymes which are involved in fatty acid oxidation and in the glyoxalate cycle (discussed further in Chapter 8) (Heath, 1976; Powell, 1976).

Some vesicles contain the hydrolytic enzyme acid phosphatase as well as other hydrolytic enzymes. Because of their distinctive contents, they are known as *lysosomes*. Lysosomes are often responsible for the controlled breakdown of cellular components. Examples include septum breakdown, which may allow nuclear migration to take place between cells (Wessels and Sietsma, 1979), breakdown of a host cell wall by a parasitic fungus (Armentrout and Wilson, 1969), and breakdown of some spore-bearing sacs (asci), which become an ooze and release the spores (Wilson and Stiers, 1970).

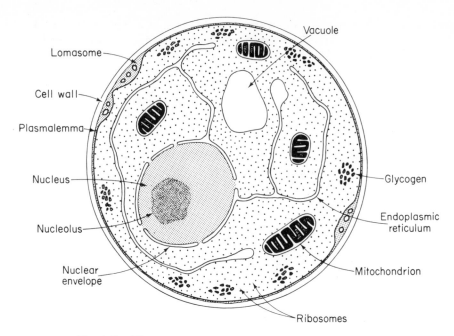

Fig. 1-7. Diagram of a cross section of a generalized fungus cell.

Fig. 1-8. Ultrastructural detail of a portion of a fungal cell. Mb = microbody, Mt = mitochondrion, G = Golgi apparatus, ER = endoplasmic reticulum, and R = ribosomes. × 63,700 [Courtesy I. B. Heath.]

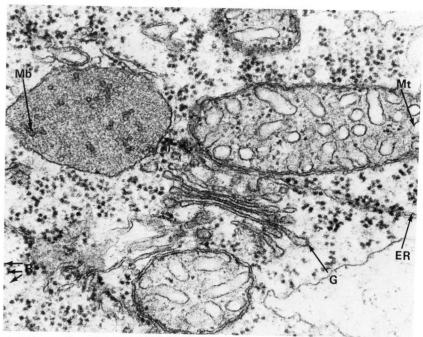

Vacuoles are often apparent in fungal cells (Fig. 1–7) and, like the vesicles, are surrounded by a single unit membrane. There is sometimes no sharp delineation between the use of the terms *vesicle* and *vacuole* for a given structure. Usually the term vacuole is reserved for structures that appear to be conspicuously large and have light contents. The vacuoles may originate from the enlargement and fusion of small vesicles. Vacuoles become increasingly numerous and enlarge in maturing cells. Fungal vacuoles do not, however, achieve the prominence of the large vacuoles in mature plants. The vacuolar contents are aqueous and may include amino acids, pigments, or hydrolytic enzymes. In the "inky cap" mushroom, in which the gills undergo autolysis and drip away as an ooze (Fig. 5–43), the hydrolytic enzyme chitinase has been found in the vacuoles and is apparently responsible for the breakdown of the gills (Iten and Matile, 1970).

Mitochondria occur in the cytoplasm of fungal cells (Figs. 1–7 and 1–8). Fungal mitochondria are generally similar to those which occur in other organisms. They usually possess platelike cristae. Some yeast cells may possess a single, highly branched mitochondrion.

An unusual feature of fungal cells is the widespread occurrence of membranous configurations that are sometimes elaborate. These include the *lomasomes,* membranous configurations associated with the fungal cell wall (Fig. 1–9). They consist of membranous tubules, vesicles, or parallel sheets which occur in a matrix between the plasmalemma and the cell wall. The matrix may be part of the cell wall itself or of

Fig. 1–9. A complex membranous configuration (mc) containing both tubules and vesicles is adjacent to the fungal cell wall (w). Wall material is apparently permeating the spaces between the vesicles so that they become embedded within the wall as a lomasome (l). × 112,700. [From I. B. Heath and A. D. Greenwood, 1970, *General Microbiology* **62**:129–137.]

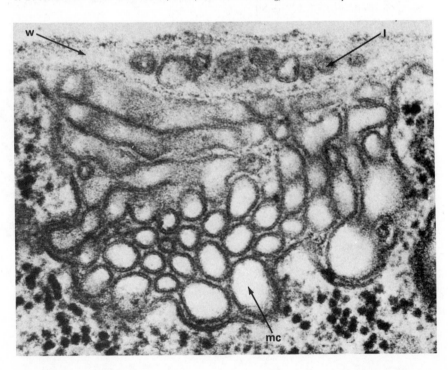

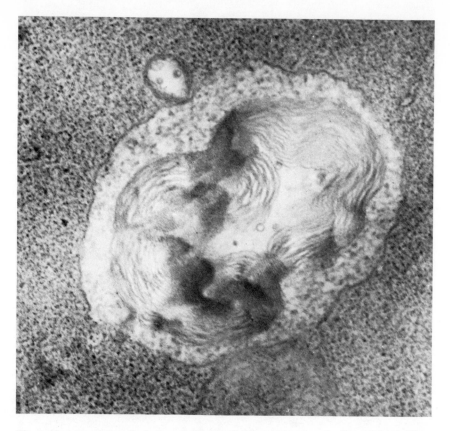

Fig. 1-10. A complex whorl of myelinlike membranes within a fungal cell. × 120,000. [From E. Moore-Landecker, 1971, *Cytologia* **4**:563-574.]

some other composition. The lomasomes may occur singly or in groups. Usually the plasmalemma assumes a domelike configuration over the lomasomes, but it may protrude inward to form a pouch. Sometimes the membranes within the lomasome may be continuous with that of the plasmalemma. Membranous configurations which resemble the lomasomes are frequently observed in the cytoplasm. Some of these membranous configurations appear to be attached to the plasmalemma, while others appear to be isolated within the cytoplasmic matrix or within vacuoles. Many of these have concentric whorls of membranes, which resemble myelin figures (Fig. 1-10) and are sometimes associated with an osmiophilic material. The significance of these various membrane configurations is not known and could vary according to the species or physiological condition of the fungus involved. At least some of the lomasomes and membranous configurations appear to be normal constituents of the cell (Heath and Greenwood, 1970; Marchant and Moore, 1973) and, because of their frequent association with the walls of young cells, have had various roles suggested for them which relate to the normal development of either the plasmalemma or the cell wall (Duncan, 1974; Marchant and Moore, 1973; Weisberg and Turian, 1974).

An Introduction to the Fungi   **13**

NUCLEI AND NUCLEAR DIVISION

Fungal nuclei are very small and are near the limit of resolution with the light microscope. In spite of their small size, they share many similarities with nuclei of other organisms and have a nucleolus, nucleoplasm, and nuclear envelope. Mitotic division is unique, however, and usually differs markedly from that observed in higher organisms. Futhermore, details of mitosis vary from fungus to fungus.

One of the most conspicuous features of mitosis is that the nuclear envelope usually does not disappear but instead constricts in a dumbbell-like fashion and eventually separates into daughter nuclei (Fig. 1–11). In some species, there may be extremely large gaps in the nuclear envelope. Similarly, the nucleolus is usually retained during mitosis and may be stretched out and divided between the daughter nuclei. In some fungi, the nucleolus either is cast out of the nucleus, disperses and disappears within the nucleus, or disperses within the nucleus but remains detectable during mitosis.

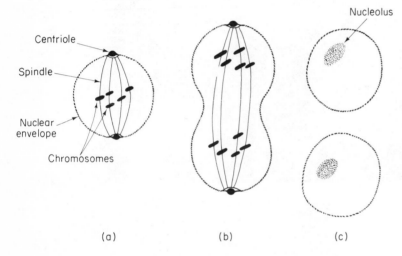

Fig. 1–11. Some stages in fungal mitosis: (a) Metaphase; (b) Telophase; (c) Interphase.

Centrioles or *spindle pole bodies,* may be associated with the nucleus during interphase. Centrioles contain distinct triplets of microtubules, while the spindle pole bodies lack microtubule triplets but rather are platelike, globular, or of some other form. During nuclear division, the centrioles or spindle pole bodies will replicate and migrate to opposite poles of the nucleus, will lie next to or within the nuclear envelope, and will organize the microtubules comprising the spindle. Spindle formation may be initiated from the centrioles or spindle pole bodies either while they are migrating or after they have reached their final position at opposite poles of the nucleus. The spindle typically lies within the nuclear envelope. Some of the microtubules of the spindle are attached to chromosomes. The chromosomes are small, and often they are not observed at all or appear only as patches of chromatin. In a few fungi, the chromosomes are aligned at metaphase on an equatorial plate, as in other organisms. However, in most fungi the chromosomes are randomly distributed during metaphase and therefore do not form an equatorial plate (Figs. 1–11 and 1–12).

Meiosis conforms more closely to the norm for other organisms than mitosis does. In meiosis both the nucleolus and nuclear envelope degenerate, usually near the end of prophase, although the nucleolus may degenerate later. As in mitosis, the chromosomes are usually irregularly distributed on the spindle at metaphase, but rarely they may be aligned on an equatorial plate. Sometimes the chromosomes are so small and poorly defined even in meiosis that studies of cytology may be extremely difficult. This has led to some confusion in understanding certain fungal life cycles if the time of meiosis cannot be precisely determined by conventional cytological techniques. Other methods, such as quantitative analysis of DNA, may be required to pinpoint the place or time of meiosis.

In addition to their role in nuclear division, centrioles are important in motile cells formed by members of the Division Mastigomycota. This topic is discussed further on page 31.

Fig. 1-12. Early metaphase of mitosis in *Saprolegnia ferax*. The centrioles (at the large arrows) lie outside the nuclear envelope at the spindle poles. Some of the spindle fibers terminate at kinetochores and their chromatin (small arrows). The bar equals 1 micron. [From I. B. Heath, 1978, *in:* I. B. Heath, Ed., *Nuclear division in the fungi,* Academic Press, New York, pp. 89–176.]

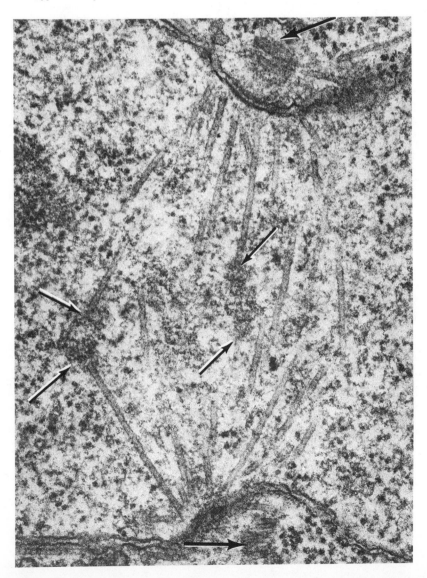

REPRODUCTION

Sexual reproduction may involve the union of hyphae, motile gametes, differentiated multinucleate male and female organs *(gametangia),* or a female gametangium with a motile or nonmotile male gamete. All these structures have haploid *(N)* nuclei. Typically these nuclei are of a single type, and the gamete or gametangium is *monokaryotic.* When the male and female elements unite, their protoplasts mingle. This stage in which union of the cytoplasm takes place is termed *plasmogamy.* There are now two nuclear types within the same cell, so that cell is *dikaryotic,* but the nuclei are still haploid. The nuclear state of such a dikaryon may be designated as *(N + N).* A dikaryotic phase occurs in probably all organisms in which sexual reproduction takes place, but usually this stage is very transient. In most fungi, however, the dikaryotic phase occupies the major part of the life cycle and is prolonged for several cell generations. Eventually, the two haploid nuclei in a dikaryotic cell fuse (undergo *karyogamy),* and a diploid *(2N)* zygote is formed. Either that diploid zygote or its diploid progeny undergoes meiosis to restore the haploid state. Although fungal sexual reproductive cycles typically progress through the haploid, dikaryotic, and diploid phases, the relative length of each varies with the species (Fig. 1–13).

Some fungi do not produce either gametes or gametangia, and it is often not known whether or not plasmogamy between hyphae has actually taken place. In many of these fungi, karyogamy and meiosis do take place just as in other fungi which produce conspicuous gametes or gametangia and in which plasmogamy is known to occur. Sexual reproduction is said to have taken place in such fungi, even in the apparent absence of sexual organs or plasmogamy. The use of the term sexual reproduction in conjunction with fungi implies only that karyogamy and meiosis occur.

Not all fungi reproduce sexually and some must rely entirely upon *nonsexual reproduction.* Nonsexual reproduction may take place during the haploid, dikaryotic, or diploid phases of the sexual cycle or in organisms which apparently lack a sexual cycle. Unlike sexual reproduction, karyogamy does not take place at a precise stage in the life cycle (it may not occur at all), and meiosis never takes place in nonsexual reproduction. This topic is discussed further in Chapter 9.

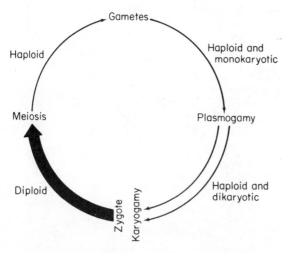

Fig. 1–13. Succession of events and nuclear stages in a typical sexual reproductive cycle in fungi. The relative lengths of these stages vary from fungus to fungus. Gametes and plasmogamy may be absent, and a true dikaryon may not be produced if different nuclear types are not brought together in plasmogamy.

Fig. 1–14. Section of a spore of a terrestrial fungus. The endospore (En) appears as a broad light band next to the cytoplasm, and it is bordered by the epispore (E) which appears as a thinner and darker band. Both are surrounded by the perispore, which is a thick ribbed band. Organelles within the cytoplasm are the nuclei (N), and mitochondria (M). × 5,600. [Courtesy R. J. Lowry and A. S. Sussman. From A. S. Sussman and H. O. Halvorson, 1966, *Spores—their dormancy and germination,* Harper and Row, Publishers, New York.]

    A typical reproductive unit in the fungi is the *spore,* which may contain one or more nuclei derived from either sexual or nonsexual reproduction. A spore is a structure that can be separated from its parent thallus and disseminated. Spores produced by aquatic fungi are usually capable of swimming (Chapter 2), while those produced by terrestrial fungi are surrounded by a thick wall (thicker than that of the hyphae). This wall may be composed of an outer coat, the *epispore,* and an inner coat, the *endospore,* which surrounds the protoplasm. The entire spore may be surrounded by a *perispore,* an external layer(s) that can be removed without impairing the functions of the spore (Fig. 1–14). Cytological characteristics shared both by terrestrial and aquatic spores are the conspicuous scarcity of the endoplasmic reticulum and fewer and larger mitochondria as compared with the assimilative or vegetative cells. Nuclei, vacuoles, and other inclusions (particularly lipid droplets) are present within the cytoplasm of the spore. Unlike the seeds produced by higher plants, spores do not contain an embryo. When the spore germinates, it gives rise to a unicellular protoplast or alternatively to a *germ tube* that grows and differentiates into a new thallus.

A variety of nonsexual spore types are formed by the fungi and each type will be described more completely in subsequent chapters. Briefly, swimming *zoospores* (Chapter 2) and nonmotile *sporangiospores* are formed endogenously within a saclike cell. The *conidium* is an extremely common type of nonsexual spore, is usually formed apically or laterally upon its sporogenous cell, and is readily deciduous at maturity. In some fungi, the hyphae may fragment into sporelike cells, the *oidia,* or cells of the hyphae may round up and become thick-walled, forming a *chlamydospore* (Chapters 4, 5, and 6). Both oidia and chlamydospores are considered to be specialized types of conidia, a topic which is discussed further in Chapter 6. The different types of sexual spores will be introduced in the appropriate chapters.

Spores may be produced upon or within hyphae similar to the vegetative hyphae or within isolated single cells. In many of the more complex fungi, multicellular reproductive bodies are formed and bear the spores. These reproductive bodies are separated spatially from the mycelium that assimilates nutrients and often occur above the surface of the substratum. A relatively large multicellular structure bearing spores is often called a *sporocarp.* An example of a sporocarp with which you might be familiar is the colorful mushroom that you might see on a hike through the forest. The visible mushroom is a sporocarp because it bears spores, but it is nourished by the mycelium, which is invisible to the naked eye and is buried in the soil.

CLASSIFICATION

The intent of any biological classification scheme is to group closely related species or other taxa above the species level together, and conversely, to separate unrelated species or higher taxa. Classification of fungi is based to a large extent on the particular life cycle involved and the morphology of the spores and sporocarps produced. To a lesser extent, morphology of tissues or other sterile structures may be used. In general the classification systems which we are currently using are based on the classical morphological studies of past centuries. Recently scientists have been utilizing data based on physiology or biochemistry to help determine relationships among the fungi. Such techniques include comparison of the genomes represented by base composition and sequence within DNA, determination of the serological response elicted by fungal proteins when used as antigens, comparison of electrophoretic properties of enzymes and other proteins (Bradford et al., 1975), and comparison of the enzymatic ability to utilize certain carbohydrates or other nutrients. Further, mathematical analyses may be utilized to compare the characters found in different fungi and to analyze the degrees of similarities and differences. These techniques are especially helpful if there is limited morphology to study in the case of single-celled or otherwise simple fungi or if it is difficult to make valid decisions based on morphological evidence alone. The present classification systems are being constantly modified as our knowledge and understanding of the fungi increases. In general, such information supplements rather than replaces taxonomic systems based on morphological studies.

As a group, the fungi have been classified within different kingdoms. These have been the plant kingdom in the older classification schemes, which were based on only two kingdoms (plants and animals), and then later in the Kingdom Protista with other relatively simple organisms. Whittaker (1969) argued that the fungal mode of nutrition, which is based on extracellular digestion and absorption, is fundamentally different from that of autotrophic plants. Further, the eukaryotic structure of the fungal cell also makes the fungi fundamentally different from the prokaryotic bacteria and blue-green algae, while their walled cells make them different from most protozoa. Whittaker devised a five-kingdom classification system in which the fungi are assigned to their own kingdom. This reflects his belief that the primitive protozoan groups gave rise to three evolutionary lines, which are distinguished primarily by their mode of nutrition. These lines are the plants, animals, and fungi.

The slime molds have sometimes been included within the fungi, but these organisms have many characters which are unlike those of true fungi. For example, they have a naked cell or mass of protoplasm which has animallike motility and may creep about. They also feed by engulfing and ingesting their food, as do protozoa. When they reproduce, they form spores with thick walls that may be wind-blown, as do most of the fungi. Although Whittaker (1969) includes these organisms in the Kingdom Fungi, he states, ''The slime molds cross the distinctions of the kingdoms in both nutrition and organization, and offer a free choice of treatment as aberrant fungi, eccentric protists, or very peculiar animals.'' Because these organisms are not clearly related to the vast majority of the fungi, they will not be treated further in this textbook.

We shall classify the fungi in the Kingdom Myceteae (Fungi) and recognize the subdivisions proposed by Ainsworth (1973). Characters important in delimiting the following taxa include the presence or absence of a swimming stage (zoospores) and their mode of sexual reproduction. We shall explore these characters in subsequent chapters. The classification follows:

Kingdom Myceteae
(Nutrition by extracellular digestion and absorption: eukaryotic, vegetative cells typically with a thick wall and usually elongated).

Division Mastigomycota (Forming swimming zoospores. The following classes are based on the morphology of zoospores and are explained in Chapter 2).
Class Chytridiomycetes
Class Hyphochytridiomycetes
Class Plasmodiophoromycetes
Class Oomycetes
Division Amastigomycota (Nonmotile, not forming zoospores)
Subdivision Zygomycotina (forming zygospores as the sexual spore)
Subdivision Ascomycotina (forming ascospores as the sexual spore)
Subdivision Basidiomycotina (forming basidiospores as the sexual spore)
Form-subdivision Deuteromycotina (lacking sexual spores)

References

Ainsworth, G. C. 1973. Introduction and keys to higher taxa. *In:* Ainsworth, G. C., F. K. Sparrow, and A. S. Sussman, Eds., The Fungi—an advanced treatise. Academic Press, New York. 4A:1–7.

Armentrout, V. N., and C. L. Wilson. 1969. Haustorium-host interaction during mycoparasitism of *Mycotypha microspora* by *Piptocephalis virginiana.* Phytopathology 59:897–905.

Aronson, J. M. 1965. The cell wall. *In:* Ainsworth, G. C., and A. S. Sussman, Eds., The fungi—an advanced treatise. Academic Press, New York. 1:49–76.

Barnett, J. A. 1977. The nutritional tests in yeast systematics. J. Gen. Microbiol. 99:183–190.

Bartnicki-Garcia, S. 1968. Cell wall chemistry, morphogenesis, and taxonomy of fungi. Botan. Rev. **22**:87-108

——, J. Ruiz-Herrera, and C. E. Bracker. 1979. Chitosomes and chitin synthesis. *In:* Burnett, J. H. and A. P. J. Trinci, Eds., Fungal walls and hyphal growth, Cambridge Unversity Press, Cambridge, pp. 149-168.

Beckett, A. 1976. Fibrous and tubular inclusions in four *Xylariaceous* fungi. Protoplasma **89**:279-290.

Benedict, G. R. 1970. Chemotaxonomic relationships among the Basidiomycetes. Adv. Appl. Microbiol. **13**:1-23.

Bertoldi, Marco de. 1976. New species of *Humicola:* an approach to genetic and biochemical classification. Can. J. Botany **54**:2755-2768.

Bracker, C. E. 1967. Ultrastructure of fungi. Ann. Rev. Phytopathology **5**:343-374.

——, and E. E. Butler. 1963. The ultrastructure and development of septa in hyphae of *Rhizoctonia solani.* Mycologia **55**:35-58.

——, and E. E. Butler. 1964. Function of the septal pore apparatus in *Rhizoctonia solani.* J. Cell Biol. **21**:152—157.

Bradford, L. S., R. J. Jones, and E. D. Garber. 1975. An electrophoretic survey of fourteen species of the fungal genus *Ustilago.* Botan. Gaz. **136**:109-115.

Burnett, J. H. 1979. Aspects of the structure and growth of hyphal walls. *In:* J. H. Burnett and A. P. J. Trinci, Eds. Fungal walls and hyphal growth. Cambridge University Press, Cambridge. pp. 1-25.

Campbell, I. 1974. Methods of numerical taxonomy for various genera of yeasts. Adv. Appl. Microbiol. **17**:135-156.

Choinski, J. S., and J. T. Mullins. 1977. Ultrastructural and enzymatic evidence for the presence of microbodies in the fungus *Achlya.* Am. J. Botany **64**:593-599.

Clémençon, H. 1975. Ultrastructure of hymenial cells in two Boletes. Nova Hedwigia Beih. **51**:93-98.

Dowding, E. S., and A. H. R. Buller. 1940. Nuclear migration in *Gelasinospora.* Mycologia **32**:471-488.

Duncan, E. J. 1974. Studies on the vegetative hyphae of *Corticium areolatum.* Trans. Brit. Mycol. Soc. **63**:115-120.

Dykstra, M. J. 1974. Some ultrastructural features in the genus *Septobasidium.* Can. J. Botany **52**:971-972.

Flegler, S. L., G. R. Hooper, and W. G. Fields. 1976. Ultrastructural and cytochemical changes in the basidiomycete dolipore septum associated with fruiting. Can. J. Botany **54**:2243-2253.

Fletcher, J. 1973. The distribution of cytoplasmic vesicles, multivesicular bodies and paramural bodies in elongating sporangiophores and swelling sporangia of *Thamnidium elegans* Link. Ann. Botany **37**:955-961.

——, 1973. Ultrastructural changes associated with spore formation in sporangia and sporangiola of *Thamnidium elegans* Link. Ann. Botany **37**:963-971.

Frederick, S. E., F. J. Gruber, and E. H. Newcomb. 1975. Plant microbodies. Protoplasma **84**:1-29.

Fuller, M. S. 1976. Mitosis in fungi. Intern. Rev. Cytol. **45**:113-153.

Giesey, R. M., and P. R. Day. 1965. The septal pores of *Coprinus lagopus* in relation to nuclear migration. Am. J. Botan. **52**:287-293.

Grove, S. N., and C. E. Bracker. 1970. Protoplasmic organization of hyphal tips among fungi: vesicles and spitzenkörper. J. Bacteriol. **104**:989-1009.

——, ——, and D. J. Morré. 1968. Cytomembrane differentiation in the endoplasmic reticulum-Golgi apparatus-vesicle complex. Science **161**:171-173.

——, ——, and ——. 1970. An ultrastructural basis for hyphal tip growth in *Pythium ultimum.* Am. J. Botany **57**:245-266.

Gull, K., and R. J. Newsam. 1975. Ultrastructural organization of cystidia in the basidiomycete *Agrocybe praecox.* J. Gen. Microbiol. **91**:74-78.

Hall, R. 1969. Molecular approaches to taxonomy of fungi. Botan. Rev. **35**:285-304.

Hammill, T. M. 1974. Septal pore structure in *Trichoderma saturnisporum.* Am. J. Botan. **61**:767-771.

Heath, I. B. 1976. Ultrastructure of freshwater Phycomycetes. *In:* Jones, E. B. G., Ed., Recent advances in aquatic mycology. John Wiley & Sons, New York. pp. 603-650.

——, 1978. Experimental studies of mitosis in the fungi. *In:* Heath, I. B., Ed., Nuclear division in the fungi. Academic Press, New York. pp. 89-176.

——, and A. D. Greenwood. 1970. The structure and function of lomasomes. J. Gen. Microbiol. **62**:129-137.

Hemmes, D. E., and S. Bartnicki-Garcia. 1975. Electron microscopy of gametangia interaction and oospore development in *Phytophthora capsici*. Arch. Mikrobiol. **103**:91–112.

Hill, T. W. 1977. Ascocarp ultrastructure of *Herpomyces* sp. (Laboulbeniales) and its phylogenetic implications. Can. J. Botany **55**:2015–2032.

Isaac, P. K. 1964. Cytoplasmic streaming in filamentous fungi. Can. J. Botany. **42**:787–792.

Iten, W., and P. Matile. 1970. Role of chitinase and other lysosomal enzymes of *Coprinus lagopus* in the autolysis of fruiting bodies. J. Gen. Microbiol. **61**:301–309.

Khan, S. R. 1976. Ultrastructure of the septal pore apparatus of *Tremella*. J. Gen. Microbiol. **97**:339–342.

Kubai, D. F. 1978. Mitosis and fungal phylogeny. *In:* Heath, I. B., Ed., Nuclear division in the fungi. Academic Press, New York. pp. 177–229.

Laseter, J. L., W. M. Hess, J. D. Weete, D. L. Stocks, and D. J. Weber. 1968. Chemotaxonomic and ultrastructural studies on three species of *Tilletia* occuring on wheat. Can. J. Microbiol. **14**:1149–1154.

Lawson, J. A., J. W. Harris, and S. K. Ballal. 1975. Application of computer analysis of electrophoretic banding patterns of enzymes to the taxonomy of certain wood-rotting fungi. Econ. Botany **29**:117–125.

Madhosingh, C., and V. R. Wallen. 1968. Serological differentiation of the *Ascochyta* species on peas. Can. J. Microbiol. **14**:449–454.

Marchant, R., and R. T. Moore. 1973. Lomasomes and plasmalemmasomes in fungi. Protoplasma **76**:235–247.

McLaughlin, D. J. 1972. Ultrastructure of sterigma growth and basidiospore formation in *Coprinus* and *Boletus*. Can. J. Botany **51**:145–150.

——. 1974. Ultrastructural localization of carbohydrate in the hymenium and subhymenium of *Coprinus*—evidence for the function of the Golgi apparatus. Protoplasma **82**:341–364.

Mims, C. W., F. Seaburg, and E. L. Thurston. 1975. Fine structure of teliospores of the cedar-apple rust *Gymnosporangium juniperi-virginiae*. Can. J. Botany **53**:544–552.

Mollenhauer, H. H., D. J. Morré, and A. G. Kelley. 1966. The widespread occurrence of plant cytosomes resembling animal microbodies. Protoplasma **62**:44–52.

Moore, R. T. 1965. The ultrastructure of fungal cells. *In:* Ainsworth, G. C., and A. S. Sussman, Eds., The fungi—an advanced treatise. Academic Press, New York. **1**:95–118.

——, and R. Marchant. 1972. Ultrastructural characterization of the basidiomycete septum of *Polyporus biennis*. Can. J. Botany **50**:2463–2469.

Moore-Landecker, E. 1971. Ultrastructural observations on lamellar and tubular membrane configurations in fungi. Cytologia **36**:563–574.

Moss, S. T. 1975. Septal structure in the Trichomycetes with special reference to *Astreptonema gammeri* (Eccrinales). Trans. Brit. Mycol. Soc. **65**:115–127.

Mosse, B. 1970. Honey-coloured, sessile *Endogone* spores III. Wall structure. Arch. Mikrobiol. **74**:146–159.

Nehemiah, J. L. 1973. Localization of acid phosphatase activity in the basidia of *Coprinus micaceus*. J. Bacteriol. **115**:443–446.

Powell, M. J. 1974. Fine structure of plasmodesmata in a chytrid. Mycologia **66**:606–614.

——. 1976. Ultrastructure and isolation of glyoxysomes (microbodies) in zoospores of the fungus *Entophlyctis* sp. *Protoplasma* **89**:1–27.

Price, C. W., G. B. Fuson, and H. J. Phaff. 1978. Genome comparison in yeast systematics: delimitation of species within the genera *Schwanniomyces, Saccharomyces, Debaryomyces,* and *Pichia*. Microbiol. Rev. **42**:161–193.

Raudaskoski, M. 1970. Occurrence of microtubules and microfilaments, and origin of septa in dikaryotic hyphae of *Schizophyllum commune*. Protoplasma **70**:415–422.

Reid, I. D., and S. Bartnicki-Garcia. 1976. Cell-wall composition and structure of yeast cells and conjugation tubes of *Tremella mesenterica*. J. Gen. Microbiol. **96**:35–50.

Robb. J., A. E. Harvey, and M. Shaw. 1973. Ultrastructure of hyphal walls and septa of *Cronartium ribicola* on tissue cultures of *Pinus monticola*. Can. J. Botany **51**:2301–2305.

Robinow, C. F., and A. Bakerspigel. 1965. Somatic nuclei and forms of mitosis in fungi. *In:* Ainsworth, G. C., and A. S. Sussman, Eds., The fungi—an advanced treatise. Academic Press, New York. **1**:119–142.

Shannon, M. C., S. K. Ballal, and J. W. Harris. 1973. Starch gel electrophoresis of enzymes from nine species of *Polyporus*. Am. J. Botany **60**:96–100.

Thielke, C. 1972. Zisternaggregate bei höheren Pilzen. Protoplasma **75**:335–339.

Tyrrell, D. 1969. Biochemical systematics and fungi. Botan. Rev. **35**:305–316.

Weisberg, S. H., and G. Turian. 1974. The membranous type of lomasome (membranosome) in the hyphae of *Aspergillus nidulans*. Protoplasma **79**:377–389.

Wells, K. 1978. The fine structure of septal pore apparatus in the lamellae of *Pholiota terrestris*. Can. J. Botany **56**:2915–2924.

Wergin, W. P. 1973. Development of Woronin bodies from microbodies in *Fusarium oxysporum* f. sp. *lycopersici*. Protoplasma **76**:249–260.

Wessels, J. G. H., and J. H. Sietsma. 1979. Wall structure and growth in *Schizophyllum commune*. *In:* Burnett, J. H., and A. P. J. Trinci, Eds., Fungal walls and hyphal growth. Cambridge University Press,, Cambridge. pp. 27–48.

Whittaker, R. H. 1969. New concepts of kingdoms of organisms. Science **163**:150–161.

Willetts, H. J., R. J. W. Byrde, and A. H. Fielding. 1977. The taxonomy of the brown rot fungi *(Monilinia* spp.) related to their extracellular cell wall-degrading enzymes. J. Gen. Microbiol. **103**:77–83.

Wilson, C. L., and D. L. Stiers. 1970. Fungal lysosomes or sphaerosomes. Phytopathology **60**:216–227.

TWO

Division Mastigomycota

The Division Mastigomycota contains those fungi that produce nonsexual motile *zoospores*. The zoospores help make it possible for these fungi to live in aquatic environments such as oceans or in fresh water bodies, including bogs, ponds, lakes, and streams. Many of these fungi occur in damp soil, where the zoospores can swim in accumulated water.

These fungi may frequently be collected from their natural habitats by "baiting" techniques. Many fungi will grow preferentially on substrata such as fruits, twigs, seeds, feathers, hair, cellophane, pollen grains, and insects. Water, soil, or debris may be collected in the field and placed in small dishes to which a little bait is added. The choice of bait depends on the type of fungus being sought, but half of a boiled hemp seed or a few dead fruit flies are often used. Usually some fungi will colonize the bait within a few days, and then they can be studied further. Some can be directly isolated by the inoculation of agar with soil, water, or other debris. For further information on baiting and culturing, consult Emerson (1958) and Sparrow (1960).

ZOOSPORE MOVEMENT AND STRUCTURE

The zoospores are propelled by one or two *flagella* (Fig. 2–1). A flagellum is a hairlike structure that undulates. The path travelled by zoospores is determined by movement of the flagella. The zoospores may travel in a nearly straight or in a circular path with an even gliding motion without rotations, in either a spiraled or more or less straight

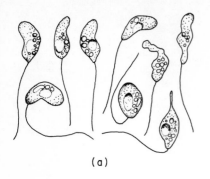

(a)   (b)

Fig. 2-1. (a) Swimming zoospores showing amoeboid movement of zoospore body and flagellum; (b) Diagram of the cross section of a flagellum showing the two central microtubules and nine pairs of microtubules near the periphery. [(a) from J. S. Karling, 1948, *Am. J. Botany* **35**:503–510. (b) from I. R. Gibbons and A. V. Grimstone, 1960, *J. Biophys. Biochem. Cytol.* **7**:697–716.]

path accompanied by clockwise rotations, or in an irregular path accompanied by hopping and darting. Zoospores having a single flagellum swim by transmitting a succession of waves, more or less in a single plane, through that flagellum. When zoospores have two flagella, one flagellum extends anteriorly and the other extends posteriorly while swimming. In this case, each flagellum undulates in a more or less single plane that is different from that of the other flagellum. Undulation is often more prominent in the anterior flagellum, which has led to suggestions that the anterior flagellum actually propels the zoospore.

Structurally zoospores are highly specialized cells and have some components not found in vegetative cells. The membrane surrounding the flagellum is continuous with the plasmalemma of the cell. The flagellum may be smooth (*whiplash type*) or may bear filaments extending laterally from the main axis (*tinsel type*) (Fig. 2-2). Like flagella in higher plants and animals, the flagella of the Division Mastigomycota consist of a core of two central microtubules that are surrounded by nine pairs of microtubules (Fig. 2-1). These microtubules are contractile and are responsible for flagellar movement. Near the point of attachment of the flagellum to the zoospore, the two central microtubules disappear and the outer microtubules extend into the *kinetosome*, a centriolelike body with nine groups of microtubules arranged in triplets and lying near the nucleus (Fig. 2-3). The point of juncture between the flagellum and kinetosome may be marked by a platelike structure, the *terminal plate*. In some zoospores, the kinetosome is connected to the nucleus by a large, fibrous *rhizoplast* consisting of fused microtubules or strands (Fig. 2-5). Structures similar to the rhizoplast, the *striated rootlets*, may be present and extend to other organelles such as the centriole, second kinetosome, or mitochondria. Individual microtubules may also be prominent as part of the rootlet system. Anchoring of the flagellum to cellular structures by rootlets is important, as otherwise the vigorous movement of the flagellum would tear it away from the cell. Uniflagellate zoospores swim more vigorously than biflagellate zoospores, and predictably the flagellum of a uniflagellate zoospore is anchored to a much larger mass of cellular organelles and by a more extensive rootlet system than are the flagella of biflagellate zoospores. This presumably allows the more vigorous swimming of the uniflagellate zoospores (Heath, 1976).

The structures mentioned in the preceding paragraph were related to flagellation. Other organelles are important in the zoospore (Figs. 2-4 and 2-5). Zoospores may contain a single giant mitochondrion enveloping the kinetosome and the smaller end of the nucleus or a number of smaller mitochondria scattered throughout the cytoplasm. Ribosomes may be scattered throughout the cytoplasm, attached to the endoplasmic reticulum, or present as a *nuclear cap*. A nuclear cap is a loose aggregation or membrane-enclosed aggregation of ribosomes surmounting the large end of the nucleus. Lipid, which provides energy to the zoospore, may be dispersed in the cytoplasm in several droplets or present as a single large droplet. If the species possesses Golgi bodies in the vegetative cells, one or two Golgi bodies are normally found in the zoospore. Uniflagellate zoospores may contain a *rumposome,* a highly organized disklike structure of which the outer surface is appressed to the cell membrane while the inner surface is connected to the nuclear cap by two or three lamellae extending through the cytoplasm. The rumposome does not occur in vegetative cells, and its function in the zoospore is unknown. *Contractile* vacuoles that function in osmoregulation by expelling water have been observed in some biflagellate zoospores and may originate as vesicles produced by the Golgi bodies. Some species have a *side body,* a simple pleomorphic organelle which resembles microbodies and is bounded by a double membrane. *Gamma particles* have been observed in zoospores of *Blastocladiella emersonii* (Fig. 2-21), but not of other species. The gamma particle is a cup-shaped organelle consisting of an inner, electron-dense core surrounded by a single unit membrane. DNA, RNA, and the enzyme chitin synthetase have been found in the gamma particles. Chitin synthetase is involved in cell wall synthesis, and it has been suggested that gamma particles may play a role in the synthesis of a new wall when the zoospore encysts (Truesdell and Cantino, 1970).

Fig. 2-2. Portions of a whiplash flagellum (top) and tinsel flagellum (bottom). × 1,250. [Courtesy A. P. Kole and K. Horstra, 1959, *Konikl. Ned. Akad. Wetenschappen, Proc.,* ser. C. **62**:404-408.]

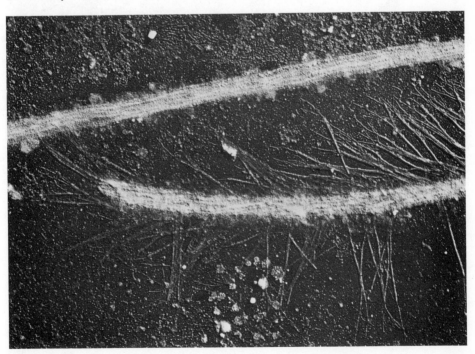

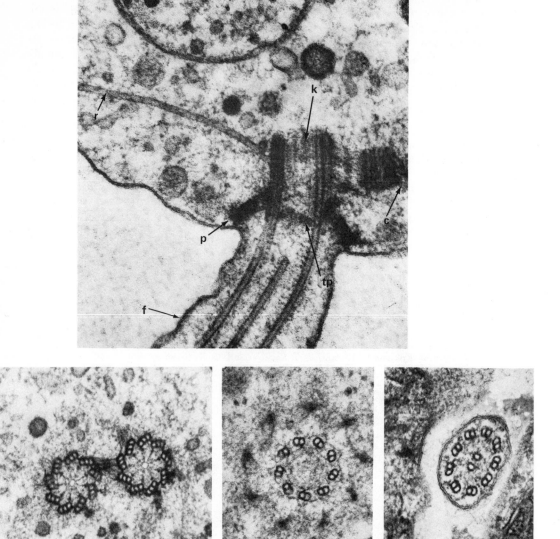

Fig. 2-3. Top: Base of a zoospore and its flagellum. The flagellum (*f*) is arising from the kinetosome (*k*). Note that the central microtubules are visible only within the flagellum and not within the kinetosome. A rootlet (*r*) is visible at the left and extends from the kinetosome to other organelles. Both a terminal plate *(tp)* and props *(p)* are present at the kinetosome base. A nonfunctional centriole *(c)* is present at the right. × 77,000; Bottom: A cross-sectional view of the kinetosome and nonfunctional centriole (left), the flagellum in the vicinity of the props (center), and the flagellum that is exterior to the body of the zoospore (right). Note the transition from the spiral of nine triplets of microtubules in the kinetosome to the ring of nine doublets in the flagellum. [From D. J. S. Barr and V. E. Hadland-Hartmann, 1978, *Can. J. Botany* **57**:887–900.]

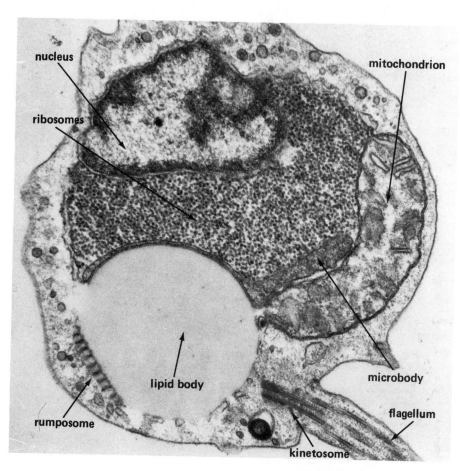

nucleus

mitochondrion

ribosomes

lipid body

microbody

rumposome

flagellum

kinetosome

Fig. 2-4. An electronmicrograph of a zoospore of *Rhizophydium subangulosum* (Chytridiales). This zoospore contains a large single mitochondrion, ribosomes clustered into an area near the nucleus, a pleomorphic microbody, a single large lipid body, and a rumposome adjacent to the lipid body. The kinetosome and flagellum are at the lower right. × 32,400. [Courtesy D. J. S. Barr. From D. J. S. Barr and V. E. Hadland-Hartmann, 1978, *Can. J. Botany* **56**:2380–2404.]

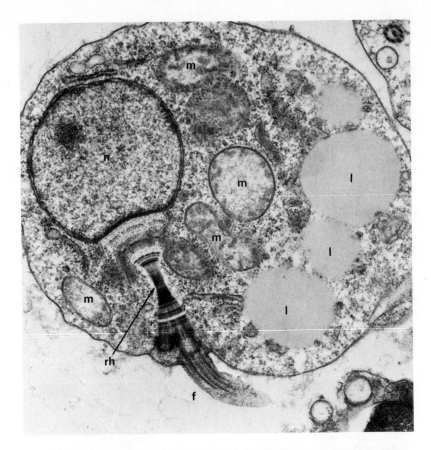

Fig. 2–5. An electronmicrograph of the zoospore of *Rhizophlyctis rosea* (Chytridiales). Compare this zoospore with that in Fig. 2-4 and the flagellar apparatus with that in Fig. 2-3. This zoospore has several small mitochondria (*m*), scattered ribosomes throughout the cytoplasm, several clustered lipid bodies (*l*); it lacks a rumposome and has a conspicuous rhizoplast (*rh*) extending from the nucleus (*n*) to the flagellum (*f*). × 22,000. [Courtesy D. J. S. Barr. From D. J. S. Barr and V. E. Hartmann, 1977, *Can. J. Botany* **55**:1221–1235.]

ZOOSPOROGENESIS

Zoospores are formed within a saclike structure, the *zoosporangium* (Fig. 2-6). Zoosporogenesis has been studied in detail in a number of fungi (for example, Hoch and Mitchell, 1972; Lessie and Lovett, 1968). At the beginning of zoosporogenesis, the zoosporangium contains cytoplasm and various cytoplasmic organelles and is multinucleate. Cleavage of the multinucleate cytoplasm into smaller uninucleate units follows and may be accomplished in one of two ways.

The first method involves the formation of vesicles, which may arise from the Golgi apparatus or possibly from some other source (Fig. 2-7). The vesicles surround a uninucleate portion of the cytoplasm, and the membranes of the vesicles fuse with each other to form a double unit membrane. This membrane becomes the plasma-

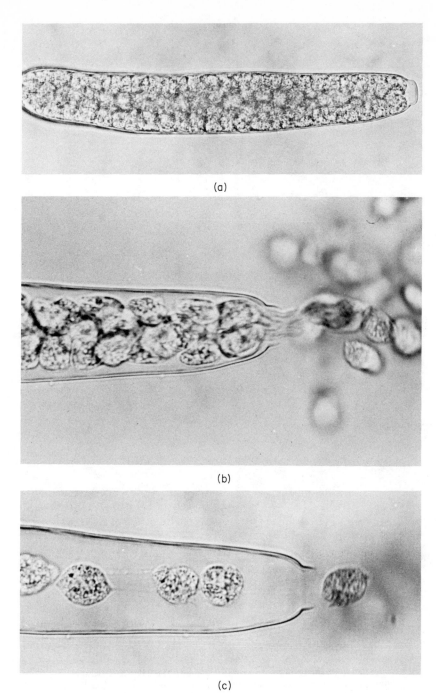

(a)

(b)

(c)

Fig. 2-6. Zoosporangia and zoospores of *Saprolegnia*. (a) Zoospores packing a mature zoosporangium; (b) Zoospores escaping from the zoosporangium through an apical opening; (c) A nearly emptied zoosporangium containing a few zoospores. [Courtesy Carolina Biological Supply Company.]

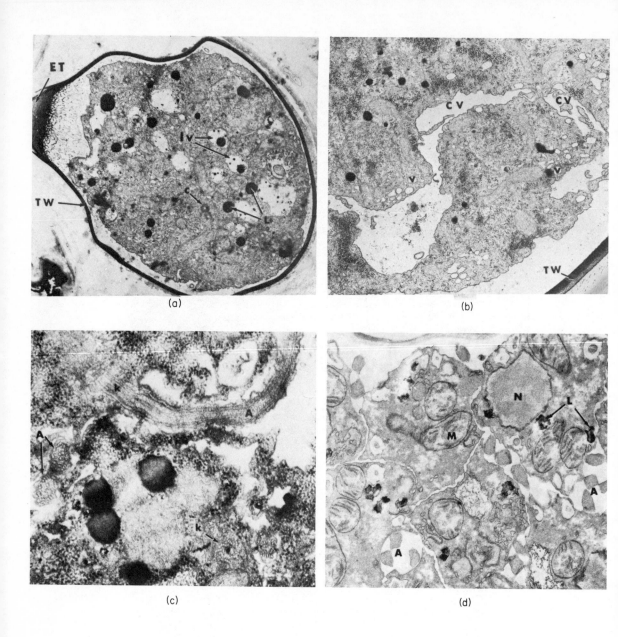

Fig. 2-7. Zoosporogenesis in *Olpidium brassicae*. In this species zoosporogenesis occurs by fusion of vesicles. (a) Zoosporangium before cleavage; (b) Part of zoosporangium showing cleavage of zoospores; (c) Flagellum in cytoplasm of zoosporangium; (d) Part of mature zoosporangium. The flagella are wrapped around the zoospores. Key to labeling: *A* = axoneme (shaft) of flagellum, *CV* = cleavage vacuole, *ET* = exit tube, *L* = lipid globule, *M* = mitochondrion, *N* = nucleus, *TW* = thallus wall, *V* = vacuole. [Courtesy J. H. M. Temmink and R. M. Campbell, 1968, *Can. J. Botany* **46**:951–956.]

lemma of the zoospore and cleaves the zoospore, which is then free within an essentially empty zoosporangium.

In the second method, a large central vacuole is formed in the zoosporangium. This vacuole increases in size both by the synthesis of additional membrane material and by absorption of water from the cytoplasm, and the enlarging vacuole extends in various directions. Meanwhile the plasmalemma is pulling inward, and extensions of the plasmalemma meet and fuse with the vacuolar membrane. These membranes delimit the uninucleate zoospores and become the plasmalemma of the zoospores.

Development of the flagellum is an important phase of zoosporogenesis and occurs concurrently with cleavage (Heath, 1976). Two centrioles lie in the vicinity of each nucleus in the uncleaved zoosporangium. Each centriole contains nine groups of triplet fibers, and a centriole may elongate to form a kinetosome. The fibers of the kinetosome in turn elongate to form the fibers in the remainder of the flagellum. Once the kinetosome is formed in the uncleaved zoosporangium, it may lie at rest until cleavage is completed and the young zoospore is surrounded by a plasmalemma. The elongating flagellum is then covered by a membrane which is an extension of the plasmalemma. Alternatively, the kinetosome may not rest but rather extend into the flagellum while cleavage is taking place. If this happens, the flagellum lies within one of the vesicles surrounding the uninucleate mass of cytoplasm that will become the zoospore, and the membrane surrounding the flagellum is derived from some unknown source, possibly the vesicles. In biflagellate fungi, both centrioles form flagella. In uniflagellate fungi, the second centriole either does not undergo change or elongates into a kinetosome that does not undergo further development.

ZOOSPORE DISCHARGE AND GERMINATION

Zoospores may be released from the zoosporangium in a number of ways. In the simplest cases, the zoosporangial wall simply disintegrates. Usually, however, a localized nipplelike protrusion, the *papilla,* is formed on the surface of the zoosporangium or at the end of an elongated *exit tube.* The papilla may or may not have a more or less distinct plug within it. Enzymatic breakdown of the papilla causes it to open. The plug may also be disintegrated or pushed out. The zoospores then escape through the opening.

When a zoospore has completed its swarming stage, it is ready to settle down, germinate, and produce a vegetative thallus. In most cases, the flagellum is withdrawn into the zoospore and subsequently broken down and dispersed within the cell. The zoospore becomes converted into a nonmotile walled cell, the cyst. The wall material is initially contained in vesicles, which may arise from the Golgi bodies. The vesicles release the wall material that forms the outermost wall of the cyst. The inner wall of the cyst is subsequently deposited from the cytoplasm. After encystment occurs, the organelles undergo a transition to the form typical for vegetative structures. The rootlets are lost within a few minutes, the pear-shaped nucleus becomes spherical, the ribosomes in the nuclear cap become dispersed, the Golgi body moves away from its typical position adjacent to the nucleus or vacuole, and any specialized structures such as the rumposome break down. If a single large mitochondrion is present initially, it fragments into smaller mitochondria. In most cases, when the cyst germinates, the inner wall of the cyst extends and ruptures through the outer wall to form an extension that will become the germ tube. The germ tube will give rise to the vegetative thallus.

CLASSIFICATION

The Division Mastigomycota may be divided into four classes on the basis of the number, position, and type of flagella (Sparrow, 1973-a, 1976). The uniflagellate fungi are separated into the Class Chytridiomycetes (posterior flagellum) and the Class Hyphochytridiomycetes (anterior flagellum). The members of the two remaining classes are biflagellate. In the Class Plasmodiophoromycetes both flagella are of the whiplash type, while a tinsel and a whiplash flagellum occur in the Class Oomycetes. Further division into orders is based primarily on thallus structure and type of sexual reproduction. The taxa follow:

Division Mastigomycota
Class Chytridiomycetes
    Order Chytridiales
    Order Blastocladiales
    Order Monoblepharidales
    Order Harpochytriales
Class Hyphochytridiomycetes
    Order Hyphochytriales
Class Plasmodiophoromycetes
    Order Plasmodiophorales
Class Oomycetes
    Order Saprolegniales
    Order Lagenidiales
    Order Peronosporales

Class Chytridiomycetes

Fungi with posteriorly uniflagellate zoospores are grouped within a single class, the Chytridiomycetes. The single flagellum is of the whiplash type. These are small, comparatively simple fungi. Principal differences among members of this group are to be found in complexity of thallus structure and the type of sexual reproduction.

ORDER CHYTRIDIALES

Members of the Chytridiales, commonly known as "chytrids," are small, relatively simple fungi that produce zoospores with a single posterior flagellum and usually contain a conspicuous refractive lipid globule. The chytrids are also characterized by formation of a simple coenocytic thallus either entirely or partially within the substratum upon which they are feeding.

Chytrids occur principally in fresh water, although they also occur in marine waters, moist terrestrial environments such as bogs, and even dry, hot soils. Fossil chytrids have been recovered from 50-million-year-old shale (Bradley, 1967). Chytrids apparently live most often as saprophytes, some may live either as saprophytes or parasites, and some seem to live only as parasites. They occur on or in algae, other chytrids or aquatic fungi, microscopic animals such as rotifers or mosquito larvae, eggs of microscopic animals, cast insect exoskeletons, roots of higher plants, and plant debris. One can recover chytrids by collecting and examining the above materials or by baiting water samples with bits of hair, snake skin, cellophane, filter paper, or boiled pollen or plant stems. Chytrids which were parasitic within their algal hosts were often mistakenly thought by early biologists to represent part of the algal life cycle. If environmental conditions are favorable in lakes, parasitic chytrids may multiply so quickly that natural populations of their host are infected in epidemic proportions, which results in the decline of the populations. Roots of terrestrial plants may be parasitized by chytrids. *Olpidium brassicae* may damage grass roots, and *Synchytrium endobioticum* causes the black wart disease of potato tubers.

Thallus. The morphology of the chytrids is extremely varied, and many distinctive features may occur in various combinations (Figs. 2–8 to 2–12). In some species, the thallus may be elongated or ellipsoidal and without *rhizoids,* while in others it may have rhizoids. Rhizoids are determinant, tapering, threadlike extensions of the thallus. They may arise from a broad globose or turbinate part of the thallus, the *apophysis.* Also, the rhizoids may be unbranched, slightly branched, or extensively branched. They absorb nutrients from the food source. The thallus may occur entirely within the cell of the living or dead substratum in which case it is termed *endobiotic.* In other species, the thallus is partly within and partly outside the substratum, and that portion outside the substratum is *epibiotic.* In still other species, only a few rhizoids actually penetrate the substratum and the reproductive structures are not in contact with the substratum (in this case the thallus is *interbiotic*). The vegetative thallus is coenocytic and usually appears lustrous because it contains numerous vacuoles and refractive globules (probably lipid) (Sparrow, 1960). When reproduction is to take place, the thallus of some endobiotic chytrids will become entirely converted into a single reproductive structure (*holocarpic* type). A holocarpic thallus lacks rhizoids.

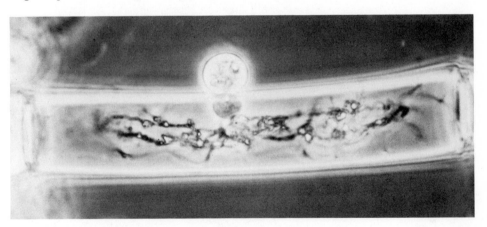

Fig. 2–8. A chytrid, *Chytridium lagenaria,* growing as a saprophyte on a dead algal cell. × 2000. [Courtesy D. J. S. Barr.]

Other chytrids have a thallus that only partially converts into reproductive structures (*eucarpic* type); these are the endobiotic species or species with epibiotic reproductive structures and rhizoids. In at least some chytrids, it is known that the portion of the thallus that will become a reproductive structure enlarges, is delimited from the vegetative part of the thallus by a septum, and continues to differentiate. Most of the chytrids form only a single reproductive structure on a thallus (are *monocentric*), while some produce several at different locations on a thallus (are *polycentric*). A polycentric thallus typically has a more extensively branched system of rhizoids than does a monocentric thallus (Figs. 2–9 and 2–10).

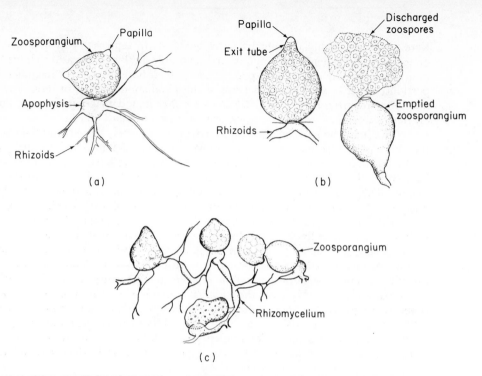

Fig. 2-9. Vegetative and nonsexual reproductive structures in the chytrids. (a) Monocentric thallus of *Rhizidium braziliensis;* (b) Monocentric thallus of *Rhizidium Nowakowskii;* (c) Polycentric thallus of *Nowakowskiella elegans.* All the above chytrids are eucarpic. [(a) and (b) from J. S. Karling, 1944, *Am. J. Botany* **31**:254–261. (c) from T. W. Johnson, Jr., 1973, *Mycologia* **65**:1337–1355.]

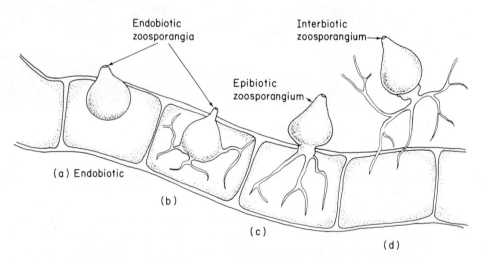

Fig. 2-10. Thallus types in the chytrids. All are monocentric types (a polycentric type is illustrated in Fig. 2-9c). (a) Holocarpic thallus showing complete conversion into zoosporangium; (b–d) Eucarpic thalli showing zoosporangia with rhizoids.

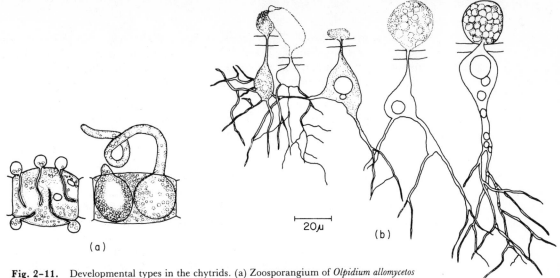

(a)

20μ

(b)

Fig. 2-11. Developmental types in the chytrids. (a) Zoosporangium of *Olpidium allomycetos* in algal host cell; (b) From left to right, successive developmental stages of *Entophlyctis confervae-glomeratae* f. *marina;* (c) Successive stages in development from left to right of *Chytridium lagenaria* on its algal host; (d) Development of *Rhizidium braziliensis*. Successive stages in the germination of the zoospore are at the left. A zoosporangium with two exit tubes is free in the water and is attached to rhizoids that have penetrated the substratum wall (indicated by a horizontal line) in the center. Right, a resting spore has germinated by producing a zoosporangium. [(a) from J. S. Karling, 1948, *Am. J. Botany* **35:**503–510. (b) from Y. Kobayashi and M. Ôkubo, 1954, *Bull. Natl. Sci. Museum (Tokyo)* **1:**62–71. (c) from J. S. Karling, 1936, *Am. J. Botany* **23:**619–627. (d) from J. S. Karling, 1944, *Am. J. Botany* **31:**254–261.]

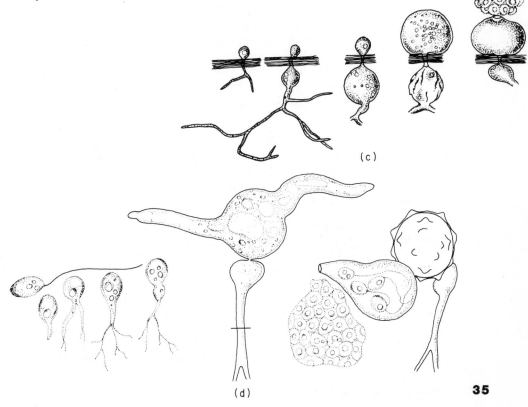

(c)

(d)

An eminent mycologist, F. K. Sparrow, recognized four basic groups into which the chytrids may be divided on the basis of their development (Sparrow, 1960). Each developmental type is designated by the name of a genus in which it is typical, although it may occur in members of other genera. The developmental types are:

1. *Olpidium* type. A zoospore withdraws its flagellum, encysts on the substratum surface, and a penetration tube enters the cell. The cytoplasm of the zoospore passes through the penetration tube and is deposited within the substratum cell. A wall forms around the cytoplasm, which has become a spherical or ellipsoidal thallus without rhizoids. At maturity, the entire thallus converts into a single reproductive structure (endobiotic and holocarpic) [Fig. 2-11(a)].

2. *Entophylictis* type. The zoospore encysts on the substratum surface, producing a penetration tube that develops into a branching rhizoidal system within the substratum. Usually the zoospore and penetration tube become emptied of cytoplasm and disintegrate. Limited portions of the thallus lying just beneath the substratum wall expand to form the rudiments of the reproductive structures, which then continue to expand and differentiate. Either one or several reproductive structures form on a single thallus, and the rhizoidal system remains distinctly vegetative throughout. The cytoplasm from the rhizoids travels back and drains into the reproductive structures, which are then separated by septa from the empty rhizoids. These chytrids are therefore endobiotic, eucarpic, and either mono- or polycentric [Fig. 2-11(b)].

3. *Chytridium* type. An encysted zoospore on the substratum surface produces a penetration tube that enters the substratum. The penetration tube develops into rhizoids or remains as an unbranched peg and meanwhile utilizes some of the cytoplasm from the zoospore. After the rhizoids are established, a segment of the thallus immediately below the cyst may become enlarged as the apophysis (the apophysis is formed by some species only). Cytoplasmic contents from the rhizoids and apophysis then flow back into the cyst, which enlarges as the cytoplasm from the vegetative system is concentrated within it. The cyst is then separated from the emptied vegetative thallus by a septum and subsequently differentiates into a reproductive structure. This chytrid type is eucarpic and the reproductive structures are epibiotic. [Fig. 2-11(c)].

4. *Rhizidium* type. Chytrids of this type are unique because the encysted zoospore does not necessarily rest directly on the substratum but germinates in the water and produces one or more rhizoids. These rhizoids eventually penetrate the substratum. The rhizoids become established, develop further, and convey materials back to the cyst. The cyst enlarges and becomes the reproductive structure. This type is eucarpic and interbiotic, as reproductive structures are not directly situated on the substratum surface [Fig. 2-11(d)].

Nonsexual Reproduction. The reproductive structures mentioned above are almost always zoosporangia, which are nonsexual. Zoosporangia occur in a variety of shapes, such as globose, ellipsoidal, tubular, and pyriform. Crowding of zoosporangia within the substratum may alter zoosporangial shape (Roane, 1973). In nature, the size of zoosporangia of a given species may vary, possibly because of competition for available nutrients, space available, or nature of the substratum (Sparrow, 1960). In his study of *Endochytrium operculatum,* Karling (1937) noted that size and shape of the zoosporangia, opercula, and resting spores vary on different host species. Studies of chytrids in culture under carefully controlled conditions show that zoosporangial size can be modified by pH, temperature, vitamins, and nutrients provided (Miller, 1976).

Zoospores are cleaved from the cytoplasm within the zoosporangium. The size of the zoosporangium can influence the number of zoospores formed within it. In *Endochytrium operculatum,* the number of zoospores in a single zoosporangium ranges from fewer than 20 in the smallest to an estimate of thousands in the larger (Karling, 1937). Once the zoospores have matured, they may be discharged immediately or may be retained within the zoosporangium for a relatively long period of time. Sometimes an abrupt change in surrounding conditions, such as temperature fluctuation or transfer to clean water with a higher level of oxygen, will stimulate zoospore discharge (Sparrow, 1960).

In a few chytrids (e.g., some *Rhizophydium* species), the entire zoosporangial wall may deliquesce or split to release the zoospores (Barr, 1978; Sparrow, 1960). In most chytrids, the zoospores are released through one or more *discharge papillae,* each a lens- or nipple-shaped protrusion. The discharge papillae may occur directly on the surface of the zoosporangium or at the end of an elongated exit tube. A hyaline, refractive, apparently gelatinous mass occurs inside the apical wall of the papilla. This gelatinous mass remains distinct from the remainder of the cytoplasm where the zoospores are formed (Figs. 2–12, 2–13, and 2–14). Eventually an opening is formed in the wall of the discharge papilla. The mechanism for forming the opening varies, and three major types are recognized. In the *operculate* type, a lidlike flap of wall material (the *operculum*) is formed and is thrown aside to leave an opening through which the zoospores escape.

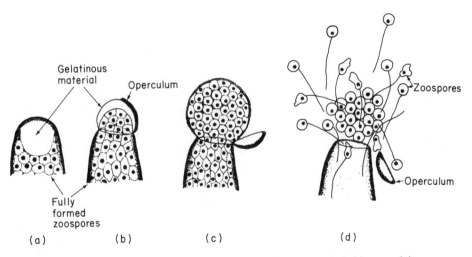

Fig. 2–12. Exooperculation. (a) Fully developed discharge papilla; (b) Dehiscence of the discharge papilla; (c) A slightly later stage showing the envelope of gelatinous substance surrounding the passive zoospores; (d) Zoospores swimming away, with detached operculum nearby. [Figures from I. J. Dogma, Jr., 1973, *Nova Hedwigia* **24:**393–411.]

In turn, the operculate type may be subdivided into the *exooperculate* and *endooperculate* types, depending upon whether the operculum is formed on the surface of the papilla or within the exit tube. The *inoperculate* zoosporangia lack an operculum and open by a simple pore. Dogma (1973) has described development of discharge papillae and has distinguished four types. These include exooperculation, two types of endooperculation, and inoperculation. We shall consider three of these further.

Exooperculation. In this type of mechanism, (Fig. 2–12), the wall at the tip of the discharge papilla develops a line of weakness which delimits a circular cap, the operculum. This type of operculum has been called an *exooperculum*, and has been considered by some mycologists to be the ''true'' operculum. When the operculum dehisces, the gelatinous mass is extruded outward, forming an envelope around the zoospores which are passively released [Fig. 2–12(b, c)]. The zoospores rest quietly at the mouth of the papilla for a brief period, and then begin to dart away. The lidlike operculum may simply be bent back, remaining attached to one edge of the papilla as though it were on hinges, or may be pushed off by the emerging mass of zoospores.

Endooperculation. In endooperculation, (Fig. 2–13), the operculum is formed within the papilla and is sometimes called an *endooperculum*. The endooperculum may be formed either near the tip of the papilla or near its base. In the type in which the operculum is to be formed within the exit tube near the tip, the apex of the papilla inflates slightly and the wall dissolves [Fig. 2–13(a)]. The resulting pore is then plugged with a gelatinous mass. Beneath the plug, the cytoplasm becomes modified to form the operculum, which is continuous with the wall of the papilla [Fig. 2–13(b)]. A second gelatinous plug is formed inside the operculum [Fig. 2–13(c)]. Dehiscence of the operculum occurs, and the gelatinous plug is exuded and surrounds the zoospores, which are passively released. The operculum is carried away by the discharged mass of zoospores [Fig. 2–13(f)].

Inoperculation. In this mechanism, (Fig. 2–14), the wall at the tip of the discharge papilla becomes thin, and eventually the tip ruptures to form a pore. The hyaline, gelatinous mass is expelled [Fig. 2–14(b)], expands, and surrounds the

Fig. 2-13. Endooperculation. (a) Papilla with gelatinizing apex; (b) First gelatinous plug in the pore; and the operculum having formed within the papilla; (c) After the disappearance of the first gelatinous plug, and appearance of the second gelatinous plug; (d, e, f) Successive stages in zoospore release. [Figures from I. J. Dogma, Jr., 1973, *Nova Hedwigia* **24**:393–411.]

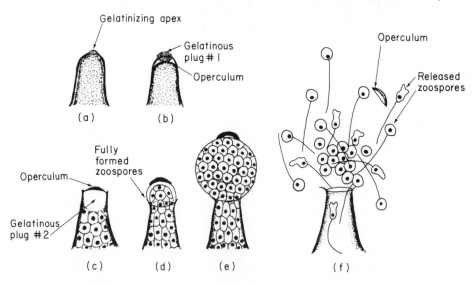

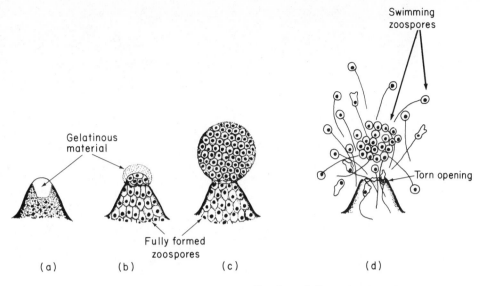

Fig. 2–14. Inoperculation. (a) Discharge papilla; (b, c, d) Successive stages in zoospore discharge. A torn opening remains behind in (d). [Figures from I. J. Dogma, Jr., 1973, *Nova Hedwigia* **24**:393–411.]

zoospores, which are passively ejected [Fig. 2–14(c)]. After a brief period of in-activity, the zoospores swim away [Fig. 2–14(d)]. After the zoospores have left the zoosporangium, the pore has jagged, irregular edges.

Once the zoospores are released, they swim with a hopping or jerking motion, sometimes abruptly pausing and then becoming amoeboid in shape. Each zoospore is capable of producing a thallus if it lands on a suitable substratum.

Sexual Reproduction. Sexual reproduction has been found in comparatively few chytrids and is probably absent in many chytrid species. When it occurs, plasmogamy has been reported to involve fusion of gametes, gametangia, or rhizoids of thalli. Morphological variations are numerous, and sexual structures in a given chytrid species may be unique for that species as far as is known. Fusion of motile isogametes is illustrated in the life cycle of *Synchytrium brownii* in Fig. 2–17, and gametangial fusion occurring in *Polyphagus euglenae* is illustrated in Fig. 2–18. The gametangia in *Polyphagus euglenae* are not necessarily typical for chytrids. For example, encysted zoospores may function as gametangia in species of *Rhizophlyctis* and allied genera (Dogma, 1974); furthermore in *Dangeardia mammillata* the male gametangium strongly resembles an immature zoosporangium while the female gametangium resembles a mature zoosporangium (Canter, 1946). Fusion of rhizoids as the sexual mechanism is exemplified by *Chytriomyces hyalinus* (Fig. 2–15). In *C. hyalinus,* the rhizoids of two thalli approach each other and anastomose. At the point of fusion, a zygote is initiated and the contents of the contributing thalli flow through the rhizoids into the developing zygote. The zygote enlarges rapidly and at first has two nuclei. The two nuclei move to the central position in the cell and undergo karyogamy (Moore and Miller, 1973; Miller, 1977).

Some chytrid species may be heterothallic, as gametes of two genetically dif-ferent strains are required for fusion to take place (Mullins, 1961).

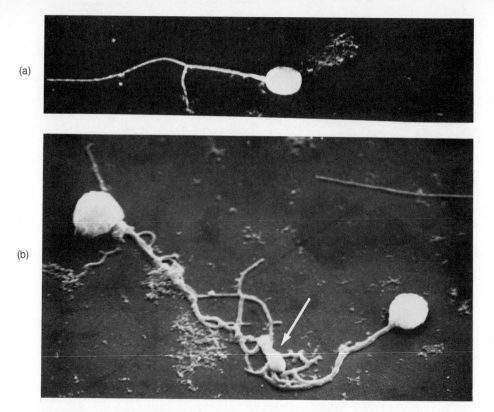

(a)

(b)

Fig. 2-15. Sexuality in the chytrid *Chytriomyces hyalinus*. (a) Germinated zoospores, × 2750; (b) Fusion between the rhizoids of two thalli. A zygote that will also function as a resting spore is being initiated at the point of anastomosis (arrow). × 3,000; (c) The developing resting body is enlarged (arrow) while the two contributing thalli have become empty. × 3,600. [From C. E. Miller, 1977, *Bull. Soc. Botany France* **124**:281–289.]

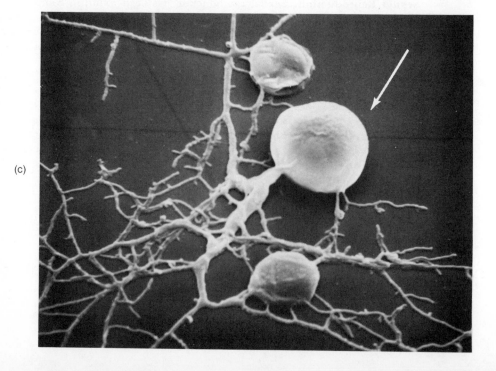

(c)

The Resting Spore. After the zygote is formed, it is converted into a *resting spore.* Resting spores may also be formed from zoosporangia under unfavorable conditions. The resting spore is characterized by its guttulate contents and thick wall, which may be smooth or ornamented with spines, blunt knobs, lobes, or undulations. It may be colorless or may range from light to dark shades of yellow or brown. Upon germination, the resting spore cracks open or forms a pore and its contents emerge. In some chytrids, a thin-walled zoosporangium is formed which protrudes through the opening in the resting spore wall (Fig. 2–16). Subsequently the protoplasm is emptied from the resting spore into the zoosporangium in which the zoospores are formed. In a few chytrids the zoospores are formed within the resting spore (which is therefore functioning as a zoosporangium) and are then released. Each zoospore may give rise to a new thallus.

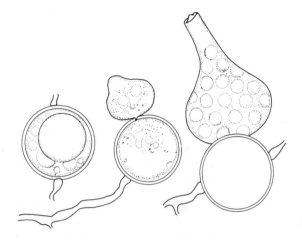

Fig. 2–16. Successive stages in germination of the smooth resting spore of *Nowakowskiella macrospora,* showing emergence of a zoosporangium (center) followed by cleavage of zoospores. [Figures from J. S. Karling, 1954, *Am. J. Botany* **32**:29–35.]

Representative Chytrids. We will discuss the life cycles of two chytrids, *Synchytrium brownii* and *Polyphagus euglenae.* Both these chytrids have a well-established sexual cycle.

Synchytrium. There are about 150 species in this genus. These holocarpic chytrids are parasites of algae, mosses, ferns, and flowering plants (Karling, 1964). These fungi live within the plant cells, typically causing gall formation on the diseased parts. Galls are enlarged structures consisting primarily of host cells that have been stimulated to enlarge or divide abnormally or both. The "black wart" disease of potato is an example of these diseases and is caused by *Synchytrium endobioticum.* In this disease pulpy irregular protuberances are formed on the underground parts of the plant. These galls blacken with age owing to the maturation of the resting spores of the parasites that they contain.

We will examine the life cycle of *Synchytrium brownii,* which occurs on the leaves of *Oenothera laciniata.* Although this particular *Synchytrium* is of little or no economic importance, its life cycle has been carefully studied and is well known (Lingappa, 1958-a, 1958-b, 1958-c). Lingappa (1958-a) believes that *S. brownii* may be the same organism that Kusano (1930) described as *S. fulgens* and in which he discovered a sexual cycle.

SYNCHYTRIUM BROWNII. *S. brownii* has nonsexual and sexual cycles that may occur concurrently, and perhaps even in the same host cell.

The nonsexual cycle begins when a zoospore settles down on the surface of the plant and becomes firmly attached (Fig. 2–17). The zoospores then produce a narrow protoplasmic extension, which enters the epidermal cell, and the protoplasm from the zoospore flows into the plant, leaving an emptied membrane on the outside. The thallus, now within a host cell, develops into a *prosorus*. During this development the thallus becomes located near the base of the host cell, enlarges, and undergoes cytological changes including vacuole formation. Development can be completed in as little as 6 days at 30°C but requires twice as long at 20°C. The mature prosori are usually spherical or subspherical in shape and do not completely fill the host cell. The prosorus will then give rise to a *sorus,* a cluster of zoosporangia. In this process, the cytoplasm of the prosorus flows slowly out through a papilla into a vesicle. Vesicle formation requires about 6 hours. The single nucleus from the prosorus elongates and passes through the papilla into the vesicle, which will become the sorus. Successive nuclear divisions take place until eventually more than 100 nuclei are formed. After these nuclear divisions have started, a plug forms between the emptied prosorus and the developing sorus, perhaps to prevent the protoplasm from flowing back into the prosorus. Cleavage of the protoplasm into approximately 40–60 multinucleate segments takes place, beginning at the periphery of the sorus and progressing inward. Each segment acquires a hyaline wall and matures into a zoosporangium. Maturation of the zoosporangium requires about 8 hours. These zoosporangia may be forcibly discharged from the galls and may be scattered on the leaf surface in nature. There is some question about whether these exposed zoosporangia germinate. However, zoosporangia under appropriate conditions will cleave internally, and zoospores are released about an hour after zoosporogenesis begins. The zoospores are released and are then able to reinfect the host and produce another nonsexual generation involving the formation of prosori, sori, and zoosporangia. A complete developmental cycle takes about 7 days at 30°C, but requires a longer time at lower temperatures.

The sexual cycle is initiated when some of the zoospores become gametes. The gametes are morphologically similar but differ slightly in behavior. One gamete becomes sedentary and absorbs its flagellum; then a second, more active, flagellate gamete comes into contact with it. Plasmogamy occurs about 1 to 3 hours after pairing, and karyogamy occurs within another half hour. The zygote has the flagellum of the active gamete for a while but is nonmotile. The zygote penetrates the plant and forms an intercellular thallus in the same manner as described above for infection initiated by a zoospore. The thallus enlarges and develops into a resting spore. Maturation of the resting spore requires about 12 days at 27°C and involves the development of a thick, orange outer wall and oily contents. The contents of the host cell disintegrate and may appear as residue on the resting spore. The resting spores must go through a dormancy period before germinating. Upon germination, the resting spores function as the prosori. Each forms an exit tube that penetrates through the host wall and produces a sorus in essentially the manner described above. The single diploid nucleus in the resting spore migrates into the developing sorus. This nucleus undergoes a division, presumably meiotic, and several mitotic nuclear divisions follow. Cleavage into multinucleate zoosporangia, zoosporogenesis, and release of zoospores follow.

Zoospores derived from the nonsexual or sexual cycle are indistinguishable from each other. Presumably any of these zoospores can function as gametes or can initiate a nonsexual cycle in *S. brownii.*

POLYPHAGUS EUGLENAE. *Polyphagus euglenae* parasitizes cells of species of *Euglena* that have encysted and are nonmotile. Like motile cells of *Euglena,* the zoospores of *P.*

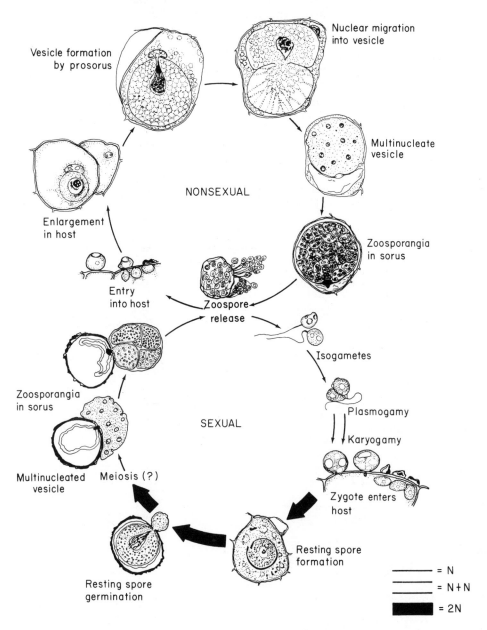

Nuclear migration
into vesicle

Vesicle formation
by prosorus

Multinucleate
vesicle

NONSEXUAL

Enlargement
in host

Zoosporangia
in sorus

Entry
into host

Zoospore
release

Isogametes

Plasmogamy

Karyogamy

Zoosporangia
in sorus

SEXUAL

Multinucleated
vesicle

Meiosis (?)

Zygote enters
host

Resting spore
formation

Resting spore
germination

———— = N

———— = N+N

■■■■ = 2N

Fig. 2-17. Life cycle of *Synchytrium brownii*. [Figures from
B. T. Lingappa, 1958, *Am. J. Botany* **45**:116–123, 613–620.]

euglenae are positively phototropic, swimming towards a lighted area. The positive phototactic response of *P. euglenae* is advantageous, as the zoospores move into regions where the host cells have also accumulated.

This species follows the *Rhizidium* type of development. The zoospore germinates in the water and forms a profuse, interbiotic thallus. Only a few of the rhizoidal tips penetrate an individual *Euglena* cell, and a single thallus may be in contact with as many as 50 *Euglena* cells. After the *Euglena* cell is infected, the chlorophyll changes in color from green to yellow and eventually the protoplasm disintegrates. The encysted zoospore functions as a *prosporangium* from which the zoosporangium develops. The prosporangia lie free in the water and are usually fusiform in shape. The zoosporangium forms at the side of the prosporangium, is usually elongated and tubular in form, and is inoperculate. From a single zoospore to hundreds are cleaved out.

Sexual reproduction is initiated when the ends of the thalli become differentiated into elongated gametangia (Fig. 2-18). Antheridia and the female gametangia are formed on different thalli. The gametangia may be tubular or irregular in form, or alternatively the antheridium may be spherical or club-shaped while the female gametangium is saclike. Cytoplasm from the thallus accumulates in the gametangium.

Fig. 2-18. Progressive sexual stages in *Polyphagus euglenae*. Top: The antheridium is on the left and the female gametangium is on the right. They are separated by a very long conjugation tube, the tip of which is just beginning to swell to form the resting spore. Middle: The male nucleus has begun its migration from the antheridium into the conjugation tube. Bottom: the male nucleus has reached the resting spore initial, and the female nucleus is ready to begin its migration. [From H. Wager, 1913, *Ann. Botany London* **27**:173–202.]

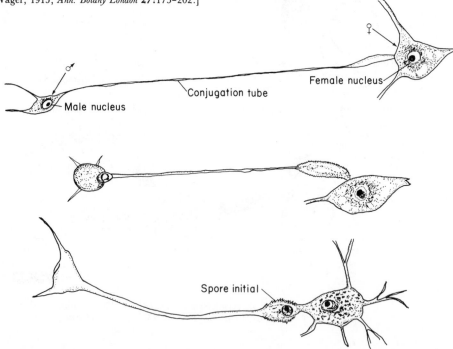

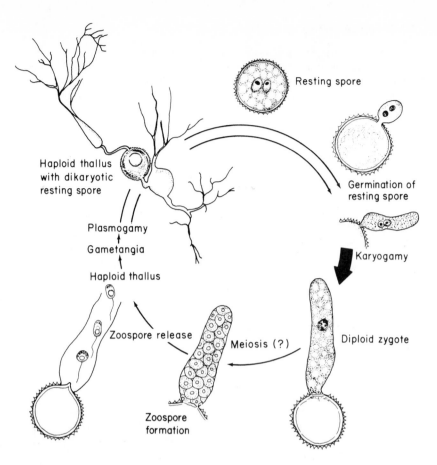

Fig. 2-19. Sexual phase of life cycle of *Polyphagus euglenae.* [Figure of thallus from A. F. Bartsch, 1945, *Mycologia* **37**:553–570; remaining figures adapted from H. Wager, 1913, *Ann. Botany London* **27**:173–202.]

An elongated conjugation tube is formed from the antheridium, which then makes contact with the female gametangium. Each gametangium contains a single nucleus that functions as the gamete. The nucleus from the antheridium and a second larger nucleus from the female gametangium pass into the conjugation tube. These nuclei do not fuse but lie near each other in the distal portion of the conjugation tube, which will undergo transformation to form a resting spore (Fig. 2-19). The male nucleus enlarges until it is the same size as the female nucleus and the nuclei move apart to opposite sides of the cell, which now becomes covered with a thin inner wall and a thick spiny outer wall and becomes the resting spore. When the resting spore germinates, the thick outer wall cracks and the inner wall emerges through the crack and becomes a zoosporangium. The two nuclei move out into the zoosporangium and undergo karyogamy to form a diploid nucleus, the zygote. Meiosis is presumed to take place immediately. The resulting nuclei, now haploid, multiply to form a multinucleate zoosporangium. Zoospores are formed within the zoosporangium and released. It has been suggested that the zoospores ultimately formed in the zoosporangium will produce male and female thalli in equal numbers.

ORDER BLASTOCLADIALES

Most members of this order are saprophytes in the soil or in aquatic environments and have a relatively well-developed thallus (Fig. 2–20). Thallus morphology ranges from naked, endobiotic thalli without rhizoids in *Coelomomyces,* a parasite of arthropods, to mycelial, branching thalli in *Allomyces.* Most of the thalli have well-developed tapering and branching rhizoids, much like those formed by the chytrids. The rhizoids not only anchor the thallus but also gather food from the substratum.

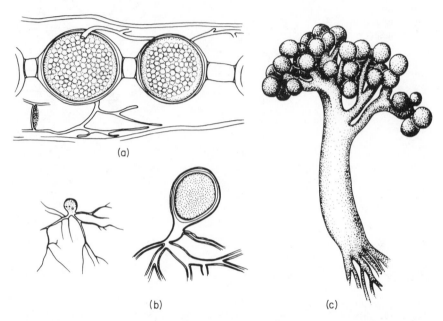

(a)

(b)   (c)

Fig. 2–20. Representative members of the Blastocladiales: (a) Rhizoids and resting sporangia of *Catenaria allomycis* in host, *Allomyces;* (b) Thallus of *Blastocladiella cystogena;* (c) Thallus of *Blastocladia pringsheimii* with resting sporangia. [(a) From J. N. Couch, 1945, *Mycologia* **37**:163–193. (b) From J. N. Couch and A. J. Whiffen, 1942, *Am. J. Botany* **29**:582–591. (c) From R. Thaxter, 1896, *Botan. Gaz.* **21**:45–52.]

Those characteristics unifying the members of this group are (1) production of zoospores with a single posterior whiplash flagellum and a prominent nuclear cap, (Fig. 2–21), and (2) formation of resting spores in either nonsexual or sexual cycles. Resting spores are ovoid, spherical, or pyriform. In *Allomyces neo-moniliformis,* Skucas (1967) demonstrated that the thick outer wall layer contains melanin and lipid in addition to chitin and other components that normally occur in the hyphal walls. Unlike the other portions of the thallus, the resting spores can withstand unfavorable conditions such as freezing or drying. When the resting spore germinates, the outer wall cracks open and the zoospores escape.

Zoospores are also formed in colorless, thin-walled zoosporangia. In several species of this order, but apparently not in other orders, a refractive, pulley-shaped mass of pectic material plugs the discharge papilla in the thin-walled zoosporangia. When the zoospores are about to be discharged, first the lower portion and then the

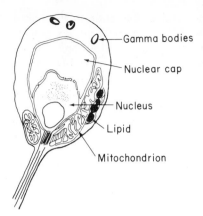

Gamma bodies

Nuclear cap

Nucleus

Lipid

Mitochondrion

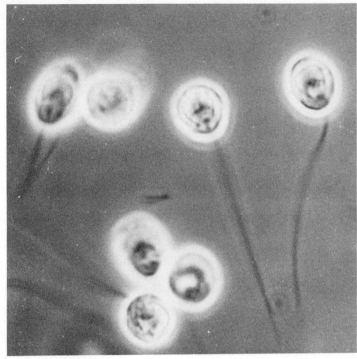

Fig. 2-21. Zoospores of *Blastocladiella emersonii,* a member of the Blastocladiales. Note the gamma bodies and large nuclear cap (the nuclear caps are visible in the photograph). [Diagram adapted from E. C. Cantino, J.S. Lovett, L. V. Leak, and J. Lythgoe, 1963, *J. Gen. Microbiol.* **31**:393–404: Photograph courtesy of L. C. Truesdell and E. C. Cantino.]

upper portion of the plug are digested (probably by enzymatic action). The upper portion is digested from the inside, while the remaining outer shell expands to form a vesicle. The vesicle bursts and leaves an opening through which the zoospores can escape. This entire process takes about 30 seconds (Skucas, 1966). Typically, the zoospores are spherical when they escape from the zoosporangium but become ovoid as they swim. Sometimes the zoospores produce pseudopods and move in an amoeboid fashion. The nuclear cap is visible in the swimming zoospores as a turbinate or pyramidal structure which gleams faintly. A side body is also present in some zoospores and is likewise visible as a faintly gleaming structure.

Sexual reproduction is not known in all members of the Blastocladiales. When nonsexual reproduction takes place, it occurs through the union of motile isogametes or anisogametes. Alternation of morphologically similar gametophytic (haploid) and sporophytic (diploid) thalli occur in some life cycles (for example, in species of *Coelomomyces* and *Allomyces*).

Coelomomyces. Species of *Coelomomyces,* representing the family Coelomomycetaceae, are obligate parasites within the body cavity of arthropods. There they form a coenocytic mycelium that is either irregularly or dichotomously branched and lacks a cell wall. The thallus is holocarpic and also lacks septa that separate rhizoids. At least some species that have been studied in detail are unique, as they are the only fungi known that require two arthropod host species to complete their life cycle. The gametophytic thallus occurs in a copepod, while the sporophytic thallus occurs in a mosquito (Whisler, 1979).

Infection of the hosts in the life cycle was first established in *Coelomomyces psorophorae* (Whisler, 1979). Let us begin with a haploid zoospore which is swimming about in the water. The zoospore invades the copepod and establishes the naked, gametophytic thallus. At first, the copepods swim actively and do not seem to be affected by the fungal invasion. They then slow down abruptly, fall to the bottom of the water, and die. Within a few minutes, the entire body cavity is swarming with the motile, uniflagellate isogametes. Union of isogametes takes place. The zygote, now bi-flagellate, swims away and its protoplasm must invade a mosquito larva to complete the life cycle. The zygote comes into contact with a mosquito larva and becomes attached to the cuticle with a gluelike substance which is released from cytoplasmic vesicles. The zygote then encysts, forms an appressorium from which a narrow tube extends, and penetrates the cuticle. The protoplasm passes from the cyst into an epidermal cell, and eventually the fungus completes its development in the hemocoel. The sporophytic thallus is morphologically similar to the naked, coenocytic gameto-phyte. Diploid resting spores are eventually formed. Upon germination, the resting spore undergoes meiosis and cracks along a preformed groove; the inner wall and con-tents protrude, and the four haploid zoospores formed as a result of meiosis are re-leased. The zoospores are capable of infecting a copepod but are not capable of infect-ing another mosquito larva.

Blastocladiella. Members of this genus typically are soil inhabitants in warm climates such as that of Texas or Mexico but have been found in Great Britain and even in Iceland. The morphology of members of this genus was described by Couch and Whiffen (1942). The relatively simple thallus consists of a basal cell anchored by branched rhizoids at one end and supporting a single reproductive cell from the other end. The sporophytic thallus (presumably diploid) is observed most frequently. The sporophytic thallus produces a terminal zoosporangium which has a thin wall and is colorless or orange. The zoosporangium discharges its zoospores through one or more pores produced upon deliquescence of the papilla. Zoospores formed from these thin-walled zoosporangia are only capable of forming additional sporophytes upon ger-mination. Resting spores with thick, dark, sculptured walls may be formed by the sporophyte. Upon germination, the resting spore releases zoospores. The behavior of the zoospores varies depending upon the species. In *B. emersonii,* the zoospores simply form additional sporophytic plants. In *B. cystogenes,* a zoospore encysts and upon ger-mination produces four haploid isogametes. Fusion occurs only between isogametes from different cysts. These isogametes fuse in pairs after the release, and the resulting biflagellate zygote germinates to form a sporophytic thallus. In *B. variabilis,* the presumably haploid zoospores settle down and each germinates into a gametophytic thallus. The male thallus bears a single orange gametangium while the female thallus bears a single colorless gametangium. The gametes are released and fuse in pairs, and the resulting zygote germinates to form a sporophyte.

Blastocladiella emersonii has been the subject of extensive physiological studies by Cantino and his associates (e.g., Cantino, 1966). For example, we mentioned on page 25 that gamma particles are known only in this fungus. Because of the extensive work by Cantino we have some insight into the manner in which metabolic changes are cor-related with morphological differentiation in *B. emersonii.* This topic is discussed fur-ther in Chapter 8.

Allomyces. The genus *Allomyces* is of special interest to us because some members of the subgenus *Euallomyces* are among the few fungi known to have alternative genera-tions of sporophytic and gametophytic thalli (Fig. 2–22) and have been the subjects of extensive genetic and cytological studies. The diploid and haploid thalli are mor-phologically similar, consisting of a stout, trunklike portion anchored to the sub-stratum by rhizoids and bearing slender branches. Branching of the thallus is more or

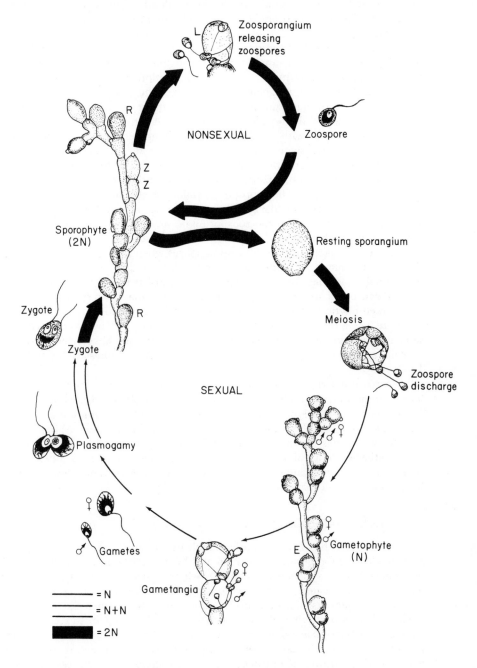

Zoosporangium
releasing
zoospores

L

NONSEXUAL

Zoospore

R

Z
Z

Sporophyte
(2N)

Resting sporangium

Zygote

R

Meiosis

Zygote

Zoospore
discharge

SEXUAL

Plasmogamy

♀
♂

Gametes

Gametophyte
(N)

E

Gametangia

_____ = N
_____ = N+N
▬▬▬▬ = 2N

Fig. 2-22.   Life cycle of *Allomyces arbuscula*.

less dichotomous. Pseudosepta with large triangular openings occur at intervals throughout the thallus, which appears to be jointed because it is constricted at each pseudoseptum. Reproductive organs are formed at the tips of the hyphal branches and are separated from the remainder of the thallus by true septa. These reproductive structures may be zoosporangia, gametangia, or resting spores.

The gametophytic thallus bears orange, globular male gametangia (*antheridia*) and colorless, somewhat cylindrical or ovoid female gametangia at the tips of the branches. The male gametangia may be immediately above the female gametangia as in *A. macrogynus,* while in *A. arbuscula* the position is reversed. Both the male and female gametes are uniflagellate single cells. The male gametes produced in the antheridium are orange and are smaller and more active than the colorless female gametes produced in the female gametangium. After their release from the gametangia, the gametes pair with the flagellated ends together, and their cytoplasm mingles. The united gametes swim about and then settle down, their nuclei fuse, and the resulting zygote develops into a sporophytic thallus. The first germ tube formed from the zygote develops into rhizoids, while the second germ tube develops into the remainder of the thallus. The sporophytic thallus is similar to the gametophytic thallus except that it produces dark-walled resting spores and colorless zoosporangia. Zoospores are produced in the zoosporangia, and each zoospore may develop into a new sporophytic thallus, thus allowing many generations of nonsexual development to occur. Meiosis takes place in the resting spores when they germinate. The walls, consisting of thin outer and inner layers and a thick middle layer, crack open, a prominent papilla is formed from the inner wall as it pushes through the thick wall, and four haploid zoospores are released. A haploid zoospore may then settle down and give rise to a gametophytic thallus. The female gamete, the diploid zoospore, and the haploid zoospore are morphologically indistinguishable, but differ in their origin and subsequent behavior and development. Gamma particles have been observed in the haploid zoospore but not the diploid zoospore at the ultrastructural level (Olson, 1973).

A natural series of polyploids occurs within this genus. Isolates of *A. macrogynus* have chromosomes in multiples of 14 (14, 28, and probably 56), while isolates of *A. arbuscula* represent a polyploid series with a multiple of 8 chromosomes (8, 16, 24, and 32). In addition, there is a naturally occurring hybrid (*A. javanicus*) between these two species, which has chromosome numbers ranging from 13 to 21. This hybrid also bears antheridia both above and beneath the female gametangia, unlike the parental types, which bear the antheridia either above (*A. macrogynus*) or below (*A. arbuscula*) the female gametangia (Emerson and Wilson, 1954).

ORDER HARPOCHYTRIALES

This order has been recently established to include only a few genera, *Harpochytrium and Oedogoniomyces* (Emerson and Whisler, 1968). The thallus is typically an unbranched single, coenocytic filament, which is usually curved in an arc or spiral. The thallus is attached to the substratum by an acellular, excreted holdfast, which is usually globular, discoid, or cup-shaped but may have other shapes [Fig. 2–23(b)]. Cellular holdfasts or rhizoids, found in some other fungi in this class, are lacking. Members of *Oedogoniomyces* are epiphytic on shells of living snails and on dead or living plant material such as fruits or twigs. *Harpochytrium* species occur as epiphytes on living algae. The substratum is not penetrated by the holdfast, and these fungi apparently obtain their nutrients in the form of a solution from the surrounding water. These

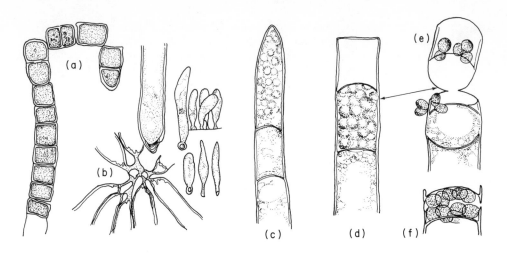

Fig. 2-23. *Oedogoniomyces lymnaeae.* (a) Chain of gemmae breaking apart; (b) Basal parts of the thalli, showing an especially elaborate holdfast and germinated zoospores with discoid holdfasts; (c) Apical zoosporangium with zoospores. The next zoosporangium is delimited but without zoospores; (d) Zoosporangia formed later. The top zoosporangium has been ruptured to release zoospores, and the line of cleavage is forming in the second one at the arrow; (e) The same zoosporangium dehisces to release zoospores; (f) an especially short zoosporangium in the process of dehiscence. [From Y. Kobayashi and M. Ôkubo, 1954, *Bull. Natl. Sci. Museum (Tokyo)* **1**:59–66.]

fungi are found in fresh water and have been recovered from ponds that undergo periodic drying.

In members of the Order Harpochytriales, parts of the vegetative filamentous thallus function as zoosporangia and the cytoplasm cleaves into spores. In *Harpochytrium,* zoospores are formed only in the upper half of the thallus, which is now functioning as a zoosporangium. The zoospores are released when an operculum is formed by a circular rupture of the zoosporangial wall just beneath the apex. In contrast, the young vegetative thallus of *Oedogoniomyces* cleaves into separate cells, each functioning as a zoosporangium. An elongated apical cell is first delimited and forms zoospores, and meanwhile a second cell is being delimited by septum formation beneath it [Fig. 2-23(c)]. Zoosporangium delimitation continues in this basipetal direction until approximately one-half the thallus has been converted into zoosporangia. The apical zoosporangium of *Oedogoniomyces* opens by an operculum, as does *Harpochytrium*. All zoosporangia beneath the apex open by a unique mechanism [Fig. 2-23(e)]. A transverse cleavage line is formed around the wall near the upper part of the zoosporangium. This cleavage line dehisces and thereby releases a tubular section of the thallus consisting of segments of walls of two adjacent zoosporangia and the septum that originally divided them. The zoospores escape from the broken zoosporangium, and the zoosporangium immediately beneath it is ready to form zoospores.

Once released, the zoospores swim with a jerky motion. The zoospore swarms, encysts, and germinates. Upon germination, the encysted zoospore produces a single germ tube, which usually remains unbranched and develops into the single vegetative filament. Sexual reproduction is unknown.

No resistant bodies have been detected for *Harpochytrium*. One species of *Oedogoniomyces, O. lymnaeae,* however, produces chains of gemmae which are shorter and thicker-walled than the zoosporangia and are cylindrical cells which appear cuboidal from the side (Fig. 2-23). At maturity, they separate, fall into the mud, and overwinter. A gemma can germinate when conditions are favorable by releasing zoospores through a slit in its wall (Kobayashi and Ôkubo, 1954).

ORDER MONOBLEPHARIDALES

The Order Monoblepharidales is comparatively small, containing only three genera, *Gonapodya, Monoblepharis,* and *Monoblepharella* (Sparrow, 1960, 1973-a). *Monoblepharis* is most frequently recovered from dead, submerged twigs collected from cool, freshwater habitats and incubated in a cool environment. Species of *Gonapodya* are also recovered from twigs but are found most frequently on submerged fruits such as those of the apple or rose. *Monoblepharella* can be recovered from tropical soils by baiting water cultures with plant material. None of these fungi are of economic importance but they have fascinated mycologists because of their unique mode of sexual reproduction. Sexual reproduction is *oogamous,* involving the fertilization of a nonflagellated egg by flagellated, motile male gametes. We will examine *Monoblepharis* as an example of this order as it perhaps is the best-known genus.

Monoblepharis. These fungi form a eucarpic thallus with distinct hyphae. Zoospore germination is bipolar, with one germ tube giving rise to the rhizoids, which function as holdfasts, and the second germ tube giving rise to the remainder of the thallus. The thallus may be branched or unbranched, depending upon the species, and is nonseptate. The thallus typically has a foamy appearance because there are numerous vacuoles and refractive granules regularly distributed through the cytoplasm, producing a characteristic lustrous quality.

Usually nonsexual and sexual reproductive structures are not found on the same thallus, as their formation is controlled by temperature. They are found together only if the thallus has grown within a narrow temperature range allowing formation of both types.

Zoosporangia are delimited by septa from the terminal portions of the thallus. The zoosporangia are usually elongated, cylindrical, and slightly wider than the hyphae from which they were formed. A circular pore is formed by deliquescence at the tip of the zoosporangium, allowing the posteriorly uniflagellate zoospores to escape. The zoospores creep through this pore in an amoeboid fashion, remain temporarily caught in the vicinity of the pore by their tangled flagella, jerk away, and then swim away with a gliding motion. The zoospores have refractive granules that move to the anterior end of the zoospore as it begins to swim. Once a zoosporangium has been emptied of its zoospores, a new zoosporangium may form within the walls of the old one (this process is called *internal proliferation*).

Sexual reproduction involves the formation of antheridia and oogonia. The arrangement of the antheridia and oogonia varies among the species. In some species, including *Monoblepharis insignis* (Fig. 2-24), the antheridium is formed when a terminal portion of the hypha is cut off by a septum. Following this, a section of the hypha beneath the antheridium enlarges to form an intercalary oogonium. At maturity, the small antheridium appears as if it is inserted on the oogonium. In some other species, the oogonia are first formed in a terminal position and this is followed by formation of the antheridium from a hyphal protrusion below. In still other species, antheridia may be in the terminal position on some branches while oogonia are terminal on other

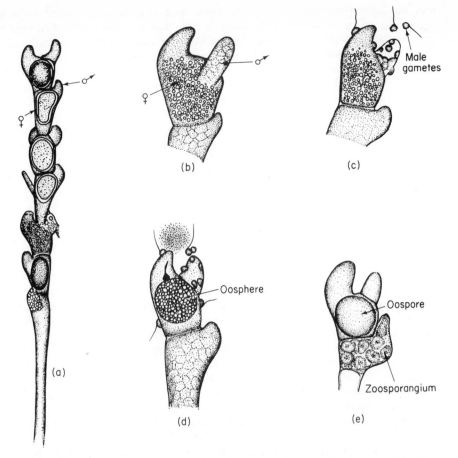

Fig. 2-24. *Monoblepharis insignis:* (a) Hypha bearing terminal oogonia with antheridia; (b) Immature oogonium and antheridium; (c) Antheridium has matured and some male gametes have escaped. Oogonium is yet immature. (d) Male gametes entering oogonium to unite with the differentiated oosphere; (e) Oospore within the oogonium. Immediately beneath the oogonium is a zoosporangium in which zoospores are differentiating. [From R. Thaxter, 1895, *Botan. Gaz.* **20**:433–440.]

branches. Additional gametangia may be produced beneath those already formed so that eventually a chainlike succession of gametangia results.

Four to eight sperm are formed within each antheridium and escape through a pore, which opens by deliquescence. The sperm are morphologically similar to the zoospores except that they are somewhat smaller and usually move in an amoeboid fashion.

As the uninucleate oogonium matures, scattered small oil droplets coalesce into large droplets, which move to the center of the oogonium. Each oogonium produces a single, uninucleate oosphere. The oosphere moves towards the apex of the oogonium, and the nucleus becomes located especially close to the apex. When the oosphere is ready for fertilization, the oogonium dilates and opens at its tip, and a mucilaginous material is present that is more or less continuous with the cytoplasm of the oosphere. When the sperm reaches the oogonium, it rests momentarily and is then engulfed by the mucilaginous material; subsequently, both the mucilaginous material and sperm appear to be absorbed by the oosphere. After plasmogamy has been completed, the oosphere may remain within the oogonium in some species, while it is extruded to the outside in others. If the oosphere is extruded, it remains attached to the oogonium by a narrow hyaline collar. A thick wall with regularly spaced protuberances eventually develops around it, converting the oosphere into an *oospore*. The male and female nuclei remain side by side until the thickening of the oosphere wall reaches an advanced stage, and then karyogamy occurs. The mature oospore is uninucleate, and the single nucleus is presumably diploid. Oospores germinate by forming a germ tube (this process has rarely been observed). The timing of meiosis has not been conclusively established.

Class Hyphochytridiomycetes

ORDER HYPHOCHYTRIALES

All fungi with a single tinsel flagellum, which is inserted anteriorly, belong to the Hyphochytridiomycetes. This class contains a single order, the Hyphochytriales. In all major respects other than flagellation, the small number of fungi in this group parallel the members of the Chytridiales in their structure and life cycle.

Although nonsexual reproduction is essentially like that in the chytrids, an important difference among members of this group is that in some the zoospores are completely differentiated in the zoosporangium, while in others the undifferentiated protoplasm is extruded through the orifice of the zoosporangium and the differentiation of the zoospores is completed outside the sporangium.

Sexual reproduction is generally unknown in this group, and resting spores have been found only in representatives of certain genera.

Hyphochytrium catenoides. One of the best known fungi in this group is *Hyphochytrium catenoides*. Karling (1939) has described the development of this fungus, which occurs as a weak parasite and saprophyte in root hairs and parenchyma cells of corn.

Within the corn cells, the thallus consists of hyphalike tubular segments connecting intercalary enlargements and oval or spherical zoosporangia. In agar cultures, the intercalary enlargements later mature into zoosporangia. The connecting tubular segments vary in length from approximately 10 to 138 microns but are relatively constant in width. The thallus type, therefore, is usually polycentric and eucarpic.

When the zoosporangia mature, they form one to four extremely long exit tubes that may be branched, straight, curved, coiled, or irregular. The tip of the exit tube eventually deliquesces and the protoplasm streams out, taking about 15 to 32 seconds to emerge. The protoplasm is naked and is apparently undifferentiated at first. About 120 seconds after the cytoplasm emerges, cleavage furrows that will eventually delimit zoospores are first evident. The partially delimited cytoplasmic segments begin to move individually and glide over each other. Meanwhile, the flagella are formed and become active, making the entire mass of protoplasm vibrate. Eventually the fully formed zoospores separate from each other and swim away; this occurs about 8 minutes after the entire process started. Although zoospores are most often formed outside the zoosporangium, in unusual cases they may be partially or entirely delimited within the zoosporangium.

The zoospores are oval and somewhat flattened, have numerous conspicuous vacuoles and granules, and have a single anterior flagellum. The flagellum moves rapidly, with the shaft of the flagellum extending more or less straight in front and the tip lashing back and forth. The period of activity usually lasts 20 to 80 minutes.

The zoospores round up and germinate by producing a germ tube that penetrates the host cell and then swells to form an extension [Fig. 2–25(b)], which will enlarge even further [Fig. 2–25(c)] and eventually form hyphalike extensions that will develop into the remainder of the thallus [Fig. 2–25(d)]. The original swelling will eventually become a zoosporangium. As the thallus develops, intercalary enlargements that will become zoosporangia are formed and the cytoplasm from the hyphae accumulates within them, leaving most of the hyphal segment devoid of cytoplasm [Fig. 2–25(d)].

Fig. 2-25. *Hyphochytrium catenoides.* (a) Zoospore; (b–e) Successive developmental stages in the hair of corn; (b) Initial invasion of the hair has occurred; (b–c) Approximately 5 hours later, the original enlargement of the thallus has formed a hyphalike extension, which is forming an additional enlargement; (d) Approximately 32 hours after the original infection, the first enlargement has expanded into a zoosporangium with a short exit tube, and its contents have undergone cleavage internally into zoospores. Below, an oval zoosporangium has discharged its undifferentiated protoplasm through a large, curving exit tube; (e) Zoospores are becoming delimited in a mass of exuded protoplasm, and their flagella are beating. [Figures from J. S. Karling, 1939, *Am. J. Botany* **26**:512–519.]

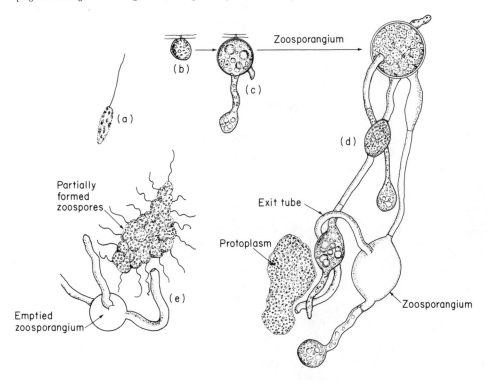

Class *Plasmodiophoromycetes*

ORDER PLASMODIOPHORALES

Fungi with two unequal whiplash flagella are classified in the Plasmodio-phoromycetes. Approximately 35 species are known and are contained within a single family, the Plasmodiophoraceae (Karling, 1968). These fungi are obligate parasites and most often occur within cells of higher plants, but some also attack algae or aquatic fungi. Usually these parasites cause marked enlargement of infected parts of the host. This may result simply from abnormal enlargement of the host cells, from stimulated division of the host cells, or from a combination of these events. Two serious diseases caused by members of this group are powdery scab of potato (caused by *Spongospora subterranea*), in which scabby lesions form on the tubers and galls form on the roots and stems, and clubroot of crucifers (caused by *Plasmodiophora brassicae*), in which roots become shortened and distorted from the formation of tumorlike galls.

The Plasmodiophorales are further characterized by producing a naked *plasmodium* within the host cells (Fig. 2–26). A plasmodium is a multinucleate mass of protoplasm that lies embedded within the host protoplasm, from which it is separated only by its plasmalemma. Once initiated, the plasmodium increases in size by growth that is associated with nuclear divisions. The nuclear divisions associated with vegetative growth of the plasmodium have a unique crosslike appearance, leading to the term *cruciform division*. In cruciform divisions, the nucleolus elongates and divides into two parts. While the nucleolus is stretched out into a dumbbell-like shape, it intersects a ring of chromosomes lying on the equatorial plane, giving the appearance of a cross. The nuclear envelope persists throughout metaphase and anaphase (Braselton et al., 1975). Similar division figures have been noted in protozoa but not

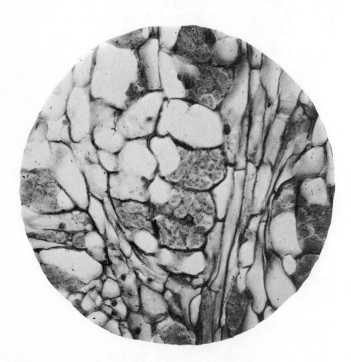

Fig. 2–26. Amoeboid cells of *Plasmo-diophora brassicae* in enlarged cells of cabbage root. [Courtesy Plant Pathology Department, Cornell University.]

in other fungi. Following the last cruciform division in the plasmodium is the *akaryote* phase, during which the nuclei cannot be stained and therefore seem to temporarily disappear.

Members of the Plasmodiophorales apparently share some common features in their life cycles, although many details about the life cycles are poorly understood (Karling, 1968). Two types of plasmodia are formed: (1) those that will form globular, thin-walled zoosporangia, and (2) those that will form thick-walled resting spores. Let us begin with zoospores that were released from a zoosporangium. The zoospores usually become intermittently amoeboid and move about in the water or water film until they come in contact with a host plant. After penetrating the plant, the zoospore divides and initiates formation of a new plasmodium. Although many details of subsequent events are uncertain, mycologists have sometimes suggested that the invading cell may fuse with a young plasmodium or that small plasmodia may coalesce to form a single larger plasmodium. Nevertheless, growth of the plasmodium, involving the cruciform divisions and akaryote phase, occurs. After the akaryote phase, the plasmodium is cleaved into zoosporangia and zoospores. The zoospores escape from the zoosporangium through a pore or exit papilla. Apparently any number of zoosporangial generations may occur before the second plasmodial type is formed.

Eventually a plasmodium is formed that cleaves into resting spores, which lie loose within the host cell or else are clustered together in a *cystosorus*, or loose mass. The resting spores are released into the water or soil by decay of the host plant. They may remain viable for several years. Upon germination each resting spore releases a single zoospore, which may directly infect a host plant or may become amoeboid before infecting the plant. After the zoospore enters the plant, it forms a plasmodium in the manner described above, and also the typical cruciform divisions and final akaryote phase occur before sporulation takes place.

Sexual reproduction has been reported in only a few species, and details concerning sexuality are controversial because chromosome counts are generally lacking and meiosis has not been conclusively demonstrated. Some investigators maintain that some of the zoospores function as isogametes, fusing to form a diploid zygote either outside or within the plant. Presumably the zygote then divides to form a plasmodium. Other investigators maintain that mating occurs when the nuclei within a haploid plasmodium fuse. In either case, a diploid plasmodium would be eventually formed. Meiosis is presumed to occur either during the last two nuclear divisions in the diploid plasmodium, before it cleaves to form resting spores, or during the actual cleavage into resting spores.

Plasmodiophora brassicae. *Plasmodiophora brassicae* is perhaps the most economically important member of this group because it causes the clubroot disease of cabbage and its relatives. Clubroot occurs worldwide. Entire crops may be lost in badly infested fields, which may be virtually useless for cultivation of crucifers for several years because of the longevity of the resting spores (which may remain viable for up to 8 years).

The roots of diseased plants have galls that may be as large as a human fist. Water transport in these plants is affected, and often the first indication of disturbance is that leaves of the infected plants wilt on hot and dry days. As the disease progresses, the leaves turn yellow and the entire plant appears stunted.

Many aspects of the life cycle of *P. brassicae* are poorly understood, although there is general agreement on many morphological details. The resting spore germinates to form a single zoospore. This zoospore comes to rest on the root hair, rounds up and encysts, and then quickly penetrates the root hair. The penetration process is rather intricate—a flattened, adhesive ending of a tubular extension of the zoospore is applied to the surface of the root hair, and then the wall of the root hair is pierced by a specialized bullet-shaped organelle that is ejected through this tubular extension. Subsequently, the cytoplasm from the zoospore passes into the host cell (Aist and Williams, 1971). Once within the root hair, a plasmodium with more than 100 nuclei develops. The plasmodium contains numerous large lipid droplets, mitochondria, Golgi dictyosomes, endoplasmic reticulum, and ribosomes (Williams and McNabola, 1967). The plasmodium cleaves into uninucleate units, the zoosporangia. The nucleus in each zoosporangium divides mitotically, and at maturity about 4 to 10 zoospores are formed. The zoospores are discharged to the outside or remain within the root. Some investigators (e.g., Tommerup and Ingram, 1971) believe that these zoospores function as gametes, undergoing plasmogamy. Nevertheless, additional plasmodia occur in the root cortex. The origin of these particular plasmodia is uncertain. It has been variously suggested that they could result from infection by zoospores emitted from the zoosporangia, by invasion of the zygote, or by *in situ* encystment and germination of the zoospores. At maturity, the plasmodium cleaves into uninucleate segments of cytoplasm that develop into resting spores. The resting spores are cleaved out of the plasmodium by coalescence of vacuoles and subsequent deposition on the cell wall (Tommerup and Ingram, 1971). Meiosis presumably occurs at this stage.

A number of cellular changes occur in the infected host cells (Williams, 1966; Yukawa et al., 1966; Karling, 1968). The invaded host cells divide normally at first, although they eventually lose their ability to divide but continue to enlarge. The nucleus of the host cell enlarges as DNA synthesis is stimulated by the infection. Starch accumulates temporarily in infected cells. Cell and nuclear enlargement and starch accumulation all reach approximately 10 times normal levels. The nucleoli increase in number.

Class Oomycetes

The Class Oomycetes is a large, heterogenous group of fungi. Morphologically, the thalli range from simple holocarpic, endobiotic forms to eucarpic, freely branching mycelial forms. Habitats are equally diverse. The Oomycetes occur as parasites within or on simple animals such as rotifers, simple plants such as algae or fungi, or higher plants or animals. Many of these fungi live as saprophytes. They occur predominantly in aquatic environments or moist soil, but some also occur on the aerial parts of plants.

The Oomycetes are characterized by (1) *oogamy,* in which nonmotile eggs are formed, and (2) biflagellate zoospores. One flagellum is of the tinsel type and is directed forward while swimming; the second flagellum is of the whiplash type and is directed backward. Two morphological types of zoospores occur, depending upon the particular life cycle. These are zoospores with flagella inserted anteriorly, which are typically pip- or pear-shaped when they swim, and zoospores with the flagella inserted laterally (Fig. 2–27). Usually the laterally flagellate zoospores are shaped like a bean (reniform) or grape seed and have a long groove along the side from which the flagella emerge.

An individual member of the Oomycetes may lack sexual reproduction but be classified in this group because its biflagellate zoospores resemble those of other Oomycetes. Alternatively, some members may lack motile zoospores but have oogamous sexual reproduction as do others in this class.

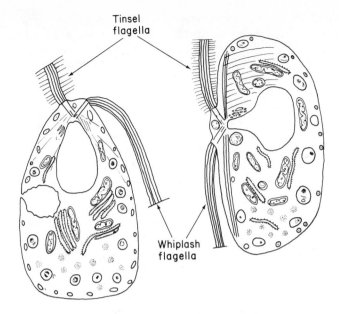

Tinsel flagella

Whiplash flagella

Fig. 2-27. Oomycete zoospores. The apically flagellate *primary* type is at the left and the laterally flagellate *secondary* type is at the right. [Figures from S. A. Holloway and I. B. Heath, 1977, *Can. J. Botany* **55**:1328–1339.]

Unlike the majority of the fungi, several Oomycetes have walls which contain cellulose rather than chitin. Both chitin and cellulose have been found in walls of some of the members (Aronson and Lin, 1978).

Sexual Reproduction. The female gametangium is the *oogonium,* typically a spherical or pear-shaped organ. The antheridium may occur in various forms, depending upon the species, but is most often an elongated organ formed at the tip of hyphal branches. The antheridium may be on the same thallus producing the oogonium or on a different thallus.

Fig. 2-28. Sexual reproductive structures formed by *Saprolegnia litoralis:* (a) Young terminal oogonium. × 310; (b) Terminal oogonium with oospheres. The antheridia have not yet discharged their contents. × 310; (c) Oogonium with mature oospores. × 310; (d) Intercalary oogonium. × 170. [From W. C. Coker, 1923, *The Saprolegniaceae,* University of North Carolina Press, Chapel Hill, N.C.]

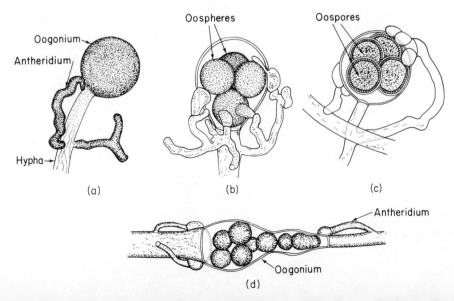

Oospheres

Oospores

Oogonium

Antheridium

Hypha

(a)

(b)

(c)

Antheridium

Oogonium

(d)

One or more *oospheres* in the oogonium will function as female gametes (Fig. 2–28). The oosphere consists of a haploid nucleus surrounded by a small amount of cytoplasm. Depending upon the species, the oosphere may be either a distinct membrane-bounded cell that lies free within the oogonium or a naked gamete that lies within differentiated surrounding protoplasm, the *periplasm*. The oospheres may develop without fertilization into thick-walled *oospores*. Alternatively, the oospheres may be fertilized by nuclei from the antheridium and then mature into oospores. During maturation, a reorganization of the protoplasmic contents occurs. This especially involves lipid drops and reserve globules, which are of poorly known composition but apparently contain reserve food. The oospore may fill the oogonium entirely, or there may be a space left between the oospore and the oogonial wall. If periplasm is present, it disappears as the oospore matures. The oospore then functions as a resting spore (Dick, 1969).

Although the above general characteristics apply to the entire Class Oomycetes, we know comparatively little about sexuality in many members of this group. This is especially true of the reduced chytridlike forms. Studies of sexuality have concentrated on the larger, filamentous fungi in the families Leptomitaceae and Saprolegniaceae in the Order Saprolegniales and the Order Peronosporales. Further discussion of sexuality will be based on these better-known fungi.

Until the early 1960s, mycologists generally believed that meiosis occurred in the oospore before germination to form a haploid thallus. In 1962, Sansome and Harris published a paper reporting that they had induced polyploids in three members of this group and had observed multivalent chromosomal configurations typical of meiosis in the gametangia. Therefore, the thallus would be diploid, contrary to the former concept. As in other fungi, the chromosomes are extremely small in the Oomycetes, which generally makes cytological analysis difficult and contributed to the early misinterpretations. Following the publication by Sansome and Harris (1962), other investigators have reexamined numerous members of the two major Oomycete orders, the Saprolegniales and Peronosporales. Cytological investigations showed that meiosis occurred in the gametangia of all those fungi which were critically reexamined (Dick and Win-Tin, 1973). Light microscope studies were supported by ultrastructural observations of meiosis in gametangia (e.g., Ellzey, 1974; Howard and Moore, 1970). This interpretation has been supported by microspectrophotometric analysis of changing DNA levels during the life cycle. In microspectrophotometric studies of two members of the Saprolegniales, the DNA content of nuclei just before division in the gametangia was shown to be four times that of the nuclei resulting from division and twice that of nuclei in the vegetative hyphae. These data are consistent with the interpretation that gametangial division is meiotic (Bryant and Howard, 1969; Howard and Bryant, 1971). Genetic data have been obtained which can best be interpreted as involving segregation of characters in a Mendelian fashion at meiosis during gametogenesis, with the resulting production of a diploid thallus (Dick and Win-Tin, 1973). We may conclude that a predominantly diploid life cycle (Fig. 2–30), including a diploid thallus, therefore occurs widely among the Oomycetes, at least within the Saprolegniales and Peronosporales.

Numerous investigators have studied the morphology of the sexual organs and their functions in the Saprolegniaceae (Order Saprolegniales) and Peronosporales (Figs. 2–28 and 2–29). Light microscope observations have been supplemented by recent electron microscopic observations, and it has been established that these organisms generally share a common pattern of reproduction (Dick, 1969; Dick and Win-Tin, 1973), which is described below.

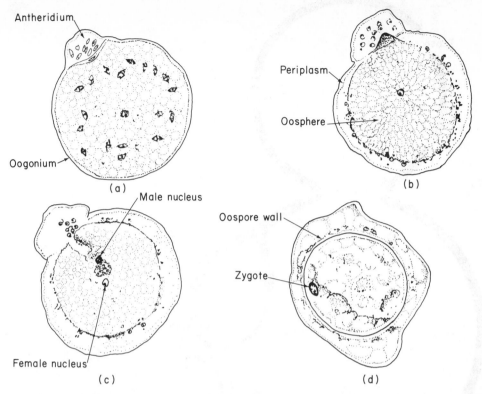

Antheridium

Oogonium

(a)

Periplasm

Oosphere

(b)

Male nucleus

Female nucleus

(c)

Oospore wall

Zygote

(d)

Fig. 2-29. Fertilization in an Oomycete. (a) Final nuclear division in both the antheridium and oogonium. Within the oogonium, the nuclei have started to migrate towards the periphery; (b) The periplasm is delimited, and nuclear disintegration is beginning to occur; (c) The male nucleus is passing into the oosphere; (d) Oospore wall is beginning to form, and karyogamy has presumably occurred. [Figures from K. M. Safeulla, M. J. Thirumalachar, and C. G. Shaw, 1963, *Mycologia* **55**:819–823.]

    The oogonium and antheridium originate as multinucleate swellings on a diploid thallus. Each gametangium is eventually separated from the thallus by a septum. The gametangia are packed with nuclei and cellular organelles, including the endoplasmic reticulum, mitochondria, Golgi bodies, and various types of vesicles. These vesicles include some which contain lipid or dense bodies. Meiosis occurs more or less simultaneously in the gametangia. Some of the excess nuclei may abort before, during, or after meiosis.

    The antheridium and oogonium come into contact with each other, forming a more or less distinct adhesion zone at the point of contact (Hemmes and Ribeiro, 1977), perhaps at an early stage of development. Microbodies have been observed in the cytoplasm of both gametangia near the point of contact (Howard and Moore, 1970) or in the antheridium but not the oogonium (Hemmes and Ribeiro, 1977). Plasmogamy is initiated when a pore is formed between the two organs or when the antheridium penetrates the oogonial wall with one or more elongate *fertilization tubes*. Howard and Moore (1970) suggested that the microbodies may contain enzyme(s) that are responsible for development of a pore through which the fertilization tube passes. A fertilization tube reaches each egg, and a male nucleus is deposited in the

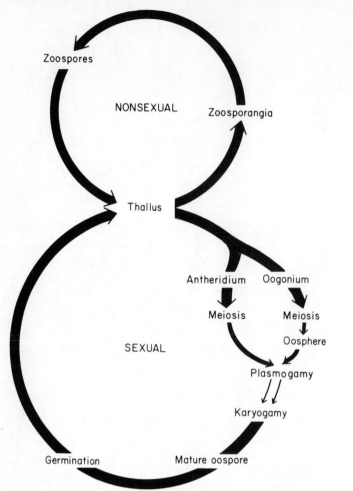

Zoospores

NONSEXUAL

Zoosporangia

Thallus

Antheridium

Oogonium

Meiosis

Meiosis

Oosphere

SEXUAL

Plasmogamy

Fig. 2-30. A generalized Oomycete life cycle.

Karyogamy

Germination

Mature oospore

vicinity of the egg to accomplish plasmogamy. The antheridium may be either partially or completely emptied.

The oogonium is meanwhile undergoing internal changes which lead to the formation of one or more eggs. There are two oogonial types. In one type, the mature oogonium will have the eggs differentiated as discrete, membrane-bounded oospheres floating within an oogonium which appears to be empty. The entire oosphere will form an oospore. In the second type of oogonium, a naked egg is surrounded by *periplasm*, which will not be incorporated into the oospore. Let us examine ultrastructural observations which exemplify each type.

1. *Saprolegnia* species form numerous free oospheres in oogonia. Within the young oogonium, a large vacuole is formed when numerous vesicles containing dense bodies enlarge and their membranes fuse. The vacuole expands further by taking up water from the surrounding cytoplasm, which simultaneously decreases the volume of the cytoplasm. The vacuole eventually forms furrows that extend through the peripheral cytoplasm and join with the plasmalemma. The oospheres are cleaved out by this action, and the tonoplast of the vacuole becomes the

plasmalemma of the oospheres. After plasmogamy, a wall is laid down around each oosphere to form the oospore (Gay et al., 1971; Howard and Moore, 1970).

2. *Pythium* species have oogonia in which a single naked egg lies within the periplasm. During the early stages, the oogonium has continuous cytoplasm, with the organelles more or less evenly distributed. The large central vacuole is absent. The organelles, including the nuclei, move to the peripheral cytoplasm in the oogonium. One of these nuclei will function as the female gamete. Plasmogamy occurs, and the periplasm becomes delimited at approximately the time of plasmogamy; therefore an oosphere is not discernible. The periplasm contains vacuoles and a great deal of endoplasmic reticulum. The oospore wall begins to form at the inner face of the periplasm as soon as plasmogamy occurs, delimiting the oospore from the periplasm. While the oospore wall is forming, the vacuole in the periplasm increases and all other organelles disintegrate. The zone between the oospore wall and oogonium eventually becomes empty, and the disintegrated periplasm may form a thin covering over the oospore and inside the oogonium (Haskins et al., 1976; McKeen, 1975).

After plasmogamy in species of *Pythium* and *Saprolegnia,* the male nucleus lies near the female nucleus, and the developing oospore is binucleate. The timing of karyogamy varies, depending upon the particular strain or species. The male and female nuclei may undergo karyogamy soon after plasmogamy or it may be delayed until the oospore has developed its thickened wall or, rarely, until the oospore begins to germinate.

A number of changes occur inside the oosphere as it matures into an oospore. There are several types of vesicles present in the oosphere. Some of these contain electron-dense material (a lipid-phosphoglucan complex), while others are microbodies (presumably glyoxysomes) that are associated with lipid (Beakes, 1980-b). The vesicles containing the electron-dense material coalesce to form a reserve globule. There is partial utilization of the neutral lipid and reorganization, so that at maturity the reserve dense globules and a variety of vesicles lie near the periphery of the oospore. Meanwhile many cytoplasmic organelles are lost.

The oospore undergoes a resting period and then germinates, usually by forming a germ tube terminated by a small functional zoosporangium. In his ultrastructural studies of oospore germination in *Saprolegnia ferax,* Beakes (1980-a) noted that the inner oospore wall becomes thinner while a new wall is secreted around the protoplast. The germ tube forms from the oospore. Germination may occur while the oospore is still in the oogonium, and then the elongating germ tube ruptures the oogonial wall. Meanwhile a large central vacuole has been forming in the oospore as a result of cytoplasmic breakdown (Beakes, 1980-a). The dense reserve globule disintegrates and the reserve glucans and lipids are utilized. Small granules from the disintegrating reserve globule are released into the central vacuole and peripheral vesicles, many of which become associated with mitochondria. Nuclear division occurs and the nuclei migrate into the elongating germ tube. Some of the protoplasmic constituents required for the early growth of the germ tube are probably derived by autolysis of the oospore contents by lysosomes (Beakes, 1980-b).

ORDER SAPROLEGNIALES

The Saprolegniales may be divided into six families (Sparrow, 1976). These fungi share at least some, but not necessarily all, the following characteristics: (1) they have both apically and laterally biflagellate zoospores or equivalent stages; (2) they form one to several oospores that lie free in the oogonium and lack periplasm; (3) the oospores usually have one or several reserve globules in the peripheral position (Sparrow, 1976).

The most conspicuous fungi are in the families Saprolegniaceae and Leptomitaceae. These fungi are often called *water molds* and are commonly found as saprophytes on plant and animal debris in water and soil. Although they typically inhabit fresh, unpolluted water, some members of the Leptomitaceae thrive in waters high in organic wastes, such as those polluted by industrial wastes. Although the majority are saprophytes, some species are parasitic on aquatic animals or on the roots of higher plants. Members of *Saprolegnia* are sometimes parasites on fish, causing superficial lesions and possibly damage to the viscera.

The thallus of members of the Saprolegniaceae is a well-developed mycelium consisting of a highly branched mass of hyphae of indeterminate growth. The hyphae lack constrictions and are typically nonseptate except where reproductive structures are delimited. Frequently the hyphae are visible macroscopically as a pustule or dense filamentous mass on the substratum after sufficient growth has occurred (Fig. 2-31).

Fig. 2-31.   A colony of *Achlya* growing in water culture.
[Courtesy Carolina Biological Supply Company.]

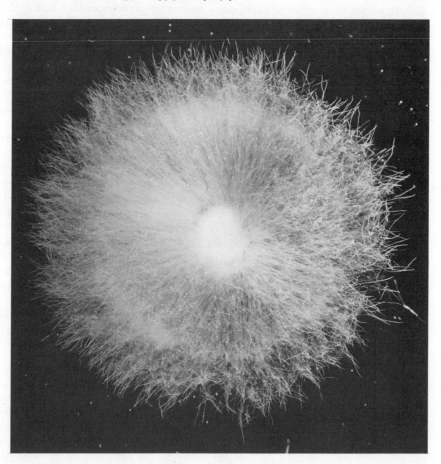

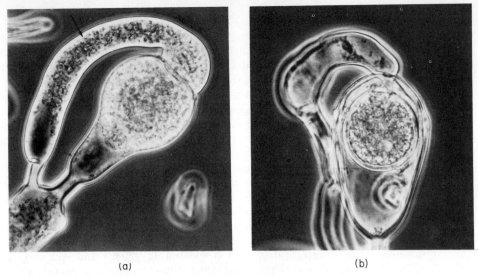

(a)   (b)

Fig. 2-32. Gametangia of *Sapromyces androgynus,* a member of the Leptomitaceae; (a) Plasmogamy between an antheridium (arrow) and an oogonium. Note the constrictions at the base of the antheridium and oogonium; (b) After the completion of plasmogamy, an oospore formed within the oogonium. [Courtesy D. Gotelli and K. Mitchell. *In:* Fuller, M. S., Ed. *Lower fungi in the laboratory.*]

Fig. 2-33. *Aphanomycopsis sexualis,* a member of the Ectrogellaceae. (a) Holocarpic thallus that converted into a zoosporangium which almost entirely fills a midge egg. × 235; (b) Branched discharge tubes arising from the zoosporangium. × 160; (c) Antheridium in contact with two oogonia. Each oogonium contains an oospore. × 360. [(a) Courtesy W. W. Martin; (b, c) from W. W. Martin, 1975, *Mycologia* **67:**923-933.]

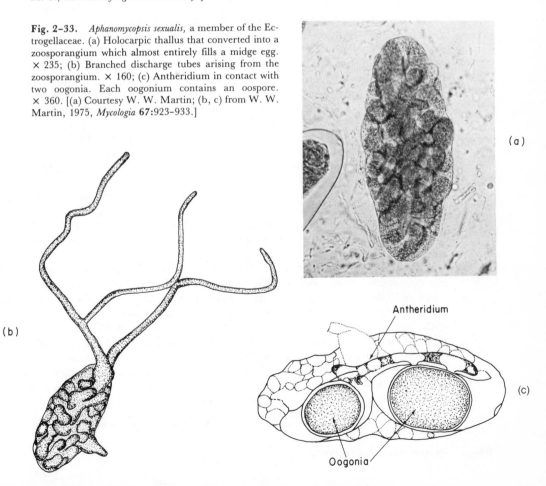

The thallus of the Leptomitaceae is generally similar to that of members of the Saprolegniaceae except that pseudosepta are formed at more or less regular intervals. The pseudosepta are partially plugged by a refractive material, *cellulin*. Because these hyphae are constricted at the sites where pseudosepta occur, the hyphae have a characteristic jointed and segmented appearance (Fig. 2–32). These hyphae are coenocytic with continuous cytoplasm.

Members of the remaining families are simple, chytridlike fungi that may lack sexual reproduction (Fig. 2–33). The best known among these simple fungi are those in the family Ectrogellaceae. Species in this family are endobiotic within algae (especially diatoms) or other fungi. The thallus is unicellular and unbranched or sparingly branched. It may develop to the extent that it pushes the valves of a diatom host apart. At maturity, the holocarpic thallus develops into a zoosporangium. Sexuality in this family is poorly known but may occur (Fig. 2–33).

Sexual Reproduction. There is a great deal of morphological variation among the gametangia and oospores in members of the Saprolegniales. This variation is of taxonomic importance and is used in delimiting genera and species (e.g., Dick, 1973-a, 1973-b; Johnson, 1956; Scott, 1961). Plasmogamy, karyogamy, and a general life cycle are described on pages 59–63.

The antheridium may simply be a cell beneath the oogonium [Fig. 2–34(a)]. Usually the antheridial apparatus consists of a branched or unbranched filament terminated by an antheridial cell. An antheridium may occur on the same thallus producing its oogonium and may arise from the oogonium itself, the stalk supporting the oogonium [Fig. 2–34(c)], or a vegetative hypha supporting the oogonium [Fig. 2–34(b)]. In some species, the antheridium arises from a different hypha than that supporting the oogonium [Fig. 2–34(d)].

In its simplest form, the oogonium may simply be similar to a hyphal segment. The oogonium may occur in an intercalary position (within the hypha) [Fig. 2–34(e)], on a short lateral branch [Fig. 2–34(d,e)], or terminating the main hypha [Fig. 2–34(f,g)]. The shape of the oogonium may vary from elongate to angular [Fig. 2–34(g)]. Usually the oogonia are pyriform [Fig. 2–34(b,f)] or spherical [Fig. 2–34(a,c)]. Oogonial walls may be smooth [Fig. 2–34(a–f)] or ornamented [Fig. 2–34(d)]. Although some species produce a single oosphere in an oogonium [Fig. 2–34(d)], most produce more than one, with a range of 2 to 100 or more in a single oogonium [Fig. 2–34(g)].

Oospores have their reserve materials located in various more or less constant locations. Coker (1923) recognized two principal arrangements. These are the *centric* oospore, in which one or two layers of small droplets surround the central ooplasm [Fig. 2–34(b,f)], and the *eccentric* oospore, which has one or a few large drops on one side of the ooplasm only [Fig. 2–34(g)].

In the genera *Dictyuchus* and *Achlya,* some thalli will have only antheridia while others will have only oogonia. Therefore one thallus is "male" and the second is "female." Because fertilization must occur between two thalli, they are *heterothallic.* Most species of the Saprolegniales are *homothallic,* with fertilization occurring between gametangia on the same thallus. Homothallism and heterothallism are discussed further in Chapter 9. Differentiation of the male and female thalli in heterothallic species of *Achlya* is under complex hormonal control. This topic is also discussed in Chapter 9.

Nonsexual Reproduction. The zoosporangia are usually elongate or spherical and occur most often in a terminal position on the thallus. The zoosporangium may proliferate internally, that is, when a zoosporangium has been emptied after the discharge of its zoospores, another zoosporangium may form within it.

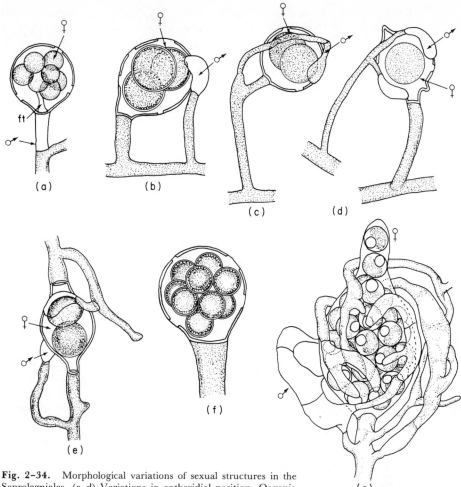

Fig. 2-34. Morphological variations of sexual structures in the Saprolegniales. (a-d) Variations in antheridial position. Oogonia are all lateral; (e-f) Variations in oogonial position, intercalary and terminal; (a-g) Variations in shape of oogonium; (g) Highly branched and swollen antheridia; (a, c-e) Oosphere stage; (b, f, g) Oospore stage; (f) Centric oospores; (g) Eccentric oospores. [Figures a-f from R. L. Seymour, 1970, *Nova Hedwigia* **19**:1-124; Figure g from T. W. Johnson, Jr., 1974, *Am. J. Botany*, **61**:244-252.]

There is a great deal of variation in the type and behavior of the zoospores. Species may have *monomorphic* zoospores, that is, form only one type of zoospores in their life cycle. Other species may have *dimorphic* (literally meaning "two-forms") *zoospores* and form anteriorly biflagellate zoospores followed later by laterally biflagellate zoospores. Because of their order of appearance, the anteriorly biflagellate zoospores are generally termed *primary* zoospores while the laterally biflagellate zoospores are termed *secondary* zoospores (Fig. 2-27).

We will first examine representatives of *Saprolegnia,* which have dimorphic zoospores. The primary zoospore squeezes through the papilla and then swims slowly for about 5 to 10 minutes (Fig. 2-6). The zoospores then withdraw their flagella, round up, and form a cyst. The cyst rests for up to 4 hours or more and then germinates. The secondary zoospore squeezes out through a pore at the mouth of an exit papilla and emerges as a cell with a much larger volume than the cyst. It remains attached to the emptied cyst for about 15 to 30 minutes and then swarms for another 30 minutes. The secondary zoospore darts and glides quickly; it is a more active swimmer and swarms longer than the primary zoospore. It then slows down and jerks, and a beadlike protrusion is formed on the flagella, which then fall off the zoospore within 1 to 2 minutes. Encystment occurs after the flagella are lost (Crump and Branton, 1966). Unlike the primary zoospore, which encysts only once, the secondary zoospore may encyst repeatedly. When a cyst formed by a secondary zoospore germinates, it

Fig. 2-35. Nonsexual reproductive structures in the Saprolegniales. (a) *Achlya*-type with encysted primary zoospores at mouth of the zoosporangium; (b) Cysts and emptied cyst cases in zoosporangia of *Dictyuchus;* (c) Row of cysts remaining behind after disintegration of zoosporangial wall in *Brevilegnia;* (d) Cysts germinating before their release from the zoosporangium in *Achlya.* [From W. C. Coker, 1927, *J. Elisha Mitchell Sci. Soc.* **42**:207–226.]

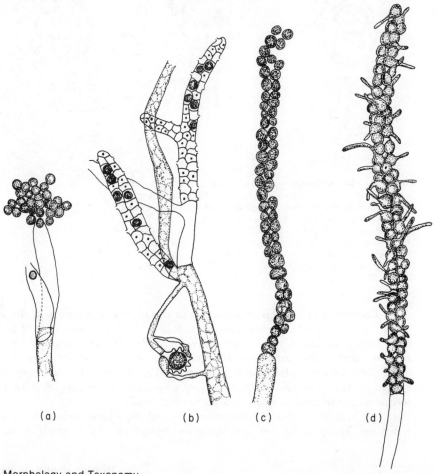

(a)   (b)   (c)   (d)

may release another secondary zoospore or may form a germ tube. At the ultrastructural level, it may be seen that there are a number of changes in the arrangement and morphology of the cellular organelles as the zoospore goes through its cycle involving the alternation of primary zoospore, cyst, secondary zoospore, cyst, and germination (Holloway and Heath, 1977). These changes include elongation of mitochondria in the motile stage and their shortening in the cysts. The cisternae of the rough endoplasmic reticulum are stacked in the primary zoospore but randomly arranged in both types of cysts and in the secondary zoospore. The organelles have a precise arrangement in both the primary and secondary zoospores but appear to be randomly arranged in the cysts.

Unlike species of *Saprolegnia*, which have pronounced swarming of both primary and secondary types of zoospores, other members of the Saprolegniales may have suppression of either one or the other of the stages. Members of the genus *Pythiopsis* have monomorphic zoospores and form only the primary type. These swarm, encyst, and germinate to form a new thallus. In species of *Achlya* and some species of *Ectrogella*, which are dimorphic, the primary zoospores typically emerge from the zoosporangium and immediately encyst at the mouth of the zoosporangium without swarming [Fig. 2-35(a)]. These cysts then germinate to release secondary zoospores that swarm. In species of *Dictyuchus*, there is further suppression and the primary zoospore stage is represented by cysts within the zoosporangium. These cysts remain in the zoosporangium and release secondary zoospores through a papilla that penetrates the zoosporangial wall and in which a pore is formed. The empty cysts, still contained within the zoosporangium, give this cell a netlike pattern [Fig. 2-35(b)]. Motile stages are regularly absent in members of one genus, *Geolegnia*, or absent under unfavorable environmental conditions in other fungi (e.g., *Achlya*). The zoospores are represented by cysts in the zoosporangium, and these cysts are usually released when the wall disintegrates [Fig. 2-35(c)]. Sometimes the cysts will begin to germinate while in the zoosporangium [Fig. 2-35(d)]. The cysts germinate by forming a germ tube.

Gemmae may be produced by some members of the Saprolegniales. Gemmae are hyphal segments which have been delimited by septa, are often irregular in form, and have dense cytoplasm. These structures may be separated from the parent mycelium and when conditions are favorable, may produce a new thallus. Gemmae germinate by forming a germ tube or by functioning as a zoosporangium or even as an oogonium.

ORDER LAGENIDIALES

The Order Lagenidiales is a group of simple, holocarpic, and endobiotic fungi which may or may not be closely related (Sparrow, 1960, 1973-c). They are facultative parasites and occur chiefly within algae and other aquatic fungi, although some are parasitic within various forms of animal life such as nematodes, daphne, mosquito larvae, and other insects. These fungi occur principally in fresh water habitats, although there are some marine forms.

The multinucleate thallus may be a single, more or less amoeboid mass as in *Olpidiopsis*, or it may be a tubular or filamentous structure as in *Lagenidium* or *Haliphthoros* (Fig. 2-36). It may be either unbranched or sparingly branched and is nonseptate at first but may later become septate. In unicellular thalli such as those formed by species of *Olpidiopsis* (Fig. 2-37), each thallus is converted into a zoosporangium or gametangium. In the filamentous or tubular thalli, each cell will convert into a zoosporangium or gametangium at maturity.

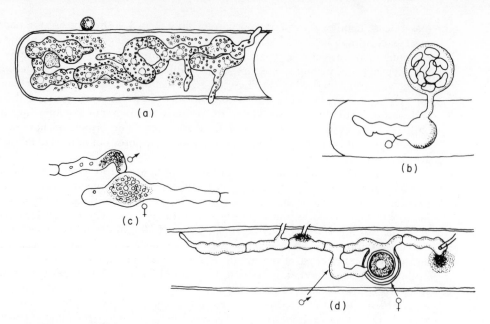

Fig. 2-36. *Lagenidium rabenhorstii,* a member of the Lagenidiales. (a) Vegetative thallus in cell of algal host; (b) An isolated zoosporangium has formed a vesicle and zoospores; (c) An early stage in union of the gametangia; (d) A thallus has become septate and divided into zoosporangia, an antheridium, and oogonium. An oospore is visible within the oogonium. [From W. Zopf, 1884, *Nova Acta Acad. Leop.* **47**:143–236, plates 12–21.]

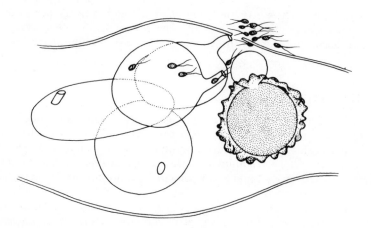

Fig. 2-37. A member of the Lagenidiales, *Olpidiopsis* within a hypha of its host. Note the zoosporangia that are discharging zoospores (left) and the small antheridium that is in contact with the spiny oogonium (right). × 447. [From W. C. Coker, 1923, *The Saprolegniaceae,* University of North Carolina Press, Chapel Hill, N.C.]

In some species, the contents of the zoosporangium cleave into biflagellate zoospores before their discharge from the zoosporangium. The zoospores exit through the usually single papilla which leads from the zoosporangium to the exterior. In contrast, others discharge the undifferentiated zoosporangial protoplasm to the exterior, where it continues its differentiation into zoospores. In some species, the outer portion of the undifferentiated protoplasm forms an evanescent, delicate bubble or saclike structure, the *vesicle*. The remainder of the protoplasm within it then continues to differentiate into zoospores, which are released when the vesicle bursts. In other members of this family, a vesicle is not formed around the extruded protoplasm.

The zoospores are somewhat variable in shape, but typically they are of the secondary type and are reniform, with the two flagella inserted in the concave elongate indentation. The zoospore is monomorphic, but it may undergo repeated encystment and emergence.

Sexual reproduction is only known to occur in some members of this group. Typically, a portion of the thallus may either function as or differentiate into gametangia. The gametangia may be alike in size or, more often, the antheridium may be smaller than the oogonium (Fig. 2–36). Plasmogamy takes place when the contents of the antheridium flow into the oogonium through either a fertilization tube or a pore. There are few records of either karyogamy or meiosis in this group. A thick, often spiny wall forms around the zygote, now an oospore. The oospore usually lies free within the gametangium. In a few species some cytoplasm surrounds the oospore after it matures (this remaining cytoplasm is the periplasm). In some instances, the oospore has been observed to germinate as a zoosporangium.

Parthenogenetic development of the thick-walled oospore is common.

ORDER PERONOSPORALES

In the Order Peronosporales, there are some fungi that inhabit water or damp soil and live as saprophytes or weak parasites (Waterhouse, 1973-b). Many of these fungi, however, have abandoned the aquatic habitat and live on the dry, aerial parts of plants as parasites. Plant parasites in this order include species of *Pythium* that cause damping-off diseases of seedlings, *Peronospora* and other genera that cause downy mildew diseases of various plants, *Albugo,* which causes the white rust disease, and *Phytophthora. Phytophthora infestans* causes "late blight of potatoes," a disease responsible for the potato famine in Ireland, and which thus changed the course of history in that country (see Chapter 12).

Characteristics delimiting this group of fungi from others are: (1) only monomorphic zoospores of the secondary type may be formed; (2) periplasm surrounds the egg in the oogonium; and (3) the oospore bears a centrally located reserve globule (Sparrow, 1976). Sexual reproduction usually involves the formation of a single oospore within an oogonium. Morphological and cytological details of sexual reproduction are essentially similar to those detailed on pages 59–63 for the Saprolegniales.

The coenocytic hyphae are more slender than those in the Saprolegniales and are freely branching. Septa may be formed to delimit reproductive organs and in older mycelia. In parasitic species, hyphae may occur principally between the cells (intercellularly) and derive their nourishment through peg-like hyphal branches which they send into the host cell. These specialized food-absorbing organs are called *haustoria.* In some members, an intracellular hypha which penetrates the cells may also be produced.

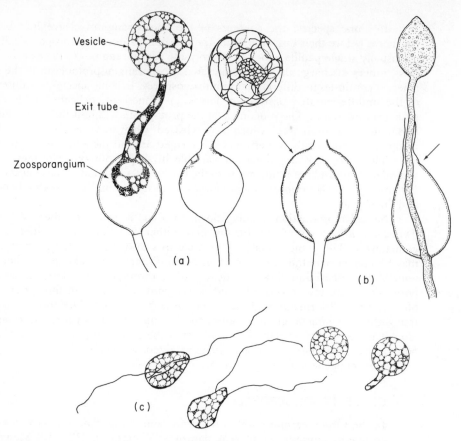

Fig. 2-38. Nonsexual reproduction in *Pythium.* (a) Zoosporangia with vesicles. A vesicle with undifferentiated protoplasm is on the left, and a later stage with delimited zoospores is on the right; (b) Two types of proliferation of zoosporangia. The original, older zoosporangium is at the arrow; (c) From left to right, swarming, cyst, and germination of zoospores. [From J. T. Middleton, 1943, *Mem. Torrey Botan. Club* **20** (1):1-171.]

Many aspects of nonsexual reproduction are unique for this order and some species may show an adaptation to the drier terrestrial environment and a requirement for spore dissemination by air rather than water. Most of the species form zoosporangia, which range from an undifferentiated hyphal filament to spherical, oval, or pyriform cells. The zoosporangia are often borne on a specialized hypha, which is morphologically different from the vegetative hyphae and is called a *sporangiophore.* The zoosporangia usually germinate when they form zoospores. Depending upon the species, the zoosporangia may: (1) release their zoospores directly into the environment; (2) release their mature zoospores into a vesicle; or (3) release apparently undifferentiated cytoplasm into the vesicle, where it then differentiates into zoospores (Fig. 2-38). Alternatively, the zoosporangium may not form zoospores at all but germinate by forming a germ tube. This latter type of germination may be either environmentally or genetically controlled. If the zoosporangium germinates by forming a germ tube, it is functioning as does a *conidium,* a wind-disseminated nonsexual spore. Conidia are common in all groups of the higher fungi, which will be discussed in the following chapters. Differences in nonsexual reproduction are of great importance in delimiting the genera in the Peronosporales. We will examine some representative genera, particularly noting nonsexual reproduction.

Pythium. *Pythium* is a large genus. Its species are distributed worldwide and include saprophytes in water or soil and parasites on algae, other fungi, or higher plants (Middleton, 1943). The coenocytic thallus consists of an irregularly branching mycelium. In higher plants, the mycelium is most often found growing within the cells (intracellularly) of roots or stems. This invasion results in the rotting of the plant part. One well-known disease of this type is damping off, which attacks the roots of emerging seedlings and may be caused by *Pythium debaryanum* (Fig. 2–39). The zoosporangia are intercalary or terminal in position. In shape, they are usually filamentous and only slightly wider than the mycelium or spherical. The papilla may be sessile or may be borne upon a discharge tube of varying length. Upon germination, a membrane-bounded vesicle is produced at the end of the discharge tube [Fig. 2–38(a)]. The undifferentiated protoplasm from the zoosporangium passes into it. Within about 15 to 20 minutes, the protoplasm differentiates into zoospores. The zoospores move about and are finally released when the vesicle ruptures. The zoospores are of the secondary, laterally biflagellate type and usually encyst only once, but they may undergo repeated swarming and encystment. If the cyst is to release a zoospore, it does so by forming a small vesicle at the mouth of an exit tube, and subsequently a single zoospore emerges from the vesicle. Proliferation of the emptied zoosporangium may occur when a new zoosporangium forms inside the old one. Sometimes the supporting hypha will grow through the emptied zoosporangium, producing another zoosporangium at its tip [Fig. 2–38(b)].

Phytophthora. The genus *Phytophthora* includes some species which are capable of living saprophytically in the soil until a host plant is available, while others are known to occur only on their host plants. Many species are able to attack the aerial parts of the plants, causing the blights or rots of affected parts and quickly killing infected tissues. *Phytophthora infestans* causes late blight of potatoes, resulting in the rotting of the tubers. The mycelium of these plant parasitic fungi may grow either within the cells (intracellularly) or between the cells (intercellularly). Haustoria are produced in intercellular mycelium. The sporangiophores are usually branched and may be formed on the surface of the soil or, in the case of fungi on plants, may emerge from the host through the epidermis or stomates. The zoosporangia are oval, ellipsoidal, or somewhat lemon-shaped. For those species occuring within the soil, the zoosporangia usually lack a papilla, undergo proliferation, and are not shed. In contrast, the zoosporangia of those species that occur aboveground on plants have a broad, conspicuous papilla at the apex, do not undergo proliferation, and are deciduous. The zoosporangia which have been shed may then be wind-disseminated. The zoosporangia are then capable of germinating in one of two ways: (1) indirectly by forming zoospores that are released without a vesicle being formed; or (2) directly by forming a germ tube. Once released, the zoospores are capable of repeated swarming and encystment stages.

The mode of germination of the zoosporangia (either direct or indirect) may be determined by age or environmental conditions. In *P. infestans,* indirect germination by zoospore formation occurs when temperatures are low (optimum at 12°C) and direct germination occurs at higher temperatures (optimum at 25°C). This differential germination contributed greatly to the Irish potato famine as there were a number of cool summers in succession, which resulted in more propagative units being formed.

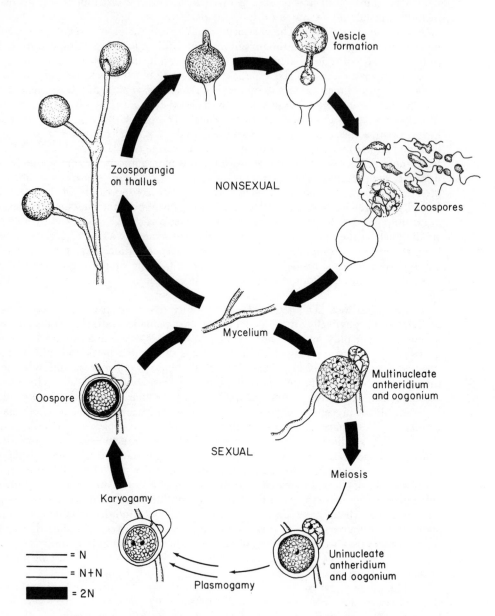

Vesicle
formation

Zoospores

Zoosporangia
on thallus

NONSEXUAL

Mycelium

Multinucleate
antheridium
and oogonium

Oospore

SEXUAL

Meiosis

Karyogamy

Uninucleate
antheridium
and oogonium

_____ = N

_____ = N+N

▬▬▬ = 2N

Plasmogamy

Fig. 2–39. Life cycle of *Pythium debaryanum,* a member of the Peronosporales.

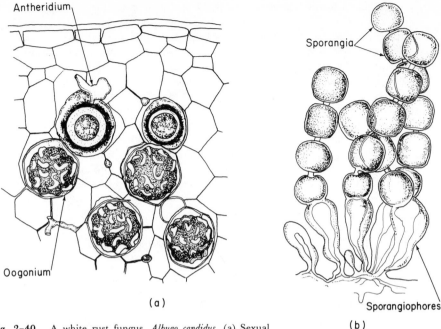

Antheridium

Sporangia

Oogonium

(a)

Sporangiophores

(b)

Fig. 2-40. A white rust fungus, *Albugo candidus*. (a) Sexual reproductive structures within the host tissue. An antheridium, oospores, and oogonia are visible; (b) Chains of sporangia on surface of the host. [From A. N. Berlese, 1898, *Icones Fungorum. Phycomycetes.* Published by author.]

Ultrastructural studies have been made of zoosporangium germination in *Phytophthora parasitica*. When the zoosporangium cleaves into zoospores in indirect germination stimulated by environmental conditions, cleavage of the zoospores by vesicles occurs. These vesicles are apparently derived from the Golgi apparatus. Cleavage of the zoospores occurs in the manner which is typical for this group (Hohl and Hamomoto, 1967). If the zoosporangium is to undergo indirect germination, formation of the flagella and the initial steps in fusion of vesicles and cleavage occur as they would for zoosporogenesis. However, this is usually followed by the abnormal disappearance of the flagella, and cleavage is rarely completed. Sometimes the zoospores may be cleaved out but not released. An additional wall is formed around the inside of the zoosporangial wall. The Golgi-derived vesicles that would ordinarily form the plasmalemma of the zoospores in indirect germination appear to be involved in the transport of precursor materials for the new wall. A germ tube is formed which penetrates through either the zoosporangial wall or the fibrous plug in the discharge tube. Lomasomes, possibly derived from the Golgi apparatus, appear to be active during the penetration of the wall by the germ tube (Hemmes and Hohl, 1969).

Albugo. Members of the genus *Albugo* are parasites of plants, causing the white rust disease. This disease is especially common on the plant called shepherd's purse (*Capsella*). The mycelium is intercellular and produces haustoria. The sporangiophores and zoosporangia* are produced in a palisadelike mass (*sorus*), which is formed be-

* These zoosporangia are called *conidia* by some mycologists. Refer to Chapter 6 for a discussion of this term.

tween the mesophyll and epidermis of the host plant and subsequently ruptures the epidermis (Fig. 2–40). The exposed pustule appears white. The sporangiophores are upright, club-shaped, and unbranched. Zoosporangia are formed at the tip of the sporangiophores in chains, with the youngest zoosporangia at the base of the chain. In this process, a zoosporangial initial buds out from the tip of the sporangiophore, and after it has achieved a certain size, it is delimited from the sporangiophore by a septum and is now a separate zoosporangium. This zoosporangium is then pushed upward when another one is delimited in the same fashion immediately under it. The zoosporangia are separated by a *disjunctor,* consisting of a double septum with an unusually thick middle wall (Khan, 1977). When the disjunctors rupture, the zoosporangia are shed. When the zoosporangia fall into water, perhaps forming an aqueous film on a plant, they germinate by cleaving into zoospores that then escape from the zoosporangium. These zoospores encyst and reinfect a host plant when they form a germ tube. The sexual stages are found within the host tissue.

Fig. 2–41.   Three members of the Peronosporales (downy mildew fungi). (a). *Basidiophora entospora,* which has simple, unbranched stout sporangiophores emerging through the epidermis of its host. An oogonium lies buried within the tissues (arrow); (b) Apical portion of a slender, dichotomously branching sporangiophore of *Peronospora potentillae;* (c) Tips of sporangiophores of *Bremia.* [From A. N. Berlese, 1898, *Icones Fungorum. Phycomycetes.* Published by author.]

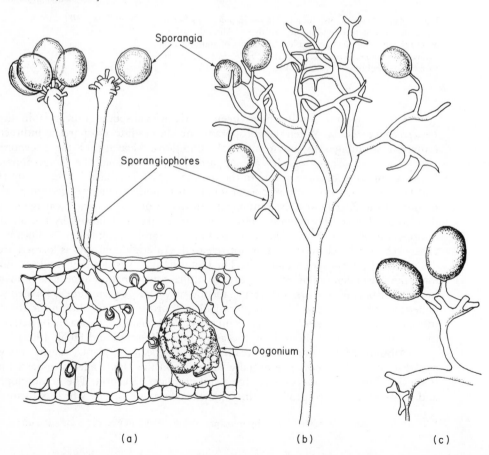

Sporangia

Sporangiophores

Oogonium

(a)   (b)   (c)

Peronospora and Other Downy Mildew Fungi. Members of *Peronospora* are among the fungi that cause downy mildew diseases of vascular plants (Waterhouse, 1973-b). These fungi are specialized parasites, apparently occurring only on vascular plants. The mycelium is intercellular and haustoria are produced. Sporangiophores protrude through stomates of the host, giving the appearance of a "down" that varies from white to gray in color. The sporangiophores branch in a way that is characteristic for the genus (Fig. 2–41). For example, the branches of *Peronospora* species are produced dichotomously and are gently curving. Zoosporangia are formed at the tips of the sporangiophores and are always deciduous and wind-disseminated. In most species, the zoosporangia cleave into zoospores when they germinate. However, in *Peronospora* and *Bremia* the zoosporangia typically germinate by producing a germ tube. In members of these two genera, the zoosporangium almost always germinates in the same manner as a conidium (a term used by many mycologists to describe these structures) and its role as a zoosporangium is essentially absent. The downy mildews have therefore bridged the gap between the aquatic and terrestrial environments as they do not occur in the aquatic habitat, and the motile zoospore stage is all but absent in species of two of the genera. The sexual stages (gametangia and oospores) are found within the host tissues (Fig. 2–41).

PHYLOGENY AND RELATIONSHIPS

The origin of the fungi in the division Mastigomycota and their evolution are not known but have been the subject of some speculation. Sparrow (1958, 1960, 1973-a, 1973-b, 1976) places great emphasis on the flagellation, and suggests that the separate classes arose from different flagellate ancestral lines. According to Sparrow (1958), the Chytridiomycetes were derived from a posteriorly uniflagellate protozoan, following a line sequentially through the Chytridiales, Blastocladiales, and Monoblepharidales. The Monoblepharidales did not give rise to additional fungi. The Hyphochytridiomycetes were derived from an anteriorly uniflagellate protozoan and presumably represent an evolutionary blind alley. Similarly, the Plasmodiophoromycetes were derived from a separate, biflagellate protozoan characterized by at least some of the Oomycetes and were presumed to originate from biflagellate algae, an argument which is bolstered by the cellulose in their walls and their coenocytic hyphae. Possible relationships and phylogeny within the Oomycetes is fairly complex, and Sparrow (1976) recognizes two groups, the Saprolegnian Galaxy which has dimorphic zoospores, one to several eggs without periplasm, and oospores with a peripheral reserve globule. The Eurychasmales and Saprolegniales belong to this galaxy. The second group is the Perosporacean Galaxy with monomorphic zoospores, periplasm surrounding a single egg, and the oospore with a single, central reserve body. The Lagenidiales and Peronosporales belong to the last galaxy.

Olson and Fuller (1968) take an opposing viewpoint and claim that there is ultrastructural evidence that the uniflagellate fungi were derived from biflagellate stock through loss of a flagellum. They consider the second centriole to be a vestigial flagellum base in uniflagellate chytrids.

Barr (1978) has considered the evolution of chytrids. Much of his evidence is based on zoospore ultrastructure. He states that the primitive type of zoospores comprises those lacking specialized organelles such as Golgi bodies or rumposomes and also lacking elaborate internal organization of the organelles (close association of organelles or connection of organelles to microtubules). According to Barr, the

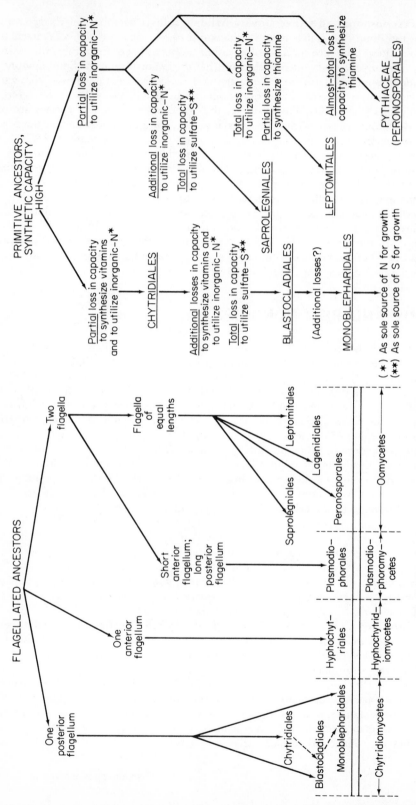

Fig. 2-42. Possible evolution of the division Mastigomycota based on morphological (left) and nutritional evidence (right). [From E. C. Cantino, 1966. *In:* Ainsworth, G. C., and A. S. Sussman, Eds. *The fungi—an advanced treatise.* Academic Press, New York. **2:**283–337.]

holocarpic and eucarpic chytrids are not closely related because the holocarpic chytrids have a fibrillar rhizoplast, unlike the microtubular rhizoplast in the eucarpic forms. Presumably the ancestral type of the eucarpic line was a monocentric chytrid in which the zoosporangial wall dissolved to release the zoospores, giving rise to an ancestral *Rhizophydium,* perhaps with multiple papilla. From this ancestor two lines diverged. One line gave rise to monocentric chytrids and finally to polycentric chytrids in which the number of papilla was reduced. The second line retained multiple papillae and gave rise to both monocentric and polycentric chytrids, culminating in the evolution of the Blastocladiales. The zoospore in the Blastocladiales is of the advanced type with a nuclear cap, side-body complex consisting of lipid globules and a microbody, and microtubules that radiate from the kinetosome. The Harpochytriales and Monoblepharidales may have been derived from polycentric chytrids.

Nutritional evidence has been used as evidence of possible relationships (Cantino, 1966). The scheme that can be derived generally supports the grouping that Sparrow has made on the basis of flagellation (Sparrow, 1958, 1960). Those fungi which are presumed to be most primitive have the capacity to utilize inorganic sulfur and inorganic nitrogen as nutrients, while more advanced members of this group require more complex forms of sulfur and nitrogen. In addition, those fungi which are assumed to be most primitive are able to synthesize all their required vitamins, while more advanced fungi have lost some of this synthetic capacity and may require an exogenous source of vitamins. Compare the evolutionary schemes based on morphology and nutrition in Fig. 2-42. Nutrition of fungi is discussed further in Chapter 7.

References

Aist, J. R., and P. H. Williams. 1971. The cytology and kinetics of cabbage root hair penetration by *Plasmodiophora brassicae.* Can. J. Botany 49:2023–2043 (Plasmodiophorales).

Aronson, J. M., and C. C. Lin. 1978. Hyphal wall chemistry of *Leptomitus lacteus.* Mycologia 70:363–369 (Saprolegniales).

Barr, D. J. S. 1975. Morphology and zoospore discharge in single-pored, epibiotic Chytridiales. Can. J. Botany 53:164–178.

——. 1978. Taxonomy and phylogeny of chytrids. Biosystems 10:153–165.

——, and V. E. Hadland-Hartmann. 1977. Zoospore ultrastructure of *Olpidium cucurbitacearum (Chytridiales).* Can. J. Botany 55:3063–3074.

——, and ——. 1978. Zoospore ultrastructure in the genus *Rhizophydium* (Chytridiales). Can. J. Botany 56:2380–2404.

Beakes, G. W. 1980-a. Electron microscopic study of oospore maturation and germination in an emasculate isolate of *Saprolegnia ferax.* 1. Gross changes. Can. J. Botany 58:182–194 (Saprolegniales).

——. 1980-b. ——. 3. Changes in organelle status and associations. Can. J. Botany 58:209–227 (Saprolegniales).

——. 1980-c. ——. 4. Nuclear cytology. Can. J. Botany 58:228–240 (Saprolegniales).

Bland, C. E., and H. V. Amerson. 1973. Observations on *Lagenidium callinectes:* isolation and sporangial development. Mycologia 65:310–320 (Lagenidiales).

Bradley, W. H. 1967. Two aquatic fungi (Chytridiales) of Eocene age from the Green River Formation of Wyoming. Am. J. Botany 54:577–582.

Braselton, J. P., C. E. Miller, and D. G. Pechak. 1975. The ultrastructure of cruciform nuclear division in *Sorosphaera veronicae* (Plasmodiophoromycete). Am. J. Botany 62:349–358.

Bryant, T. R., and K. L. Howard. 1969. Meiosis in the Oomycetes: 1. a microspectrophotometric analysis of nuclear deoxyribonucleic acid in *Saprolegnia terrestris.* Am. J. Botany 56:1075–1083 (Saprolegniales).

Canter, H. M. 1946. Studies on British chytrids. I. *Dangeardia mammillata* Schröder. Trans. Brit. Mycol. Soc. **29:**128–134. (Chytridiales).

Cantino, E. C. 1966. Morphogenesis in aquatic fungi. *In:* Ainsworth, G. C., and A. S. Sussman, Eds., The fungi—an advanced treatise. Academic Press, New York **2:**283–337.

Coker, W. C. 1923. The Saprolegniaceae with notes on other water molds. University of North Carolina Press, Chapel Hill, N.C. 201 pp.

Couch, J. N. 1924. Some observations on spore formation and discharge in *Leptolegnia, Achlya,* and *Aphanomyces.* J. Elisha Mitchell Sci. Soc. **40:**27–42 (Saprolegniales).

——. 1935. A new saprophytic species of *Lagenidium,* with notes on other forms. Mycologia **27:**376–387 (Lagenidiales).

——, and A. J. Whiffen. 1942. Observations on the genus *Blastocladiella.* Am. J. Botany **29:**582–591.

Crump, E., and D. Branton, 1966. Behavior of primary and secondary zoospores of *Saprolegnia* sp. Can. J. Botany **44:**1393–1400 (Saprolegniales).

Dick, M. W., 1969. Morphology and taxonomy of the Oomycetes, with special reference to Saprolegniaceae, Leptomitaceae and Pythiaceae. 1. Sexual reproduction. New Phytologist **68:**751–755.

——. 1973-a. Saprolegniales. *In:* Ainsworth, G. C., F. K. Sparrow, and A. S. Sussman, Eds., The fungi—an advanced treatise. Academic Press, New York **4B:**113–144.

——. 1973-b. Leptomitales. *In:* Ainsworth, G. C., F. K. Sparrow, and A. S. Sussman, Eds., The fungi—an advanced treatise. Academic Press, New York **4B:**145–158.

——, and Win-Tin. 1973. The development of cytological theory in the Oomycetes. Biol. Rev. **48:**133–158.

Dogma, I. J., Jr. 1973. Developmental and taxonomic studies on Rhizophlyctoid fungi, Chytridiales. I. Dehiscence mechanisms and generic concepts. Nova Hedwigia **24:**393–411.

——. 1974. Developmental and taxonomic studies on Rhizophlyctoid fungi. IV. *Karlingia granulata, Karlingia spinosa,* and *Karlingiomyces dubius.* Nova Hedwigia **25:**91–105.

Drechsler, C. 1930. Repetitional diplanetism in the genus *Phytophthora.* J. Agr. Res. **40:**557–573 (Peronosporales).

——. 1952. Production of zoospores from germinating oospores of *Pythium ultimum* and *Pythium debaryanum.* Bull. Torrey Botan. Club **79:**431–450 (Peronosporales).

Elliott, C. G., and D. MacIntyre. 1973. Genetical evidence on the life-history of *Phytophthora.* Trans. Brit. Mycol. Soc. **60:**311–316 (Peronosporales).

Ellzey, J. T. 1974. Ultrastructural observations of meiosis within antheridia of *Achlya ambisexualis.* Mycologia **66:**32–47.

Emerson, R. 1958. Mycological organization. Mycologia **50:**589–621.

——, and H. C. Whisler. 1968. Cultural studies of *Oedogoniomyces* and *Harpochytrium,* and a proposal to place them in a new order of aquatic Phycomycetes. Arch. Mikrobiol. **61:**195–211 (Harpochytriales).

——, and C. M. Wilson. 1954. Interspecific hybrids and the cytogenetics and cytotaxonomy of *Euallomyces.* Mycologia **46:**393–434 (Blastocladiales).

Fuller, M. S. 1977. The zoospore, hallmark of the aquatic fungi. Mycologia **69:**1–20.

——, and R. Reichle. 1965. The zoospore and early development of *Rhizidiomyces apophysatus.* Mycologia **57:**946–961 (Hyphochytridiales).

Gay, J. L., A. D. Greenwood, and I. B. Heath. 1971. The formation and behaviour of vacuoles (vesicles) during oosphere development and zoospore germination in *Saprolegnia.* J. Gen. Microbiol. **65:**233–241 (Saprolegniales).

Haskins, R. H., J. A. Brushaber, J. J. Child, and L. B. Holtby. 1976. The ultrastructure of sexual reproduction in *Pythium acanthicum.* Can. J. Botany **54:**2193–2203 (Peronosporales).

Hatch, W. R. 1938. Conjugation and zygote formation in *Allomyces arbuscula.* Am. J. Botany **2:**583–614 (Blastocladiales).

Heath, I. B. 1976. Ultrastructure of freshwater Phycomycetes. *In:* Jones, E. B. G., Ed., Recent advances in aquatic mycology. John Wiley & Sons, New York. pp. 603–650.

Held, A.. A., 1973. Development of endoparasitic, zoosporic fungi. Bull Torrey Botan. Club **100:**203–216.

——.1975. The zoospore of *Rozella allomycis:* ultrastructure. Can. J. Botany **53:**2212–2232 (Chytridiales).

Hemmes, D. E., and H. R. Hohl. 1969. Ultrastructural changes in directly germinating sporangia of *Phytophthora parasitica*. Am. J. Botany **56**:300–313 (Peronosporales).

——, and O. K. Ribeiro. 1977. Electron microscopy of early gametangial interaction in *Phythphthora megasperma* var. *sojae*. Can. J. Botany **55**:436–447.

Hoch, H. C., and J. E. Mitchell. 1972. The ultrastructure of *Aphanomyces euteiches* during asexual spore formation. Phytopathology **62**:149–160 (Saprolegniales).

Hohl, H. R., and S. T. Hamamoto. 1967. Ultrastructural changes during zoospore formation in *Phytophthora parasitica*. Am. J. Botany **54**:1131–1139 (Peronosporales).

Holloway, S. A., and I. B. Heath. 1977. An ultrastructural analysis of the changes in organelle arrangement and structure between the various spore types of *Saprolegnia*. Can. J. Botany **55**:1328–1339 (Saprolegniales).

Howard, K. L., and T. R. Bryant. 1971. Meiosis in the Oomycetes: II. A microspectrophotometric analysis of DNA in *Apodachlya brachynema*. Mycologia **63**:58–68 (Saprolegniales).

——, and R. T. Moore. 1970. Ultrastructure of oogenesis in *Saprolegnia terrestris*. Botan. Gaz. **131**:311–336 (Saprolegniales).

Johns, R. M. 1964. A new *Polyphagus* in algal culture. Mycologia **56**:441–451 (Chytridiales).

Johnson, T. W., Jr. 1956. The genus *Achlya:* morphology and taxonomy. University of Michigan Press, Ann Arbor, Mich. 180 pp. (Saprolegniales).

Karling, J. S. 1937. The structure, development, identity, and relationship of *Endochytrium*. Am. J. Botany **24**:352–364 (Chytridiales).

——. 1939. A new fungus with anteriorly uniciliate zoospores: *Hyphochytrium catenoides*. Am. J. Botany **26**:512–519 (Hyphochytriales).

——. 1964. *Synchytrium*. Academic Press, New York. 470 pp. (Chytridiales).

——. 1968. The Plasmodiophorales, Second ed. Hafner Press, New York. 256 pp.

Khan, S. R. 1977. Light and electron microscopic observations of sporangium formation in *Albugo candida* (Peronosporales; Oomycetes). Can. J. Botany **55**:730–739.

Kobayasi, Y., and M. Ôkubo. 1954. On a new genus *Oedogoniomyces* of the Blastocladiaceae. Bull Natl. Sci. Museum (Tokyo), N.S. **1**:59–66 (Harpochytriales).

Koch, W. J. 1961. Studies of the motile cells of chytrids. III. Major types. Am. J. Botany **48**:786–788 (Chytridiales).

Kole, A. P. 1965. Flagella. *In:* Ainsworth, G. C., and A. S. Sussman, Eds., The fungi—an advanced treatise. Academic Press, New York. **1**:77–93.

——, and A. J. Gielink. 1963. The significance of the zoosporangial stage in the life cycle of the Plasmodiophorales. Neth. J. Plant Pathol. **69**:258–262.

Kusano, S. 1930. The life-history and physiology of *Synchytrium fulgens* Schroet., with special reference to its sexuality. Japan. J. Botany **5**:35–132 (Chytridiales).

Lasure, L. L., and D. H. Griffin. 1975. Inheritance of sex in *Achlya bisexualis*. Am. J. Botany **62**:216–220 (Saprolegniales).

Lessie, P. E., and J. S. Lovett. 1968. Ultrastructural changes during sporangium formation and zoospore differentiation in *Blastocladiella emersonii*. Am. J. Botany **55**:220–236 (Blastocladiales).

Lingappa, B. T. 1958-a. Sexuality in *Synchytrium brownii*. Mycologia **50**:524–537.

——. 1958-b. Development and cytology of the evanescent prosori of *Synchytrium brownii* Karling. Am. J. Botany **45**:116–123 (Chytridiales).

——. 1958-c. The cytology of development and germination of resting spores of *Synchytrium brownii*. Am. J. Botany **45**:613–620 (Chytridiales).

Marchant, R. 1968. An ultrastructural study of sexual reproduction in *Pythium ultimum*. New Phytology **67**:167–171 (Peronosporales).

McKeen, W. E. 1975. Electron microscopy studies of a developing *Pythium* oogonium. Can. J. Botany **53**:2354–2360 (Peronosporales).

Middleton, J. T. 1943. The taxonomy, host range and geographic distribution of the genus *Pythium*. Mem. Torrey Botan. Club **20**:1–171 (Peronosporales).

Miller, C. E. 1976. Substrate-influenced morphological variations and taxonomic problems in freshwater, posteriorly uniflagellate Phycomycetes. *In:* Jones, E. B. G., Ed., Recent advances in aquatic mycology. John Wiley & Sons, New York. pp. 469–487.

———. 1977. A developmental study with SEM of sexual reproduction in *Chytriomyces hyalinus*. Bull. Soc. Botan. France **124:**281–289 (Chytridiales).

Moore, E. D., and C. E. Miller. 1973. Resting body formation by rhizoidal fusion in *Chytriomyces hyalinus*. Mycologia **65:**145–154 (Chytridiales).

Mullins, J. T. 1961. The life cycle and development of *Dictyomorpha* gen. nov. (formerly *Pringsheimiella*), a genus of the aquatic fungi. Am. J. Botany **48:**377–387 (Chytridiales).

Olson, L. W. 1973. The meiospore of *Allomyces*. Protoplasma **78:**113–127 (Blastocladiales).

———, and M. S. Fuller. 1968. Ultrastructural evidence for the biflagellate origin of the uniflagellate fungal zoospore. Arch. Mikrobiol. **62:**237–250.

Perrott, P. E. T. 1955. The genus *Monoblepharis*. Trans. Brit. Mycol. Soc. **38:**247–282 (Monoblepharidales).

Reichle, R. E., and M. S. Fuller. 1967. The fine structure of *Blastocladiella emersonii* zoospores. Am. J. Botany **54:**81–92 (Blastocladiales).

Roane, M. K. 1973. Two new chytrids from the Appalachian highlands. Mycologia **65:**531–538 (Chytridiales).

Sansome, E. 1963. Meiosis in *Pythium debaryanum* Hesse and its significance in the life-history of the biflagellatae. Trans. Brit. Mycol. Soc. **46:**63–72 (Peronosporales).

———. 1965. Meiosis in diploid and polyploid sex organs of *Phytophthora* and *Achlya*. Cytologia **30:**103–117.

———, and C. M. Brasier. 1974. Polyploidy associated with varietal differentiation in the *megasperma* complex of *Phytophthora*. Trans. Brit. Mycol. Soc. **63:**461–467 (Peronosporales).

———, and B. J. Harris. 1962. Use of camphor-induced polyploidy to determine the place of meiosis in fungi. Nature **196:**291–292.

Scott, W. W. 1961. A monograph of the genus *Aphanomyces*. Virginia Agr. Expt. Sta. Tech. Bull. **151:**1–95 (Saprolegniales).

Skucas, G. P. 1966. Structure and composition of zoosporangial discharge papillae in the fungus *Allomyces*. Am. J. Botany **53:**1006–1011 (Blastocladiales).

———. 1967. Structure and composition of the resistant sporangial wall in the fungus *Allomyces*. Am. J. Botany **54:**1152–1158 (Blastocladiales).

Sparrow, F. K. 1933. The Monoblepharidales Ann. Botany London **47:**517–542.

———. 1936. Evidences for the possible occurrence of sexuality in *Diplophlyctis*. Mycologia **28:**321–322 (Chytridiales).

———. 1958. Interrelationships and phylogeny of the aquatic Phycomycetes. Mycologia **50:**797–813.

———. 1960. Aquatic Phycomycetes, Second ed. University of Michigan Press, Ann Arbor, Mich. 1187 pp.

———. 1973-a. Mastigomycotina (zoosporic fungi). *In:* Ainsworth, G. C., F. K. Sparrow, and A. S. Sussman, Eds., The fungi—an advanced treatise. Academic Press, New York **4B:**61–73.

———. 1973-b. Chytridiomycetes, Hyphochytridiomycetes. *In:* Ainsworth, G. C., F. K. Sparrow, and A. S. Sussman, Eds., The fungi—an advanced treatise. Academic Press, New York **4B:**85–110.

———. 1973-c. Lagenidiales. *In:* Ainsworth, G. C., F. K. Sparrow, and A. S. Sussman, Eds., The fungi—an advanced treatise. Academic Press, New York **4B:**159–163.

———. 1976. The present status of classification in biflagellate fungi. *In:* Jones, E. B. G., Ed., Recent advances in aquatic mycology. John Wiley & Sons, New York. pp. 213–222.

Thirumalachar, M. J., M. D. Whitehead, and J. S. Boyle. 1949. Gametogenesis and oospore formation in *Cystopus (Albugo) evolvuli*. Botan. Gaz. **110:**487–491 (Peronosporales).

Tommerup, I. C., and D. S. Ingram. 1971. The life-cycle of *Plasmodiophora brassicae* Woron. in *Brassica* tissue cultures and in intact roots. New Phytologist **70:**327–332 (Plasmodiophorales).

Travland, L. B., and H. C. Whisler. 1971. Ultrastructure of *Harpochytrium hedinii*. Mycologia **63:**767–789 (Harpochytriales).

Truesdell, L. C., and E. C. Cantino. 1970. Decay of g particles in germinating zoospores of *Blastocladiella emersonii*. Arch. Mikrobiol. **70:**378–392.

Vujičič, R. 1971. An ultrastructural study of sexual reproduction in *Phytophthora palmivora*. Trans. Brit. Mycol. Soc. **57:**525–530 (Peronosporales).

Wager, H. 1913. The life-history and cytology of *Polyphagus euglenae*. Ann. Botany London **27:**173–202 (Chytridiales).

Waterhouse, G. M. 1973-a. Plasmodiophoromycetes. *In:* Ainsworth, G. C., F. K. Sparrow, and A. S. Sussman, Eds., The fungi—an advanced treatise. Academic Press, New York **4B:**75–82.

———. 1973-b. Peronosporales. *In:* Ainsworth, G. C., F. K. Sparrow, and A. S. Sussman, Eds., The fungi—an advanced treatise. Academic Press, New York. **4B:**165–183.

Whisler, H. 1979. The fungi versus the arthropods. *In:* Batra, L. R., Ed. Insect-fungus symbiosis—nutrition, mutualism, and commensalism. Allanheld, Osmun and Company, Montclair, N.J. pp. 1–32.

Williams, P. H. 1966. A cytochemical study of hypertrophy in clubroot of cabbage., Phytopathology **56:**521–524 (Plasmodiophorales).

———, and S. S. McNabola. 1967. Fine structure of *Plasmodiophora brassicae* in sporogenesis. Can. J. Botany **215:**1665–1669 (Plasmodiophorales).

Wilson, C. M. 1952. Meiosis in *Allomyces*. Bull. Torrey Botan. Club **79:**139–160.

Umphlett, C. J. 1962. Morphological and cytological observations on the mycelium of *Coelomomyces*. Mycologia **54:**540–554 (Blastocladiales).

———. 1964. Development of the resting sporangia of two species of *Coelomomyces*. Mycologia **56:**488–497 (Blastocladiales).

Voos, J. R. 1969. Morphology and life cycle of a new chytrid with aerial sporangia. Am. J. Botany **56:**898–909 (Chytridiales).

Yukawa, Y., P. H. Williams, and J. O. Strandberg. 1966. Ultrastructural studies on *Plasmodiophora brassicae*. Phytopathology **56:**907 (Plasmodiophorales).

THREE

Division Amastigomycota— Subdivision Zygomycotina

The remaining groups of fungi are in the Division Amastigomycota. Unlike the fungi in the previous taxon, the Division Mastigomycota, the members of the Amastigomycota lack a motile stage and usually are not adapted to the aquatic habitat.

The fungi in the Subdivision Zygomycotina are characterized by their sexual reproduction, which culminates in *zygospore* formation. In this type of reproduction, two gametangia fuse and subsequently develop into the zygote, which then forms thick walls to become a zygospore. Details of this type of reproduction are discussed in the section on the Order Mucorales, pages 87–89.

The typical nonsexual spore formed by members of the Subdivision Zygomycotina is the *sporangiospore*. Sporangiospores are nonmotile spores that are formed endogenously within a saclike structure, the *sporangium* (Fig. 3–1). Although most sporangia produce several sporangiospores, there are some that produce only a few or even a single one. These reduced sporangia are often called *sporangioles*. In some monosporous sporangioles, the sporangial wall remains distinct and can separate from the sporangiospore at maturity. In other sporangioles, this wall is fused with that of the sporangiospore at maturity. In still other sporangioles, the wall of the sporangiole is fused with that of a sporangiospore, which may also appear to develop exogenously, like most conidia. Although many mycologists call the latter spore type a conidium, the view has been generally supported that the monosporous sporangioles are all homologous structures (e.g., Hawker et al., 1970). An electron microscopic study of representatives of both types of monosporous sporangioles supports this view, as it showed that in each case the wall layers are deposited *within* the surrounding sporangiole wall and that the total number of wall layers is equivalent (Khan and Talbot, 1975). The sporangia are borne on specialized stalks, the *sporangiophores*.

84

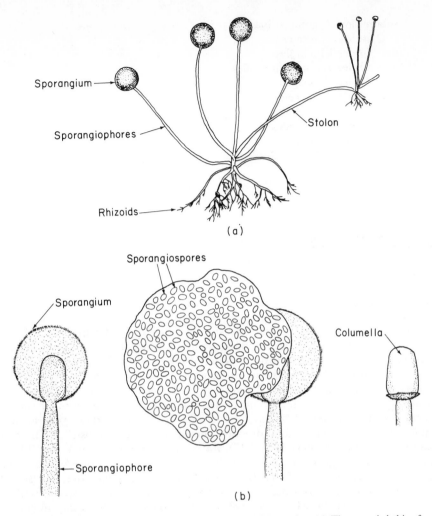

Fig. 3–1. Nonsexual reproductive structures in the Mucorales. (a) The growth habit of *Rhizopus nigricans,* showing the production of sporangiophores bearing a single multispored sporangium; (b) A single multispored sporangium of *Mucor mucedo* (left), which ruptures to release numerous sporangiospores (center) and finally leaves only the columella (right). [(a) from G. F. Atkinson, 1908, *in*: H. M. Fitzpatrick, 1930, *The Lower Fungi,* McGraw-Hill Book Company, New York, 331 pp.; (b) from O. Brefeld, 1872, *Botanische Untersuchungen über Schimmelpilze,* Heft 1, 1-64, Leipzig.]

Members of the Zygomycotina occur on or within a variety of substrata. The two classes may be distinguished on the basis of the relationship of the fungi to their substratum. The Class Zygomycetes has large numbers of saprophytes and parasites within or upon a variety of hosts. The Class Trichomycetes contains fungi that occur primarily in the gut of arthropods as commensals. We shall use the following classification (Ainsworth, 1973):

Subdivision Zygomycotina
Class Zygomycetes
Order Mucorales
Order Entomophthorales
Order Zoopagales
Class Trichomycetes
Order Harpellales
Order Asellariales
Order Eccrinales
Order Amoebidiales

CLASS ZYGOMYCETES

Order Mucorales

The majority of the Mucorales occur as saprophytes on plant debris or, to a lesser extent, on animal debris in soil. Some members cause plant or animal diseases. A few of the parasitic ones are obligate parasites on other fungi (Chapter 12). When fungi are isolated from the soil, air, or dung by using agar culture dishes, members of the Mucorales are among the most commonly isolated fungi. They are often notable in culture because of their extremely fast growth rate, a characteristic important in their ability to compete in the soil (Chapter 11). Some members not only are common soil saprophytes but also are important or are involved in our lives in other ways. Three such genera are *Mucor, Rhizopus,* and *Absidia.* Members of these genera may cause spoilage of stored grain and diseases known as "muromycosis" in humans and animals (see also Chapters 11 and 12). *Rhizopus stolonifer* is known as the "black bread mold" because it grows on moist, old bread. Some members of *Rhizopus* and *Absidia* may also cause rotting of vegetables such as sweet potatoes that are stored after harvest. On the positive side, some members of *Rhizopus* and *Mucor* as well as some other members of the Mucorales are used to produce commercially important products. These include organic acids, pigments, fermented Oriental foods, alcohols, and modified steroids. This commercial utilization is discussed further in Chapter 14. *Phycomyces* has served as an important experimental organism because of its response to light and nutritional requirements (Chapter 7). *Pilobolus* has an interesting spore discharge mechanism (Chapter 10).

These fungi have a well-developed vegetative mycelium. The mycelium is multinucleate and nonseptate when first produced, but later it becomes septate at the delimitation of reproductive structures or when aging. In members of two families, septa are regularly formed in the aerial hyphae and sporangiophores. These septa appear Y-shaped in longitudinal sections of the hyphae, with the arms of the Y pointing towards a central pore in the septum and enclosing a lens-shaped area. The pore is plugged with an electron-dense material which is apparently lipid (Saikawa, 1977). The mycelium is usually light in color and is typically submerged in the substratum with only the reproductive portions emerging. In other cases, the mycelium may form a dense, cottony growth on the surface of the substratum.

SEXUAL REPRODUCTION

The vegetative mycelium is haploid. Meiosis occurs after zygospore initiation but before or upon germination of the zygospore. Therefore the life cycle is one in which the haploid stage is emphasized (Fig. 3–2).

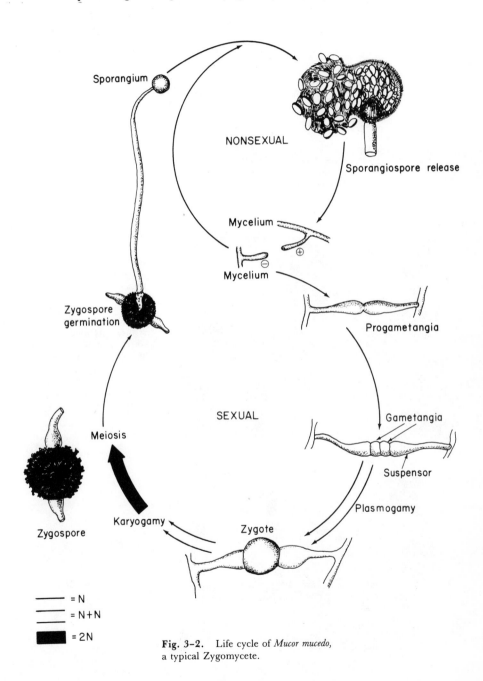

Fig. 3–2. Life cycle of *Mucor mucedo,* a typical Zygomycete.

Individual species may be homothallic, requiring only one thallus type for sexual reproduction (in this group, a single thallus only), or it may be heterothallic, requiring two thalli of opposite mating types. These mating types may be designated "plus" and "minus" (see Chapter 9 for additional information on this topic). The majority of the Mucorales are heterothallic.

Differentiation of the sexual organs is initiated when adjacent branches meet, if homothallic, or when plus and minus hyphae touch. This differentiation is controlled

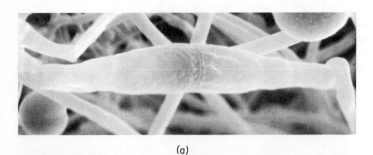

(a)

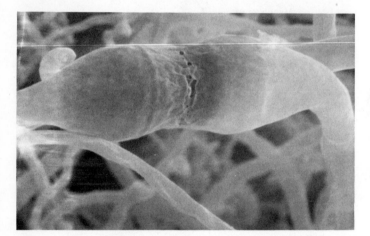

(b)

Fig. 3-3. Stages in the development of the zygospore of *Gilbertella persicaria*. (a) Progametangia from opposing hyphae have come into contact (× 810); (b) The zygote has been formed and the outer wall is beginning to rupture from the expansion of the developing zygospore (× 850); (c) Nearly mature zygospores with remnants of the outer wall (× 835). Note the prominent markings and suspensors. [Courtesy K. L. O'Donnell, J. J. Ellis, C. W. Hesseltine, and G. R. Hooper, 1977, *Can. J. Botany* **55**:662–675.]

(c)

by hormones, as discussed further in Chapter 9. Numerous light microscopic studies have been supplemented by recent ultrastructural studies (e.g., Hawker and Gooday, 1967, 1968, 1969; O'Donnell et al., 1977-a, 1978). Generally development proceeds in the following manner, although there may be minor morphological variations. Those hyphae that meet to initiate the sexual development, called *progametangia,* are multinucleate with dense cytoplasm and numerous lipid droplets. When the progametangia meet, the walls making contact fuse into a single wall, and the progametangia increase in size and gradually push apart the hyphae from which they have arisen [Fig. 3-3(a)]. The progametangia will become separated by a septum into the *suspensor,* the tapered cell adjoining the vegetative hyphae, and the terminal gametangium. Vesicles, probably arising from the endoplasmic reticulum, may be involved in formation of the wall separating the gametangium from its suspensor. The vesicles fuse, forming the plasmalemma of each cell, and wall material is laid down between the adjacent plasmalemmas (Hawker and Gooday, 1967). Cytoplasmic strands (plasmodesmata) may extend through the wall that separates the suspensor from its gametangium and may provide a means for food to be transported into the gametangium (Hawker and Gooday, 1968). The gametangia are usually equal in size [Fig. 3-3(b)], but gametangia and their supporting suspensors are unequal in some members. The common wall separating the gametangia is then digested enzymatically, beginning at the center. This allows plasmogamy to occur, and the cytoplasm from the two gametangia mingles in the common cell just formed, the *zygote.* The wall of the zygote is at first thin, as it was derived from the gametangia, but as the zygote develops into the zygospore, a thick wall is formed inside the original thin wall. Patches of new wall material form and gradually coalesce to form the zygospore wall ornamentations. As the zygospore expands, the thin outer wall is sloughed off [Fig. 3-3(c)]. The wall of the zygospore may have ornate markings such as warts or starfish-shaped plates when mature (Schipper et al., 1975). While the above processes are taking place, a few fungi such as *Phycomyces* produce ornate projections that arise from the suspensors and then surround the zygospore (Fig. 3-4).

The zygospore undergoes a period of rest before germinating. Germination typically occurs through production of a germ tube which is rather short and terminates in a sporangium. Often the sporangiospores within a single sporangium are of the same mating type, indicating that some of the products of meiosis disintegrate.

Cytological events occurring after plasmogamy follow different patterns. In *Mucor,* the nuclei within the zygote fuse in pairs and meiosis follows immediately before the zygospore becomes dormant (Fig. 3-2). In *Rhizopus,* karyogamy occurs before the onset of dormancy and meiosis occurs upon zygospore germination, while in *Phycomyces* both karyogamy and meiosis take place when the zygospore germinates. In some genera, karyogamy does not take place but development of the zygospore is apomictic.

Under some conditions, plasmogamy may fail to take place normally and a gametangium may develop parthenogenetically into an *azygospore.* Azygospores are morphologically similar to zygospores but are smaller. They may be either single or double, depending on whether one or both of the copulating gametangia developed parthenogenetically. Sometimes an isolated, noncopulated gametangium will develop into an azygospore. Azygospores are formed infrequently as a result of unfavorable cultural conditions, or they may be induced by forcing a mating between plus and minus strains of different species or genera (Blakeslee and Cartledge, 1927; Benjamin and Mehrotra, 1963). In a few species, azygospores but not zygospores are routinely formed (O'Donnell et al., 1977-b).

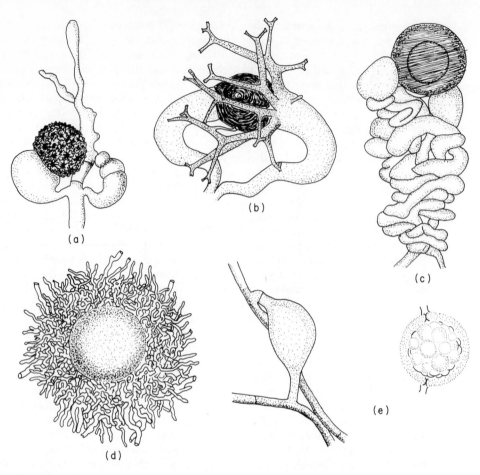

Fig. 3-4. Variation in the sexual structures of the Mucorales. (a) Unequal suspensors arising from the same hypha in *Zygorhynchus macrocarpus;* (b) Spines arising from the tong-shaped suspensors in *Phycomyces microsporus;* (c) Twined suspensors of *Choanephora conjuncta;* (d) Covering of sterile hyphae around zygospore in *Mortierella rostafinskii;* (e) Morphologically simple suspensors and hyaline, smooth zygospore of *Coemansia mojavensis.* [(a) from M. Ling-Young, 1930, *Rev. Gen. Bot.* **42**:147–158; (b) from G. A. Christenberry, 1940, *J. Elisha Mitchell Sci. Soc.* **56**:333–366. (c) from J. N. Couch, 1925, *J. Elisha Mitchell Sci. Soc.* **41**:141–150; (d) from O. Brefeld, 1877, *Botanische Untersuchungen über Schimmelpilze,* Heft 4, Teil 5, 81–96, Leipzig; (e) from R. K. Benjamin, 1958, *Aliso* **4**:149–169.]

NONSEXUAL REPRODUCTION

Nonsexual reproductive spore types formed by the Mucorales are sporangiospores and chlamydospores. Unlike the sporangiospores, which are formed in a special sac and are shed, the chlamydospores are nondeciduous spores that are thick-walled and are formed by the rounding up of vegetative cells. Chlamydospores may be formed in either a terminal or intercalary position and sometimes occur in chains.

The most commonly occurring sporangia are of the type that occurs in the Mucoraceae. In these fungi, the sporangium is large and multispored [Fig. 3-4(a,b)].

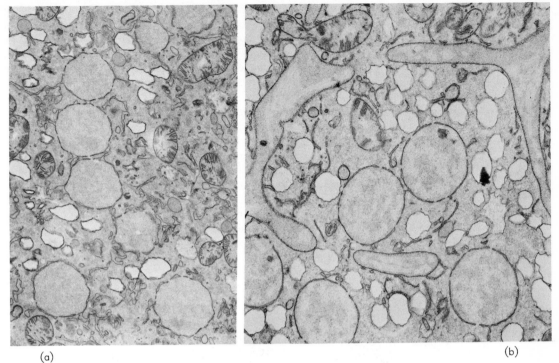

(a)

(b)

Fig. 3–5. Cleavage to form sporangiospores in the Zygomycete, *Gilbertella persicaria:* (a) Portion of sporangium before cleavage begins. × 4,000; (b) Mid-cleavage, parts of cleavage furrows are evident. × 6,200; (c) Post-cleavage, the spore initials are present as unwalled, individual cells. Walls form later. × 1,800; (d) A diagrammatic model of cleavage, resulting from fusion of vesicles in a sporangium: I, Early cleavage; II, Mid-cleavage; III, Late cleavage. [Courtesy C. E. Bracker, 1968, *Mycologia* **60**:1016–1067.]

(c)

(d)

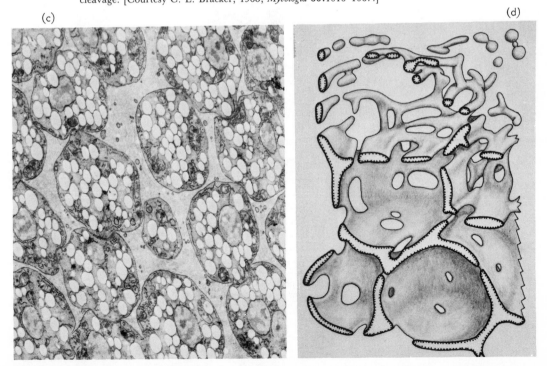

A *columella* may be present, which is a domelike extension of the sporangiophore lying within the sporangium. Details of sporangiospore formation within the sporangium are similar to those for zoospore formation. Let us consider a representative, *Gilbertella persicaria* (Bracker, 1968). The sporangium is at first a multinucleate structure with continuous cytoplasm. Vesicles are formed and are at first scattered throughout the cytoplasm [Fig. 3–5(a)]. These vesicles apparently arise from the cisternae of the endoplasmic reticulum. The membranes of these vesicles fuse and they become transformed into tubules. The tubules expand and branch and eventually extend throughout the cytoplasm as membranous cleavage furrows [Fig. 3–5(b)]. Uninucleate segments of protoplasm are cleaved out by these membranes and will become the sporangiospores. Meanwhile the columella is being delimited by a similar mechanism. Immediately after cleavage, the sporangiospore initials appear as naked protoplasts [Fig. 3–5(c)] and are surrounded by an amorphous material derived from the matrix of the cleavage apparatus. The plasmalemma of the sporangiospore is derived from the membrane of the cleavage apparatus, and the wall of the sporangiospore is then deposited outside the plasmalemma. In the Mucorales in general, the sporangial wall may either deliquesce or be persistent, requiring an outside force to rupture it. The sporangiospores are liberated when the sporangial wall breaks down by either mechanism.

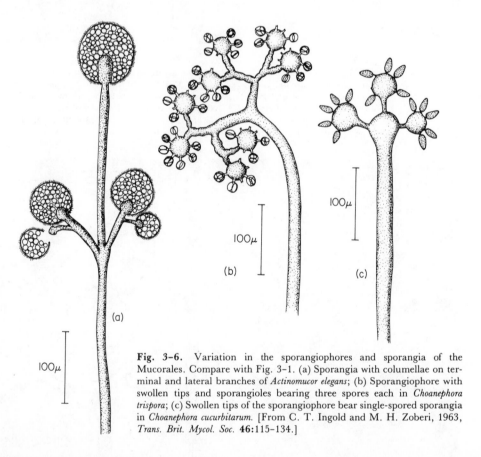

Fig. 3-6. Variation in the sporangiophores and sporangia of the Mucorales. Compare with Fig. 3-1. (a) Sporangia with columellae on terminal and lateral branches of *Actinomucor elegans*; (b) Sporangiophore with swollen tips and sporangioles bearing three spores each in *Choanephora trispora*; (c) Swollen tips of the sporangiophore bear single-spored sporangia in *Choanephora cucurbitarum*. [From C. T. Ingold and M. H. Zoberi, 1963, *Trans. Brit. Mycol. Soc.* **46**:115–134.]

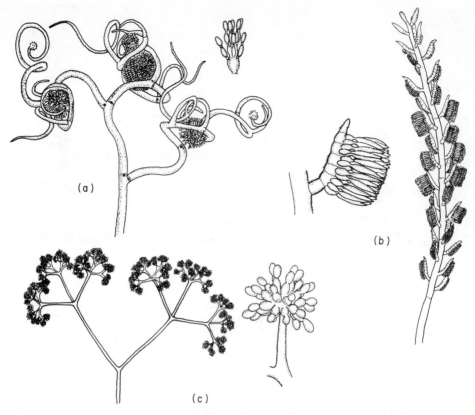

Fig. 3-7. Sporangiophores bearing elongated sporangioles, each with a larger detail of a cluster of sporangioles. (a) *Dispira cornuta,* which has both sterile and fertile branches; (b) *Coemansia mojavensis,* which produces monosporous sporangioles; (c) *Piptocephalis lepidula.* Magnifications (a) × 30 and × 1,080, (b) × 30 and × 400, (c) × 180 and × 1,700. [(a) and (c) from R. K. Benjamin, 1958, *Aliso* **4:**321-433; (b) from R. K. Benjamin, 1958, *Aliso* **4:**149-169.]

Sporangioles are produced by some members of the Mucorales. These reduced sporangia are much smaller than the sporangia described above, lack a columella, and produce only one or a few sporangiospores (Fig. 3-6). In its initial stages, the sporangiole contains continuous cytoplasm. Cleavage of a few sporangiospores in a sporangiole may occur in essentially the way described above in *Thamnidium elegans* (Fletcher, 1973). In monosporous sporangioles in *Mycotypha* and *Cunninghamella,* cleavage is absent and the cytoplasm of the sporangiole becomes that of the sporangiospore when the sporangiospore wall is deposited around it (Khan and Talbot, 1975). Sporangiola may be borne singly on lateral branches of the sporangiophore or in clusters on swollen apices of the sporangiophore. They may be sessile on the sporangiophore or borne on a short, peglike stalk.

Elongated, rodlike sporangiola are formed by some genera and are sometimes called *merosporangia* (Fig. 3-7) (Benjamin, 1959, 1966). These usually contain approximately 10 to 15 sporangiospores arranged in a linear sequence, but some of these contain only one or two spores. The sporangiola may be borne in whorls from a swollen

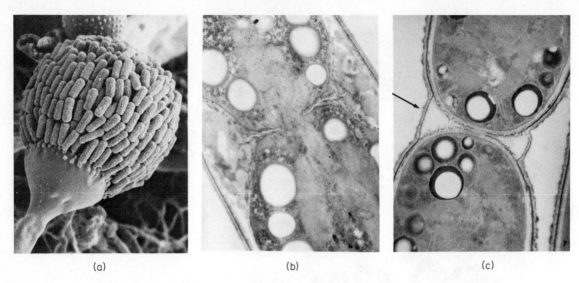

(a)   (b)   (c)

Fig. 3-8. *Syncephalis sphaerica.* (a) Swollen tip of sporangiophore bearing chains of sporangiospores after maturation is completed. × 1,700. (b) An early stage in which the plasmalemma is invaginating within the sporangiolum to delimit sporangiospores. × 19,250. (c) Completed cleavage of sporangiospores by deposition of wall material between the plasmalemma. The wall of the sporangiolum is still present (arrow). × 19,250. [Courtesy K. L. Baker, G. R. Hooper, and E. S. Beneke, 1977, *Can. J. Botany* **55**:2207–2215.]

terminal head [Fig. 3-7(a,c)] and may be either sessile on that head or borne on short, peglike projections. The most unusual arrangement is found in the family Kickxellaceae, in which the sporangiola are borne laterally on a modified sporangiophore, much like teeth on a comb [Fig. 3-7(b)]. The sporangiophore branches may break apart at a line of dehiscence to shed the sporangiola, or the sporangiola may be deciduous and drop away from the sporangiophore. In other species, the sporangiola are not shed but the sporangiolum wall fragments, leaving behind single-spored segments, which remain temporarily in chains. Lastly, the wall of the sporangiolum may simply be evanescent at maturity. Some species form a slime drop that encloses either the shed sporangiola or the single-spored fragments, while others remain dry.

Development of the elongated sporangiola has been studied ultrastructurally (e.g., Benny and Aldrich, 1975; Fletcher, 1972), and there is some variation in the manner in which development occurs. Let us consider a representative, *Syncephalis sphaerica* (Fig. 3-8) (Baker et al., 1977). In *S. sphaerica,* buds appear simultaneously over the surface of the swollen apex of the sporangiophore. The buds then elongate to form the sporangiola. Three sporangiospores are cleaved out simultaneously when the plasmalemma invaginates, moving from the periphery of the sporangiolum to the center and forming the plasmalemma of each spore. Fibrillar wall material is deposited between the plasmalemma of the sporangiospore and the merosporangial wall, thereby forming the walls of the sporangiospores. Meanwhile an electron-dense abscission zone is formed between each spore. At maturity, the wall of the sporangiolum ruptures to leave a chain of single sporangiospores.

The sporangia are borne on sporangiophores. The sporangiophores are typically stout aerial structures, usually terminating in a sporangium or a cluster of sporangioles. Sporangiophores range from relatively simple forms in which a single unbranched stalk terminates in a single sporangium (e.g., *Mucor* or *Rhizopus*) to elaborate branched types (Figs. 3-1 and 3-7). If branched, there may be a main axis with smaller lateral branches, dichotomous branching, or the formation of umbellike clusters of essentially equal branches. The branches usually bear a large single sporangium or perhaps smaller sporangioles at each end, but sometimes some of the branches are sterile. The main axis of the sporangiophore may terminate in a large multispored sporangium, and clusters of sporangioles may occur on the side branches (e.g., in *Thamnidium*). In some species, a prominent rhizoid anchors the sporangiophore to the substratum. A distinct lateral hypha (the *stolon*) may extend from the base of one sporangiophore to another.

The Mucorales may be divided into 14 families, primarily on the basis of the morphology of their nonsexual structures and habitat (Hesseltine and Ellis, 1973).

ENDOGONACEAE

An important family in the Mucorales is the Endogonaceae. These fungi have a symbiotic association with roots of higher plants and have not been cultured in the laboratory on artificial media.

Fig. 3-9. Sporocarps of *Endogone*. Top: Surface and cross-sectional views of *E. lactiflua*, × 3.2; Bottom: *E. flammicorona* attached to roots of a plant, × 1.5 [Courtesy J. W. Gerdemann and J. M. Trappe, 1974, *Mycologia Mem.* **5**:1-76.]

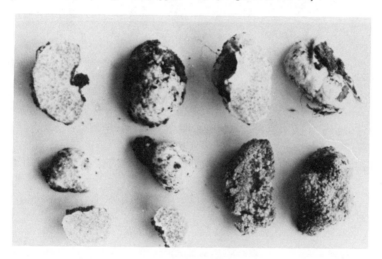

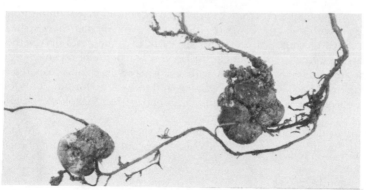

Some members of the Endogonaceae may be readily collected because they produce macroscopic sporocarps (Fig. 3-9). The sporocarps may occur above the surface of the soil, where they may be collected from litter, wood, decaying mushrooms, or the tips of moss gametophytes. Most species produce hypogeous sporocarps, that is, they occur beneath the surface of the soil. The sporocarps sometimes are found connected by hyphae to the roots with which they are associated (Fig. 3-9). The sporocarps range in diameter from approximately 1 to 25 millimeters, are sometimes brightly colored, and may or may not have a distinct covering. The shape varies from globose to ellipsoid or lobed and irregular, and they may be hollow or solid. The sporocarps contain zygospores [Fig. 3-10(a)], chlamydospores [Fig. 3-10(b)], or sporangia which are embedded within or surrounded by sterile hyphae. Only one type of spore is found within a single sporocarp (Gerdemann and Trappe, 1974; Godfrey, 1957-a). Not all species form sporocarps, and the spores may be found on isolated hyphae in the soil. These spores may be collected by a technique involving the decanting and sieving of a soil suspension.

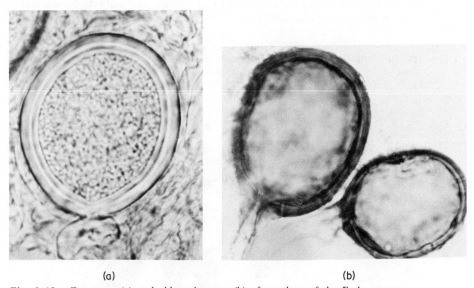

(a)   (b)

Fig. 3-10. Zygospore (a) and chlamydospores (b) of members of the Endogonaceae. × 625. [Courtesy J. W. Gerdemann and J. M. Trappe, 1974, *Mycologia Mem.* **5**:1–76.]

Species names and descriptions are based on collections of similar isolated spores or sporocarps with one type of spore. It is not known whether these spore types may represent different stages of the same life cycle in some cases or whether they are indeed unrelated fungi. For example, *Endogone* is one of the best known genera, and it usually produces either sporocarps with zygospores or chlamydospores or isolated chlamydospores. Each species of *Endogone* is described as forming only one type of spore, e.g., zygospores (Gerdemann and Trappe, 1974; Nicholson and Gerdemann, 1968).

Zygospore formation is known in some species of *Endogone*. In *Endogone*, the gametangia fuse at their tips and then a bud grows out from the point of union or from

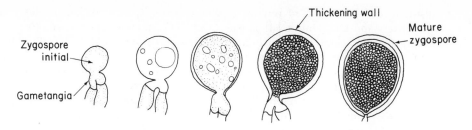

Fig. 3–11. Successive stages in zygospore formation in *Endogone incrassata*. The zygospore forms above the point of union of the two gametangia. [From J. W. Gerdemann and J. M. Trappe, 1970, *Mycologia* **62**:1204–1208.]

the larger gametangium. This bud then matures into the zygospore (Fig. 3–11). Two distinct walls are visible, the outer resulting from the expansion of the gametangial wall and the inner developing later. Typically the gametangia and suspensors eventually disintegrate, leaving the zygospores free within the sporocarp.

Azygospores are formed free in the soil by some species. Development of azygospores was studied by light and electron microscopy by Mosse (1970-a, 1970-b, 1970-c) in a species that was later identified as *Acaulospora laevis* (Fig. 3–12) (Gerdemann and Trappe, 1974). A globose vesicle appears as an enlargement at the end of a stout, funnel-shaped supporting hypha. Later an azygospore is initiated as a bud from the side of the supporting hypha, and the contents of the vesicle are transported into the developing azygospore. The vesicle then collapses. Unusual membrane-bounded pigment granules containing melanin are prominent in the vesicle and azygospore but disappear as the wall of the azygospore develops, which suggests that they play a role in wall formation. The azygospore wall is thick, consisting of six layers, and contains chitin and cellulose in separate parts of the wall. The azygospore undergoes a period of rest before germinating. Upon germination, the peripheral portion of the cytoplasm is cleaved by radial walls into multinucleate segments, and each segment germinates by extending a germ tube through newly formed splits in the azygospore wall.

The chlamydospores are produced in terminal positions on undifferentiated hyphae. Although superficially similar to the zygospores or azygospores, they may be

Fig. 3–12. Successive stages in azygospore development of *Acaulospora laevis*. [Figures from J. W. Gerdemann and J. M. Trappe, 1974, *Mycologia Mem.* **5**:1–76.]

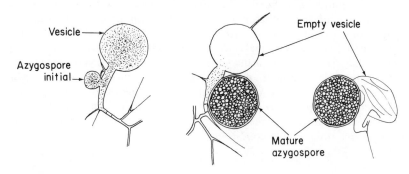

distinguished because they remain attached to the hyphae, which do not resemble suspensors and form only a single wall layer. The chlamydospores are often ovate to spherical in shape and at maturity have oily yellow contents and thick walls. They germinate by forming a germ tube.

Sporangia are thin-walled and resemble immature chlamydospores. A large number of sporangiospores may be formed within a sporangium.

Fungi with vesicles resembling the immature chlamydospores or sporangia may be found within or upon roots. These fungi have a symbiotic association, which is described as the vesicular-arbuscular type of mycorrhiza and is discussed further in Chapter 13. At least some members of the Endogonaceae, including some species of *Endogone,* can form these mycorrhizae when their azygospores, zygospores, and chlamydospores are added to the soil of potted plants. No such connection has been established for the species that produce sporangiospores. The fungi within the roots have not been isolated in pure culture, so it is not known whether all these morphologically similar fungi belong to the Endogonaceae (Gerdemann, 1968; Mosse and Bowen, 1968).

Order Entomophthorales

The members of the Entomophthorales are predominantly parasites of insects and other arthropods. Members of the genera *Entomophthora* and *Massospora* are insect parasites, and these are discussed further in Chapter 12. Those not parasitic on insects may be parasitic on desmids or fern prothalli, while some species may be saprophytes on plant debris. *Basidiobolus haptosporus* causes a disease of man in which tumorlike enlargements develop in the subcutaneous tissue, while *Entomophthora coronata* is parasitic on humans and horses as well as on insects.

The vegetative stages may consist of a coenocytic mycelium, but often the growth of a true mycelium is sparse and limited. In almost all species, the mycelium later fragments into a number of short segments, the *hyphal bodies,* which separate from each other. The hyphal bodies may serve as propagative units by budding and undergoing fission. Rhizoids may be formed by some species. The rhizoids consist of hyphae that extend from the parasitized insect and adhere to the substratum. The rhizoids may be simple or branched, with each terminating in a holdfast that secretes a viscous, sticky substance. The mycelium or hyphal bodies may produce chlamydospores by forming a thick wall, with or without enlargement of the vegetative cell.

Sexual reproduction is initiated when two specialized mycelial endings (from the same or different hyphae) or hyphal bodies function as gametangia and unite (Fig. 3–13). The gametangia may be alike or dissimilar in size and the zygospore may form either within one of the gametangia, within the conjugated cell formed by both gametangia, or in a bubblelike growth extending from one of the gametangia. Some species (such as *Entomophthora grylli*) do not reproduce sexually but instead produce azygospores that are morphologically similar to zygospores. Azygospores may arise by the parthogenetic development of hyphal bodies, by direct budding of chlamydospores, or at the tips of hyphae arising from hyphal bodies of chlamydospores. Neither zygospores nor azygospores are known in some species. The thick-walled zygospores, azygospores, and chlamydospores are sometimes termed *resting spores,* as they aid the fungus to survive unfavorable conditions.

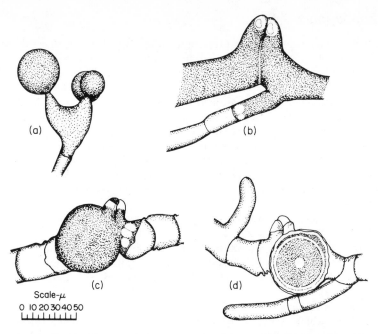

Fig. 3-13. *Basidiobolus magnus,* a member of the Entomophthorales: (a) Sporangiophore with sporangioles; (b) and (c) Successive stages in gametangial fusion; (d) Mature zygospore with a single nucleus. [From C. Drechsler,1964, *Am. J. Botany* **51**:770-777.]

Nonsexual reproduction involves the formation of unbranched or comparatively simple branched sporangiophores. The sporangiophores usually bear ovoid to spherical sporangioles, but other shapes occur. The sporangioles are monosporous, except that in *Basidiobolus* more than one sporangiospore may be formed in a sporangiole. The sporangiospores are uni- or multinucleate. Except in the case of *Massospora,* the sporangiole is discharged with great force. Mechanisms responsible for forcible discharge are considered in Chapter 10. The sporangiophores are phototropic in many species. Germination occurs by production of germ tubes or if conditions are not favorable for subsequent growth, by production of another sporangiophore and sporangiole. The shape of the secondary sporangiolum may be either like that of the primary sporangiolum or dissimilar (e.g., globose or fusiform). This sporangiole may be discharged and again germinate by producing another sporangiophore and sporangiole. Each generation becomes progressively smaller, and this process may continue until either the vitality of the fungal structures is exhausted or a suitable substratum is encountered.

A particularly interesting member of this order is *Basidiobolus ranarum,* which has an unusual life cycle. *B. ranarum* grows on excrement of frogs and lizards. Beetles crawl about over the excrement and devour the sporangia. The beetles in turn may be eaten by a lizard or frog, and while in the stomach the sporangiole cleaves into a number of sporangiospores that will undergo no further development until liberated with the feces. Once these are exposed to the air, mycelium formation follows.

Order Zoopagales

All members of the Zoopagales are predacious fungi, which means that they trap their prey (other predacious fungi are discussed in Chapter 12). There are at least 65 species in the Zoopagales and they are of widespread geographical distribution. The filamentous thallus is often at first devoid of septa, but septa are later formed and delimit degenerating portions of the mycelium from that part which is physiologically active and from the reproductive structures. Sexual reproduction does not always occur, but if it does, it involves the pairing of similar gametangia followed by zygospore formation. Nonsexual reproduction occurs through the formation of sporangiospores ranging in shape from filiform to spherical. The sporangiospores may be borne on long aerial sporangiophores or they may be nearly sessile on short *sterigmata* (peglike projections). Sporangiospores may be borne singly along the length of the sporangiophore or in long chains.

The Zoopagales occur predominantly on amoebae and other protozoa, although a few species are large enough to trap nematodes. These fungi are apparently obligate predators, requiring a living animal in order to satisfy their nutritional requirements. Often they are host specific and are restricted to a certain kind of animal. These fungi may either develop within their host (endogenously) or exist primarily outside their host (exogenously).

Endogenous Zoopagales include members of the genera *Cochlonema* (Fig. 3–14) and *Endocochlus,* which are somewhat similar. As an example of this group, we examine *Cochlonema verrucosum,* a predator on *Amoeba* spp.. Infection is initiated by a fusiform sporangiospore which is attached to the ectoplasm of the amoeba, and penetration is accomplished by a germ tube. A globular swelling develops within the amoeba at the tip of the germ tube, and both the sporangiospore and germ tube fall away, leaving only this swelling within the animal. The globular swelling develops into a minute elongate coiled thallus about 6 microns in length. During the early stages of infection, the amoeba leads a normal life, apparently unhindered by the fungus infection. As the fungus continues to grow and to feed on the endoplasm of the amoeba, however, the amoeba becomes increasingly sluggish and finally dies. The fungus continues to feed on the contents of the amoeba until only the shriveled ectoplasm remains (Fig. 3–15). Chains of sporangiospores are formed on the exterior of the amoeba before its death. Sexual reproduction occurs after the death of the host when slender club-shaped gametangia emerge from the remains of the amoeba and fuse, leading to zygospore formation.

The predacious habit is more obvious in a fungus such as *Stylopage rhynchospora,* an example of an exogenous member of the Zoopagales (Drechsler, 1959). Like the majority of the species of *Stylopage* that prey upon protozoa, *S. rhynchospora* preys upon amoebae. The sparingly septate, delicate mycelium adheres to any passing amoeba, probably by forming a sticky secretion. At the point of contact, the fungus forms a number of haustoria that penetrate the amoeba and absorb its contents. The amoeba is at first relatively unharmed, as cytoplasmic streaming continues to occur and the contractile vacuole continues to function. With further depletion of the endoplasm, the amoeba becomes rounded, all movements cease, and death follows. The fungus forms sporangiospores and thick-walled zygospores with walls marked by hemispherical warts.

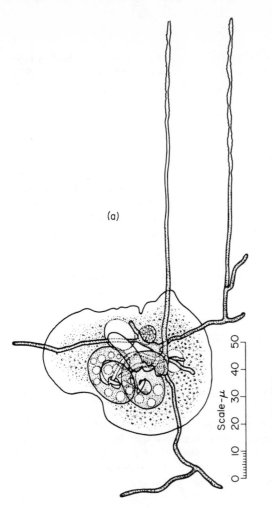

(a)

(b)

Fig. 3-14. *Cochlonema verrucosum:* (a) Conidia arising from endogenous thallus within dead amoeba; (b) Zygospores formed on disintegrating amoeba. [From C. Drechsler, 1935, *Mycologia* **27**:6-40.]

Fig. 3-15. Portion of hypha of *Stylopage rhynchospora* on which five amoebae are attached. Amoeba *e* has been captured so recently that it has not been penetrated; amoebae *a* and *d* have each been invaded by a haustorium; and amoebae *b* and *c* have been depleted of their contents. [From C. Drechsler, 1939, *Mycologia* **31**:338-415.]

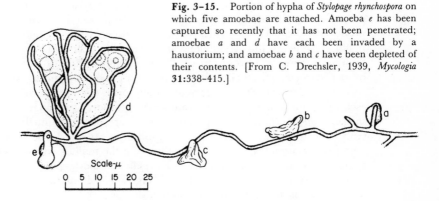

Division Amastigomycota—Subdivision Zygomycotina   **101**

CLASS TRICHOMYCETES

Members of the Class Trichomycetes inhabit a variety of arthropods, including insect larvae, crayfish, millipedes, beetles, and isopods (Lichtwardt, 1973-a, 1973-b, 1976). The arthropod hosts may live on land or in fresh water or marine waters. A particular member of the Trichomycetes may be restricted to a narrow range of host species, perhaps all within a single family. There is also a specificity for parts of the host, for example, a given fungus may be found in the rectum. The Trichomycetes do not grow independently of their hosts in nature.

The fungi almost always occur within the gut, and may be detected if the thallus protrudes from the anus of the host or if the host is dissected. Members of one genus, *Amoebidium,* occur on the outer surface. The thalli are attached to the host by a holdfast, which penetrates the lining of the gut or exoskeleton only and does not penetrate into the living tissues of the host. When the arthropod molts, the lining of the gut and exoskeleton are shed and the fungus is also expelled. These fungi are not parasitic on their hosts and live as commensals, apparently neither harming nor benefiting them (Moss, 1979). The fungi obtain their nutrients from the surrounding gut contents or water. In experiments in which spores of the fungus *Smittium* were fed to bacteria-free mosquito larvae, it was found that the fungal spores could germinate only in the gut of the larva, although the thallus could be cultured alone in defined medium. Therefore it was concluded that the spores of this Trichomycete require components of the gut contents produced by the larvae and not by bacteria for spore germination (Williams and Lichtwardt, 1972).

Only a few of these fungi have been isolated and grown in pure culture; mostly these are species of *Smittium* or *Amoebidium.* Therefore our understanding of the life cycle usually results from the dissection of numerous host specimens from a heavily infected population.

The thallus of the Trichomycetes is comparatively simple (Fig. 3–16). It is anchored by a holdfast, which may be a simple secretion or may have a more elaborate shape such as that of a bulb. The thallus may have one of two basic forms: (1) unbranched and coenocytic, or (2) branched with septa. The septa are similar to those described for members of the Kickxellaceae in the Mucorales, consisting of a bifurcated wall with a plug [Fig. 3–18(b)]. Similarity of the septa in the Kickxellaceae and Trichomycetes has been used to argue for their close phylogenetic relationship (Moss and Young, 1978).

There are four orders in the Trichomycetes. These orders are distinguished primarily by the morphology of nonsexual reproductive structures. *Arthrospores* are produced by the Order Asellariales. Arthrospores are formed when the vegetative thallus fragments into sporelike segments, which then fall apart from each other. Sporangiospores are produced by the remaining orders. In the Eccrinales, the unbranched coenocytic thallus cleaves into sporangia, beginning at the apex and proceeding downward. Each sporangium then forms a single sporangiospore, which then escapes. Thin-walled, multinucleate sporangiospores germinate in the gut, while the thicker-walled, uninucleate sporangiospores are passed from the gut and are apparently able to withstand unfavorable conditions. In the Order Amoebidiales, naked amoeboid sporangiospores are formed within the thallus, which assumes the role of a sporangium. In the last order, the Harpellales, sporangioles bearing one to several filaments are formed and disseminated (Fig. 3–17). In all cases, the spores escape from the host when the host defecates, molts, or dies.

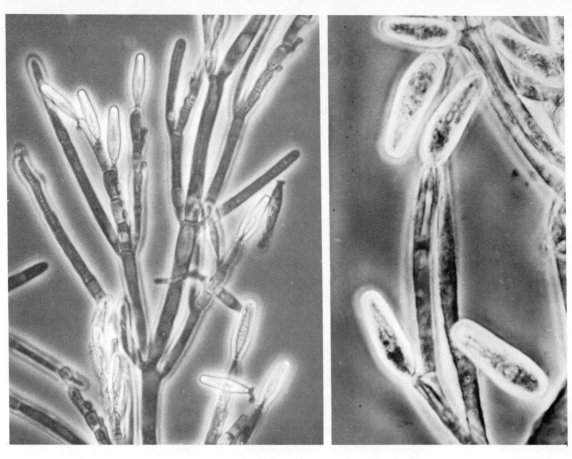

Fig. 3-16. Thalli of two members of the Trichomycetes. Left: Part of a sporulating thallus of *Smittium culisetae* showing the basipetal development of sporangioles, × 900. Right: Generative cells (arrows) and sporangioles of *Genistellospora homothallica*, × 720. [Courtesy R. W. Lichtwardt. Photograph of *G. homothallica* from 1972, *Mycologia* **64**:167–197.]

Fig. 3-17. Scanning electronmicrographs of sporangioles of *Smittium culicus.* Left: a generative cell bearing a collar and a terminal sporangiole, × 2600. Right: A released sporangiole with its collar and uncoiling single appendage, × 2650 [Courtesy S. T. Moss.]

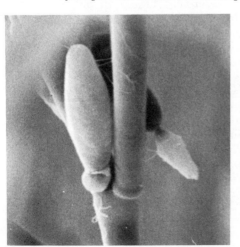
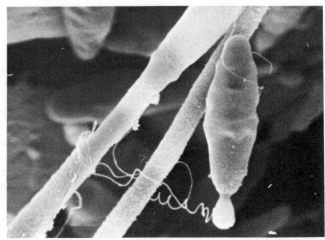

We will discuss the Order Harpellales further as it is the best-known order in the Trichomycetes and is also the only order in which sexual reproduction is known. Additional details about remaining orders may be seen in Table 3-1.

Table 3-1: Characteristics of the Trichomycete Orders

Order	Host	Thallus	Nonsexual Reproduction	Sexual Reproduction
Harpellales	Hindgut lining or peritrophic membrane of immature insects	Branched or unbranched	Sporangiospores	Zygospores
Asellariales	Hindgut lining of isopods or insects	Branched and septate	Arthrospores	Unknown
Eccrinales	Hindgut or foregut of insects, millipedes, isopods, amphipods, decapods	Unbranched or branched only at the base	Sporangiospores	Unknown
Amoebidiales	External parts of aquatic crustacea and insects	Unbranched, entire thallus becoming a sporangium	Amoeboid cells	Unknown

Order Harpellales

Members of this order occur in larvae of insects that inhabit fresh water, most usually the water of rapidly flowing streams. Some species form unbranched, coenocytic thalli in the midgut, while others form branched thalli in the hindgut (Fig. 3-16).

The unbranched coenocytic thalli become entirely separated into single cells, the *generative cells,* while the branched thalli have distinct terminal generative cells. The generative cells support the sporangioles externally (Figs. 3-16 and 3-17). The sporangioles may be borne directly upon the generative cells or may be borne on short, nonseptate lateral branches extending from the generative cells. Each sporangiole contains a single uninucleate sporangiospore and is typically elongate in shape. During development of the sporangiole in *Genistellospora homothallica* (Figs. 3-16 and 3-18), the young sporangiole is initiated as an extension of its generative cell, from which it is separated by a septum. The wall of the sporangiole and that of the generative cell are continuous and electron-opaque (Fig. 3-18). The sporangiole wall expands, and meanwhile an inner electron-transparent wall layer, representing the wall of the sporangiospore, is deposited immediately inside the sporangiole wall. Vesicles are involved in the deposition of wall material in the expanding outer wall and newly forming inner wall (Moss and Lichtwardt, 1976). The appendages are formed in *G. homothallica* as in other Trichomycetes studied, that is, they are extensions of the sporangial wall which form within the generative cell but external to the plasmalemma

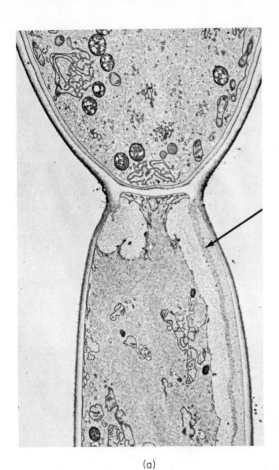

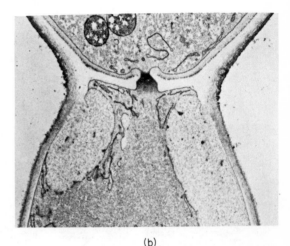

Fig. 3-18. Ultrastructural details of the point of attachment of the sporangiole to the generative cell in *Genistellospora homothallica*. In both electronmicrographs, the sporangiole is on top while the generative cell is on the bottom. (a) Note the continuity of the outermost wall layer between the sporangiole and the generative cell. An additional wall layer is present inside the sporangiole. An appendage is visible at the arrow. The septum appears continuous because this is a nonmedian section. × 7,500; (b) The septum separating the sporangiole from the generative cell in a median section. Note the characteristic forking of the septum. × 14,500. [Courtesy S. T. Moss and R. W. Lichtwardt, 1976, *Can. J. Botany* **54:**2346-2364.]

(a)   (b)

(Fig. 3-18). Vesicles are involved in the deposition of the appendages (Lichtwardt, 1976; Moss, 1979; Moss and Lichtwardt, 1976). At maturity, sporangioles are shed from the generative cell by its rupture either at the base of the sporangiole or at the base of the supporting branch (Fig. 3-17). A minute collar consisting of the remnant of the generative cell is left behind. As the sporangiole moves away from the thallus, the appendages unwind and trail behind. They are nonmotile but may be adaptive in preventing the sporangiole from floating away from a population of potential hosts and also in entangling the sporangiole in debris that might be consumed by the host. A second adaptive feature is that germination cannot occur in the water but only in the gut of the host, where the thallus can develop. Once within the host, the sporangiospore germinates. In *Smittium,* germination may occur when the sporangiospore forms a germ tube which ruptures the sporangiole wall or when the entire sporangiospore protoplast and wall escape from the sporangiole. Once the sporangiospore germinates, it attaches itself to the cuticle of the host by a holdfast. In studies of *Smittium,* germination and attachment were observed to occur within a half hour of ingestion of the spores (Williams and Lichtwardt, 1972). Fairly rapid germination and holdfast production help prevent the sporangiospores or young thallus from being expelled from the gut.

Many members of the Harpellales reproduce sexually by producing zygospores (Figs. 3-19 and 3-20). Conjugation may occur between different thalli if the thalli are unbranched and between cells of the same thallus or different thalli in the branched forms. Conjugation occurs between hyphae, and then a specialized branch (the *zygosporophore*) develops from one of the conjugated cells (Fig. 3-19) in all genera except *Glotzia*, where it arises from a conjugation tube. The zygospore develops as an outgrowth of the zygosporophore. At maturity, the zygospore is an elongated uninucleate cell terminating in two conelike ends with thickened walls (Figs. 3-19 and 3-20). The zygospore may be perpendicular to the zygosporophore or may occur in other spatial relationships to it. Cytological aspects of sexual reproduction are incompletely known, as neither karyogamy nor meiosis has been observed.

Conjugation does not occur in *Genistellospora homothallica*, and parthenogenetic azygospores are formed. The azygospores are released by rupture of the zygosporophore, and appendages similar to those formed on the sporangioles unwind and trail behind as they are dispersed (Moss and Lichtwardt, 1977).

Fig. 3-19. Zygospores of *Trichozygospora chironomidarum*, a Trichomycete. (a) The conjugating cells *(cj)* have produced a zygosporophore *(zs)* and an immature zygospore *(z)*; (b) Mature released zygospore with collar and numerous appendages. [(a) Courtesy S. T. Moss and R. W. Lichtwardt, 1977, *Can. J. Botany* **55**:3099–3110; (b) Courtesy R. W. Lichtwardt, 1976, *in:* E. B. G. Jones, ed. *Recent advances in aquatic mycology,* John Wiley & Sons, New York.]

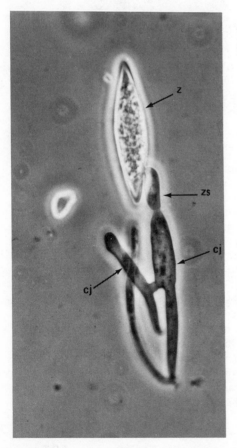

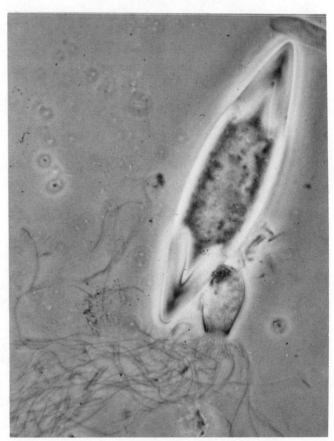

(a)   (b)

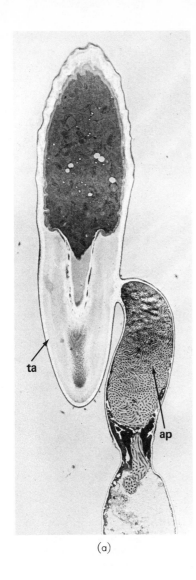

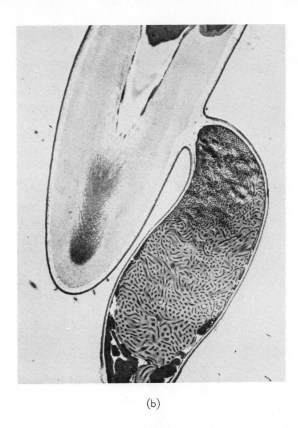

(b)

Fig. 3-20. Ultrastructural details of zygospores of *Trichozygospora chironomidarum*. (a) The zygospore borne on the zygosporophore. Both the apical thickenings (*ta*) and appendages (*ap*) of the zygospores are visible. × 2001. (b) Detail of the numerous folded zygospore appendages that lie to the outside of the plasmalemma of the zygosporophore but are within the collar region. × 3680. [Courtesy S. T. Moss and R. W. Lichtwardt, 1977, *Can J. Botany* **55**:3099–3110.]

(a)

EVOLUTION OF THE ZYGOMYCOTINA

The origin of the Subdivision Zygomycotina is unknown, and there are few speculations about its origin. The evolution of the groups within the Zygomycotina has been considered by various mycologists, including Benjamin (1959) and Hesseltine and Ellis (1973). According to Hesseltine and Ellis, the Zygomycetes constitute a natural class of related organisms, as they all have nonmotile spores, sexual reproduction as a result of gametangial fusion, and walls containing chitosan and chitin. The multi-spored sporangium is considered to be primitive, with evolution proceeding from that type to the few-spored sporangioles, to the monosporous sporangiole with a dehiscent outer wall, and finally to those monosporous sporangioles with a double, fused wall. A second major evolutionary trend was from the primitive saprophytic mode of nutrition as found in many Mucorales to the more advanced parasitic modes of forms which cannot be grown in culture, such as members of *Endogone*. The steps along this pathway proceeded from those fungi that live as saprophytes with no special nutritional requirements to some such as *Pilobolus* which are saprophytic but require a specialized growth factor (ferrichrome) found in dung, to the Entomopthorales which can be grown in culture but have complex growth requirements, and finally to those fungi in the Endogonaceae that cannot be cultured. A third major evolutionary trend was from passive spore discharge in most members of the Mucorales to the more elaborate forcible spore discharge mechanisms occurring in *Pilobolus* and the Entomophthorales. The Entomophthorales may be more advanced than *Pilobolus* in this respect, as the spore is capable of repeated discharge until it alights on a favorable substratum.

Benjamin (1959) dealt with evolution within the Mucorales, placing emphasis on the relative specialization of the vegetative thallus, number of sporangiospores, and nature of the progametangia and zygospores (Fig. 3-21). He recognized the nonseptate thallus condition as primitive and the septate condition as advanced and the large, multispored sporangia with a columella as primitive and the monosporous sporangioles as advanced. The primitive sexual apparatus involves highly differentiated progametangia, which form their zygospores on aerial hyphae. The zygospores are rough, thick-walled, and often highly pigmented. The advanced condition for the sexual apparatus involves mating between relatively undifferentiated hyphae, the zygospores are formed in or near the surface of the substratum, and the zygospore is thin walled, hyaline, and nearly smooth. Emphasizing these characteristics, Benjamin then recognized six separate lines of evolution from the ancestral type. Many genera and species are included in most of the lines, and a given species may not have the particular combination of primitive and advanced characteristics which is typical for the line in which it is included. These six lines and their advancements from the primitive condition are listed below.

1. The *Mucor* line: All primitive conditions are represented.
2. *Thamnidium-Cunninghamella* line: Reduction in the number of sporangiospores is represented. Numerous species have few-spored sporangioles as well as multispored sporangia.
3. *Choanephora* line: Reduction of sporangiospores is found here, as in the previous line, but there is a greater representation of monosporous sporangioles, and sporangia and sporangioles are borne on separate sporangiophores (Fig. 3-6). Some specialization of the gametangia is represented (tonglike, coiled suspensors) [Fig. 3-4(c)].
4. *Pilobolus* line: Thallus and sporangiospore characters are primitive, but a highly specialized sporangium discharge mechanism exists. Specialization of the gametangia is similar to that in the previous line.
5. *Mortierella-Endogone* line: The gametangia of *Mortierella* are tonglike, as in the previous two lines. However, the smooth-walled, light-colored zygospore becomes covered by a sheath of sterile hyphae [Fig. 3-4(d)], which superficially resembles the sporocarp of *Endogone*. These genera share some other similarities, including the production of chlamydospores. It is thought that the *Mortierella* line may have given rise to the *Endogone* line.
6. *Piptocephalis-Kickxella* line: All the advanced characteristics are represented in this line. The elongated sporangioles are present in all representatives (Figs. 3-7) and are laterally arranged in the most advanced types [Fig. 3-7(b)]. Septa occur

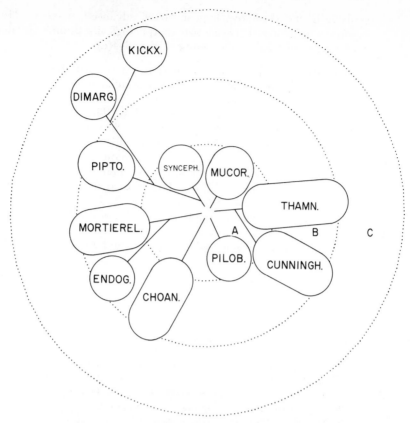

Fig. 3-21. Relationships of the families of Mucorales. Areas A, B, and C, bounded by dotted lines, are intended to represent levels of advancement of morphological characters, which may be summarized, in general, as follows:

A. Primitive

    1. Vegetative and fruiting hyphae generalized, initially nonseptate; simple imperforate septa formed to delimit reproductive structures or laid down adventitiously in aging hyphae.

    2. Sporangia relatively large, multispored, columellate.

    3. Progametangia opposed, highly differentiated morphologically; zygospores usually formed on aerial hyphae; zygosporangium thick-walled, rough, often highly pigmented.

B. Intermediate

    1. Vegetative and fruiting hyphae essentially as in A; perforate septa sometimes formed in fruiting structures.

    2. Sporangiola containing relatively small numbers of spores; columellae absent or rudimentary.

    3. Progametangia opposed, more or less differentiated; zygospores often as in A, sometimes formed in or at the surface of the substrate; zygosporangium sometimes thin-walled, hyaline, nearly smooth.

C. Advanced

    1. Vegetative and fruiting hyphae septate from the beginning; septa highly modified.

    2. Sporangiola containing one or two spores.

    3. Sexual hyphae relatively undifferentiated; zygospores usually formed in or near the surface of the substrate; zygosporangium thin-walled, hyaline, nearly smooth.

[Figure and legend from R. K. Benjamin, 1959, *Aliso* **4**:321–433. Erratum, 1960, *Aliso* **4**:531.]

regularly in the vegetative hyphae of the advanced members. Hyaline, smooth zygospores produced between two morphologically undifferentiated gametangia occur in the most advanced members [Fig. 3-4(e)].

Members of the Trichomycetes may not necessarily be closely related to each other. For example, the Amoebidiales form peculiar amoeboid sporangiospores and lack chitin or cellulose in their walls, unlike the remainder of the Trichomycetes, which have chitinous walls (Lichtwardt, 1973-a). The Trichomycetes in the Order Harpellales or closely related fungi may have close relationships with the Family Kickxellaceae in the Mucorales. These fungi all produce elongated, monosporous sporangioles, which are borne in a similar fashion on the thallus. The supporting structures associated with the sporangioles resemble each other and may be homologous (Moss and Young, 1978). The peculiar flared septum with the lenticular plug occurs in the Kickxellaceae and also in two orders of the Trichomycetes, which argues further for a possible close relationship (Moss, 1975; Moss and Young, 1978). The zygospore of *Glotzia,* which arises as a bud from the conjugation tube superficially, resembles those of *Syncephalis, Endogone,* and *Entomophthora* (Moss and Lichtwardt, 1977).

References

Ainsworth, G. C. 1973. Introduction and keys to higher taxa. *In:* Ainsworth, G. C., F. K. Sparrow, and A. S. Sussman, Eds., The fungi—an advanced treatise. Academic Press, New York. **4B**:1-7.

Baker, K. L., G. R. Hooper, and E. S. Beneke. 1977. Ultrastructural development of merosporangia in the mycoparasite *Syncephalis sphaerica* (Mucorales). Can. J. Botany **55**:2207-2215.

Benjamin, R. K. 1958. Sexuality in the Kickxellaceae. *Aliso* **4**:149-169 (Mucorales).

——. 1959. The merosporangiferous Mucorales. *Aliso* **4**:321-433.

——. 1966. The merosporangium. Mycologia **58**:1-42.

——, and B. S. Mehrotra. 1963. Obligate azygospore formation in two species of *Mucor* (Mucorales). *Aliso* **5**:235-245.

Benny, G. L., and H. C. Aldrich. 1975. Ultrastructural observations on septal and merosporangial ontogeny in *Linderina pennispora.* (Kickxellales; Zygomycetes). Can. J. Botany **53**:2325-2335 (Mucorales).

Blakeslee, A. F., and J. L. Cartledge. 1927. Sexual dimorphism in Mucorales. II. Interspecific reactions. Botan. Gaz. **84**:51-57.

Bracker, C. E. 1968. The ultrastructure and development of sporangia in *Gilbertella persicaria.* Mycologia **60**:1016-1067 (Mucorales).

Drechsler, C. 1959. Several Zoopagaceae subsisting on a nematode and on some terricolous amoebae. Mycologia **51**:787-823.

——. 1961. Two species of *Conidiobolus* often forming zygospores adjacent to antheridium-like distentions. Mycologia **53**:278-303 (Entomophthorales).

Duddington, C. L. 1973. Zoopagales. *In:* Ainsworth, G. C., F. K. Sparrow, and A. S. Sussman, Eds., The fungi—an advanced treatise. Academic Press. New York. **4B**:231-234.

El-Buni, A. M., and R. W. Lichtwardt. 1976. Spore germination in axenic cultures of *Smittium* spp. (Trichomycetes). Mycologia **68**:573-582.

Fletcher, J. 1972. Fine structure of developing merosporangia and sporangiospores of *Syncephalastrum racemosum.* Arch. Mikrobiol. **87**:269-284 (Mucorales).

——. 1973. Ultrastructural changes associated with spore formation in sporangia and sporangiola of *Thamnidium elegans* Link. Ann. Botany **37**:963-971 (Mucorales).

Gerdemann, J. W. 1968. Vesicular-arbuscular mycorrhiza and plant growth. Ann. Rev. Phytopathol. **6**:397-418.

——, and T. H. Nicholson. 1963. Spores of mycorrhizal *Endogone* species extracted from soil by wet sieving and decanting. Trans. Brit. Mycol. Soc. **46:**235–244.

——, and J. M. Trappe. 1974. The Endogonaceae in the Pacific Northwest. Mycol. Mem. 5, 1–76.

——, and ——. 1975. Taxonomy of the Endogonaceae. *In:* Sanders, F. E., B. Mosse, and P. B. Tinker, Eds., Endomycorrhizas. Academic Press, New York. pp. 35–51.

Godfrey, R. M. 1957a. Studies of British species of *Endogone* I. Morphology and taxonomy. Trans. Brit. Mycol. Soc. **40:**117–135.

——. 1957b. Studies of British species of *Endogone* III. Germination of spores. Trans. Brit. Mycol. Soc. **40:**203–210.

Hawker, L. E., and M. A. Gooday. 1967. Delimitation of the gametangia of *Rhizopus sexualis* (Smith) Callen: an electron microscope study of septum formation. J. Gen. Microbiol. **49:**371–376.

——, and ——. 1968. Development of the zygospore wall in *Rhizopus sexualis* (Smith) Callen. J. Gen. Microbiol. **54:**13–20.

——, and ——. 1969. Fusion, subsequent swelling and final dissolution of the apical walls of the progametangia of *Rhizopus sexualis* (Smith) Callen: an electron microscope study. New Phytol. **68:**133–140.

——, ——, and C. E. Bracker. 1966. Plasmodesmata in fungal cell walls. Nature **212:**635.

——, B. Thomas, and A. Beckett. 1970. An electron microscope study of structure and germination of conidia of *Cunninghamella elegans* Lendner. J. Gen. Microbiol. **60:**181–189 (Mucorales).

Hesseltine, C. W., and J. J. Ellis. 1973. Mucorales. *In:* Ainsworth, G. C., F. S. Sparrow, and A. S. Sussman, Eds., The fungi—an advanced treatise. Academic Press, New York. **4B:**187–217.

Khan, S. R., and P. H. B. Talbot. 1975. Monosporous sporangiola in *Mycotypha* and *Cunninghamella*. Trans. Brit. Mycol. Soc. **65:**29–39 (Mucorales).

King, D. S. 1976. Systematics of *Conidiobolus* (Entomophthorales) using numerical taxonomy. I. Biology and cluster analysis Can. J. Botany **54:**45–65.

——. 1977. Systematics of *Conidiobolus* (Entomophthorales) using numerical taxonomy. Can. J. Botany **55:**718–729.

——. 1979. Systematics of fungi causing Entomophthoramycosis. Mycologia **71:**731–745.

Lichtwardt, R. W. 1960. Taxonomic position of the Eccrinales and related fungi. Mycologia. **53:**410–428.

——. 1967. Zygospores and spore appendages of *Harpella* (Trichomycetes) from larvae of *Simuliidae*. Mycologia **59:**482–491.

——. 1973-a. The Trichomycetes: what are their relationships? Mycologia **65:**1–20.

——. 1973-b. Trichomycetes. *In:* Ainsworth, G. C., F. K. Sparrow, and A. S. Sussman, Eds., The fungi—an advanced treatise. Academic Press, New York **4B:**237–243.

——. 1976. Trichomycetes. *In:* Jones, E. B. G., Ed., Recent advances in aquatic mycology. Elec. Sci. London. pp. 651–671.

Moss, S. T. 1975. Septal structure in the Trichomycetes with special reference to *Astreptonema gammari* (Eccrinales). Trans. Brit. Mycol. Soc. **65:**115–127.

——. 1979. Commensalism of the Trichomycetes. *In:* Batra, L. R., Ed., Insect-fungus symbiosis—nutrition, mutualism, and commensalism. Allanheld, Osmun & Co. Publishers, Inc., Montclair. pp. 175–227.

——, and R. W. Lichtwardt. 1976. Development of trichospores and their appendages in *Genistellospora homothallica* and other Harpellales and fine-structural evidence for the sporangial nature of trichospores. Can. J. Botany **54:**2346–2364.

——, and ——. 1977. Zygospores of the Harpellales: an ultrastructural study. Can. J. Botany **55:**3099–3110.

——, ——, and J. F. Manier. 1975. *Zygopolaris,* a new genus of Trichomycetes producing zygospores with polar attachment. Mycologia **67:**120–127.

——, and T. W. K. Young. 1978. Phyletic considerations of the Harpellales and Asellariales (Trichomycetes, Zygomycotina) and the Kickxellales (Zygomycetes, Zygomycotina). Mycologia **70:**944–963.

Mosse, B. 1970-a. Honey-coloured, sessile *Endogone* spores. I. Life history. Arch. Mikrobiol. **70:**167–175.

——. 1970-b. Honey-coloured, sessile *Endogone* spores. II. Changes in fine structure during spore development. Arch. Mikrobiol. **74:**129–145.

——. 1970-c. Honey-coloured, sessile *Endogone* spores. III. Wall structure. Arch. Mikrobiol. **74:**146–159.

——, and G. D. Bowen. 1968. A key to the recognition of some *Endogone* spore types. Trans. Brit. Mycol. Soc. **51**:469–483.

Nicholson, T. H., and J. W. Gerdemann. 1968. Mycorrhizal *Endogone* species. Mycologia **60**:313–325.

——, and N. C. Schenck. 1979. Endogenous mycorrhizal endophytes in Florida. Mycologia **71**:178–198.

O'Donnell, K. L., J. J. Ellis, C. W. Hesseltine, and G. R. Hooper. 1977-a. Zygosporogenesis in *Gilbertella persicaria*. Can. J. Botany **55**:662–675 (Mucorales).

——, ——, ——, and ——. 1977-b. Azygosporogenesis in *Mucor azygosporus*. Can. J. Botany **55**:2712–2720 (Mucorales).

——, ——, ——, and ——. 1977-c. Morphogenesis of azygospores induced in *Gilbertella persicaria* (+) by imperfect hybridization with *Rhizopus stolonifer* (-). Can. J. Botany **55**:2721–2727 (Mucorales).

——, S. L. Flegler, and G. R. Hooper. 1978. Zygosporangium and zygospore formation in *Phycomyces nitens*. Can. J. Botany **56**:91–100 (Mucorales).

Saikawa, M. 1977. Ultrastructure of septa of two species of Dimargaritaceae (Mucorales). J. Japan. Botany **52**:200–203.

Sassen, M. M. A. 1962. Breakdown of cell wall in zygote formation of *Phycomyces blakesleeanus*. Proc. Koninkl. Ned. Akad. Wetenschappen Ser. C. **65**:447–452 (Mucorales).

Schipper, M. A. A., R. A. Samson, and J. A. Stalpers. 1975. Zygospore ornamentation in the genera *Mucor* and *Zygorhynchus*. Persoonia **8**:321–328 (Mucorales).

Whisler, H. C. 1963. Observations on some new and usual enterophilous Phycomycetes. Can. J. Botany **41**:887–900.

——, and M. S. Fuller. 1968. Preliminary observations on the holdfast of *Amoebidium parasiticum*. Mycologia **60**:1068–1079.

Williams, M. C., and R. W. Lichtwardt. 1972. Infection of *Aedes aegyptii* larvae by axenic cultures of the fungal genus *Smittium* (Trichomycetes). Am. J. Botany **59**:189–193.

Young, T. K. W. 1968. Electron microscopic study of asexual spores in Kickxellaceae. New Phytol. **67**:823–836.

——. 1969. Ultrastructure of aerial hyphae in *Linderina pennispora*. Ann. Botany **33**:211–216 (Mucorales).

——. 1973. Ultrastructure of the sporangiospore of *Coemansia reversa* (Mucorales). Trans. Brit. Mycol. Soc. **60**:57–63.

FOUR

Division Amastigomycota— Subdivision Ascomycotina

Members of the Subdivision Ascomycotina are often called "ascomycetes," a name known to all mycologists but without taxonomic rank. This is the largest fungal group, containing almost 2,000 genera. These include a wide range of diverse organisms such as the yeasts, powdery mildews, cup fungi, and the edible delicacies, the morels and truffles. Members of the Ascomycotina are predominantly terrestrial, although a large number live in fresh or marine waters. The majority are saprophytes on decaying plant debris. Many saprophytic members are highly specialized and are capable of growing only on decaying parts of a certain host species; they may even be restricted to a particular part of the host, such as the petiole of the leaf. Other specialized saprophytic types include those fungi that form sporocarps only where a fire recently occurred or on the dung of a certain kind of animal. Many ascomycetes are parasites on plants and, less commonly, on insects or other animals (Chapter 12).

Fungi in the Subdivision Ascomycotina are of great importance, in either a destructive or beneficial sense. Devastating plant diseases such as chestnut blight, apple scab, blackspot of rose, Dutch elm disease, and peach leaf curl are caused by ascomycetes. Beneficial members include the yeasts used in commercial production of alcohol and bread and *Claviceps purpurea,* or ergot, used as a source of medicinal agents. Manufacture of these products is discussed further in Chapter 14.

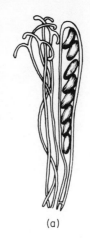

(a)

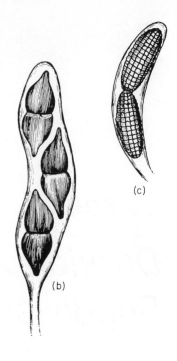

(b)

(c)

Fig. 4-1. Asci and ascospores. Note the variation in form and numbers of the ascospores. Paraphyses, discussed on page 120, are illustrated by two of the asci. [From A. Engler and K. Prantl, 1897, *Die natürlichen Pflanzenfamilien.* Engelmann, Leipzig, and H. M. Fitzpatrick, 1920, *Mycologia* **12**:206–267.]

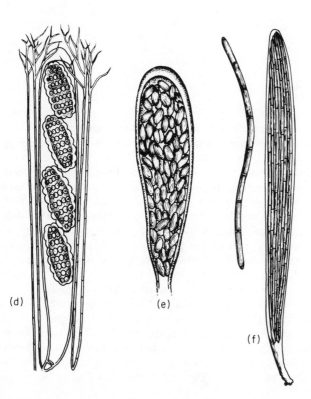

(d)

(e)

(f)

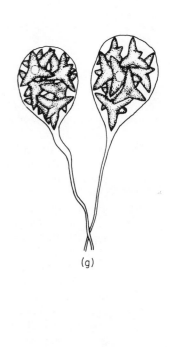

(g)

THE ASCUS AND SEXUAL REPRODUCTION

One characteristic shared by these fungi is that their sexual spores are borne within an *ascus* (plural, *asci*). The ascus is a cell that at first contains a diploid nucleus resulting from karyogamy, and this nucleus undergoes meiosis. Haploid ascospores with thick walls are formed within the ascus (Fig. 4–1).

To understand the role of the ascus in the life cycle, we begin with the ascospore. The haploid ascospore germinates, producing either individual vegetative cells or a monokaryotic mycelium. These cells or mycelia may exist for an extended period of time in the assimilative or vegetative state, perhaps reproducing nonsexually.

Plasmogamy

The sexual cycle is usually initiated when plasmogamy takes place. Plasmogamy may occur by the isogamous union of two similar cells or hyphae or by the heterogamous union of dissimilar cells, a differentiated female gametangium with a detached male cell (a *spermatium*) or a female gametangium with a male gametangium (the antheridium). The gametangia may be simple and rather similar single-celled structures, or they may be quite complex (Figs. 4–2 and 4–3). The female gametangium

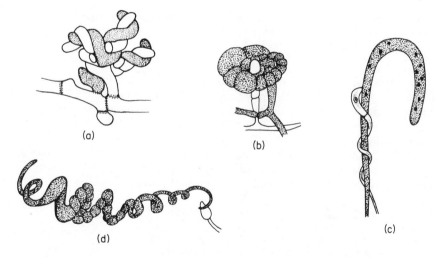

(a)

(b)

(c)

(d)

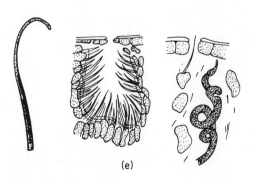

(e)

Fig. 4–2. Examples of reproductive structures in the Ascomycotina. The archicarp is shaded and the male elements are unshaded: (a) A Discomycete, *Ascodesmis nigricans;* (b) A Discomycete, *Ascobolus magnificus;* (c) A Plectomycete, *Penicillium wortmanni;* (d) A Discomycete, *Ascobolus carbonarius;* (e) A Pyrenomycete, *Polystigma rubrum.* The elongated cells in the sporocarp at the left are probably spermatia. [Adapted from: (a) P. Claussen, 1905, *Botan. Zeit.* **63:**1–27; (b) B. O. Dodge, 1920, *Mycologia* **12:**115–134; (c) P. A. Dangeard, 1907, *Botaniste* **10:**1–385; (d) B. O. Dodge, 1912, *Bull. Torrey Botan. Club* **39:**139–197; (e) V. H. Blackman and E. J. Welsford, 1912, *Ann. Botany London* **26:**761–767.]

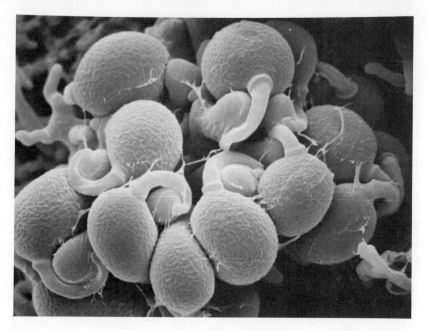

Fig. 4-3. A scanning electronmicrograph of the gametangia of *Pyronema domesticum*. The ascogonia are the balloonlike structures, and each is producing a tubular trichogyne that is in contact with a clubshaped antheridium. × 1,000. [Courtesy K. L. O'Donnell.]

(the *archicarp*) may be divided into a number of cells. Proceeding from the apex downward, these are: (1) the *trichogyne*, (2) the *ascogonium*, and (3) the basal cells which support the remainder of the archicarp. The trichogyne is the terminal cell that initially receives the male protoplasm and allows it to pass to the ascogonium. The ascogonium consists of either a single cell or a chain of cells and is the final recipient of the male protoplasm in plasmogamy (Fig. 4-4). Many ascomycetes are apogamous and the male nuclei are not transferred to the archicarp. Also, many lack sexual organs. Any exchange of nuclei must occur by hyphal anastomosis in these latter cases.

Following plasmogamy there are two basic types of behavior:

1. Karyogamy may follow immediately after plasmogamy, transforming those cells which functioned as gametes into zygotes. Either the newly formed zygote or its vegetative progeny becomes the ascus.
2. Alternatively, karyogamy may be delayed and the dikaryotic condition initiated after plasmogamy is prolonged for several cell generations. This may be accomplished by forming a succession of dikaryotic cells (as in the Taphrinales, pages 131-132) or by forming a special hyphal system, the ascogenous hyphae.

The Ascogenous Hyphae

Ascogenous hyphae are formed by the majority of the members of the Ascomycotina. The ascogenous hyphae arise from the ascogonium (Fig. 4-4) or from other cells if the ascogonium is absent. Usually the ascogenous hyphae contain haploid nuclei of both male and female origin and are therefore dikaryotic. These nuclei divide repeatedly, and the cells in the ascogenous system contain several nuclei. The ascogenous hyphae repeatedly branch, and finally the individual tips reach the site where the asci will be produced. Usually a *crozier* is formed when the tip of the ascogenous hypha recurves and forms a hooked cell. Two nuclei are present in the crozier. These two nuclei divide simultaneously, with their division figures parallel to the long axis of the crozier. Two septa form across the spindle and form two cells from the crozier. These are the apical cell, the recurved portion of the hyphal tip, and the second (penultimate) domelike cell, which will become the ascus. Sister nuclei are separated by the septa, which isolate two nuclei in the penultimate cell and a third nucleus in the apical cell; the fourth nucleus remains in the ascogenous hypha. The two nuclei in the penultimate cell fuse and form a diploid nucleus, and this cell is now a young ascus (Fig. 4–4). In those fungi forming an ascogenous system, the ascus is the only diploid stage in the life cycle, and the haploid stage is quickly regained in the meiotic division that follows.

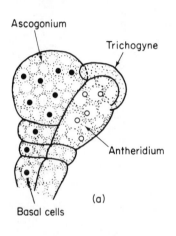

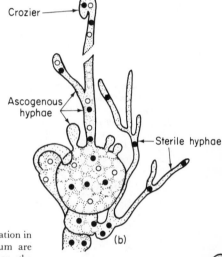

Fig. 4-4.   Sexual organs, plasmogamy, and ascus formation in *Pyronema domesticum:* (a) The archicarp and antheridium are undergoing plasmogamy. Note the continuity between the trichogyne and antheridium; (b) Ascogenous hyphae are arising from the ascogonium, and sterile hyphae are arising from the basal cells of the archicarp; (c) Binucleate crozier; (d) Mitotic division of nuclei; (e) Formation of septa to delimit penultimate cell; (f) Karyogamy; (g) Diploid nucleus in young ascus; (h) Binucleate ascus after first division of meiosis: (i) Four-nucleate ascus after second division of meiosis; (j) Mature ascus with eight ascospores after final mitotic division. [(a) and (b) adapted from E. J. Moore, 1963, *Am. J. Botany* **50:**37-44.]

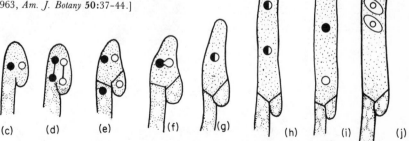

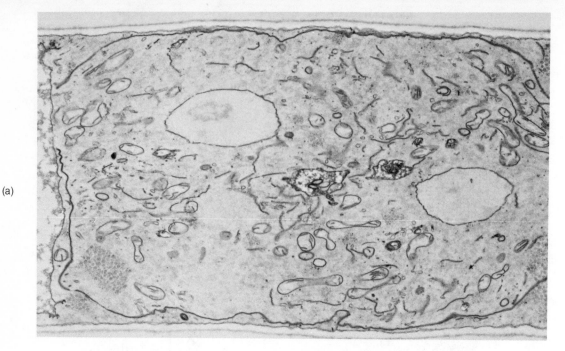

(a)

Fig. 4–5. Stages in ascospore delimitation. (a) Longitudinal section of a young ascus showing the membrane sac forming. × 9,900; (b) The membrane sac is at an advanced stage of invagination, and the ascospore initials are almost completely delimited. × 8,500; (c) Production of wall material (clear areas) has begun. × 4900; (d) A nearly mature reticulated ascospore in ascus. Note that the surrounding protoplasm has disappeared. × 12,600. [Courtesy G. C. Carroll. (a) and (b), 1967, *J. Cell Biol.* **33**:218–224. (c) and (d), 1966, from Thesis, University of Texas.]

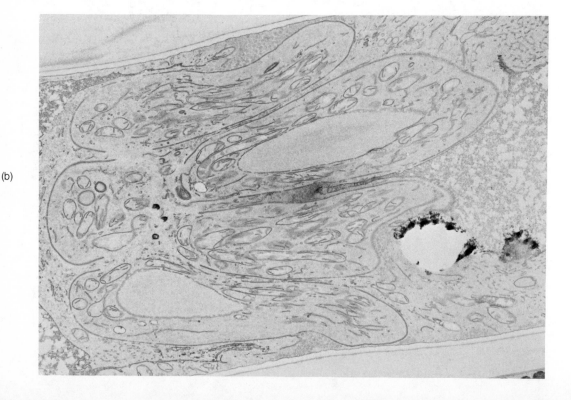

(b)

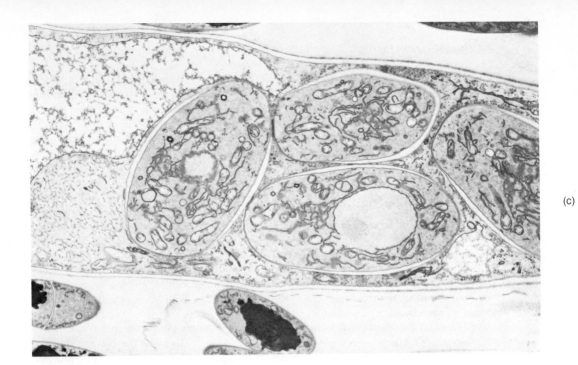

(c)

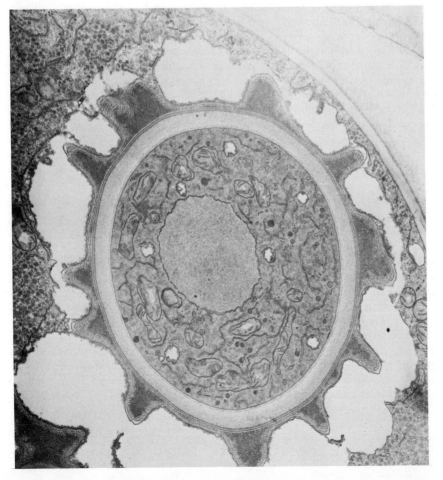

(d)

119

Ascus Formation

Beginning with the young diploid ascus, ascospore formation is similar in all the Ascomycotina. The diploid nucleus in the ascus undergoes a meiotic division, and the ascus acquires two nuclei as a result of meiosis I and four nuclei as a result of meiosis II. In some ascomycetes the asci will have no further divisions, while in most species a single mitotic division follows and the four nuclei divide to form eight nuclei. The resulting nuclei are then incorporated into ascospores (Fig. 4-4). Ascospore development has been studied extensively at the ultrastructural level, and with the exception of members of the Hemiascomycetes (page 125), virtually all members of the Ascomycotina conform to the following general pattern (e.g., Carroll, 1969; Reeves, 1967; Stiers, 1974, 1976; Tyson and Griffiths, 1976; Wells, 1972). Two continuous unit membranes form a sac near the periphery of the ascus and lie parallel to each other and the ascus wall [Fig. 4-5(a)]. These membranes comprise the *ascus vesicle*. The origin of the ascus vesicle is difficult to determine but the membranes apparently develop from a preexisting membrane system such as the plasmalemma or nuclear envelope. The two membranes then move inward, meanwhile pushing organelles into new positions near the nuclei [Fig. 4-5(b)]. Eventually each nucleus is isolated, along with its associated organelles and surrounding cytoplasm, from the remainder of the cytoplasm by two concentric membranes. Each cell isolated in this manner becomes an ascospore. The inner membrane of the concentric pair becomes the plasmalemma of the ascospore, and ascospore wall material is deposited between the two membranes [Fig. 4-5(c)]. The wall is at first uniform in appearance but as it matures, distinct layers become apparent. The number of wall layers and the sequence in which they differentiate vary from species to species (e.g., Merkus, 1976). Some cytoplasm remains in the ascus around the ascospore but it disappears as the ascus matures [Figs. 4-5(c) and (d)]. Although 8 ascospores are usually produced in each ascus, the number of ascospores may be 4 if the third division does not occur or it may be 16, 32, or any multiple of 8, depending on the number of mitotic divisions taking place (Fig. 4-1). As many as 7,000 ascospores occur in *Trichobolus* (Kimbrough and Korf, 1967). Asci may or may not be produced within or upon complex multicellular sporocarps, the *ascocarps*.

Sterile Tissues in an Ascocarp

The sequence from plasmogamy through ascospore formation represents only the sexual phase of development in a typical Ascomycotina member that forms an ascocarp. In addition to the sexual organs, ascogenous hyphae, and asci, an ascocarp usually contains a large quantity of sterile tissue. The sterile tissue arises as hyphae, usually from the basal cells supporting the gametangia, and then envelopes the sexual structures and becomes organized into tissues. In some instances, the sterile elements may be represented by only a few hyphae incompletely surrounding the developing sexual system. The sterile hyphae are not dikaryotic, as the ascogenous hyphae frequently are, but rather are monokaryotic, as they arise from mycelium or cells that are not the products of plasmogamy (Fig. 4-6).

The hyphae are intertwined so that distinct tissues are formed, and the ascocarp is often divided into anatomical regions characterized by a distinctive type of tissue. Hyphae from these tissues may terminate in sterile, hairlike processes. These are: (1) *paraphyses,* which lie among the asci and are free at their apices; (2) *pseudoparaphyses,*

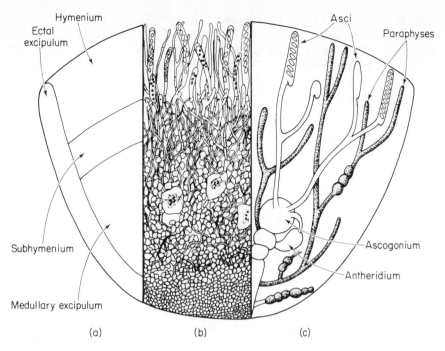

Fig. 4-6. Longitudinal section through an apothecium of *Pyronema domesticum* showing: (a) The division into major tissue zones; (b) The natural appearance of the tissues; (c) A diagrammatic representation of the origin of sexual structures (unshaded) and sterile structures (shaded) that intertwine to form the tissues. [(b) from E. J. Moore, 1962, Thesis, Cornell University.]

which grow downward from the upward surface of the ascocarp cavity and form a vertical palisade of hairs, usually becoming attached at the bottom by interweaving to form a matlike plexus; and (3) *periphyses,* which are small hairs that line a canal in the necklike extension of some ascocarps (perithecia, see below) (Fig. 4–7).

There are fundamental differences in the morphology of the ascocarp and in the origin of the surrounding tissues. Some ascomycetes form a *stroma,* a pad of vegetative tissue. *Locules* (cavities) may form within the stroma. Asci are then formed within the locules and are not surrounded by any sterile tissue that originates from the archicarp or that is not sexually stimulated. A second major type of ascocarp is the *perithecium,* which may be immersed within a stroma or free on the substratum and is character-

Fig. 4-7. Diagrammatic longitudinal sections through perithecia to show the position and nature of sterile filaments.

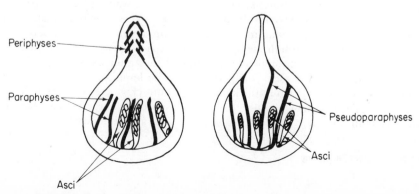

ized by an enclosed cavity bearing asci. This cavity is lined by a wall of sterile cells arising from the archicarp. The *perithecium* may be of various shapes, most often flask-shaped, and opens by forming a slit or ostiole. A third type of ascocarp is the *cleistothecium,* which is globular and without a regular means of opening. Finally, an *apothecium* is an open ascocarp that bears asci on an exposed surface. These ascocarp types are illustrated in Fig. 4–8.

The number of asci borne by an ascocarp varies from one to several. They may be borne randomly within the ascocarp cavity, or they may arise as a clustered *fascicle* from the base of the ascocarp cavity. Most often, they are borne in a palisadelike layer, the *hymenium,* which lines the base of the ascocarp cavity or an exposed part of the apothecium.

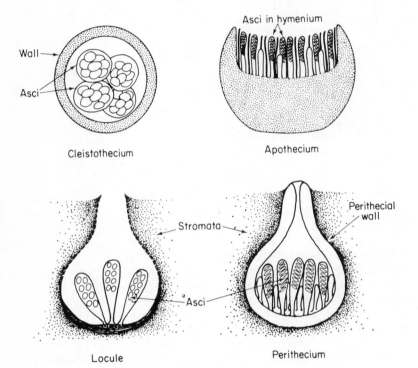

Fig. 4–8. The different types of ascocarps.

Ascus Types

Two fundamentally different types of asci occur. These are the bitunicate and unitunicate asci. The *bitunicate ascus* has a rigid outer wall and an inner wall which is flexible and can become considerably extended. When ascospore discharge takes place, the outer wall splits and may sometimes slip downward to form a collar, and the inner wall protrudes through the opening and expands (Fig. 4–9). The ascospores are expelled in succession through the opening of the inner ascus wall. The *unitunicate* ascus possesses a single ascus wall, which may be laminated. The unitunicate ascus wall either may be uniform in thickness throughout [Fig. 4–9(a)] or may have a thickened apex. The wall may rupture to allow ascospore discharge to take place, or a pore may occur in the apex. Ascospore discharge occurs simultaneously.

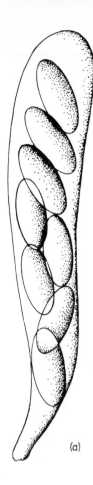

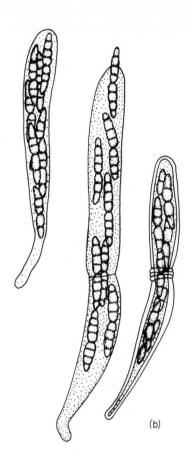

Fig. 4-9. The different types of asci: (a) A unitunicate ascus; (b) A bitunicate ascus. From left to right, a mature bitunicate ascus before spore discharge begins; the inner wall expanded in the process of spore ejaculation; and a collapsed ascus showing the thickened inner wall and the thin, rolled-back outer wall which forms the collar. [(a) from A. Engler and K. Prantl, 1897, *Die natürlichen Pflanzenfamilien.* Engelmann, Leipzig. (b) From E. S. Luttrell, 1960, *Mycologia* **52:** 64-79.]

(a)

(b)

NONSEXUAL REPRODUCTION

Many members of the Ascomycotina have nonsexual reproduction (Fig. 4–10). Therefore their life cycles consist of two alternating phases, nonsexual and sexual. This entire life cycle including both nonsexual and sexual cycles is sometimes termed the *holomorph.* The nonsexual phase of the cycle may be referred to as the *anamorph* while the sexual phase is termed the *telomorph.* (Until recently, the anamorph was called the "imperfect" stage and the telomorph the "perfect" stage: you may encounter these terms as you read other literature.) At its simplest, nonsexual reproduction may occur by budding of cells in the yeasts (this is discussed further on page 127). Nonsexual reproduction by means of conidia is very common throughout the group. Conidia are borne on specialized hyphae, the *conidiophores.* The conidiophores may be scattered loosely over the mycelium, or they may be organized into groups or into sporocarps that closely resemble ascocarps. The conidial stage (anamorph) may follow germination of ascospores. Often the conidia are borne on a mycelium separate from that producing the sexual stage, which gives a spatial separation of the anamorph and telomorph. In many ascomycete life cycles, the conidia will be produced in abundance

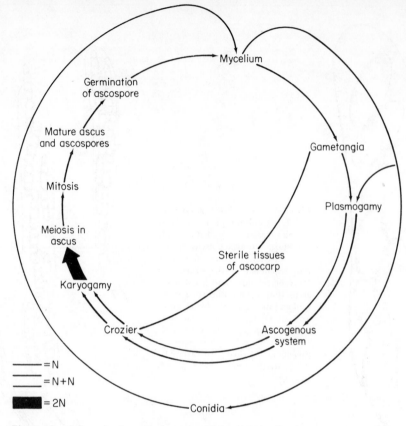

Fig. 4-10. Life cycle of a typical member of the Ascomycotina that forms a multicellular ascocarp, showing both sexual and nonsexual reproduction.

throughout the spring and summer, with the ascocarp beginning its development in the late autumn and maturing the following spring, which culminates in ascospore discharge.

An ascomycete may propagate so constantly and persistently through its non-sexual stage that mycologists are unaware that a particular conidia-producing fungus is the anamorph of an ascomycete. They may discover this relationship when they germinate ascospores of a known ascomycete and find that a conidial fungus is produced which under certain conditions produces an ascocarp.

For further information about the morphology and development of the non-sexual structures, consult Chapter 6.

CLASSIFICATION

Taxonomy of the Subdivision Ascomycotina is based largely on gross morphology, anatomy, and life history of the individual organisms. To a lesser extent, physiological characters are used and are especially important in the taxonomy of the yeasts. This subdivision may be divided into six classes (Ainsworth, 1973) based primarily on the presence or absence of an ascocarp and its morphology. Structure of the ascus also plays an important role in classification. Within the following classes, the orders are those used by Kreger-van Rij (1973), Kramer (1973), Fennell (1973), Müller and Von Arx (1973), Korf (1973), Benjamin (1973), and Von Arx and Müller (1975), respectively. Partly because of the large number of organisms in this group and their great diversity, not all mycologists agree on how they should be classified. An outline of the classification system used here follows:

Subdivision Ascomycotina
    Class Hemiascomycetes
        Order Endomycetales
        Order Protomycetales (not treated further)
        Order Taphrinales
    Class Plectomycetes
        Order Eurotiales
    Class Pyrenomycetes
        Order Erysiphales
        Order Meliolales
        Order Coronophorales
        Order Sphaeriales
    Class Discomycetes
        Order Medeolariales (not treated further)
        Order Cyttariales (not treated further)
        Order Tuberales
        Order Pezizales
        Order Phacidiales
        Order Ostropales
        Order Helotiales
    Class Laboulbeniomycetes
        Order Laboulbeniales
    Class Loculoascomycetes
        Order Dothideales

Class Hemiascomycetes

The Class Hemiascomycetes contains simple reduced members of the Ascomycotina which are either unicellular or form a limited mycelium. The unicellular organisms are commonly called "yeasts" although there is often no sharp dividing line between the unicellular and mycelial forms. The Hemiascomycetes are especially characterized by the absence of an ascocarp and of ascogenous hyphae. Either a zygotic cell, the diploid progeny developing nonsexually from that cell, or a usually isolated dikaryotic cell is transformed directly into an ascus bearing ascospores. Structurally, the ascus containing ascospores may superficially resemble a sporangium or the fused gametangia of the Zygomycotina, but this cell functions as an ascus and, after karyogamy, the nucleus undergoes two or three successive divisions with no resting stage. Ascospore formation shares many similarities with that outlined on page 120 but differs in one important feature (Ashton and Moens, 1979; Black and Gorman, 1971; Moens, 1971; Syrop and Beckett, 1976). A continuous ascus vesicle is not formed near the periphery of the ascus. Instead, discontinuous membranes arise in the vicinity of each nucleus, perhaps originating as invaginations of the plasmalemma or from the endoplasmic reticulum. The membranes enlarge and surround a nucleus and some associated cytoplasm and cytoplasmic organelles. The nucleus is isolated by the membranes, and development of the ascospore walls proceeds as described previously (page 120). Usually either four or eight ascospores are formed, but the number is variable if some nuclei degenerate after the nuclear divisions occur. In some species, more or fewer than eight ascospores may be formed. The ascospores are usually oval or spherical, but may be elliptical, reniform, needlelike, or shaped like a hat with a brim. Although the ascospores are usually smooth, some yeasts produce rough, warted ascospores.

Fungi resembling the unicellular yeasts also occur in other taxonomic groups (the Basidiomycotina and Deuteromycotina, Chapters 5 and 6). Unlike the yeasts in the Hemiascomycetes, the yeastlike fungi in other groups do not form asci and ascospores. In addition, the DNA base composition differs (for example, the mole percent of guanine and cytosine combined is lower in those Hemiascomycetes tested than in the yeastlike fungi in the Basidiomycotina). Other differences include cell-wall composition and the ultrastructure of the cell wall.

The Class Hemiascomycetes may be divided into the Orders Endomycetales, Protomycetales, and Taphrinales (Kreger-Van Rij, 1973; Kramer, 1973). The Protomycetales is a very small group of specialized plant parasites and will not be treated further.

ORDER ENDOMYCETALES

Members of the Endomycetales include those fungi that form an ascus directly from a zygote or from the diploid progeny of a zygote. The approximately 150 species include the well-known yeast of commerce *Saccharomyces cerevisiae* (see Chapter 14), in addition to many others of little or no economic importance. The yeasts are well-known for their ability to ferment sugars (Chapter 7), and the ability or inability to ferment particular sugars is useful in delimiting species and genera in this group. In nature, the yeasts are frequently found growing saprophytically in sugary materials. Yeasts and their allies may be found in nectaries of flowers; on leaves and fruits; in plant exudates; in association with insects; in soil; in fresh and marine waters; in sewage; and in foods such as syrups, brines, meats, and milk products. Some parasitic yeasts cause diseases of plants and animals. Yeasts may be distributed by being carried in the air and are probably carried on dust particles. The unicellular yeasts have been widely studied from both the biological and commercial viewpoints owing to their extensive and varying physiology, their potential contribution to basic biological problems such as cytology and phylogeny, and their ease of manipulation in culture.

Thallus. The unicellular yeasts vary from round, oval, or elongate shapes to rather peculiar dumbbell-like or triangular configurations. Macroscopically, yeasts cultured on solid media appear as slimy, mucoid colonies which are often brightly colored. These colonies strongly resemble bacterial colonies.

Cytology of the yeast cells has been studied extensively, and much of our knowledge has been derived from studies of *Saccharomyces cerevisiae* and closely related yeasts (Matile, Moor, and Robinow, 1969). A typical yeast cell will be described. The yeast cell is surrounded by a cell wall which consists of layers: the outermost layer contains mainly mannan-protein and some chitin, the middle layer consists of glucan and appears lighter than the other two layers in electronmicrographs, and the innermost layer consists of a proteinaceous material. The plasmalemma exhibits numerous invaginations. The cytoplasmic contents include glycogen, ribosomes, and mitochondria that may be spherical, rod-shaped, or threadlike and branched. Vesicles containing lipid or catalase are distributed within the cytoplasm and may have originated from the endoplasmic reticulum. The endoplasmic reticulum is sometimes attached to the plasmalemma but is continuous with the nuclear envelope. The nucleus has a large, dense cup-shaped nucleolus and more or less translucent dome-shaped nucleoplasm occupying the remainder of the nucleus (both the nucleolus and nucleoplasm may be seen with a phase contrast microscope). The most conspicuous organelle is a vacuole that is usually so large that it is visible with the light microscope. This vacuole, which has a diameter of approximately 0.3 to 3 microns, is bounded by a single membrane and has watery contents. Under certain conditions (e.g., anaerobic conditions), the vacuole may have granular inclusions. By analyzing the contents of isolated vacuoles it has been determined that they contain high levels of hydrolytic enzymes, including proteases, ribonuclease, and esterase. The vacuole apparently functions as a lysosome, representing an isolated compartment of the cell where macromolecules such as proteins and nucleic acids may be degraded by the hydrolytic enzymes before their components are incorporated into newly synthesized macromolecules. In addition to functioning as a lysosome, the vacuole serves as a storage place for cellular components such as amino acids or potassium ions that may be utilized later.

Cellular division in the unicellular yeasts occurs when nuclear division is accompanied by *fission* or *budding*. Fission is the division of the mother cell into two daughter cells and involves the formation of new wall material to divide two essentially equal protoplasts. In budding, a new cell arises as a protrusion on a narrow base from the mother cell, increases in size, and eventually breaks away [Fig. 4–11(b)]. Several cellular changes are involved in budding. In a cell preparing to form a bud, the endoplasmic reticulum fuses and forms a nearly closed envelope containing the nucleus and vacuole and subsequently produces vesicles near the site where the bud will form. The vesicles are thought to contain enzymes involved in the localized weakening of the wall that is followed by pore formation. The vesicles may also contain proteinaceous wall precursors. Production of a pore in the mother cell is accompanied by the simultaneous production of new wall materials appearing in the daughter cell. The nucleus originally lies between the vacuole and the developing bud. Nuclear division occurs by elongation and constriction and without loss of the nuclear envelope. One half of the elongated nucleus enters the bud in an amoeboid manner, temporarily leaving a long, narrow neck between it and the other half of the nucleus, which remains in the mother cell. Eventually division of the nucleus is completed. The vacuole fragments and the products of the fragmentation are distributed between the mother and daughter cells. The plasmalemma constricts inward to separate the protoplast of the bud from that of the mother cell, and deposition of two walls occurs at the point of junction. Finally, the bud separates from the mother cell, leaving a prominent scar on the mother cell and a smaller almost invisible scar on the daughter cell where the walls were connected. As many as 24 bud scars have been detected on a single yeast cell, corresponding to the number of daughter cells formed by that cell. In some yeasts, especially in older cultures, new cells formed by budding may remain together in simple or branched chains which superficially resemble a mycelium. Other genera, especially the yeast allies, form a true but limited mycelium through cell division. More than one thallus type may occur in a single species, depending partially on age and cultural conditions.

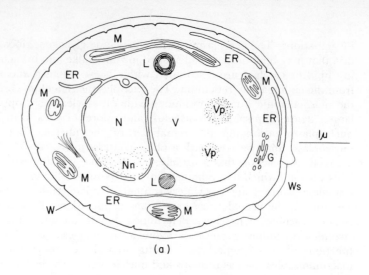

(a)

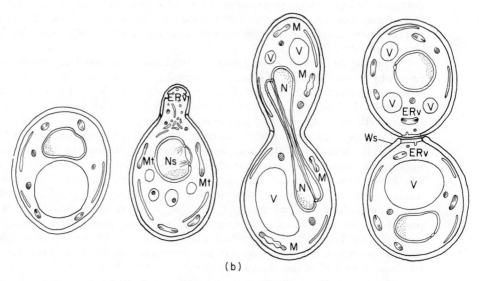

(b)

Fig. 4–11. The yeast *Saccharomyces cerevisiae*. (a) Diagram of a resting cell. Abbreviations: *ER* = endoplasmic reticulum, *ERv* = vesicles derived from endoplasmic reticulum, *G* = Golgi apparatus, *L* = lipid granule, *M* = mitochondrion, *N* = nucleus, *Nm* = nucleolus, *Ns* = spindle apparatus, *V* = vacuole, *Vp* = polymetaphosphate granule, *W* = cell wall, *Ws* = bud scar; (b) Successive stages in budding. [Adapted from P. Matile, H. Moor, and C. F. Robinow. 1969. *In:* Rose, A. H., and J. S. Harrison, eds. *The yeasts.* Academic Press, New York. **1**:219–302.]

Nonsexual Reproduction. In the unicellular yeasts, cell division or budding comprises nonsexual reproduction. Mycelial allies may reproduce nonsexually by forming yeastlike conidia, or the mycelium may fragment into a number of *oidia*. Conidia and oidia may propagate by budding or dividing for a period of time. These cells are often indistinguishable from the cells of a unicellular yeast, and they may form colonies of a similar appearance. Eventually a reversion to the mycelial type occurs.

Sexual Reproduction. Sexual reproduction occurs by conjugation of either similar or dissimilar cells or gametangia. Morphology of the reproductive structures and events involved in plasmogamy vary widely. There may be the isogamous union of haploid vegetative cells or of haploid ascospores to yield a zygote. A heterogamous union may occur between a large and small vegetative cell or between morphologically distinct gametangia. In addition to the morphological variation in reproductive structures, there is variation in the overall life cycle pattern. There may be a long haploid stage with a transient diploid stage, a long diploid stage with a transient haploid stage, or a haploid and diploid stage of approximately equal length (alternation of generations). In order to explore at least a little of this variation, we examine sexual reproduction and life cycles of three yeasts.

In *Saccharomyces cerevisiae,* similar haploid vegetative cells fuse to form a zygote. The diploid zygote typically buds and produces several generations of diploid cells. Eventually meiosis takes place in one of these cells, and the cell is converted into an ascus in which the four haploid nuclei are incorporated into ascospores. Upon their release, the haploid ascospores bud and produce several generations of haploid vegetative cells. Eventually some of these again function as gametes to initiate the cycle again. In this case, isogamous union of vegetative cells is coupled with a life cycle with both an extended haploid and diploid stage [Fig. 4–12(a)].

In *Saccharomycodes ludwigii,* isogamous union between ascospores takes place. There are four ascospores in each ascus, and they become separated into two pairs which lie at opposite ends of the ascus. The ascospores swell and form small beaks. The beaks extend to meet each other and meanwhile broaden to form a canal that eventually provides cytoplasmic continuity between the ascospores. The nuclei from the ascospores migrate into the canal and undergo karyogamy. The canal now penetrates through the ascus and forms a number of diploid vegetative cells. The cycle is completed when meiosis takes place in a diploid vegetative cell, converting it into an ascus. This life cycle combines isogamous union of ascospores with a long diploid phase and a short haploid phase [Fig. 4–12(b)].

Dipodascus aggregatus is an example of the mycelial yeast allies that form specialized gametangia [Fig. 4–12(c)]. The vegetative mycelium is haploid. Sexual reproduction is initiated when two adjacent cells in a filament form protrusions which will become an ascogonium and antheridium, respectively. In *D. aggregatus,* the ascogonium is considerably larger than the antheridium. The gametangia fuse at their apex and are separated from their parental cells by septa. Since only the tip fuses, two stalklike structures, or "feet," remain behind. Karyogamy takes place, and the ascogonium elongates into an ascus with two footlike projections. While the ascus is elongating, meiosis takes place, followed by a number of further divisions. The mature ascus is filled with numerous haploid ascospores that form a haploid mycelium upon germination. This life cycle combines heterogamous gametangial copulation with a long haploid phase and a short diploid phase. Although this fungus is a mycelial yeast ally, similar cycles occur in the unicellular yeasts.

ORDER TAPHRINALES

Members of the Taphrinales are plant parasites that cause localized deformation and swelling of the host tissues. A severe disease, peach leaf curl, is caused by *Taphrina deformans,* a member of this group. Peach leaf curl is characterized by pocketlike bulges and curled areas on the leaves. The Taphrinales differ from the Endomycetales principally in that: (1) the asci are organized into a palisadelike layer on the surface of the

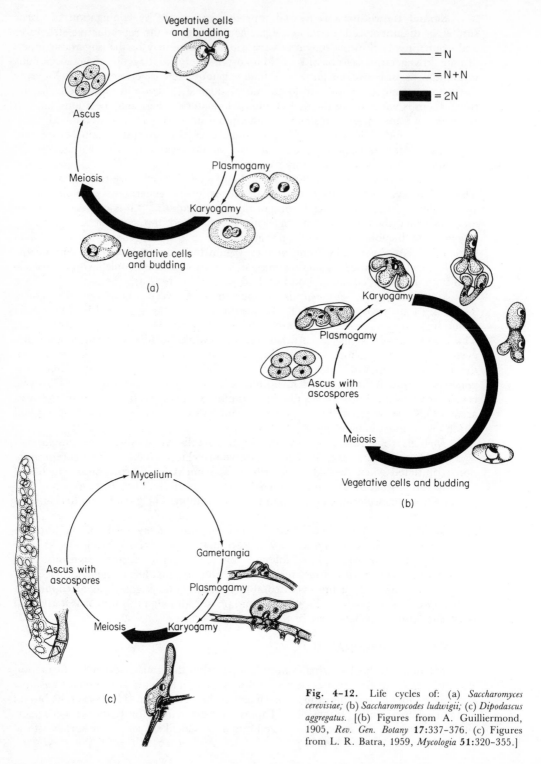

Fig. 4-12. Life cycles of: (a) *Saccharomyces cerevisiae;* (b) *Saccharomycodes ludwigii;* (c) *Dipodascus aggregatus.* [(b) Figures from A. Guilliermond, 1905, *Rev. Gen. Botany* **17**:337–376. (c) Figures from L. R. Batra, 1959, *Mycologia* **51**:320–355.]

host and (2) the ascus develops from a dikaryotic cell, not from the zygote. Otherwise, the Taphrinales share many similarities with the yeasts and their allies.

The assimilative structures of the Taphrinales are found within the host tissue, usually as a limited mycelium with dikaryotic cells in which the nuclei divide conjugately (Mix, 1935). The mycelium may occur intercellularly or subcuticularly, or it may penetrate the walls of the host where it develops. In addition to the mycelium, the vegetative structures may consist partially or completely of binucleate, separate cells distributed among host cells.

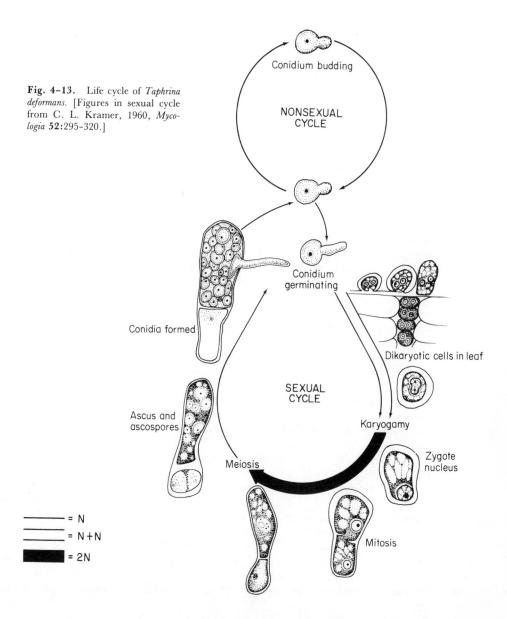

Fig. 4-13. Life cycle of *Taphrina deformans.* [Figures in sexual cycle from C. L. Kramer, 1960, *Mycologia* **52**:295–320.]

Conidium budding

NONSEXUAL CYCLE

Conidium germinating

Conidia formed

Dikaryotic cells in leaf

SEXUAL CYCLE

Ascus and ascospores

Karyogamy

Meiosis

Zygote nucleus

Mitosis

———— = N

———— = N+N

▬▬▬ = 2N

In most species of *Taphrina,* the vegetative mycelium fragments into binucleate ascogenous cells, usually immediately beneath the upper cuticle of the leaf. These binucleate cells form the asci that penetrate the leaf surface. Less commonly, the cells of the vegetative hyphae may give rise to asci without separating, or if a mycelium is not formed, the yeastlike cells will simply differentiate into asci. Karyogamy takes place in the binucleate cell, and the cell elongates and forms the ascus. The diploid fusion nucleus usually undergoes a single mitotic division; one of the daughter nuclei of this division remains in the basal portion of the developing ascus, which becomes separated as a stalk cell, and the second nucleus migrates into the tip of the young ascus. If the species does not form a stalk cell, this single mitotic division is lacking. Meiosis follows, and the single nucleus in the young ascus divides to give either four or eight ascospore nuclei (Fig. 4–13).

While in the ascus or after their release, the ascospores produce a large number of uninucleate conidia by budding in a yeastlike fashion. In culture, yeastlike colonies consist of these cells. Conidia are capable of overwintering and of initiating new infections in the spring. Apparently mycelia are incapable of overwintering. In some species (including *T. epiphylla*), conidia regularly conjugate to reestablish the dikaryon before penetration of the host occurs, but in *T. deformans* the single haploid nucleus in the conidium divides to establish a dikaryon that invades the host.

Class Plectomycetes

ORDER EUROTIALES

The Class Plectomycetes includes the simplest members of the ascocarp-forming ascomycetes. The members of this class share a combination of characteristics (Fennell, 1973). The ascocarp is a cleistothecium, an enclosed ascocarp lacking an ostiole or other regular means of opening (Figs. 4–14 and 4–15). Paraphyses are lacking and asci are scattered randomly within the cleistothecium. Typically the asci are rounded in shape, contain eight ascospores, and have thin walls that break down at maturity to release the ascospores within the cleistothecium. The ascospores are unicellular and lack germ pores or slits. The cleistothecium opens from exposure to the weather or by pressure exerted by maturation of the contents.

Many of the Plectomycetes produce a conidial stage. This anamorph may be denoted by the name of a conidial genus. This grouping is a *form-genus* because genetically diverse (distantly related) fungi may have similar anamorphs. Therefore fungi assigned to the same form-genus may or may not be related. For a further discussion of taxonomy of conidial fungi, see Chapter 6. The conidial stages of the Plectomycetes are morphologically diverse and have been assigned to a variety of form-genera.

Fennell (1973) recognizes a single order within the Plectomycetes, the Eurotiales.

The majority of the Eurotiales live as saprophytes in the soil or on substrata such as dead insects, wood, dung, hair, bones, feathers, horns, or hooves. Skin diseases of humans and other animals may be caused by parasitic species of the form-genera *Trichophyton* and *Microsporum,* which are the conidial stages with telomorphs in *Arthroderma* and *Nannizzia,* respectively. Like those Eurotiales that occur saprophytically on hair, feathers, horns, or hooves, these skin-inhabiting parasites are able to live saprophytically on such materials because they also utilize keratin as a nutrient. The causal agent of a serious and often fatal disease of humans, histoplasmosis, is

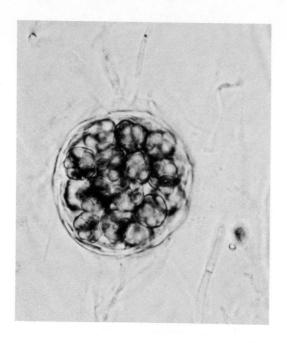

Fig. 4-14. A simple cleistothecium of *Emericellopsis microspora.* The asci are surrounded by a thin, transparent peridium. × 850. [Courtesy M. P. Backus and P. A. Orpurt, 1961, *Mycologia* **53**:64–83.]

caused by *Histoplasma capsulatum,* the anamorph of *Emmonsiella capsulata* (Kwon-Chung, 1973).

The form-genera *Penicillium* and *Aspergillus* are among the most commerically important of the conidial fungi. Sometimes a conidial culture of either of these fungi will produce cleistothecia that may be assigned to several genera in the Eurotiales (for example, *Aspergillus* may have cleistothecia in *Eurotium,* while *Penicillium* may have cleistothecia in *Talaromyces* (Benjamin, 1955: Raper and Fennell, 1965; Malloch and Cain, 1972). The telomorphs for many species of *Aspergillus* and *Penicillium* are unknown, however, and therefore these species are considered to be members of the Deuteromycotina (Chapter 6). In general, *Penicillium* and *Aspergillus* are among the most common of soil fungi (Chapter 11) and many species are important in human affairs (Chapter 14). Species of *Penicillium* are used commercially as a source of the antibiotic penicillin and in the production of Roquefort cheese. Species of *Aspergillus* are used commercially to produce citric acid, to biochemically alter steroids that will be used medically, and to manufacture certain Oriental foods. Species of both genera may be responsible for the destruction of stored foods, including grains. *Aspergillus flavus* is especially destructive as it produces aflatoxins, animal carcinogens, in stored grains. *Aspergillus* may also produce cellulolytic enzymes that destroy cloth.

The cleistothecia of some members may be collected directly on their substrata in nature. For example, species of *Onygena* form stalked cleistothecia on weathered hooves or feathers. Some keratinophilic (keratin-utilizing) species may be obtained by placing an appropriate ''bait'' such as hair in or on soil. After a short period of time, the hair will be invaded by mycelium that subsequently forms cleistothecia. The conidial stages of many members of the Eurotiales are more easily obtained than are the sexual stages but may produce cleistothecia in culture. For example, the conidial stage of *Toxotrichum cancellatum* was isolated from frozen blueberry pastries and subsequently produced cleistothecia in culture (Orr and Kuehn, 1964).

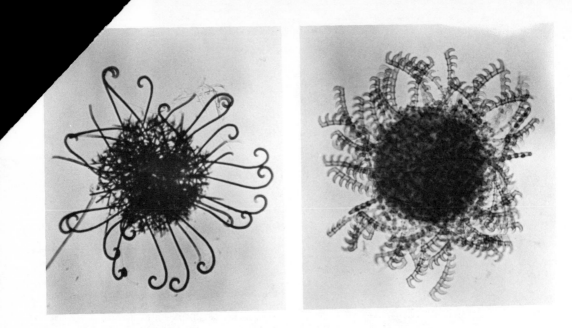

Fig. 4–15. Two simple members of the Eurotiales, *Myxotrichum chartarum* (left, × 125) and *Ctenomyces serratus* (right, × 175). Each cleistothecium bears elaborate appendages. [Courtesy R. K. Benjamin.]

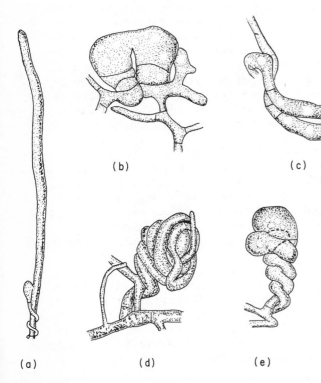

(a)

(b)

(c)

(d)

(e)

Fig. 4–16. Representative gametangia of the Eurotiales. (a) Elongate clavate ascogonium and spiraled antheridium of *Penicillium vermiculatum.* The ascogonium may reach a length of 250 microns; (b) Similar clavate gametangia of *Myxotrichum conjugatum* have fused; (c) Similar clavate gametangia of *Arachniotus reticulatus* are loosely entwined. Fusion has not been observed between these gametangia; (d) The archicarp of *Myxotrichum emmonsii* has formed a tight coil around a clavate antheridium; (e) The gametangia of *Myxotrichum uncinatum* are coiled together. [b–e, × 1,300).] [(a) from C. W. Emmons, 1935, *Mycologia* **27**:128–150; (b) from H. H. Kuehn, 1955, *Mycologia* **47**:878–890; (c) from H. H. Kuehn, 1957, *Mycologia* **49**:55–67; (d) and (e) from H. H. Kuehn, 1955, *Mycologia* **47**:533–545.]

The cleistothecia are typically globose in shape, although stalked cleistothecia may be formed by a few genera (e.g., *Onygena*). The cleistothecia are comparatively small (usually only 5 millimeters or less in diameter), and may be hyaline or some shade of yellow, orange, red, brown, or black. Most species produce isolated cleistothecia surrounded only by the cleistothecial wall (the *peridium*). A few species will produce confluent cleistothecia, or cleistothecia embedded within a stroma. The peridium shows a great deal of variation, ranging from a netlike weft of hyphae that barely covers the asci in some species of *Gymnoascus* to a compact covering consisting of interwoven hyphae or pseudoparenchymatous cells (e.g., in *Eurotium*). Some species have characteristic appendages protruding from the cleistothecium (Fig. 4-15). Among the various types of appendages are some that are spinelike and unbranched, some that have numerous small lateral branches so that each resembles a comb, and some that are unbranched and coil at the tips. Hülle cells, chlamydosporelike enlargements formed on some of the hyphae surrounding the cleistothecia, may occur in species of *Emericella*.

Developmental studies of cleistothecia of various species have been made by several mycologists (e.g., Benjamin, 1955; Dale, 1903; Emmons, 1935; Kuehn, 1957). When cleistothecia are formed from the vegetative mycelium, differentiated gametangia may be lacking. In these cases, it is not known whether plasmogamy occurred between hyphae. Differentiated gametangia are formed by the majority of the species and may be either clavate or coiled (Fig. 4-16). The antheridium is most often a straight, clublike cell. Usually the archicarp forms a flat circular coil in a single plane or a symmetrical cylindrical spiral around the antheridium, or it may become entwined with the antheridium. Fusion occurs when a pore is formed between the trichogyne and the antheridium but has not been observed in some species that form an antheridium. Ascogenous hyphae arise from the ascogonium and grow in various directions and lengths. Some species form croziers prior to ascus formation, while

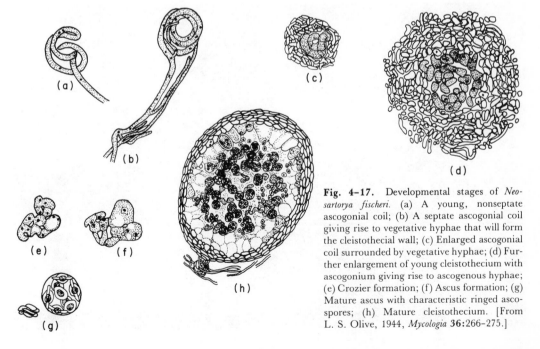

Fig. 4-17. Developmental stages of *Neosartorya fischeri*. (a) A young, nonseptate ascogonial coil; (b) A septate ascogonial coil giving rise to vegetative hyphae that will form the cleistothecial wall; (c) Enlarged ascogonial coil surrounded by vegetative hyphae; (d) Further enlargement of young cleistothecium with ascogonium giving rise to ascogenous hyphae; (e) Crozier formation; (f) Ascus formation; (g) Mature ascus with characteristic ringed ascospores; (h) Mature cleistothecium. [From L. S. Olive, 1944, *Mycologia* **36**:266–275.]

others do not. In those species that do not form croziers, the ascogenous hyphae may become septate and the separate cells enlarge and differentiate into the asci. Sometimes the asci are borne in short chains or are formed singly on short lateral branches. The asci occur randomly within the cleistothecium. Early during the above developmental sequence the gametangia become surrounded by hyphae arising from the basal cells of the archicarp, from the vegetative mycelium, or from both. These sterile hyphae will form the peridium (Fig. 4–17).

Class Pyrenomycetes

The Pyrenomycetes comprise the largest fungal class, more than 6,000 representatives being known. Not all authorities agree on which fungi should be included in this class or on the manner in which they should be divided into orders. Müller and Von Arx's (1973) treatment will be followed here.

In the Pyrenomycetes, the ascocarp is an enclosed structure. Usually the ascocarp is a perithecium having a necklike extension opening by a canal, the ostiole. Periphyses line the ostiole. Some Pyrenomycetes form a cleistothecium without an ostiole or an ascostroma with locules. The asci are typically arranged in a basal layer, the *hymenium,* within the ascocarp (in contrast to the random arrangement in the Plectomycetes). The asci are unitunicate.

The Pyrenomycetes contain the orders Erysiphales, Meliolales, Coronophorales, and Sphaeriales (Müller and Von Arx, 1973).

Fig. 4–18. A powdery mildew fungus, *Erysiphe graminis* on a leaf. The dark cleistothecia are immersed within the light mycelium and conidia. Approximately × 10. [Courtesy Plant Pathology Department, Cornell University.]

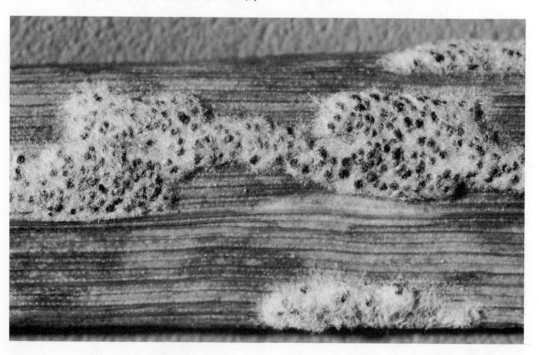

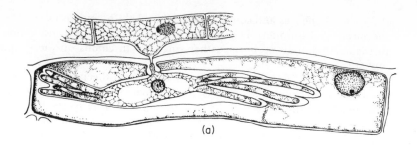

(a)

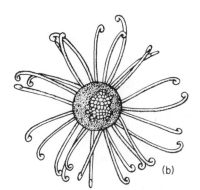

(b)

Fig. 4-19. Structures produced by powdery mildew fungi: (a) A haustorium with fingerlike projections in host cell; (b) A cleistothecium with coiled appendages; (c) and (d) Dichotomously branched appendages; (e) Appendage with a bulbous base. [(a) from G. Smith, 1900, *Botan. Gaz.* **29**:153–184. (b), (c), and (d) from E. S. Salmon, 1900, *Mem. Torrey Botan. Club* **9**:1–292. (e) From D. H. Linder, 1943, *Mycologia* **35**:465–468.]

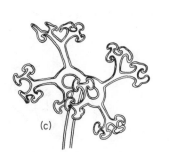

(c)

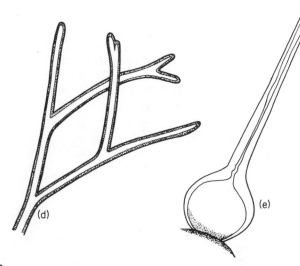

(d)

(e)

ORDER ERYSIPHALES

The Erysiphales contain a single family, the Erysiphaceae. There are at least 90 species in this group, and all of them are obligate parasites, causing the diseases known as the "powdery mildews." The diseased plant has a whitish, powdery appearance owing to the superficial mycelium and conidia growing on its surface. Powdery mildew diseases differ greatly in their damage to the host (Fig. 4-18).

Infection may be initiated by overwintered ascospores in the early spring or by conidia during the late spring and summer. The mycelium is rather delicate, consisting of white or colorless septate hyphae that branch and interweave freely. The mycelium is usually superficial, occurring only on the surface of the leaf, and at intervals it sends a projection, a *haustorium,* into the epidermal cells of the leaf. The haustoria anchor the mycelium to the leaf and absorb food from it. The form of the haustorium varies and may be peglike, globular with a narrow neck connecting it to the hypha, or lobed with many fingerlike projections (Fig. 4-19).

Conidiophores and conidia are produced abundantly on the surface of the leaf in the spring and summer. The conidia are colorless or white and somewhat cylindrical in shape; they may be borne either singly or in chains on the simple, erect conidiophore.

Near the end of the summer, spherical cleistothecia begin to form on the superficial mycelium (Fig. 4–18). The cleistothecia are small and brown or black; they occur superficially on the surface of the leaf and may be easily scraped off. The cleistothecial walls consist of many layers of tightly woven tissue and bear long conspicuous appendages. The appendages may be simple and myceloid, have a bulbous base, or have either coiled or dichotomously branched tips (Fig. 4–19). The appendage form is one of the criteria used in delimiting genera. There may be one or several asci, and if there are several, they are formed in a group at the base of the cleistothecium. Their organization within the cleistothecium differs sharply from the scattered arrangement found in the Eurotiales.

Specific events involved in cleistothecial formation vary from organism to organism, but they comply with the general type for this group. An ovoid or club-shaped ascogonium pairs with a more elongate and slender antheridium [Fig. 4–20(a)]. These organs may be laterally appressed or they may coil together. Meanwhile cells supporting the gametangia give rise to hyphae which will develop into the wall of the cleistothecium. It is uncertain whether or not plasmogamy occurs in all species [Fig. 4–20(b)] but, in any event, the ascogonium divides into three or more cells. The penultimate cell (the second cell from the apical end) contains two nuclei and the remainder of the cells contain a single nucleus [Fig. 4–20(c)]. If the fungus forms a single ascus [Fig. 4–20(e)], the penultimate cell of the ascogonium enlarges and the two nuclei then fuse and undergo three successive divisions, thus converting that cell into an ascus [Fig. 4–20(d)]. If many asci form, the central ascogonial cell

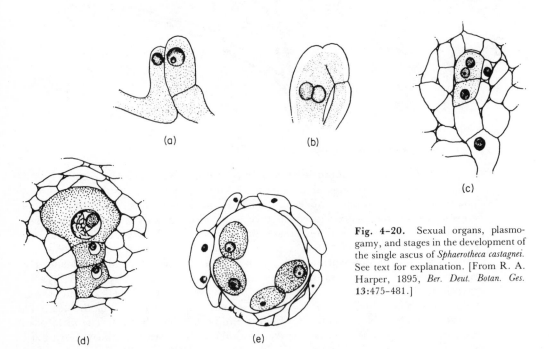

(a)

(b)

(c)

(d)

(e)

Fig. 4–20. Sexual organs, plasmogamy, and stages in the development of the single ascus of *Sphaerotheca castagnei.* See text for explanation. [From R. A. Harper, 1895, *Ber. Deut. Botan. Ges.* **13**:475–481.]

gives rise to ascogenous hyphae that form asci without crozier formation. The asci form an irregular layer across the middle of the perithecium. Owing to the tremendous expansion of the asci, some of the surrounding cells are crushed and form the cavity that will be totally occupied by the asci. The perithecium lacks an ostiole and opens at maturity when the wall splits. The asci are uniformly thin-walled and discharge their ascospores forcibly through a slit that forms in the ascus apex.

ORDER MELIOLALES

The Order Meliolales contains approximately 1,000 known species, called the black mildews because of their dark superficial growth on the surface of their hosts (Doidge, 1942; Hansford, 1961, 1963; Stevens, 1927, 1928). These fungi occur primarily in moist tropical regions but may also occur in the warmer regions of the temperate zone. They are obligate parasites on plants and occur primarily on the leaves but may also occur on twigs and petioles. Although these fungi may be parasitic on commercially important plants, they are not considered to be serious pests. The parasites may cause a reduction in the chlorophyll content of the host cells immediately beneath the colony but rarely kill the cells.

The dark brown to black superficial mycelium is thick-walled, is of uniform diameter, has cells of a uniform length, and may have a characteristic undulation and branching typical for a given species. Specialized branches, the *hyphopodia,* are produced from the hyphae [Fig. 4–21(a)]. Some of the hyphopodia consist of a single flask-shaped cell oriented at right angles to the leaf, standing erect. The function of these flask-shaped hyphopodia is not known. A second type of hyphopodium is that with two cells. The terminal cell is usually swollen and pressed against the surface of the leaf and gives rise to a simple ovate, one-celled haustorium which penetrates an epidermal cell of the host. In the genus *Meliola,* bristlelike setae may project from the mycelium or occur at the base of the ascocarp. The setae are usually straight and unbranched but may have various forms. Usually the colonies formed by the mycelium, hyphopodia, and setae can be traced to growth from one ascospore that germinated on the surface of the leaf. The colonies may have thin, sparse growth spreading over the entire surface of the leaf, or they may be comparatively dense, more or less circular, and sharply delimited. In some species, the hyphal cells may be so short and thickly packed that they form a continuous solid, thalluslike plate. Sometimes the colonies appear to be woolly or velvety, especially if the setae are numerous.

The ascocarps are either true cleistothecia, lacking an ostiole, or are perithecia and have a rudimentary neck with an ostiole. These ascocarps are dark brown to black and are generally globose, but they may be flattened and disklike. The peridium is composed of thick-walled, relatively large, dark cells. In a few genera, the ascocarps may possess appendages of various forms. Paraphyses may be present or absent within the ascocarps. The asci usually contain two or, rarely, four ascospores (more ascospores are initiated but fail to mature) and have evanescent walls that disintegrate at maturity. The dark brown ascospores are usually divided into four or five cells. The ascocarp becomes irregularly broken at the apex at maturity [Fig. 4–21(b)], allowing the ascospores to escape.

Details concerning sexuality and development of the ascocarps have not been extensively studied and are poorly known. From his study of the development of *Meliola circinans,* Graff (1932) concluded that there are many similarities to the development of members of the Erysiphales. However, a conidial stage is lacking, unlike species in the Erysiphales.

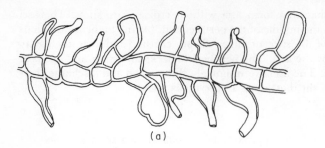

Fig. 4-21. Structures formed by members of the Meliolales. (a) A segment of a hypha produced by *Meliola ptaeroxyli* showing the two types of hyphopodia; (b) A portion of a colony of *Asterina ferruginosa* showing a cleistothecium that has opened in a stellate manner. × 460. [Extracts from Bothalia **4**:193–217 and 273–420, reproduced under Copyright Authority 6684 of 5-10-80 of the Government Printer of the Republic of South Africa.]

(a)

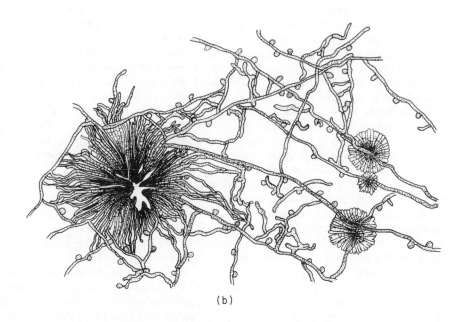

(b)

ORDER CORONOPHORALES

The Order Coronophorales is a comparatively small order containing only about 20 known species, many first described by Fitzpatrick (1923). These fungi occur as saprophytes on wood, forming their ascocarps on the surface of weathered old wood or within the bark (Fig. 4-22). A few species are parasitic on other Pyrenomycetes (Müller and Von Arx, 1973).

The mycelium is typically buried within the substratum although a weblike *subiculum* of superficial hyphae may be visible beneath the ascocarp. The superficial hyphae are dark brown to black and have a characteristic metallic iridescence. The ascocarps are ascostromata containing locules. They are carbonaceous, dark, and surrounded by a distinct peridium. The ascostromata are globose or turbinate in shape and are often solitary and free from each other. In some species, several ascostromata form an aggregate and may be seated upon a common subiculum or a compact stroma. The ascostromata are originally solid structures, and locules are formed by breakdown of some of the internal tissue. Asci then develop and eventually fill the locule. The asci are numerous and are irregularly distributed throughout the cavity.

The asci have thin walls, are unitunicate, and have long, slender stalks. The ascospores may be hyaline or brown and may have one to four cells. The asci are evanescent, disintegrating to release the ascospores into the locule at maturity. The ascostromata are closed throughout most of their development as they lack ostioles. Eventually the ascostromata open when the apices disintegrate. Disintegration is accomplished when a specialized pad of tissue at the apex undergoes gelatinization. At maturity, the ascostromata collapse to become cupulate.

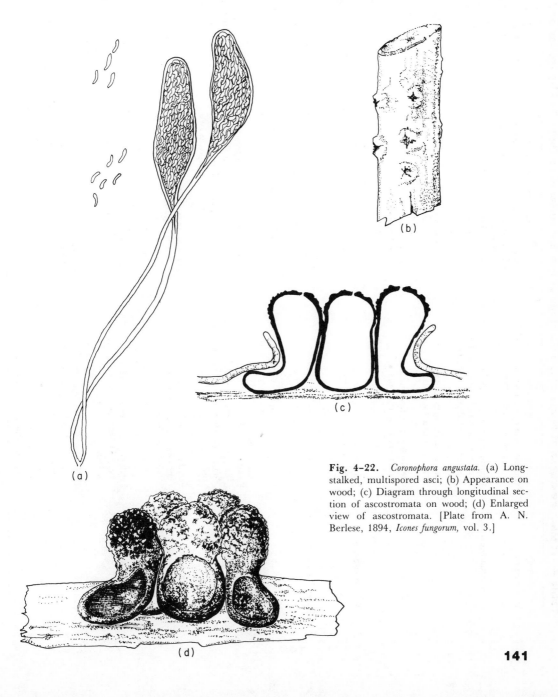

Fig. 4-22. *Coronophora angustata.* (a) Long-stalked, multispored asci; (b) Appearance on wood; (c) Diagram through longitudinal section of ascostromata on wood; (d) Enlarged view of ascostromata. [Plate from A. N. Berlese, 1894, *Icones fungorum,* vol. 3.]

Table 4-1: Main Features of Some of the Families of the Sphaeriales as Delimited by Müller and Von Arx (1973).

Family	Ascocarp	Asci and Ascospores	Representative Genera
Melanosporaceae	Cleistothecia or perithecia, usually not in a stroma	Asci evanescent, freeing ascospores in ascocarp, clavate, obovate, or spherical	Thielavia Chaetomium
Ophiostomataceae	Cleistothecia or perithecia; perithecia may have a long beak and be partially embedded in a stroma	Asci spherical, evanescent, and freeing ascospores in ascocarp	Ceratocystis
Coryneliaceae	Ascostromata with locules; dark, opening usually clefts	Asci evanescent, usually stalked and clavate, sometimes forming only two ascospores	Corynelia
Sordariaceae	Cleistothecia or perithecia; dark, lacking a stroma	Asci persistent, clavate, simple apex; ascospores usually dark with germ pores or slits and often with mucous sheaths or ornamented	Sordaria Neurospora Gelasinospora
Polystigmataceae	Perithecia immersed in host tissue or stroma; mostly parasitic on higher plants	Asci persistent with simple apex; ascospores lacking germ pores or slits, often forming appressoria upon germination	Glomerella Phyllachora Polystigma
Hypocreaceae	Perithecia often seated upon a stroma; brightly colored and fleshy	Asci persistent, simple apex, clavate; ascospores lacking germ pores or slits	Nectria Gibberella Hypomyces Pseudomeliola
Sphaeriaceae	Perithecia often upon or within a stroma; brown or black and carbonaceous	Asci persistent, simple apex, clavate; ascospores lacking germ pores or slits	
Diaporthaceae	Perithecia embedded in host tissue or stroma, sometimes covered by a shieldlike clypeus; stroma light or dark	Asci sometimes loosening from the ascogenous hyphae and forming a slimy mass; apex of ascus with an apical ring usually visible as two refractive bodies	Gnomonia Endothia Diaporthe
Clavicipitaceae	Perithecia usually immersed in a stroma, stroma bright or dark, fleshy or horny, in various forms including crustlike, cushion-shaped, or stalked	Elongate persistent asci, usually with a slightly thickened apical cap perforated by a narrow canal, often plugged with a chitinoid body; ascospores needlelike, multicellular	Epichloë Claviceps Balansia Cordyceps
Diatrypaceae	Perithecia usually immersed in a large stroma, sometimes solitary or arranged in groups	Asci persistent, clavate with long stalks; thickened apex with simple amyloid ring; ascospores one-celled and allantoid	Diatrype Eutypa
Xylariaceae	Cleistothecia or perithecia; free or immersed in host tissues or stroma; stroma dark throughout or with a dark crust, carbonaceous; stroma in various forms such as crustlike, cushion-shaped, or upright	Asci either persistent or evanescent, large with amyloid ring or system of rings, often with chitinoid pulvillus stained by cotton blue; ascospores one-celled, dark, usually with germ slit	Xylaria Hypoxylon Daldinia

ORDER SPHAERIALES

The Sphaeriales constitutes the largest group of the Pyrenomycetes, as there are at least 5,000 known members of this order. The fungi are extremely diverse morphologically. Because of the large numbers and morphological diversity, the classification of this group has undergone many changes and usually the fungi have been divided into at least four orders. However, there has been little agreement on the disposition of many borderline fungi into one order or another. Luttrell (1951) traces many of the changes in the classification of some of the fungi. Müller and Von Arx (1973) maintain that there are numerous intergrading fungi that span the more traditional orders, and also that sometimes fungi belonging to the same genus differ so much from each other that they would have to be classified in more than one of the traditional orders if certain fundamental characters such as ascus and ascospore morphology are considered. Therefore Müller and Von Arx maintain that these fungi cannot be divided into discrete orders and instead place them into a single large order. We shall follow this treatment here. See Table 4-1 for a synopsis of some of the families.

Many of the Sphaeriales are saprophytes and occur on a wide variety of substrata such as soil, dung, wood, and decaying leaves or petioles. Some occur in unusual environments, for example, on wood submerged in marine water. Many of the Sphaeriales are parasites on a wide range of organisms, including the marine red algae, lichens, other fungi, insects, and higher plants. Many of the most devastating plant diseases are caused by members of this order. *Ceratocystis ulmi* causes Dutch elm disease, which has resulted in the loss of large numbers of ornamental elm trees in the United States. Similarly, *Endothia parasitica* is responsible for the chestnut blight that almost eliminated chestnut trees in the United States. Some species of *Ceratocystis* cause a blue stain of the wood in which they grow, greatly decreasing its commercial value. Some members of the Sphaeriales are symbionts with algae, forming lichens (Chapter 13). Scientists use species of *Neurospora* as a research tool to investigate genetic problems.

The Perithecium. Ascocarps may be ascostromata with locules, cleistothecia, or perithecia. Only a few members, *Corynelia* and its allied genera (in the Coryneliaceae), form ascostromata with locules. The vast majority of the members form a true perithecium, the typical ascocarp for this group (Fig. 4-23). Other than lacking an opening, the cleistothecial members are similar to many perithecial members, sharing such important characters as ascus and ascospore morphology or developmental features (Malloch and Cain, 1973; Von Arx, 1975). Experimental evidence confirms that the cleistothecial forms are closely allied to the perithecial forms in this order. For example, *Chaetomium deceptivum* normally forms perithecia, but if cultured on certain media, will form cleistothecia that closely resemble those formed by the cleistothecial genus *Thielavia* (Malloch and Benny, 1973). A strain of *Gelasinospora calospora* underwent spontaneous mutation and subsequently formed cleistothecia rather than perithecia in culture. If the cleistothecial strain was mated with a normal perithecial strain, the ascospores segregated in approximately a 1:1 ratio in their ability to form cleistothecia or perithecia (Maniotis, 1965).

The perithecium is usually flask-shaped and has a neck surmounted by a pore. Periphyses line the ostiole within the neck. The perithecial neck varies greatly in its length and may range from extremely short, barely protruded forms to those that are so long that they resemble a bristle. In some species, the neck may be lacking and the pore seated directly within the main part of the perithecial wall. Although most perithecia are flask-shaped, the perithecium may be somewhat globular. The wall of

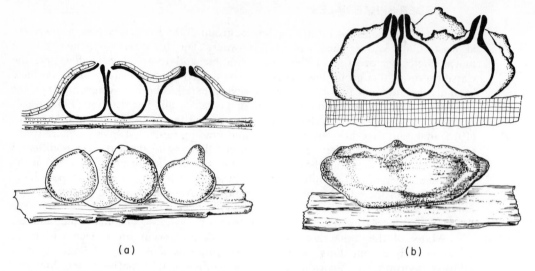

(a)   (b)

Fig. 4-23. Perithecia. (a) Free perithecia without a stroma. A longitudinal section is shown above and a habit sketch showing the entire perithecium is below; (b) Similar diagrams of perithecia embedded in a stroma. Note the prominent walls separating the perithecia from the stroma; (c) A longitudinal section through perithecia of *Sordaria.* [(a) and (b) from A. N. Berlese, 1900, *Icones fungorum* vol. 3; (c) Courtesy Carolina Biological Supply Company.]

(c)

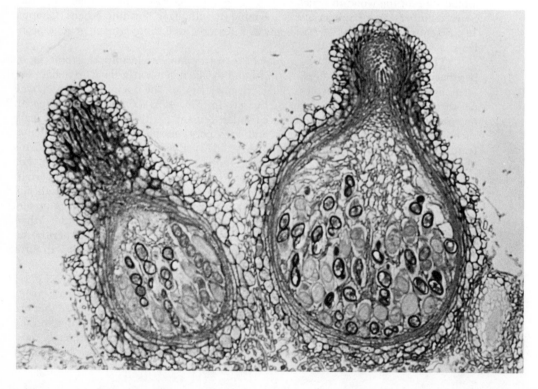

the perithecium may be either brightly colored or dark and may be somewhat fleshy, membranous, or carbonaceous in texture. Usually, bright colors are associated with a fleshy texture while the darker colors are associated with the drier textures, but this correlation is not absolute. Perithecia may occur free on the surface of the substratum or alternatively may be seated upon a pad of stroma, immersed within the host tissue, or partially or completely embedded within a stroma consisting of fungal tissue. If the perithecia are embedded within a stroma, the stroma may be brightly colored (a shade of yellow, pink, orange, red, blue, or violet) or subdued (usually dark brown or black). Although there are some exceptions, the brightly colored stromata are fleshy in consistency while the dark ones are dry and carbonaceous. There is considerable variation in the manner in which the perithecia are arranged in the stroma (Fig. 4–24).

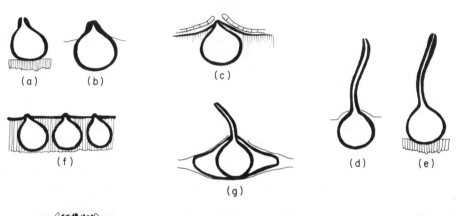

Fig. 4–24. Longitudinal sections of perithecia (heavy outline) and stromata of the Sphaeriales. (a) Solitary, free on surface of substratum; (b) Similar, but immersed in substratum; (c) Immersed in host with a shield-like layer of stromal tissue; (d) Long-necked perithecium immersed in substratum; (e) Similar, but on surface of substratum; (f) Perithecia immersed in a single layer of a crustlike stroma; (g) Perithecium in a leaf, surrounded by a circular stroma; (h) Perithecia grouped between a thin layer of stroma tissue and the substratum; (k–m) Stromata assuming various shapes. All with perithecia embedded in a single layer except in (j). [From A. N. Berlese, 1894, *Icones fungorum,* vol. 1.]

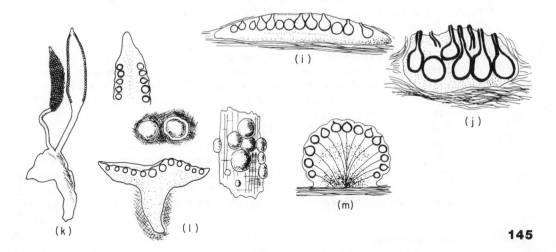

Within the perithecium, hymenial paraphyses or pseudoparaphyses may be present in addition to the asci. Some species lack both paraphyses or pseudoparaphyses, some have only paraphyses, and others have only pseudoparaphyses.

The Asci. The asci, in combination with the pseudoparaphyses or paraphyses, more or less fill the cavity that occurs within the perithecial wall, and these elements within the cavity are collectively termed the *centrum*. The asci are variable in shape and may be spherical but are usually clavate or cylindrical and are almost always borne within a distinct hymenium. In some members (e.g., *Ceratocystis ulmi* and *Chaetomium*) the asci will deliquesce and the ascospores will ooze out of the perithecium in a slime. These deliquescent asci have simple walls that are uniformly thin and lack a distinct pore. The remainder of the fungi in the Sphaeriales have persistent asci that remain attached to the perithecium and forcibly discharge their ascospores through the ostiole (Chapter 10). There is considerable morphological variation of the ascus apex among those species with persistent asci (Fig. 4–25). For example, the apical pore may be simple and lie within a thickened apical region of the ascus wall. Alternatively, the apical pore may be surrounded by distinct ringlike structures that lie within the ascus wall and are refractive or become blue when stained with iodine; the apical pore may contain a plug that blues with iodine; or the apical pore may be recessed within a funnel-shaped depression formed by the ascus apex. There are additional variations that are visible with the light microscope. The ascus apices of several species have been studied with the electron microscope (Griffiths, 1973; Reeves, 1971; Stiers, 1977). Basically, the apices consist of an electron-transparent ground material within which electron-dense granules are arranged as a ring or plate; the granules occur in combination with longitudinally oriented fibrils in similar configurations. The pore is delimited by these rings of somewhat different materials. The bottom edge of the apex, bounded by the inner wall layer of the ascus, projects downward as a shoulderlike ring into the cytoplasm of the ascus (Fig. 4–26). Electron-dense material or fibrils may be in this ring. According to Stiers (1977), some of the variation seen with the light microscope might be attributed to the degree to which the shoulderlike ring extends downward into the ascus lumen, its width, and the extent to which it is flared. It is generally difficult to precisely correlate the image of an apex seen with the light microscope with that seen with the electron microscope. However, differences among the ascus apices are considered to be extremely important in establishing relationships within the group (Müller and Von Arx, 1973). Usually eight ascospores are formed within each ascus. Ascospore morphology is also varied and considered to be important taxonomically. Germ pores or slits may be present or absent, the ascospores may be hyaline or colored and uni- to multicellular, and they may vary in shape from spherical to elongated, needlelike forms. Ornate appendages occur on the ascospores of a few species.

Ascocarp Development. Numerous studies of sexuality and perithecium development in members of the Sphaeriales have been made. Some of the more important studies include those by Ellis, 1960; Goos, 1959; Hanlin, 1961, 1963, 1964; Huang, 1976; Mai, 1977; Nelson and Backus, 1968; Rosinski, 1961; Samuels, 1973; Whiteside, 1961, 1962; Uecker, 1976.

Perithecium formation is often initiated by formation of an archicarp (Figs. 4–27 and 4–28). Typically the archicarp is coiled and may be either unicellular or multicellular. There are four patterns into which sexuality may fall. First, in many of the species investigated, no evidence exists for formation of male gametes or gametangia, plasmogamy is lacking, and further development of the ascogenous system is apogamous. This type of development is extremely widespread and has been

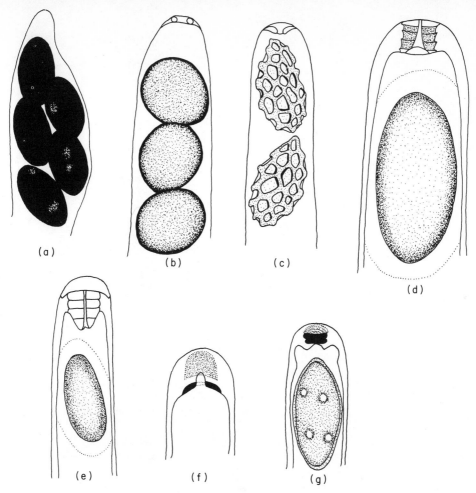

Fig. 4-25. Variation in ascus apices among members of the Sphaeriales. (a) Simple apex; (b) Slightly thickened apex with a single ring; (c) Funnellike indentation, surrounded by a thickened ring; (d) Thickened apex with three rings in a cylinder, becoming blue in iodine; (e) Cylindrical apex with four rings, some of which project downward into the ascus. The rings stain very faintly with iodine; (f) Apex with pulvillus (stippled) and a ring that blues in iodine; (g) Apex with pulvillus and two rings, blueing in iodine; [(a) from R. F. Cain, 1962, *Can. J. Botany* **40**:447–490; (b) from R. F. Cain, 1957, *Can. J. Botany* **35**:255–268; (c) from R. F. Cain, 1950, *Can. J. Res.* **C 28**:566–576; (d) adapted from J. C. Krug and R. F. Cain, 1974, *Can. J. Botany* **52**:589–605; (e) from J. C. Krug and R. F. Cain, 1974, *Can. J. Botany* **52**:809–843; (f) and (g) from R. A. Shoemaker et al., 1966, *Can. J. Botany* **44**:247–254.]

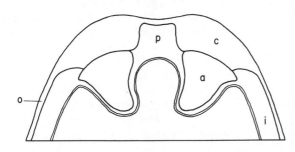

Fig. 4-26. Diagrammatic longitudinal section of ascus apex of *Hypoxylon fragiforme,* prepared from an electronmicrograph. *a* = annulus of electron-dense material, *c* = annular region of slightly electron-dense wall material, *i* = internal wall layer, *o* = external wall layer, *p* = electron-transparent apical region of ascus wall. [From G. N. Greenhalgh and L. V. Evans, 1967, *Trans. Brit. Mycol. Soc.* **50**:183–188.]

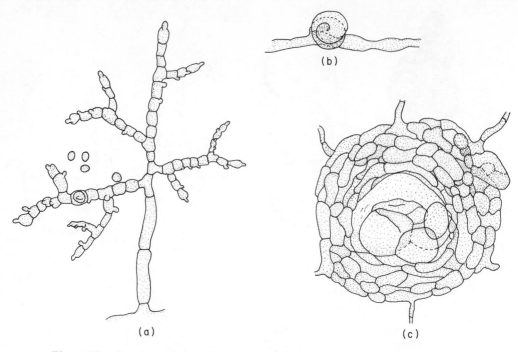

Fig. 4–27. Sexual structures of *Gelasinospora adjuncta*. All × 900. This fungus is heterothallic and requires two genetically compatible cultures to produce a fertile cross. (a) Branched spermatiophore with short cells and small projections from which the spermatia are budded; (b) Coiled archicarps; (c) Unfertilized archicarp within a young perithecium. This will not develop further without fertilization. Plasmogamy occurs by transfer of spermatia to archicarps of a second, genetically compatible strain. Spermatia are sometimes absent, and hyphal fusion between compatible cultures will take place and produce fertile perithecia. [From R. F. Cain, 1950, *Can. J. Res.* **C 28:**566–576.]

observed in representatives of several families (e.g., it has been described in species of *Chaetomium* by Whiteside, 1961, 1962). Second, many fungi produce *spermatia,* discrete sporelike cells that function as male gametes (Fig. 4–27). The spermatia, like spores, are formed on specialized hyphae, can be disseminated from the thallus producing them to other thalli, and may be responsible for cross or outbreeding. After dissemination, a spermatium fuses with a trichogyne of the receptive thallus, and its nucleus travels to the ascogonium. Spermatization occurs in several fungi, including species of *Neurospora* and *Gelasinospora* (Fig. 4–27) (Nelson and Backus, 1968; Dodge, 1932). Third, an antheridium may be formed that fuses with the trichogyne. This apparently occurs infrequently but has been reported in *Gelasinospora calospora* (Ellis, 1960). Fourth, gametangia have not been observed in some species, and in these plasmogamy presumably occurs by hyphal fusion. Sometimes hyphal fusion may occur in fungi, forming an archicarp that normally participates in plasmogamy (e.g., *Gelasinospora* and *Glomerella*) (Cain, 1950 and McGahen and Wheeler, 1951, respectively).

After the archicarps complete their maturation and possibly plasmogamy, they produce ascogenous hyphae. In most species, the ascogenous hyphae form typical croziers from which asci develop (croziers are absent in a few of the fungi, including some species of *Chaetomium*) (Whiteside, 1961). The asci are usually borne in a hymenium near the base of the perithecial cavity (Fig. 4–28).

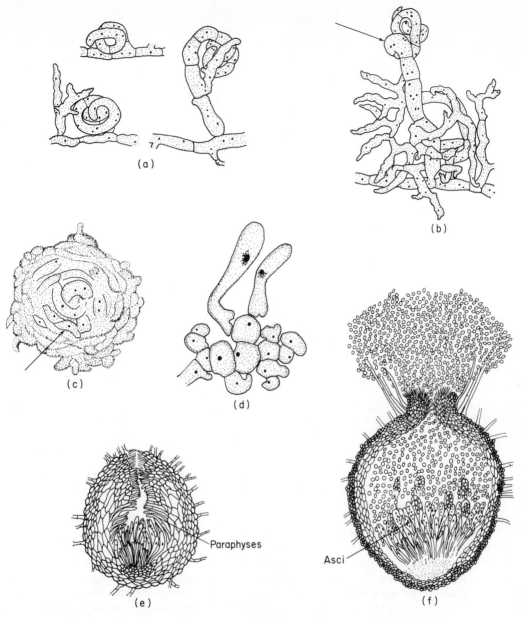

Fig. 4-28. Developmental stages of *Chaetomium globosum*. The ascogonium undergoes parthenogenetic development, and the centrum has the *Xylaria*-type of development. (a) Various configurations of the coiled archicarp; (b) Archicarp (arrow) is becoming surrounded by sterile hyphae arising from its supporting hypha; (c) Archicarp (arrow) is surrounded by sterile tissue originating from the hyphae in (b), and which will differentiate into the perithecial wall and centrum; (d) Ascogenous hyphae showing the absence of croziers and two young asci; (e) Section through a perithecium at the time that the cavity and asci are beginning to form; (f) Mature perithecium extruding ascospores released by deliquescence of the asci. [(a-d) × 1,000; (e-f) × 170]. [From W. C. Whiteside, 1961, *Mycologia* **53**:512-523.]

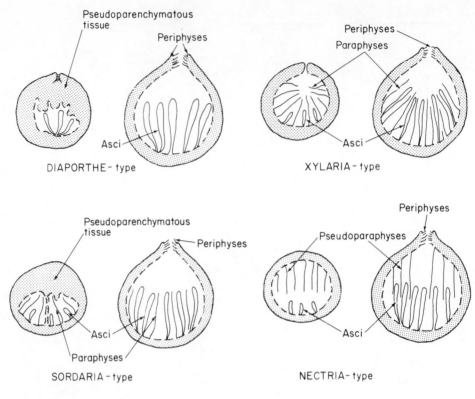

Fig. 4-29. Centrum developmental types in the Sphaeriales. An early stage with an enclosed perithecium and young asci is on the left, while a more mature, ostiolate perithecium is on the right.

The sterile hyphae that will comprise the perithecial wall usually start to appear after the archicarp has been differentiated. If the perithecium will be embedded within a stroma, the archicarp typically appears within the stroma and sterile hyphae are then formed from the basal cells of the archicarp. These sterile hyphae will develop into a discrete perithecial wall. At maturity, such a perithecial wall originating from the archicarp or from a sexual stimulus is distinct from the tissue of the stroma, which originated by modification of vegetative hyphae. Whether the perithecium is borne free or within a stroma, the hyphae initiating wall formation typically enclose the gametangia and continue to enlarge and differentiate into a distinct wall (often finally consisting of pseudoparenchymatous tissue). Meanwhile the centrum differentiates within the expanding perithecial wall. Centrum development is a critical part of development, as it may proceed in different ways. Closely related species share a similar type of development, and therefore the mode of centrum development is important in establishing relationships among these fungi. Luttrell (1951) described developmental patterns that occur among this group, including the *Diaporthe, Xylaria,* and *Nectria* types examined below (Fig. 4-29). The *Sordaria* type was added as result of subsequent research (Huang, 1976). Each type is named for an important genus in which it occurs but actually occurs in several genera. The types are as described below.

1. *Diaporthe*-type centrum: *Diaporthe* and its allies lack both paraphyses and pseudoparaphyses. The centrum is at first filled with pseudoparenchyma. Formation of the perithecial cavity is accomplished by expansion and disintegration of the pseudoparenchymatous cells. While the pseudoparenchyma is disintegrating, the growing asci push into it as they elongate. Eventually the asci form a layer lining the base of the perithecial cavity. Of the representative fungi that we consider on pages 152–154 this type of development also occurs in *Gnomonia* and *Hypomyces*. (Luttrell, 1951).

2. *Xylaria*-type centrum: These fungi form paraphyses but not pseudoparaphyses. After the perithecial wall is initiated and encloses the gametangia, paraphyses begin to grow from the inner surface of the wall at both the base and sides. The paraphyses grow upward and inward, exerting pressure by pressing against those growing in the opposite direction, and finally creating a central cavity. The developing asci push their way up through the paraphyses. In some species, the paraphyses are evanescent and will later disappear. Of those fungi discussed on pages 152–154, *Xylaria, Claviceps* (Luttrell, 1951), *Neurospora* (Nelson and Backus, 1968), and *Chaetomium* (Whiteside, 1961) have this type of development.

3. *Sordaria*-type centrum: These, like the *Xylaria* type, have paraphyses and lack pseudoparaphyses. However, development proceeds more or less as in the *Diaporthe* type, as the centrum is filled with pseudoparenchyma. The paraphyses differentiate from some of the pseudoparenchymatous cells in the centrum (Huang, 1976). This type also occurs in *Gelasinospora* (Mai, 1977).

Fig. 4–30. Perithecia of *Hypomyces trichothecoides* having the *Nectria*-type of centrum development. Left: A young stage showing the downward growth of a layer of pseudoparaphyses (arrow), × 480; Right: A mature perithecium showing asci with the ascospores (dark) interspersed among the pseudoparaphyses. Note that the ostiole has elongated also. × 410. [Courtesy R. Hanlin.]

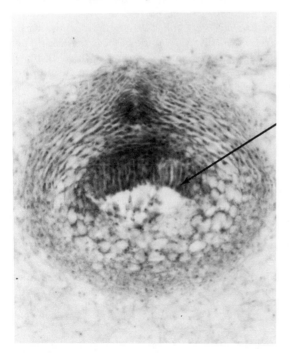 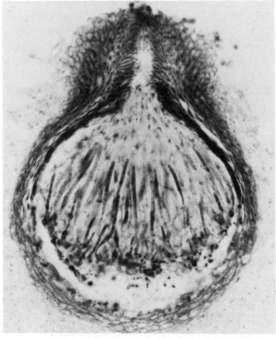

4. *Nectria*-type centrum: Species of *Nectria* and other fungi having this type centrum do not form basal paraphyses but rather have pseudoparaphyses. After the perithecial wall is initiated, there is a downward growth of pseudoparaphyses from a specialized meristematic area lying in the inner wall near the apex. The perithecial cavity is created from a combination of the pressures exerted by growing pseudoparaphyses and by further growth of the perithecial wall. The pseudoparaphyses remain free at their lower ends (Fig. 4–30). The asci form in a concave layer at the base of the cavity, and grow upward among the pseudoparaphyses. This centrum type occurs in *Hypomyces* (Fig. 4–30), discussed further on page 152 (Hanlin, 1964; Samuels, 1973) as well as many other allies of *Nectria* (Rogerson, 1970).

Anamorphs. Conidial states are known for many of the Sphaeriales. Most often, the conidial states belong to the Form-Order Moniliales, but sometimes they belong to the Form-Orders Melanconiales or Sphaeropsidales. Usually the conidia have the blastic type of development. Chlamydospores may be formed by some species. Further clarification of these form-orders and the blastic type of development may be found in Chapter 6.

Representative Members. We shall examine some members of the Sphaeriales so that some of the variation within this group may be explored. Many members of the Sphaeriales have dark perithecial walls and possibly dark stromata. One of the simplest members is *Thielavia,* which forms small isolated cleistothecia on the surface of plant roots on which it lives parasitically. *Chaetomium* is similar to *Thielavia,* but differs in having an almost globose perithecium opening by an ostiole and in bearing extremely long olive brown to black bristles on its outer surface. Perithecia of *Chaetomium* may be found on decaying plant material or dung in nature. Another common dung inhabitant is *Sordaria,* which forms black perithecia with an elongated neck so that the overall form is pear-shaped (Fig. 4–23). The isolated perithecia may be found partially immersed in dung in nature. *Neurospora* forms perithecia much like those found in *Sordaria* when cultured, but in nature the conidial state may be found on plant material that has been charred by fire.

Unlike the above representatives that lack a stroma, two common genera with a stroma are *Hypoxylon* and *Xylaria.* In *Hypoxylon* (Fig. 4–31), the globose to flask-shaped perithecia are embedded within a common carbonaceous stroma, which is black, brown, or reddish in color and may vary in shape from a hemispherical cushion to a thin spreading crust. The stromata are found on dead wood or bark. Species of *Xylaria* form an erect, stalked, club-shaped or cylindrical stroma. The stromatal tissue is white at first but as it matures, the surface becomes crustlike and black. The perithecia are embedded in a single layer beneath the surface, and their ostioles protrude through the black surface crust. One common species, *X. polymorpha,* is known as ''dead man's fingers'' because the cluster of stromata arises from the base of tree stumps and has an eerie resemblance to fingers.

Among the fleshy members of the Sphaeriales, we find some species of *Balansia* that form black or yellow stromata (Fig. 4–32). Species of *Balansia* occur as parasites on grasses or sedges. The remaining representatives with fleshy walls or stromata that we will discuss are brightly colored. Members of *Hypomyces* often parasitize other fungi, occurring on the basidiocarps of bracket fungi or mushrooms (Chapter 12). The yellow to red perithecia of *Hypomyces* may occur, densely crowded but separate, in a common web on the surface of the substratum (decaying basidiocarps) or the ground where infected basidiocarps have disintegrated, or they may be embedded within the host tissues of living basidiocarps. Some species of *Nectria* may also form scattered

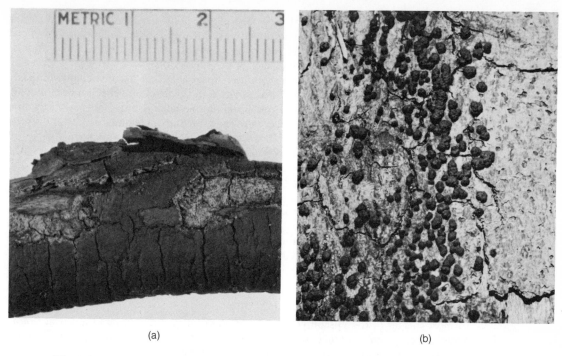

(a)   (b)

Fig. 4–31. Two species of *Hypoxylon* (Sphaeriales). The perithecia are embedded within the hard, black stromata: (a) Approximately × 2; (b) Approximately natural size. [(a) Courtesy Plant Pathology Department, Cornell University; (b) from E. J. Moore and V. N. Rockcastle, 1963, *Fungi,* Cornell Science Leaflet.]

Fig. 4–32. Fleshy stalked stromata of *Balansia claviceps.* × 9. [Courtesy U.S. Department of Agriculture.]

perithecia on the wood or bark of their host plants, but many form yellow, coral, or red stroma that are cushion-shaped and break through the bark or other superficial tissues of the host. The perithecia occur in clusters upon these stromata and may be partially immersed in the stromal tissue. Some species of *Nectria* are weakly parasitic and cause dieback disease. *Endothia parasitica* (causal agent of chestnut blight) similarly has perithecia embedded within a reddish stroma in the form of a cushion or elongated crust erumpent through cracks of the bark of the host plant. Both *Cordyceps* and *Claviceps* produce comparatively large stromata having an elongate, upright sterile stalk surmounted by a larger head in which the perithecia are immersed. Species of *Cordyceps* may be found as parasites on subterranean fungi or on insects (Chapter 12). The stromata of *Cordyceps militaris* are up to 5 centimeters in height, are orange-red to red, and have a club-shaped head merging with the stem. The stromata of *Claviceps purpurea* are somewhat smaller, are pale purple, and have a globose head about 2 millimeters in diameter. *C. purpurea* causes a disease of rye plants in which the developing rye grain is converted into a hardened black body (the *ergot*) that is actually a sclerotium and in turn produces the stromata after overwintering. Consumption of the ergots may result in poisoning (Chapter 14).

Class Discomycetes

In most species of the Discomycetes, the hymenium partially covers the surface of an open ascocarp, the apothecium. This group is commonly called the "cup fungi" because many apothecia are cup-shaped with the hymenium lining the inner surface of the cup. Many apothecia are not cupulate but may be flat, convex, or stalked, with an

Fig. 4–33. Longitudinal section through an apothecium showing the anatomical zones. [Adapted from K. S. Thind and K. S. Waraitch. 1971. *Res. Bull. Panjab Univ., (N.S.)* **22**:109–123.]

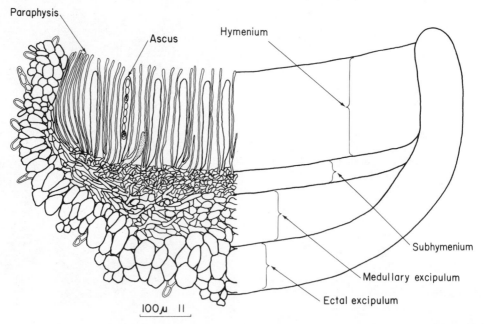

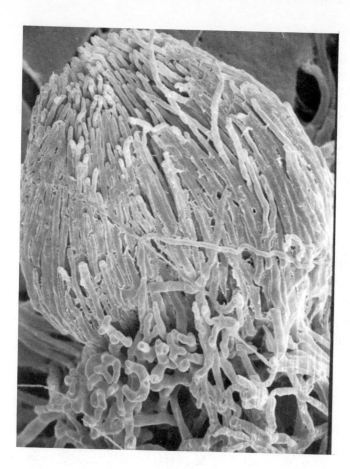

Fig. 4-34. A scanning electron-micrograph of an apothecium of the Discomycete *Pyronema domesticum.* This apothecium is pulvinate in form, and the visible surface tissue is the ectal excipulum. The tips of the paraphyses are visible at the top. × 560. [Courtesy K. L. O'Donnell.]

enlarged, club- or saddle-shaped fertile apex partially covered by a hymenium. Apothecia occur in other forms but typically share the common characteristic of having a hymenium lining a surface that is exposed at maturity. Apothecia range in color and include shades of white, buff, pink, orange, red, green, brown, and black. Characteristically they are fleshy and moist, but some form a hardened stroma resembling that of some of the Pyrenomycetes. Their size varies considerably, and they may be barely visible to the naked eye or up to several centimeters in diameter.

Apothecia are divided into distinct anatomical regions that are readily visible in longitudinal sections. The regions often differ in the structure of the tissue (Fig. 4-33). The hymenium bears a palisade layer of asci and paraphyses and is most often on the upper surface of the apothecium. Beneath the hymenium is the *subhymenium,* a relatively thin zone of tissue giving rise to the asci and paraphyses. The remaining tissues in an apothecium comprise the excipulum. The excipulum may be divided into a more or less distinct *ectal excipulum,* the outermost sterile layer of the apothecium, and the *medullary excipulum,* the central enclosed mass of sterile tissue (Fig. 4-34). In some fungi, parts of the excipulum may appear jellylike owing to the secretion of mucilage, which then accumulates among the hyphae and absorbs water [Fig. 1-1(e)].

THE ASCUS

Asci in species of the Discomycetes are usually cylindrical or clavate but occasionally may be spherical. Ultrastructural studies show that the ascus wall is laminated, usually having two or three lamina. Unlike the bitunicate ascus, which has two distinct walls, these lamina comprise only a single wall and remain permanently attached to each other. When mounted in a reagent containing iodine, the asci of some species will turn entirely blue or turn blue at the apex only. This blueing reaction is apparently due to the reaction of a minute covering of mucilaginous material (Samuelson, 1978). One of the most fundamental differences among asci is whether they are *operculate* or *inoperculate* (Fig. 4-35).

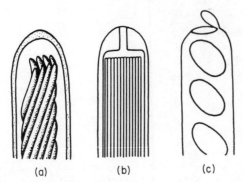

(a)   (b)   (c)

Fig. 4-35. Types of ascus openings: (a) No regular means of opening; (b) Pore in tip of the ascus. Both (a) and (b) are inoperculate; (c) An operculate ascus which opens by throwing back a lidlike operculum.

The operculate asci open by throwing back a lidlike *operculum*, which may remain attached on one side, or ejecting a lidlike operculum from the ascus (Fig. 4-36). Usually the operculum is centered at the apex of the ascus, but in some species, it is obliquely located on one side of the apex. The ascus wall consists of at least two layers, a thin outer layer and a thicker, less rigid inner layer (Fig. 4-37). The wall near the apex of the ascus may be thinner (Samuelson, 1978) or thicker than the lateral walls (Van Brummelen, 1978). Parts of the wall near the apex or at the apex may become strengthened and rigid, possibly by undergoing localized thickening or the addition of wall layers. Other parts of the adjacent wall may become weakened, perhaps by a ringlike indentation of the wall form. In the mature ascus, a swollen or indented ring or rings and possibly conspicuously thickened apices may be present. The weakened area is the dehiscent zone, which breaks open by fracturing, undergoing wall disintegration, or gelatinizing (Van Brummelen, 1978). Frequently the dehiscent zone will show different staining reactions than the remainder of the apex (Samuelson, 1975, 1978). The operculum is that part of the apex enclosed within the dehiscent zone. It may be thrown off entirely or may remain attached to the ascus by one or more wall layers. Van Brummelen (1978) distinguishes several types of operculate asci; some are illustrated in Fig. 4-37.

The inoperculate ascus may have no definite means of opening and may rely upon animals to break it and allow the spores to escape or may open by a slit or pore at maturity. The apices of the inoperculate asci may be uniformly thin or have various degrees of thickening. A distinct plug may be present in the thickened apex. When mounted in a reagent containing iodine, there may be a blueing of the entire ascus apex, of a ring around the plug, or of the plug itself. An ultrastructural study was made of the ascus of *Ciboria acerina*, which forms a cylindrical pore with a plug in an

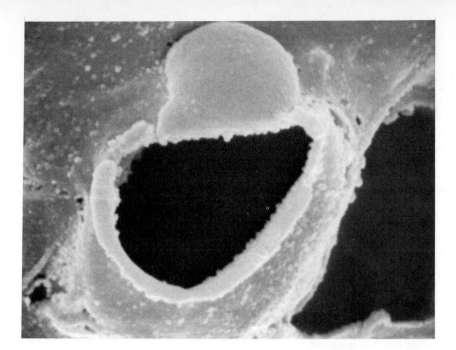

Fig. 4-36. A scanning electronmicrograph of the apex of an ascus showing the operculum, which has been thrown aside from the opening. × 11,250. [Courtesy K. L. O'Donnell.]

Fig. 4-37. Diagrams of electronmicrographs of operculate asci, both before and after opening; (a) *Pyronema omphalodes:* the operculum is not sharply delimited, and there is no indentation or ring. Irregular tearing of the operculum is due to the absence of strengthening: (b) *Thecotheus* sp. illustrates the most usual type of ascus apex. The large operculum is sharply delimited both by strengthening and by an indented ring. Right: detail of wall; Center: the entire ascus tip when closed; Bottom: the opened ascus; (c) *Sarcoscypha coccinea* has an extremely thick operculum that is sharply delimited, especially in the inner wall. The inner layer is greatly enlarged, forming an opercular plug, which will be ejected. Abbreviations: *C* = cleavage of ascus wall; *F* = line of fracturing; *I* = indentation of ascus wall; *o* = operculum; *OP* = opercular plug; *SL* = sublayering of ascus wall. [Adapted from J. Van Brummelen, 1978, *Persoonia* **10**:113–128.]

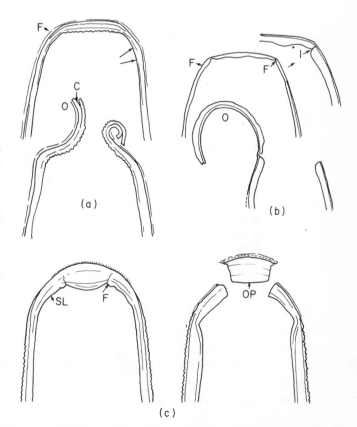

apical thickening (Corlett and Elliott, 1974). In this fungus, the outer ascus wall appears to curve inward to line the cylindrical pore. The pore is filled with a granular, amorphous material separated from the cytoplasm of the ascus by the inner wall layer but is not bounded on the outer surface. A similar amorphous material forms the thickened apex by lying between the two wall layers of the apex. The ascospores exit through the pore, and after spore discharge, the pore is empty and the outer wall that lined the cylinder is everted and pressed against the top of the ascus (Fig. 4–38). Asci of other inoperculate Discomycetes were studied by Bellemere (1977) and Schoknecht (1975) and the results are generally similar to those reported for Pyrenomycetes (page 146) and above.

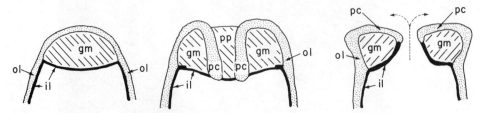

Fig. 4-38. Development of the ascus apex of *Ciboria acerina*. (a) An early stage in which the ground material (*gm*) has accumulated at the ascus apex, pushing the inner layer (*il*) downward; (b) The outer wall of the ascus has invaginated to form a pore cylinder (*pc*), which is filled with the granular material (*pp*); (c) A mature ascus in which the granular material has lysed and the pore cylinder has everted. [Figures from M. Corlett and M. E. Elliott, 1974, *Can. J. Botany* **52**:1459–1463.]

ASCOCARP DEVELOPMENT

An archicarp and fertilizing male elements have not been observed in the majority of the Discomycete species and are thought to be absent. Some of the Discomycetes having an archicarp develop apogamously. If sexual organs are present, however, they are dissimilar, making sexual reproduction heterogamous. The male element may be either a spermatium or antheridium. *Pyronema domesticum,* used to illustrate sexual reproduction in the ascomycetes (Figs. 4–3, 4–4), is a member of the Pezizales, a Discomycete order.

Once plasmogamy has been accomplished, if it occurs, the ascogenous hyphae develop. The sterile hyphae develop simultaneously, and are intertwined with ascogenous hyphae. The tissues formed by the sterile hyphae and ascogenous hyphae in combination comprise most, if not all, of the medullary excipulum and the subhymenium. The paraphyses are formed by the sterile endings of some of these hyphae. Usually the paraphyses appear before the asci, and the croziers lie among the bases of the paraphyses. The developing asci push their way up through the paraphyses to form the hymenium. There is some variation among members of the Discomycetes in regard to the exposure of the hymenium during development. Van Brummelen (1967) distinguishes two basic types of apothecia on this basis. These types are the *cleistohymenial* and *gymnohymenial*. In the cleistohymenial type of development, the hymenium is enclosed within the excipulum of the developing ascocarp. This enclosure may be permanent in some species that form cleistothecia, but in most the hymenium will be exposed when the developing asci and paraphyses rupture the excipulum. The hymenium may become exposed after the paraphyses are formed but

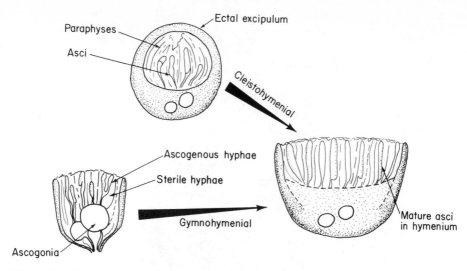

Fig. 4-39. The two major developmental types in the Discomycetes.

before croziers develop or at any point up to that at which mature ascospores are present. In contrast to the cleistohymenial development, the hymenium is exposed throughout its development in the gymnohymenial type (Fig. 4-39).

VEGETATIVE STRUCTURES

In addition to forming ascocarps, some Discomycete species produce vegetative structures. These may be *sclerotia* or stromata. Both a sclerotium and stroma have a dark outer rind surrounding a pale inner medulla and may include host tissue in addition to fungus tissue. A sclerotium is a hardened, resistant structure capable of overwintering or surviving unfavorable conditions. The sclerotium is determinant in size and is usually rounded in shape (Fig. 4-40). Upon germination, the sclerotium may give rise to vegetative hyphae or to an apothecium. Like the sclerotium, the Discomycete stroma is a resistant structure of approximately the same internal structure, but it differs primarily in that it is somewhat indeterminant in size, it may be irregular in shape, and it is often flattened or padlike. If a stroma is formed in a leaf or wood, a stromatized area of host cells together with mycelium is delimited by a border of darkened cells. A stroma may form within floral parts and fruits, and the entire plant part may become invaded by the fungal hyphae and converted into a stroma. The fruit becomes dry, hardened, and shriveled (Fig. 4-41). A stroma gives rise to an apothecium directly; this apothecium either may arise above the stroma or may be embedded within it.

TAXONOMY

Seven orders are represented in this group (Korf, 1973). Two of these orders have only a single genus, each a rather unique plant parasite group. These orders, the Medeolariales and Cyttariales, will not be discussed further. The remaining orders, which we shall consider further, are the Phacidiales, Ostropales, Helotiales, Pezizales, and Tuberales.

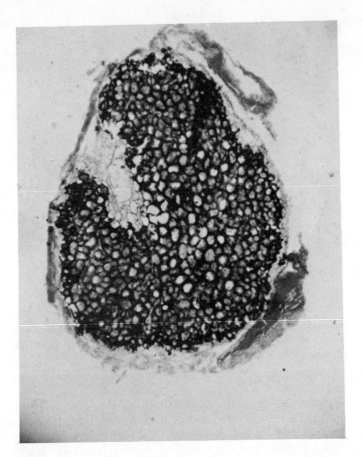

Fig. 4-40. A section through a sclerotium of a Discomycete *Pyronema domesticum*. × 440. [From E. J. Moore, 1962, *Mycologia* **54**:312–316.]

Fig. 4-41. Early stages of apothecium formation by *Monilinia fructicola* from peach fruits that have been converted to stromata. Approximately natural size. [Courtesy Plant Pathology Department, Cornell University.]

Order Phacidiales. Members of the Phacidiales are parasitic on plants, usually occurring on conifer leaves (*needles*), on which they cause diseases known as "needle blights." Species of *Rhytisma* may invade leaves of deciduous plants, causing black, shiny spots to develop that gives the disease the common name of "tar spot" (Fig. 4–42).

Fig. 4-42. A tar-spot fungus, *Rhytisma andromedae,* on its host. × 1.5 [Courtesy U.S. Department of Agriculture.]

Generally, these parasites may be divided into two broad groups. First, many of the fungi are weakly parasitic and invade injured or weakened needles, perhaps near the end of the season, and cause them to drop from the trees. These particular fungi are also called "needle-cast" fungi and include members of the largest genus, *Lophodermium.* The needle-cast fungi continue to live saprophytically on the fallen leaves and produce their ascocarps during the following spring or summer. The second group includes the majority of the members of this order and are strong parasites. They invade relatively healthy needles and do not cause them to be shed but instead form their ascocarps within the needles while they are still on the tree. They also require at least a year for the maturation of the ascocarp (Darker, 1967).

The ascocarp consists of a hymenium embedded in a stroma (Fig. 4–43). The stroma is embedded within the superficial tissues of the leaf (just beneath the cuticle, epidermal, or hypodermal tissues). The leaf tissue may become discolored and inseparable from the stroma because of the penetration by mycelium extending from the stroma. The stroma is typically elongate and is usually somewhat linear or boat-shaped. Its size varies from a fraction of a millimeter to 4 centimeters or more. These stromata are usually black but may be brown to almost colorless. According to Darker (1967), the color is related to water supply. He noted in a collection of *Lophodermium pinastri* that needles caught in a small depression filled with water were almost colorless, while normally the needles that are on the drier forest floor have black stromata. In addition, the texture of the stroma varied with the availability of water. The substance that makes the stroma dark seems to originate within the hyphae and then eventually lies in the interhyphal spaces (Silverberg and Morgan-Jones, 1974).

Each stroma contains at least one hymenium (Fig. 4–44). The hymenium contains both paraphyses which are free at the apices, and asci. The asci have an apical thickening, and sometimes the plug or surrounding pore becomes blue in iodine. The asci contain four or eight ovoid to filiform ascospores. The ascospores have a gelatinous sheath and may be single-celled or multiseptate. The hymenium is seated upon a relatively thin pad of pseudoparenchymatous tissue, which in turn is in contact with the host tissues. The cells of the host may be discolored beneath this fungal tissue.

Fig. 4-43. *Phacidium nigrum,* a member of the Phacidiales. Left: Stromata of *P. nigrum* on its host. × 1.25; Right: Ruptures of host epidermis reveal the hymenium. × 10. [From. E. K. Cash, 1942, *Mycologia* **34**:59–63; Courtesy U.S. Department of Agriculture.]

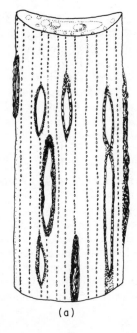

(a)

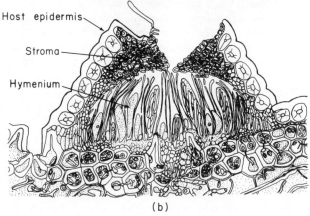

Host epidermis

Stroma

Hymenium

(b)

Fig. 4-44. *Bifusella striiformis* on a conifer. (a) Macroscopic appearance of boatlike apothecia on host; (b) Longitudinal section through an apothecium. [From G. D. Darker, 1932, *Contrib. Arnold Arboretum Harvard Univ.* **1**:1–131.]

According to Uecker and Staley (1973), who studied *Lophodermella morbida,* and Campbell and Syrop (1975), who studied *Lophodermella sulcigena,* only the basal pseudo-parenchymatous layer and the covering black shield are present in the early stages of development. The pseudoparenchymatous basal layer grows more rapidly than the covering shield, causing the basal layer to push deeper into the host tissue and meanwhile creating a cavity. Both layers of the stroma remain permanently attached to each other at the edges. Archicarps are lacking, and meanwhile some of the pseudoparenchymatous cells give rise to the ascogenous hyphae while others give rise to the paraphyses. The developing hymenium occupies the cavity created by differential growth.

After the asci have matured, the dark stromatal shield ruptures. The split may be lengthwise or stellate. If more than one hymenium is formed, there will be a split above each hymenium. The splitting sometimes occurs along the line of greatest stress, resulting perhaps from the pressure of the underlying hymenium. In some species, there is a differentiated dehiscence zone. This may consist in some cases of a band of cells within the stroma that rupture easily or in other cases of cells that become gelatinous to form the slit. The gelatinous cells may also seal off the hymenium under adverse conditions. Some species have a thickened, liplike band of tissue at the edge of the opening. After the stroma splits, the edges may only form a narrow opening or they may fold back to form most or all of the hymenium. As in the typical apothecium formed by the Discomycetes, the hymenium is exposed at maturity. Unlike the usually dark stroma, the hymenium may be light in color.

These fungi form a nonsexual stage. Conidia are formed in *acervuli* or *pycnidia,* more or less rounded or elongate dark structures formed in the superficial layers of the leaf. The anamorph is usually intermixed with the ascocarps, but the acervuli or pycnidia mature earlier than the ascocarps. Some investigators believe that these conidia may function as spermatia and be responsible for fertilization of archicarps found in some species (Darker, 1967). Some species that form conidia do not form archicarps (Uecker and Staley, 1973) while other species may form neither archicarps nor conidia (e.g., Thyr and Shaw, 1966; Woo and Partridge, 1969).

Order Ostropales. The Ostropales contain a single small family, characterized by their ascus and ascospore structure. The asci are elongate, rather cylindrical in form, and have a thickened apex penetrated by a pore. The ascospores are elongate and filiform or threadlike, and at maturity they break crosswise into a number of short, cylindrical segments. The apothecia are only a few millimeters in diameter and light but drab in color. Their form varies from a stipitate apothecium with a convex pileus or a sessile, shallow cup in *Vibrissea* to a partly submerged shallow cup in *Stictis.* These fungi are saprophytes and are often found on woody substrata submerged in a quickly running stream or semiaquatic habitats. The ascospores are discharged while submerged and have a tendency to adhere to the apothecium after their discharge. The cluster of ascospores vibrates or oscillates about in the water, which has given rise to the generic name *Vibrissea* (Latin, *vibrare,* to oscillate or vibrate) (Sanchez and Korf, 1966).

Order Helotiales. The Helotiales are a very diverse group of fungi, containing over 1,500 species. Important characteristics of the group are found in the asci and ascospores. The asci are inoperculate and have an apical pore through which ascospore discharge will occur. There may or may not be a blueing reaction with iodine. The asci are cylindrical to clavate in form. The ascospores are short and ellipsoid to needlelike and do not fragment at maturity (unlike those of the Ostropales). The apothecium may be seated within the host tissue or entirely superficial. Unlike the

(a)   (b)

Fig. 4–45. Cupulate representatives of the Helotiales: (a) *Dasyscyphus* on dead herbaceous stems collected under leaves. Approximately × 2. (b) *Sclerotinia sclerotiorum.* This species forms yellowish-brown cups borne on a downy stalk. Approximately natural size. [(a) Courtesy Plant Pathology Department, Cornell University. (b) From E. J. Moore and V. N. Rockcastle, 1963, *Fungi,* Cornell Science Leaflet.]

members of the Phacidiales, the apothecium is not immersed within a stroma. Some members of the Helotiales form a sclerotium that gives rise to the apothecium. Nonsexual stages are common and are usually Hyphomycetes.

The majority of these organisms are saprophytes on plant debris and may be collected from dead plant parts such as grass blades, leaves, and twigs. Comparatively few occur directly on soil or dung. The fungi occur in abundance in relatively moist environments such as the forest floor, the edge of streams, or in bogs. Some are virulent and important plant parasites. The plant diseases may be caused either by the apothecial stage (as in the needle blight diseases) or by the nonsexual stage (as in the brown rot of peaches, discussed below).

The majority of the apothecia formed by members of the Helotiales are comparatively small, often only a few millimeters in width, and are typically cupulate or pulvinate. The apothecium bears the hymenium on its upper surface, which may be concave or convex. The cup may be seated directly on the substratum, or it may be supported by a stemlike portion, the *stipe.* The texture of the apothecium may be soft and fleshy, somewhat waxy, or gelatinous. Colors range from pallid earth colors to bright yellow, orange, red, or green. Common genera include *Dasyscyphus,* which forms small stipitate apothecia (about 1 to 2 millimeters in diameter) with hairs on the excipulum (Fig. 4–45), and *Bisporella. Bisporella citrina* forms bright, citron yellow apothecia that are only up to 3 millimeters in diameter but are relatively noticeable because they occur in dense swarms on the surface of logs.

Some of the Helotiales are parasitic on conifers, where they cause a needle blight disease similar to that caused by members of the Phacidiales. The fungi are members of the Hemiphacidiaceae, and they form relatively simple fleshy, usually orange to yellow-brown apothecia beneath the epidermis or hypodermis of the host plant. The apothecia form on either living or recently killed leaves that are still attached to the plant. Unlike those of the Phacidiales, the apothecia are not covered by fungal tissue but are covered simply by the host tissue, which eventually tears to expose the apothecium (Korf, 1962).

Perhaps the best known and most important group of this order belongs to the family Sclerotiniaceae, which consists almost entirely of parasitic fungi. Important plant parasites include *Sclerotinia whetzelii,* which occurs on eastern poplar, *Sclerotinia sclerotiorum* (Fig. 4–45), which occurs on vegetables such as lettuce, *Monilinia fructicola,* which causes brown rot of peaches, and *Botryotina fuckeliana,* which occurs on grapes.

These fungi have a long stalk with a cuplike or funnellike fertile region, are usually somewhat waxy in texture, and are some shade of yellow, tan, or brown. The apothecia are larger than those previously described, as the fertile region may be about 10 to 20 millimeters in diameter.

The life history of many members of the Sclerotiniaceae is known much better than is that of other members of the order. Spermatia are common in the Sclerotiniaceae but are probably either rare or absent in the remainder of the Helotiales. The spermatia have been demonstrated to be functional in some members, uniting with the apothecial fundaments (the female elements) arising from the sclerotia or stromata. After plasmogamy occurs, the apothecial fundaments complete their development and form the mature apothecia.

Let us briefly examine the life history of *Monilinia fructicola* (Honey, 1936). *Monilinia* overwinters as a mycelium within a hardened and blackened fruit (a stroma or "mummy") produced by the diseased host (Fig. 4–41). In the spring, spermatia are produced from short, flask-shaped spermatiophores. The pillar-shaped fundaments of the apothecium arise from the mummy and continue their development into apothecia, presumably after fertilization by spermatia. The apothecia discharge ascospores that infect the young flowers or leaves of the peach tree. During the spring and summer, hyaline chains of conidia are produced on the leaves and young twigs. These conidia are responsible for initiating many secondary infections throughout the summer.

Fig. 4–46. Apothecia of *Leotia stipitata.* Approximately ⅔ natural size. [Courtesy Plant Pathology Department, Cornell University.]

Some of the most conspicuous members of this group are known as "earth-tongues." These are members of *Geoglossum* and closely related fungi. They occur on the ground in forests and bogs. These fungi are tall (about 2 to 8 centimeters) and have an elongate stipe bearing a slightly expanded, somewhat tonguelike region at the apex which is covered by the hymenium. The earthtongues may be dark or shades of bright orange or red. A rubbery gelatinous fungus, *Leotia,* is closely related to the earthtongues but differs in having a caplike structure at the apex of the stipe (Fig. 4-46). The hymenium covers the upper surface of the cap only.

Order Pezizales. At least 500 species comprise the Pezizales. All members of the Discomycetes with operculate asci belong to this group. Some of the fungi have asci with subapical opercula, and coupled with this character, apothecia with a car-tilaginous or leathery texture. The remaining members have centrally located opercula, and all form brittle and rather fleshy apothecia.

These fungi occur in diverse habitats as saprophytes on dung, soil, wood, or plant debris. They may be readily collected on cut banks of roads and in forests. Some members will form apothecia only on recently burned substrata (e.g., the *Pyronema,* whose name literally means "fire thread"). A few members form their apothecia beneath the soil's surface.

There is a great deal of morphological variation in this group. Some of the dung-inhabiting members simply form a mycelial mat or tuft (e.g., species of *Ascodesmis*) or a small cleistothecium. Only one ascus may occur in some of these cleistothecial dung-inhabitants, such as *Lasiobolus monascus* (Kimbrough, 1974).

Fig. 4-47. Apothecia of *Sarcoscypha coccinea,* the scarlet elf cup, a member of the Pezizales. These apothecia are conspicuously colored, with an orange-red hymenium and a white exterior. About natural size. [Courtesy Plant Pathology Department, Cornell University.]

The majority of the fungi in the Pezizales produce a larger, more complex apothecium. In some, such as *Pyronema domesticum,* the apothecium may be only a few millimeters in diameter, but in the majority it is over 1 centimeter and apothecia as large as 10 centimeters are common. Most members, including species of *Peziza* and *Sarcoscypha* (Fig. 4-47) form a cupulate apothecium, with the hymenium lining the inside of the cup. Pulvinate apothecia with a convex hymenium are also common (Fig. 4-34). The apothecia may be sessile or have a stipe of varying lengths. Many members of the Pezizales form apothecia that are noncupulate but bear fertile regions of varying forms at the apex of the stipe. The fertile region may be saddle-shaped as in *Helvella* species, spoon-shaped as in *Wynnea* species, funnel-shaped as in *Urnula* species, or deeply convoluted as in the edible morels (*Morchella* species) (Fig. 4-48).

Developmental studies have been made on numerous members of this group and have been especially numerous on species of *Pyronema* (illustrated in Figs. 4-3 and 4-4 to show many general features for the ascomycetes). *Pyronema* has received so much attention because it forms unusually large club-shaped antheridia and globose ascogonia, which may be up to 50 microns in diameter, and also because it will develop in culture (Moore, 1963). As indicated, the antheridium is usually functional and fuses with the trichogyne in *Pyronema.* However, even though some of the antheridia are functional, ascogonia in the same culture may not undergo fusion and may produce ascogenous hyphae without apparent fertilization (Moore-Landecker, 1975). Some other members of the Pezizales form antheridia and ascogonia, but their morphology varies greatly. For example, the ascogonium is a multicellular coil fertilized by a threadlike antheridium in *Anthracobia melaloma* (Rosinski, 1956). Some of the Pezizales have archicarps fertilized by spermatia; an example is *Ascobolus stercorarius,* which forms a coiled archicarp and spermatia (these are discussed further in Chapter 9). Some species form archicarps but not spermatia or antheridia (e.g., *Peziza quelepidotia* described by O'Donnell and Hooper, 1974). Sexual structures are unknown for the vast majority of the Pezizales. Subsequent development of the ascogenous system and sterile hyphae generally occurs as described on pages 116–120.

Order Tuberales. The Tuberales is a small order containing about 140 species. These fungi are commonly known as ''truffles'' and many species have been widely esteemed as a delicacy since ancient times, when they were imported from North Africa by the Romans (Trappe, 1971). The early Greeks not only relished the truffles but speculated about their nature (Chapter 14). In nature, the ascocarps are often associated with vascular plants, with which they establish a mycorrhizal relationship (Chapter 13). Efforts to culture the truffles as a food crop have been futile, perhaps because of their intricate biological requirements, met by their mycorrhizal association but not easily duplicated in cultivation (Hawker, 1955).

The apothecia usually occur beneath the soil's surface at varying depths up to approximately 8 to 10 centimeters, but they also may be partially buried in the litter on the forest floor. Macroscopically, the apothecium may appear to be cupulate or globose (Figs. 4-49 and 4-50). The exterior is often convoluted or irregularly lobed. The size of the apothecia ranges from a few millimeters to more than 10 centimeters, and they may be white, yellow, brown, black, rose, or purple in color.

The simplest members form hollow, cupulate apothecia opening by a pore or slit. In these, the hymenium occurs as a single layer lining the inside of the cup (Fig. 4-49). Unlike members of the Pezizales, the paraphyses extend beyond the asci and their tips fuse to form a tissue (the *epithecium*) that encloses the asci. In more complex forms, there is a folding inward of the excipulum to form shelf- or lobelike projections extending into the central cavity. These projections may even anastomose when

(a)

(b)

(c)

(d)

Fig. 4–48.   Noncupulate members of the Pezizales. (a) *Urnula geaster;* (b) *Helvella infula;* (c) *Wynnea americana;* (d) *Morchella esculenta.* [(a) Courtesy U.S. Department of Agriculture; (b), (c), and (d) courtesy Plant Pathology Department, Cornell University.]

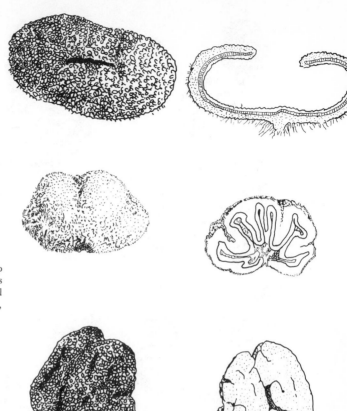

Fig. 4-49. Representative ascocarp forms in the Tuberales. A habit sketch is on the left, accompanied by a longitudinal section on the right. [From L. E. Hawker, 1954, *Phil. Trans B* **237**:429–546.]

meeting an opposing fold. The particular degree and pattern of infolding is important in distinguishing genera and species within this group (Fig. 4-49). Apothecia with a great deal of infolding tend to have extremely small cavities and are essentially solid. Individual chambers may be isolated within the apothecium by the folds, or alternatively, there may be a network of canals that wind through the apothecium. These chambers and canals become lined with a hymenium containing both asci and paraphyses. The paraphyses often extend beyond the asci, and the tips of the paraphyses fuse to form an epithecium. In some species, the chambers or canals become filled with loose hyphae. The individual chambers may eventually open to the outside. Some species do not form a hymenium but rather form asci scattered within the sterile tissue.

Members of the Tuberales do not forcibly discharge their ascospores. The ascospores may be released when the ascocarp decays or when the apothecium is broken or carried away by an animal. Usually the ripe apothecium emits a strong, sometimes unpleasant odor attracting animals that dig them up for food. If you wish to collect truffles, search for woodland sites where it appears that a small animal may have been digging recently, and then search carefully through the surrounding layer of soil.

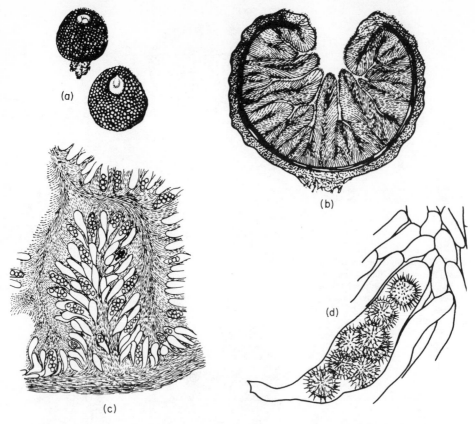

Fig. 4-50. A member of the Tuberales, *Pachyphloeus:* (a) The ascocarps as they would appear in nature; (b) Longitudinal section through the ascocarp; (c) A detail of the ascocarp as it would appear near the wall; (d) An ascus. [From A. Engler and K. Prantl, 1897, *Die natürlichen Pflanzenfamilien.* Engelmann, Leipzig.]

Class Laboulbeniomycetes

ORDER LABOULBENIALES

Members of the Laboulbeniales are minute insect parasites, ranging in size from approximately 0.1 millimeter to slightly over 1 millimeter. At least 1,500 of these tiny fungi have been described (Thaxter, 1896–1931). They penetrate the chitinous exoskeleton of the insect and form their thalli outside the insect's body. The thallus appears to the naked eye as a dark or yellowish bristle, and the insect will appear furry if it is heavily infected. The insects are apparently either unharmed or only slightly harmed by these fungi. The infections are contagious and may be spread by contact of one insect with another; infections are especially severe in dense insect populations (Benjamin, 1973). The vast majority of the Laboulbeniales are highly host-specific: they live on only certain insect species and often inhabit a certain part of the insect, such as the mouth parts, the anterior legs, or the wings. In addition, the fungus may be restricted to the right or left side, to the upper or lower surfaces, and perhaps even to male or female insects (Benjamin and Shanor, 1952). Some specificity for a particular position may be correlated with mating behavior or other behavioral patterns accounting for contact of certain parts of the insect's body (Whisler, 1968). A few species have comparatively broad host ranges. These fungi have not been successfully cultured apart from the host.

The thallus is attached to its host by a darkened foot cell that typically forms a small haustorium penetrating the host's integument. The haustorium absorbs nutrients from the insect, as a mycelium is absent. The foot cell is immediately beneath the main body of the thallus, the *receptacle.* The receptacle consists of a column of cells that may be more or less definitely arranged in one or more rows or may be somewhat parenchymatous. The receptacles may bear one or more sterile hairlike appendages, which may be unicellular or multicellular and possibly branched. The receptacle also bears the male and female reproductive structures, as well as a perithecium. Some species are monoecious and form both male and female reproductive structures on the same thallus. Other species are dioecious and form separate male and female thalli occurring in close proximity to each other in pairs (Fig. 4-51).

The male gametes are spermatia. Spermatia may be borne exogenously as cells formed from some appendages similar to sterile appendages. Usually the spermatia are formed endogenously within a flask-shaped cell, the antheridium, and escape through a neck region [Fig. 4-52(b, c)]. The antheridium is most often borne individually on the side of an appendage but may also be formed from one or more cells of the appendage itself. In some species, a compound structure consisting of fused antheridia is formed and the spermatia are discharged into a common cavity before being finally released.

The perithecium and its contents are derived from a single cell of the receptacle or one of its branches [Fig. 4-52(b)]. At maturity, the perithecium consists of three basal cells supporting a cellular perithecial wall. The perithecial wall consists of a double layer of cells regularly arranged in both horizontal and vertical rows. Usually there are four vertical rows and either four or five horizontal rows. The perithecium lacks paraphyses, pseudoparaphyses, and periphyses but contains the ascogenous system.

Development of the Perithecium. The ontogeny of some species was first described by Thaxter (1896), and some later studies include those of Batra (1963), Benjamin and Shanor (1950), Shanor (1952), and Tavares (1966). A typical sequence follows. Development begins when the two-celled ascospore lands on the surface of the host and germinates. The ascospore becomes attached by a blackened extremity that forms from the lower cell of the ascospore [Fig. 4-52(a)]. This blackened structure is the foot cell. After the foot cell is formed, the remainder of the ascospore differentiates into the receptacle. Development is characterized by apical growth of an elongating column of cells, all formed in a very precise manner. The antheridia and spermatia are differentiated before the female structures are and arise from the receptacle. The perithecium and archicarp primordia develop from a single initial cell [Fig. 4-52(b)] that divides, the lower cell giving rise to the surrounding perithecium while the upper cell gives rise to the archicarp [Fig. 4-52(b, c)].

During development of the perithecial wall, the lower cell resulting from the division of the basal cell again divides by forming an oblique septum, becoming two cells. Each of these two cells will remain as stalk cells supporting the perithecium. In turn, the stalk cells produce a total of three cells that will become the basal cells of the perithecium [Fig. 4-52(d)]. The basal cells produce a total of eight upgrowths. Four of these are external and give rise to the outer perithecial wall, while four are internal and give rise to the inner perithecial wall. The continued terminal growth and septa-

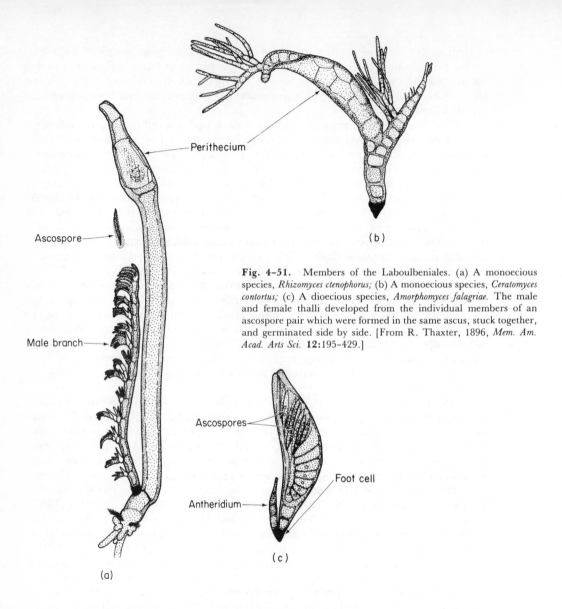

Perithecium

Ascospore

Male branch

(a)

(b)

Fig. 4-51. Members of the Laboulbeniales. (a) A monoecious species, *Rhizomyces ctenophorus;* (b) A monoecious species, *Ceratomyces contortus;* (c) A dioecious species, *Amorphomyces falagriae.* The male and female thalli developed from the individual members of an ascospore pair which were formed in the same ascus, stuck together, and germinated side by side. [From R. Thaxter, 1896, *Mem. Am. Acad. Arts Sci.* **12**:195–429.]

Ascospores

Foot cell

Antheridium

(c)

tion of the cells will result in formation of the sheathlike perithecium which envelopes the asci at maturity [Fig. 4-52(h, i)].

Meanwhile the sexual system continues its development. Let us return to the division of the single initial cell [Fig. 4-52(b)] and recall that the upper cell resulting from this division is the archicarp primordium [Fig. 4-52(c)]. The archicarp primordium divides to form three uninucleate cells; proceeding from the bottom and progressing upward, these are the *carpogenic* cell, an intermediate cell (the *trichophoric* cell), and the trichogyne in the terminal position [Fig. 4-52(e)]. In some species, the trichogyne may consist of more than one cell. The developing perithecium surrounds the basal two cells (the carpogenic and intermediate cells), while the trichogyne is either partially or wholly exposed. Spermatization occurs at about this time [Fig.

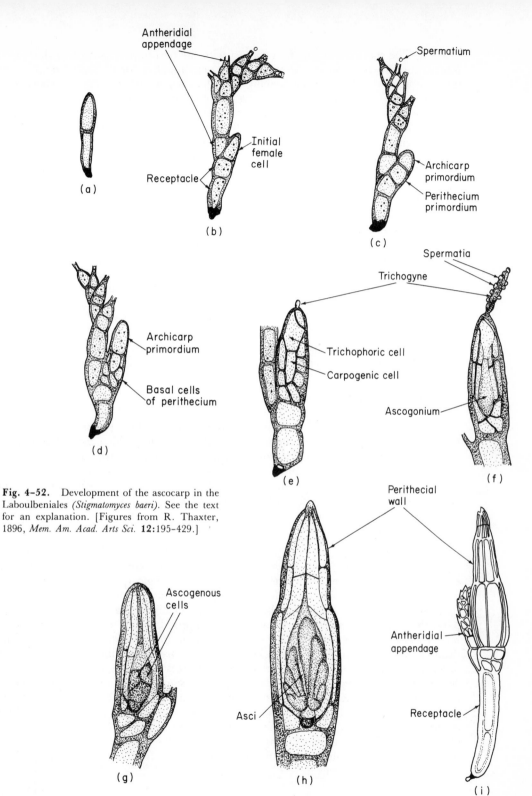

Fig. 4-52. Development of the ascocarp in the Laboulbeniales *(Stigmatomyces baeri)*. See the text for an explanation. [Figures from R. Thaxter, 1896, *Mem. Am. Acad. Arts Sci.* **12**:195–429.]

4–52(f)]. The spermatia come into contact with the trichogyne, possibly owing to the insect's movement or to the presence of a water film. Eventually the trichogyne and the trichophoric cell disappear, leaving only the carpogenic cell. The carpogenic cell divides again to form an ascogonium and supporting cells [Fig. 4–52(f)]. The ascogonium undergoes a division, and sequential divisions may occur, resulting in a precise number and arrangement of cells that is constant for a particular species. These divisions result in the production of ascogenous cells [Fig. 4–52(g)]. Each ascogenous cell produces asci budding from it both in an upward and downward direction and will produce a succession of asci [Fig. 4–52(h)]. Cytological observations of karyogamy and meiosis are lacking for most species. The asci generally become free in the cavity of the perithecium as their supporting cells disintegrate. Also, the inner wall of the perithecium disappears to provide room for the expanding asci.

Ascospores are formed with the typical pattern described for the ascomycetes (Hill, 1977). Usually, four ascospores remain after the disintegration of four nuclei. Eventually the ascus wall disintegrates, freeing the ascospores within the perithecium. The ascospores are embedded in a slime that may aid in lubricating their exit from the perithecium. The slime originates in some cells of the perithecium and is secreted by vesicles into the perithecial cavity (Hill, 1977). The ascospores tend to cling together in pairs owing to their sticky sheaths. In this fashion, ascospores land together, adhere to a surface, and germinate side by side. In dioecious species, typically one member of the ascospore pair develops into the male plant while the other, larger ascospore develops into the female plant.

Class Loculoascomycetes

Members of the Class Loculoascomycetes are characterized by: (1) the possession of a bitunicate ascus, and (2) the production of a stroma within which asci are produced. This is the only group of fungi having a bitunicate ascus. Ascostromata with asci also occur in the Coronophorales and Coryneliaceae but are formed in conjunction with a unitunicate ascus. Von Arx and Müller (1975) recognize a single order, the Dothideales, for which they delimit 34 families (Table 4-2). We shall follow Von Arx and Müller's treatment here.

ORDER DOTHIDEALES

Approximately 2,000 species are known. Numerous members occur as saprophytes on dead leaves, herbaceous stems, wood, plant debris, and dung. Species of *Myriangium* are parasitic on scale insects, which in turn occur on host plants. Other fungi are parasitic on other fungi, lichens, mosses, or higher plants. Some of the plant parasitic fungi will form ascostromata on or immersed within the living plant, while others will invade the living plant, kill it, and then form ascostromata on or within dead plant parts (Figs. 4-53 and 4-54). Such fungi are responsible for a wide variety of plant diseases usually characterized by fruit rots or leaf spots. An example is *Mycosphaerella tulipiferae,* which causes leaf spot and some defoliation of the tulip poplar in the Middle Atlantic and Gulf states. The mycelium kills the host cells, and conidiophores and conidia develop on the lower surface of the leaf. Development of the ascocarp is initiated in the autumn in leaves that are still attached to the tree but is not completed until the following spring in the fallen leaves. *Venturia inaequalis,* the causal agent of apple scab which results in leaf spots and fruit rot of the apple, undergoes a similar sequence of events.

Table 4–2: Some Important Characteristics of Representative Families of the Dothideales (as Delimited by Von Arx and Müller, 1975).

Family	Ascocarp	Asci and Ascospores	Representative Genera
Dothideaceae	Locules in stromata; stromata small to large, often erumpent, shapes include crustlike, pulvinate, and spherical; locules opening by fissures or dehiscence; pseudoparaphyses absent	Asci in fascicles or parallel at base, sometimes on column of tissue; clavate to cylindrical, with 8 or more spores; ascospores uni- to multiseptate	Dothidea Plowrightia
Myriangiaceae	Often parasitic on scale insects; stroma of various shapes (pulvinate, discoid, spherical, crustlike); small to medium; pseudoparaphyses absent	Asci borne singly in small locules, often at different levels, spherical to broadly clavate; ascospores usually both longitudinally and transversely septate	Myriangium Elsinoë
Hysteriaceae	Saprophytic on wood; stroma superficial or immersed at the base, elongated, elliptical, linear, or boat shaped; opening by a longitudinal cleft; pseudoparaphyses often branched	Asci parallel, clavate, or nearly cylindrical; apical thickening of ascus; ascospores sometimes with many cells, relatively wide	Hysterium
Pleosporaceae	Ascocarp usually a pseudothecium, conical, spherical, or flask-shaped; sometimes immersed in host tissue or forming locules in an erumpent stroma; medium to large; pseudoparaphyses present	Asci formed in a parallel array at the base of the locule, cylindrical or baglike, thickened apically	Pleospora
Mycosphaerellaceae	Pseudothecia small, spherical, and smooth, immersed in host tissue or forming locules in a stroma which may be erumpent; pseudoparaphyses absent	Asci in a fascicle or parallel at the base of the locule, ovate to cylindrical with thickened apex; ascospores elongated, narrow, uni- to triseptate	Mycosphaerella
Capnodiaceae	Epiphytic, forming superficial dark mycelium ("sooty molds"); ascocarps usually spherical; immersed in a subiculum or covered with hyphae; pseudoparaphyses absent	Asci ovoid or baglike; ascospores usually many septate	Capnodium
Microthyriaceae	Superficial, often shield-shaped with a covering wall of radially arranged cells; pseudoparaphyses present	Asci arranged in a ring with their tips converging to a pore; cylindrical or baglike; ascospores usually uniseptate	Microthyrium

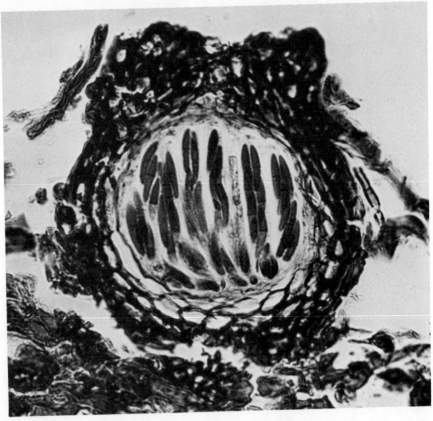

Fig. 4-53. Pseudothecium of a member of the Dothideales, *Mycosphaerella* sp. in the leaf of its host. [Courtesy Plant Pathology Department, Cornell University.]

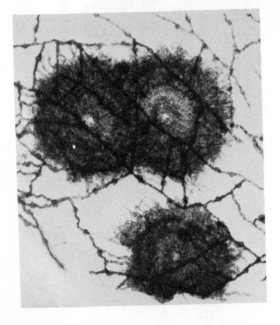

Fig. 4-54. Flattened shieldlike ascocarps of *Platypeltella angustispora,* a member of the Dothideales. × 100. [Courtesy M. L. Farr and F. G. Pollack, 1969, *Mycologia* **61**:191–195.]

Although many species of *Leptosphaeria* cause a variety of leaf spot diseases, one species, *L. sengalensis,* is responsible for a disease of humans (El-Ani, 1966). The so-called ''sooty molds'' occur as harmless epiphytes on plants, forming their dark superficial mycelium, conidia, and ascostromata on the surface of plants. Actually the sooty molds are not parasitizing the plants but feeding upon the honeydew secreted from the alimentary canal of scale insects. The honeydew contains a variety of nutrients, including sugars, amino acids, proteins, and minerals (Hughes, 1976). Some members form lichens (Chapter 13).

The Ascocarp. The ascostromata form locules that are not separated by a distinct wall from the stromal tissue. This character separates these fungi from the majority of the species in the Sphaeriales, in which a distinct wall is visible. The asci may exist individually in a small space that is essentially an individual locule (Fig. 4–55), but usually larger locules are formed in which several asci occur (Figs. 4–53 and 4–54). These larger locules resemble the cavities of the perithecia described on pages 121–122. Each ascostroma may have from one to several locules. In some species, there is a single locule surrounded by only a small amount of stroma tissue; such an ascostroma superficially resembles a perithecium and is sometimes called a *pseudothecium* (Fig. 4–53). Sometimes when mature pseudothecia and perithecia are compared, it is difficult to distinguish between the two ascocarp types. They do differ, however, in their mode of development and ascus type (unitunicate or bitunicate). Other than having a peritheciumlike form, the ascocarp shows numerous morphological variations. The shape may be lenslike owing to the formation of upper and lower plates of radiate tissue (superficially resembling the ascocarps of the Meliolales) (Fig. 4–54), elongated and boatlike and also opening by a slit, or superficially resembl-

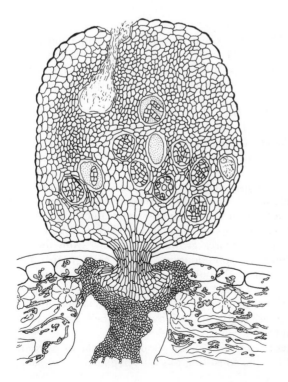

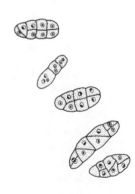

Fig. 4–55. A longitudinal section through the stroma of *Diplotheca* showing scattered locules, each containing a single ascus with septate ascospores. Details of the ascospores are at the top right, while details of the asci are at the bottom right. [From B. O. Dodge, 1939, *Mycologia* **31**:96–108.]

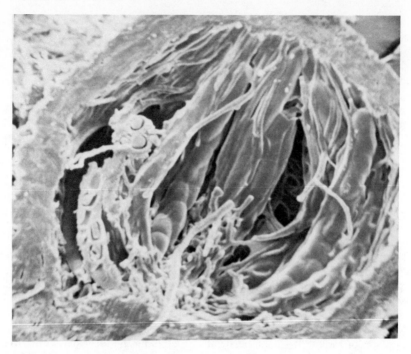

Fig. 4-56. The centrum of *Sporomiella australis,* a member of the Pleosporales. Note the thin, threadlike pseudoparaphyses among the wider asci. [Courtesy R. Blanchard, 1972, *Am. J. Botany* **59**:537-548.]

ing an apothecium (Fig. 4-55). The ascocarps range in size from comparatively small (40 to 150 microns) to quite large (1 to 5 millimeters) but are solitary. Even larger crusts may be formed if the ascocarps fuse together at their sides. The ascocarps may be brightly colored or dark. In some species, the ascocarp may be gelatinous or dissolve into a slimy mass.

Asci are usually borne at the base of the locule in a group. They may occur in a fanlike cluster (fascicle) or in a basal layer. Asci range in shape from globose to elongated cylindrical or clavate types. Unlike the ascus apex of many fungi with unitunicate asci, that of the bitunicate ascus is comparatively simple and more or less invariable and does not blue with iodine. The ascospores are usually septate and may show a number of other variations, such as a gelatinous sheath or appendages. The asci may be the only contents of the locule or they may be interspersed among pseudoparaphyses (Fig. 4-56). Hymenial paraphyses and apical paraphyses are not formed in this group. Like the apical paraphyses in the Pyrenomycetes, the pseudoparaphyses are hairlike structures growing downward from the apex of the locule, but unlike the former, the pseudoparaphyses in this group almost always become attached at the bottom of the locule when their tips recurve and become entangled in the resulting mass of tissue or push into the stromal tissue at the base of the locule (Luttrell, 1965). Growth of the pseudoparaphyses is largely intercalary, and they may also be branched (Corlett, 1973). Periphyses may be formed in those species which form a neck with an ostiole.

Ascocarp Development. In tracing the development of the ascostroma, we may begin with a relatively small initial consisting of sterile hyphae, the *stromal initial.* The stromal initial will continue to grow and will develop into all sterile elements of the ascocarp.

Apparently most species form differentiated archicarps. One or more archicarps may be formed in each stromal initial and lie embedded within the stromal tissue. The archicarps are often coiled and multicellular. In some species, plasmogamy may occur between the archicarp and spermatia: this occurs in species of *Mycosphaerella* (Fig. 4-57) (Higgins, 1936) and *Cochliobolus sativus* (Shoemaker, 1955). In others, including *Venturia inaequalis,* the archicarp undergoes plasmogamy with a differentiated antheridium (Killian, 1917). Alternatively, male gametes or gametangia may be lacking and the archicarp is then not involved in plasmogamy. In observations on *Trichometosphaeria turica,* there was no morphological evidence for plasmogamy involving the archicarp, although genetic evidence indicated that plasmogamy had occurred, perhaps involving hyphal fusion (Luttrell, 1964). Some species, including *Pleospora herbarum,* lack differentiated gametangia. In *P. herbarum,* two undifferentiated hyphae undergo fusion prior to pseudothecium formation.

Luttrell (1951) describes three major types of development (Fig. 4-58) which he considers to be typical for orders that he recognizes (1973). Because we are recognizing only a single order, each developmental type will be designated by the name of the genus on which the name of Luttrell's order is based. These types are described below.

1. *Myriangium* type: The globose asci occur individually in the tissue of the stroma and may occur throughout or in restricted fertile regions of those stromata superficially resembling apothecia. One or more archicarps are formed and produce ascogenous hyphae that penetrate through the stromal tissue, producing asci at various intervals. Although the stromal tissue is not greatly altered, it may sometimes be slightly compressed and disintegrated by the pressure of the expanding asci. When the asci are to discharge their ascospores, they penetrate individually through the overlying stromal tissue, which has become softened or crumbles.

2. *Dothidea* type: Species of *Dothidea* and their allies produce asci in fascicles in small perithecium-shaped locules in ascostromata and do not form pseudoparaphyses. This developmental sequence begins when ascogonia are formed within the stromal initial, which continues to enlarge and differentiate further. While embedded in the pseudoparenchymatous tissue, the ascogonium produces a short, radiating mass of ascogenous hyphae. This particular configuration of ascogenous hyphae ultimately produces the fanlike fascicle of asci. The elongating asci push up into the pseudoparenchyma, which becomes crushed and disintegrates to form the locule. At maturity, the locule is an unwalled cavity that is almost completely filled by asci. In some species, partially disintegrated pseudoparenchymatous tissue may remain behind as strands. A pore is formed when some tissue at the apex dissolves. This type of development also occurs in *Mycosphaerella* (Luttrell, 1951) illustrated in Fig. 4-57, and also in a few species of sooty molds (Corlett, 1970; Reynolds, 1975).

3. *Pleospora* type: Pseudoparaphyses are formed in this type of development. *Pleospora* and its allies form a pseudoparenchymatous stroma in which the ascogonia are embedded. Pseudoparaphyses appear in the vicinity of the ascogonia, grow downward, and become attached at the bottom of the locule, meanwhile forming a locule by exerting outward pressure. Elongation of the pseudoparaphyses produces a flask-shaped locule within the stroma. Asci develop among the pseudoparaphyses, forming a concave layer (hymenium) across the base of the locule. Finally a pore is dissolved in the stroma at the apex of the locule. This type also occurs in *Leptosphaeria avenaria* (Corlett, 1966-a) and *Didymosphaeria sadasivanii* (Luttrell, 1975).

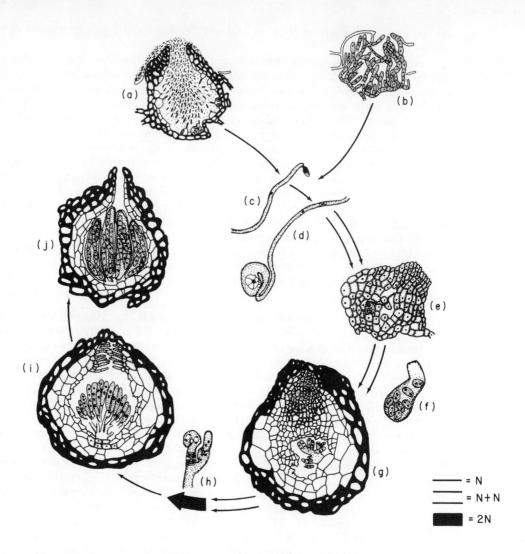

Fig. 4-57. Sexual cycle of *Mycosphaerella tulipiferae* which has spermatization and the *Dothidea*-type of development. (a) Mature spermagonium within which spermatia are produced. The spermagonium, like the pseudothecium, is formed within the leaf tissue; (b) Young stromal initial developing, with an archicarp at the lower center and with a trichogyne growing toward the surface; (c) Part of the threadlike trichogyne extending above the leaf surface with a spermatium attached to its tip; (d) Portion of an archicarp in which the male nucleus is migrating down the trichogyne. The base of the archicarp is swollen, and the tip of the trichogyne is not shown; (e) Later stage in which the ascogonium is binucleate near the base, and the trichogyne is disintegrating. Note the further development of the surrounding stroma; (f) Ascogonium with eight pairs of conjugate nuclei, each pair held in a mass of denser protoplasm; (g) Later stage of pseudothecium with branched ascogonium; (h) Tip of ascogenous hyphae with a crozier (left) and a young ascus (right); (i) Nearly mature pseudothecium with asci in various stages of development and ascogonium and ascogenous hyphae still visible. Note the formation of the neck with an ostiole and periphyses, and also the relatively conspicuous locule; (j) Mature pseudothecium with asci and ascospores. [Adapted from B. B. Higgins, 1936, *Am. J. Botany* **23**:598–602.]

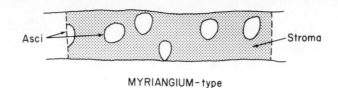

MYRIANGIUM-type

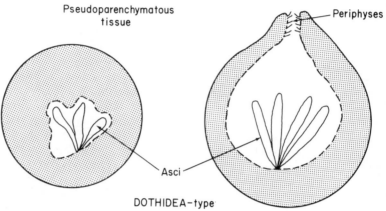

Pseudoparenchymatous tissue

Periphyses

Asci

DOTHIDEA-type

Fig. 4-58. Centrum development in the Loculoascomycetes. A younger, nonostiolate pseudothecium is on the left, showing young asci. The more mature stage with an ostiole is on the right.

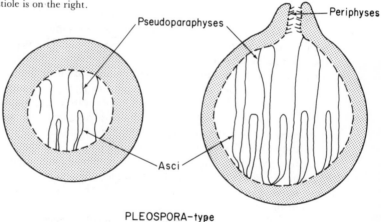

Pseudoparaphyses

Periphyses

Asci

PLEOSPORA-type

ORIGIN OF THE ASCOMYCOTINA

There have been two widely opposing views of the origin of the Division Ascomycotina: (1) that they originated from primitive fungi, and (2) that they originated from the red algae through loss of chlorophyll.

One of the proponents of the origin of ascomycetes from primitive fungi was Atkinson (1915), who suggested that the ancestor was similar to the present-day Mucorales. This ancestral fungus presumably formed a coenocytic zygote, and the ascus would have originated by reduction of the coenocytic zygote to a zygote with a single nucleus or through a progressive splitting of the coenocytic zygote until a cell with a single zygotic nucleus remained. A transitional ascomycete could have been *Dipodascus,* or a fungus similar to *Dipodascus,* which superficially resembles a multinucleate zygote. Atkinson then proposes the origin of the yeasts through reduction from *Dipodascus* and the development of many separate, diverging lines of advancement from *Dipodascus* to give the higher ascomycetes. The lines of evolution which presumably gave rise to the higher ascomycetes were those giving rise (1) to the Eurotiales, (2) to the remainder of the Plectomycetes, (3) to the Loculomycetes, Pyrenomycetes, and bitunicate series, and (4) to the Discomycetes. The Laboulbeniales may represent a separate line of evolution.

Dodge (1914), as well as many other mycologists, maintained that the ascomycetes originated from red algae through the degenerative loss of chlorophyll. His argument is based on the close similarity of nonmotile archicarps with a trichogyne fertilized by spermatia found in many fungi and red algae. He maintained that it is improbable that such similar structures could arise through two unrelated evolutionary lines. According to Dodge, the greatest obstacle to accepting this concept is that nothing resembling cleavage of ascospores from the cytoplasm of the ascus—or anything morphologically similar to the ascus—is found among the red algae. Recent proponents of the origin of the Ascomycotina from red algae are Denison and Carroll (1966) and Kohlmeyer (1973, 1975), who note many morphological similarities between the red algae and some ascomycetes in addition to similarities of the sexual structures. These include their lack of motility, the presence of a septal pore, the organization and branching of the thallus, and the superficial resemblance of the red algal pericarp to the ascomycetous perithecium.

Kohlmeyer (1973, 1975) has described some fungi in the genus *Spathulospora* which are parasitic on red algae in the Southern Hemisphere. These fungi resemble the Laboulbeniales in that they form a trichogyne and spermatia, have deliquescent asci, and lack mycelium or an ascogenous system in the form of hyphae. They differ from the Laboulbeniales but resemble the Pyrenomycetes by having a perithecium with an ostiole and thick wall typical for the fungi in the Sphaeriales and also by forming a stroma between the cells of the host or a crustlike pad. The perithecia are not immersed in the stroma. Kohlmeyer suggests that the ancestral ascomycete was similar to *Spathulospora* and was derived from red algae that had become reduced in form and lived as symbionts or parasites on other red algae. Presumably this ancestral type gave rise: (1) to the Laboulbeniales, first occurring on marine arthropods and later moving onto land along with their hosts; (2) to parasites on plants similar to the present-day *Spathulospora* group; and (3) then to other Pyrenomycetes.

An alternative argument is that morphological similarities of some ascomycetes and red algae resulted from convergent evolution. As a result of an ultrastructural study on a member of the Laboulbeniales, Hill (1977) noted that many cellular details (including the presence of a Woronin body in the septal pore, the absence of a Golgi apparatus, and the formation of the ascus vesicle and entire mode of ascospore formation) are in very close agreement with the characteristics of the remainder of the Ascomycotina but are quite different from those of the red algae. He concludes that cellular structure is more conservative than thallus morphology and that the Laboulbeniales are not necessarily closely related to the red algae. Although we will never know whether the Ascomycotina evolved from primitive fungi or the red algae, it is generally agreed that the ascomycetes represent many separate lines of evolution or are polyphyletic (Atkinson, 1915; Korf, 1958; Luttrell, 1955).

References

Ainsworth, G. C. 1973. Introduction and keys to higher taxa. *In:* Ainsworth, G. C., F. K. Sparrow, and A. S. Sussman, Eds., The fungi—an advanced treatise. Academic Press, New York. **4B:**1-7.

Ajello, L., E. Varavsky, O. J. Ginther, and G. Bubash. 1964. The natural history of *Microsporum nanum. Mycologia* **56:**873-884.

Ashton, M. L., and P. B. Moens. 1979. Ultrastructure of sporulation in the Hemiascomycetes *Ascoidea corymbosa, A. rubescens, Cephaloascus fragrans,* and *Saccharomycopsis capsularis.* Can. J. Botany **57:**1259-1284.

Atkinson, G. F. 1915. Phylogeny and relationships in the Ascomycetes. Ann. Missouri Botan. Garden **2:**315-376.

Batra, L. R. 1978. Taxonomy and systematics of the Hemiascomycetes (Hemiascomycetidea). *In:* Subramanian, C. V., Ed., Taxonomy of fungi. Proc. Intern. Symp. Taxonomy Fungi. Univ. Madras. pp. 187-214.

Batra, S. W. T. 1973. Some Laboulbeniaceae (Ascomycetes) on insects from India and Indonesia. Am. J. Botany **50:**986-992.

Bellemere, A. 1977. L'appareil apical de l'asque chez quelques discomycetes: Étude ultrastructural comparative. Rev. Mycol. **41:**233-264.

Benjamin, C. R. 1955. Ascocarps of *Aspergillus* and *Penicillium. Mycologia* **47:**669-687.

Benjamin, R. K. 1965. A new genus of the Gymnoascaceae with a review of the other genera. *Aliso* **3:**310-328.

——. 1973. Laboulbeniomycetes. *In:* Ainsworth, G. C., F. K. Sparrow, and A. S. Sussman, Eds., The fungi—an advanced treatise. Academic Press, New York. **4A:**223-246.

——, and L. Shanor. 1950. The development of male and female individuals in the dioecious species *Laboulbenia formicarum* Thaxter. Am. J. Botany **37:**471-476 (Laboulbeniales).

——, and ——. 1952. Sex of host specificity and position of certain species of *Laboulbenia* on *Bembidion picipes.* Am. J. Botany **39:**125-131 (Laboulbeniales).

Black, S. H., and C. Gorman. 1971. The cytology of *Hansenula.* III. Nuclear segregation and envelopment during ascosporogenesis in *Hansenula wingei.* Arch. Mikrobiol. **79:**231-248 (Endomycetales).

Butler, G. M. 1966. Vegetative structures. *In:* Ainsworth, G. C., and A. S. Sussman, Eds., The fungi—an advanced treatise. Academic Press, New York. **2:**83-112.

Cain, R. F. 1950. Studies of coprophilous ascomycetes I. *Gelasinospora.* Can. J. Res. **C28:**566-576 (Sphaeriales).

Campbell, R., and M. Syrop. 1975. Light and electron microscope study of hysterothecium development in *Lophodermella sulcigena.* Trans. Brit. Mycol. Soc. **64:**209-214 (Phacidiales).

Carmo-Sousa, L. D. 1969. Distribution of yeasts in nature. *In:* Rose, A. H., and J. S. Harrison, Eds. The yeasts. Academic Press, New York. **1:**79-105.

Carroll, G. C. 1969. A study of the fine structure of ascosporogenesis in *Saccobolus kerverni.* Arch. Microbiol. **66:**321-339 (Pezizales).

Conti, S. F., and H. B. Naylor. 1959. Electron microscopy of ultrathin sections of *Schizosaccharomyces octosporus* I. Cell division. J. Bacteriol. **78:**868-877 (Endomycetales).

——, and ——. 1960. Electron microscopy of ultrathin sections of *Schizosaccharomyces octosporus* III. Ascosporogenesis, ascospore structure, and germination. J. Bacteriol. **79:**417-425 (Endomycetales).

Corlett, M. 1966-a. Perithecium development in *Leptosphaeria avenaria* f. sp. *avenaria.* Can. J. Botany **44:**1141-1149 (Dothideales).

——, 1966-b. Perithecium development in *Chaetomium trigonosporium.* Can. J. Botany **44:**155-162 (Sphaeriales).

——, 1970. Ascocarp development of two species of sooty molds. Can. J. Botany **48:**991-995 (Dothideales).

——, 1973. Observations and comments on the *Pleospora* centrum type. Nova Hedwigia **24:**347-360 (Dothideales).

——, and M. E. Elliott. 1974. The ascus apex of *Ciboria acerina.* Can. J. Botany **52:**1459-1463.

Corner, E. J. H. 1929, 1930. Studies in the morphology of Discomycetes. I to III. Trans. Brit. Mycol. Soc. **14:**263-275; **14:**275-291; **15:**107-120.

Dale, E. 1903. Observations on Gymnoasceae. Ann. Botany 17:571–596.

Darker, G. D. 1967. A revision of the genera of the Hypodermataceae. Can. J. Botany 45:1399–1444 (Phacidiales).

Denison, W. C., and G. C. Carroll. 1966. The primitive Ascomycete: a new look at an old problem. Mycologia 58:249–269.

Dennis, R. W. G. 1978. British Ascomycetes. J. Cramer, Lehre. 585 pp.

Dodge, B. O. 1914. The morphological relationships of the Florideae and the Ascomycetes. Bull. Torrey Botan. Club 41:157–202.

——, 1932. The non-sexual and the sexual functions of microconidia of *Neurospora*. Bull. Torrey Botan. Club 59:347–360 (Sphaeriales).

Doidge, E. M. 1942. A revision of the South African Microthyriaceae. Bothalia 4:273–420.

El-Ani, A. S. 1966. A new species of *Leptosphaeria*, an etiologic agent of mycetoma. Mycologia 58:406–411 (Dothideales).

Ellis, J. J. 1960. Plasmogamy and ascocarp development in *Gelasinospora calospora*. Mycologia 52:557–573 (Sphaeriales).

Emmons, C. W. 1935. The ascocarps in species of *Penicillium*. Mycologia 27:128–150 (Eurotiales).

Fennell, D. I. 1973. Plectomycetes; Eurotiales. *In:* Ainsworth, G. C., F. K. Sparrow, and A. S. Sussman, Eds., The fungi—an advanced treatise. Academic Press, New York. 4A:45–68.

Fergus, C. L., and R. M. Amelung. 1971. A new thermotolerant species of *Chaetomium* from mushroom compost. Mycologia 63:1212–1217 (Sphaeriales).

Fitzpatrick, H. M. 1923. Monograph of the Nitschkieae. Mycologia 15:45–67 (Coronophorales).

Funk, A., and R. A. Shoemaker. 1967. Layered structure in the bitunicate ascus. Can. J. Botany 45:1265–1266.

Gilkey, H. M. 1939. Tuberales of North America. Oregon State Monographs (Studies in Botany) 1:1–63.

——, 1961. New species and revisions in the order Tuberales. Mycologia 53:215–220.

Goos, R. D. 1959. Spermatium-trichogyne relationship in *Gelasinospora calospora* var. *autosteira*. Mycologia 51:416–428 (Sphaeriales).

——, 1974. A scanning electron microscope and *in vitro* study of *Meliola palmicola*. Proc. Iowa Acad. Sci. 81:23–27 (Meliolales).

Graff, P. W. 1932. The morphological and cytological development of *Meliola circinans*. Bull. Torrey Botan. Club 59:241–266 (Meliolales).

Griffiths, H. B. 1973. Fine structure of seven unitunicate Pyrenomycete asci. Trans. Brit. Mycol. Soc. 60:261–271.

Guilliermond, A. 1940. Sexuality, developmental cycle and phylogeny of yeasts. Botan. Rev. 6:1–24 (Endomycetales).

Hanlin, R. T. 1961. Studies in the genus *Nectria*. II. Morphology of N. *gliocladiodes*. Am. J. Botany 48:900–908 (Sphaeriales).

——. 1963. Morphology of *Neuronectria peziza*. Am. J. Botany 50:56–66 (Sphaeriales).

——. 1964. Morphology of *Hypomyces trichothecoides*. Am. J. Botany 51:201–208 (Sphaeriales).

Hansford, C. G. 1961. The Meliolineae—a monograph. Beih. Syd. Ann. Mycol. Ser. 2, 2:1–806 (Meliolales).

——, 1963. Iconographia Meliolinearum. Beih. Syd. Mycol. Ser. 2, vol. 5, plates I - CCLXXXV (Meliolales).

Hawker, L. E. 1955. Hypogeous fungi. Biol. Rev. 30:127–158.

Higgins, B. B. 1936. Morphology and life history of some Ascomycetes with special reference to the presence and function of spermatia. III. Am. J. Botany 23:598–602.

Hill, T. W. 1975. Ultrastructure of ascoporogenesis in *Nannizzia gypsea*. J. Bacteriol. 122:743–748 (Endomycetales).

——, 1977. Ascocarp ultrastructure of *Herpomyces* sp. (Laboulbeniales) and its phylogenetic implications. Can. J. Botany 55:2015–2032.

Honey, E. E. 1936. North American species of *Monilinia*. I. Occurrence, grouping, and life-histories. Am. J. Botany 23:100–106 (Helotiales).

Huang, L. H. 1976. Developmental morphology of *Triangularia backusii* (Sordariaceae). Can. J. Botany **54**:250–267 (Sphaeriales).

Hughes, S. J. 1976. Sooty moulds. Mycologia **68**:693–820 (Dothideales).

Kamat, M. N., and A. Pande-Chiplonkar. 1971. Patterns of spermatization in the Ascomycetes. Rev. Mycol. **36**:257–276.

Killian, K. 1917. Über die Sexualität von *Venturia inaequalis* (Cooke) Ad., Zeitschr. Botanie **9**:353–398.

Kimbrough, J. W. 1963. The development of *Pleochaeta polychaeta* (Erysiphaceae). Mycologia **55**:608–626.

——, 1970. Current trends in the classification of Discomycetes. Botan. Rev. **36**:91–161.

——, 1974. Structure and development of *Lasiobolus monacus,* a new species of Pezizales, Ascomycetes. Mycologia **66**:907–918.

——, and R. P. Korf. 1967. A synopsis of the genera and species of the tribe Theleboleae (= Pseudoascoboleae). Am. J. Botany **54**:9–23 (Pezizales).

Kohlmeyer, J. 1973. Spathulosporales, a new order and possible missing link between Laboulbeniales and Pyrenomycetes. Mycologia **65**:614–647.

——, 1975. New clues to the possible origin of Ascomycetes. Bioscience **25**:86–93.

Korf, R. P. 1958. Japanese Discomycete notes I–VIII. Sci. Rept. Yokohama Natl. Univ. II **7**:7–35 (Discomycete anatomy).

——, 1962. A synopsis of the Hemiphacidiaceae, a family of the Helotiales (Discomycetes) causing needle-blights of conifers. Mycologia **54**:12–33.

——, 1973. Discomycetes and Tuberales. *In:* Ainsworth, G. C., F. K. Sparrow, and A. S. Sussman, Eds., The fungi—an advanced treatise. Academic Press, New York. **4A**:249–319.

Kramer, C. L. 1960. Morphological development and nuclear behavior in the genus *Taphrina.* Mycologia **52**:295–320 (Taphrinales).

——. 1973. Protomycetales and Taphrinales. *In:* Ainsworth, G. C., F. K. Sparrow, and A. S. Sussman, Eds., The fungi—an advanced treatise. Academic Press, New York. **4A**:33–41.

Kreger-Van Rij, N. J. W. 1969. Taxonomy and systematics of yeasts. *In:* Rose, A. H., and J. S. Harrison, Eds., The yeasts. Academic Press, New York. **1**:5–78.

——. 1973. Endomycetales, Basidiomycetous yeasts, and related fungi. *In:* Ainsworth, G. C., F. K. Sparrow, and A. S. Sussman. Eds., The fungi—an advanced treatise. Academic Press, New York. **4A**:11–32.

Kuehn, H. H. 1957. Observations on the Gymnoasceae. V. Developmental morphology of two species representing a new genus of the Gymnoasceae. Mycologia **49**:694–706 (Eurotiales).

——, and G. F. Orr. 1959. Observations on Gymnoasceae. VI. A new species of *Arachniotus.* Mycologia **51**:864–870 (Eurotiales).

Kwon-Chung, K. J. 1973. Studies on *Emmonsiella capsulata* I. Heterothallism and development of the ascocarp. Mycologia **65**:109–121 (Eurotiales).

Lodder, J., and N. J. W. Kreger-Van Rij. 1952. The yeasts—a taxonomic study. John Wiley & Sons, Inc. New York. 713 pp.

Luttrell, E. S. 1951. Taxonomy of the Pyrenomycetes. Univ. Missouri Studies **24**, no. 3:1–120.

——, 1955. The ascostromatic Ascomycetes. Mycologia **47**:511–532.

——. 1960. The morphology of an undescribed species of *Dothiora.* Mycologia **52**:64–79 (Dothideales).

——. 1964. Morphology of *Trichometasphaeria turica.* Am. J. Botany **51**:213–219 (Dothideales).

——. 1965. Paraphysoids, pseudoparaphyses, and apical paraphyses. Trans. Brit. Mycol. Soc. **48**:135–144.

——. 1973. Loculoascomycetes. *In:* Ainsworth, G. C., F. K. Sparrow, and A. S. Sussman, Eds., The fungi—an advanced treatise. Academic Press, New York. **4A**:135–219.

——, 1975. Centrum development in *Didymosphaeria sadasivanii* (Pleosporales). Am. J. Botany **62**:186–190 (Dothideales).

Mai, S. H. 1977. Morphological studies in *Sordaria fimicola* and *Gelasinospora longispora.* Am. J. Botany **64**:489–495 (Sphaeriales).

Malloch, D., and G. L. Benny. 1973. California Ascomycetes: four new species and a new record. Mycologia **65**:648–660 (Eurotiales).

——, and R. F. Cain. 1972. The Trichocomataceae: Ascomycetes with *Aspergillus, Paecilomyces,* and *Penicillium* imperfect states. Can. J. Botany **50**:2613–2628 (Eurotiales).

——, and ——. 1973. The genus *Thielavia.* Mycologia **65**:1055–1077.

Maniotis, J. 1965. A cleistothecial mutant of the perithecial fungus *Gelasinospora calospora*. Mycologia **57**:23–35 (Sphaeriales).

Matile, P., H. Moor, and C. F. Robinow. 1969. Yeast cytology. *In:* Rose, A. H., and J. S. Harrison, Eds., The yeasts. Academic Press, New York. **1**:219–302.

McClary, D. O. 1964. The cytology of yeasts. Botan. Rev. **30**:168–225 (Endomycetales).

McGahen, J. W., and H. E. Wheeler. 1951. Genetics of *Glomerella*. IX. Perithecial development and plasmogamy. Am. J. Botany **38**:610–617 (Sphaeriales).

Merkus, E. 1976. Ultrastructure of the ascospore wall in Pezizales (Ascomycetes)—IV. Morchellaceae, Helvellaceae, Rhizinaceae, Thelebolaceae, and Sarcoscyphaceae. General discussion. Persoonia **9**:1–38.

Mix, A. J. 1935. The life history of *Taphrina deformans*. Phytopathology **25**:41–66 (Taphrinales).

Moens, P. B. 1971. Fine structure of ascospore development in the yeast *Saccharomyces cerevisiae*. Can. J. Microbiol. **17**:507–510 (Endomycetales).

Moore, E. J. 1963. The ontogeny of the apothecia of *Pyronema domesticum*. Am. J. Botany **50**:37–44 (Pezizales).

——. 1965. Ontogeny of gelatinous fungi. Mycologia **57**:114–130 (Helotiales).

Moore-Landecker, E. 1975. A new pattern of reproduction in *Pyronema domesticum*. Mycologia **67**:1119–1127 (Pezizales).

Mrak, E. M., and H. J. Phaff. 1948. Yeasts. Ann. Rev. Microbiol. **2**:1–46 (Endomycetales).

Müller, E., and J. A. Von Arx. 1973. Pyrenomycetes: Meliolales, Coronophorales, Sphaeriales. *In:* Ainsworth, G. C., F. K. Sparrow, and A. S. Sussman, Eds., The fungi—an advanced treatise. Academic Press, New York. **4A**:87–132.

Nannfeldt, J. A. 1976. Iodine reactions in ascus plugs and their taxonomic significance. Trans. Brit. Mycol. Soc. **67**:283–287.

Nelson, A. C., and M. P. Backus. 1968. Ascocarp development in two homothallic *Neurosporas*. Mycologia **60**:16–28 (Sphaeriales).

O'Donnell, K. L., and G. R. Hooper. 1974. Scanning ultrastructural ontogeny of paragymnohymenial apothecia in the operculate Discomycete *Peziza quelepidotia*. Can. J. Botany **52**:873–876. (Pezizales).

Orr, G. F., and H. H. Kuehn. 1964. A re-evaluation of *Myxotrichum spinosum* and *M. cancellatum*. Mycologia **56**:473–481 (Eurotiales).

Pfister, D. J. 1979. A monograph of the genus *Wynnea* (Pezizales, Sarcoscyphaceae). Mycologia **71**:144–159.

Raper, K. B., and D. I. Fennell. 1965. The genus *Aspergillus*. Williams and Wilkins. Baltimore. 686 pp. (Eurotiales).

Reeves, F., Jr. 1967. The fine structure of ascospore formation in *Pyronema domesticum*. Mycologia **59**:1018–1033 (Pezizales).

——, 1971. The structure of the ascus apex in *Sordaria fimicola*. Mycologia **63**:204–212 (Sphaeriales).

Reynolds, D. R. 1975. The centrum of the sooty mold ascomycete *Limacinula samoensis*. Am. J. Botany **62**:775–779 (Dothideales).

Rogerson, C. T. 1970. The hypocrealean fungi (Ascomycetes, Hypocreales). Mycologia **62**:865–910 (Sphaeriales).

Rosinski, M. A. 1956. Development of the ascocarp of *Anthracobia melaloma*. Mycologia **48**:506–533 (Pezizales).

——, 1961. Development of the ascocarp of *Ceratocystis ulmi*. Am. J. Botany **48**:285–293 (Sphaeriales).

Samuels, G. J. 1973. Perithecial development in *Hypomyces aurantius*. Am. J. Botany **60**:268–276 (Sphaeriales).

Samuelson, D. A. 1975. The apical apparatus of the suboperculate ascus. Can. J. Botany **53**:2660–2679.

——, 1978. Asci of the Pezizales. VI. The apical apparatus of *Morchella esculenta, Helvella crispa*, and *Rhizina undulata*. General discussion. Can. J. Botany **56**:3069–3082 (Pezizales).

Sánchez, A., and R. P. Korf. 1966. The genus *Vibrissea*, and the generic names *Leptosporium, Apostemium, Apostemidium, Gorgoniceps* and *Ophiogloea*. Mycologia **58**:722–737 (Ostropales).

Schoknecht, J. D. 1975. Structure of the ascus apex and ascospore dispersal mechanisms in *Sclerotinia tuberosa*. Trans. Brit. Mycol. Soc. **64**:358–362 (Helotiales).

Shanor, L. 1952. The characteristics and morphology of a new genus of the Laboulbeniales on an earwig. Am. J. Botany **39**:498–504.

Shoemaker, R. A. 1955. Biology, cytology, and taxonomy of *Cochliobolus sativus*. Can. J. Botany **33:**562-576 (Pleosporales).

Silverberg, B. A., and J. F. Morgan-Jones. 1974. *Lophodermium pinastri:* certain fine-structural features of ascocarp morphology. Can. J. Botany **52:**1993-1995 (Phacidiales).

Staley, J. M., and H. H. Bynum. 1972. A new *Lophodermella* on *Pinus ponderosa* and *P. attenuata*. Mycologia **64:**722-726 (Phacidiales).

Stevens, F. L. 1927, 1928. The Meliolineae. I and II. Ann. Mycol. **25:**405-469; Ann. Mycol. **27:**165-383 (Meliolales).

——, and M. H. Ryan. 1939. The Microthyriaceae. University Ill. Biol. Monograph **17** no. 2:1-138.

Stiers, D. L. 1974. Fine structure of ascospore formation in *Poronia punctata*. Can. J. Botany **52:**999-1003 (Sphaeriales).

——, 1976. The fine structure of ascospore formation in *Ceratocystis fimbriata*. Can. J. Botany **54:**1714-1723 (Sphaeriales).

——, 1977. The fine structure of the ascus apex in *Hypoxylon serpens, Poronia punctata, Rosellinia aquila,* and *R. mammiformis*. Cytologia **42:**697-702 (Sphaeriales).

Syrop, M., and A. Beckett. 1976. Leaf curl disease of almond caused by *Taphrina deformans*. III. Ultrastructural cytology of the pathogen. Can. J. Botany **54:**293-305 (Taphrinales).

Tavares, I. I. 1966. Structure and development of *Herpomyces stylopygae* (laboulbeniales). Am. J. Botany **53:**311-318.

Thaxter, R. 1896-1931. Contributions towards a monograph of the Laboulbeniaceae. I-V. Mem. Am. Acad. Arts Sci. **12:**187-429; **13:**217-469; **14:**309-426; **15:**427-580; **16:**1-435 (Laboulbeniales).

Thyr, B. D., and C. G. Shaw. 1966. Ontogeny of the needle cast fungus, *Hypodermella arcuata*. Mycologia **58:** 192-200 (Phacidiales).

Trappe, J. M. 1971. A synopsis of the Carbomycetaceae and Terfeziaceae (Tuberales). Trans. Brit. Mycol. Soc. **57:**85-92.

Tyson, K., and D. A. Griffiths. 1976. Developmental morphology and fine structure of *Placoasterella baileyi*. II. Ascus and ascospore development. Trans. Brit. Mycol. Soc. **66:**263-279 (Microthyriales).

Uecker, F. A. 1976. Development and cytology of *Sordaria humana*. Mycologia **68:**30-46 (Sphaeriales).

——, and J. M. Staley. 1973. Development of the ascocarp and cytology of *Lophodermella morbida*. Mycologia **65:**1015-1027 (Phacidiales).

Van Brummelen, J. 1967. A world-monograph of the genera *Ascobolus* and *Saccobolus*. Persoonia Suppl. **1:**1-260 (Pezizales).

——, 1975. Light and electron microscopic studies of the ascus top in *Sarcoscypha coccinea*. Persoonia **8:**259-271 (Pezizales).

——, 1978. The operculate ascus and allied forms. Persoonia **10:**113-128.

Von Arx, J. A. 1975. On *Thielavia* and some similar genera of Ascomycetes. Studies Mycol. **8:**1-29 (Sphaeriales).

——, L. R. DeMiranda, M. T. Smith, and D. Yarrow. 1977. The genera of the yeasts and yeast-like fungi. Studies Mycol. **14:**1-42 (Endomycetales).

——, and E. Müller. 1975. A re-evaluation of the bitunicate Ascomycetes with keys to families and genera. Studies Mycol. **9:**1-159.

Wells, K. 1972. Light and electron microscope studies of *Ascobolus stercorarius*. II. Ascus and ascospore ontogeny. Univ. Calif. Publ. Botany **62:**1-33 (Pezizales).

Whisler, H. C. 1968. Experimental studies with a new species of *Stigmatomyces* (Laboulbeniales). Mycologia **60:**65-75.

Whiteside, W. C. 1961, 1962. Morphological studies in the Chaetomiaceae. I, II. Mycologia **53:**512-523; Mycologia **54:**152-159 (Sphaeriales).

Wickerham, L. J. 1952. Recent advances in the taxonomy of yeasts. Ann. Rev. Microbiol. **6:**317-332 (Endomycetales).

Woo, J. Y., and A. D. Partridge. 1969. The life history and cytology of *Rhytisma punctatum* on bigleaf maple. Mycologia **61:**1085-1095 (Phacidiales).

Yarwood, C. E. 1957. Powdery mildews. Botan. Rev. **23:**235-300 (Erysiphales).

——, 1973. Pyrenomycetes: Erysiphales. *In:* Ainsworth, G. C., F. K. Sparrow, and A. S. Sussman, Eds., The fungi—an advanced treatise. Academic Press, New York. **4A:**71-86.

FIVE

Division Amastigomycota— Subdivision Basidiomycotina

The Subdivision Basidiomycotina contains about 15,000 known species. We may refer to these fungi informally as the "basidiomycetes," a term which is commonly used but without taxonomic status. Fungi such as the jelly fungi, rusts, smuts, bracket fungi, mushrooms, and puffballs belong to this group. Many of the common names refer to the visible part of the fungus, which is a conspicuous reproductive body, the *basidiocarp*. The basidiocarp is supported by an extensive assimilative mycelium penetrating the soil or plant material and thus deriving the nutrients that supply the basidiocarp.

THE BASIDIUM AND SEXUAL REPRODUCTION

All members of the Basidiomycotina produce *basidia* (singular, *basidium*). Basidia are organs homologous to and presumably derived from the ascus. Like the typical ascus, the basidium usually originates as a binucleate structure and is the site of karyogamy and meiosis. Unlike the ascus, the basidium bears its basidiospores on the outside of the basidium.

In order to understand the role of the basidium in the Basidiomycotina life cycle, we now explore a typical life cycle, which is characterized by an extremely short diploid stage, occurring in the basidium only, and a prolonged dikaryotic stage. With the exception of the Uredinales and Ustilaginales, almost all basidiomycetes have virtually the same life cycle, which follows (Fig. 5-1). Some variations are described on pages 192-193.

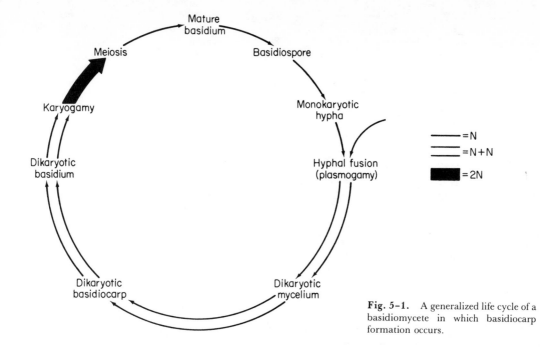

Fig. 5–1. A generalized life cycle of a basidiomycete in which basidiocarp formation occurs.

Labels in the figure:
- Mature basidium
- Meiosis
- Basidiospore
- Karyogamy
- Monokaryotic hypha
- Dikaryotic basidium
- Hyphal fusion (plasmogamy)
- Dikaryotic basidiocarp
- Dikaryotic mycelium

$——— = N$
$=== = N + N$
$▬▬▬ = 2N$

The basidiospore usually contains a single haploid nucleus. When the basidiospore germinates, it initiates a haploid monokaryotic mycelium. The monokaryon is at first nonseptate but later becomes divided into a number of uninucleate cells. This phase of the life cycle is usually of short duration.

Plasmogamy then takes place. Sexual organs are lacking in the basidiomycetes (except in the Uredinales), and plasmogamy is usually accomplished by the fusion of two monokaryotic hyphae. When the monokaryotic hyphae fuse, the nuclei of one hypha flow into the other. Usually this exchange of nuclei is reciprocal between the fusing hyphae. Invading nuclei may travel many millimeters or centimeters before stopping. The hyphae now have nuclei of two genetic types and are dikaryotic. Each cell of this dikaryon has two haploid nuclei $(N + N)$.

The dikaryon formed after plasmogamy continues to proliferate, meanwhile maintaining the binucleate, dikaryotic condition. Usually *clamp connections* are formed to maintain the binucleate condition (Fig. 5–2). The young clamp begins to form as a bulging pocket from the hyphal wall between the two nuclei (a and b) of the terminal dikaryotic cell. Both nuclei divide mitotically. The division figure of the terminal nucleus a lies obliquely along the main axis of the clamp and, at the completion of its division, one daughter nucleus a'' remains in the clamp, while the second daughter nucleus a' remains in a terminal position. The division figure of the second nucleus, b, lies along the main axis of the hypha, and its one daughter nucleus b'' remains in a posterior position, while the second daughter nucleus b' is in a terminal position near a'. A transverse septum is laid down and separates the new terminal binucleate cell $(a'b')$. The cell remaining behind still has only one nucleus (b'') and the other nucleus (a'') is still isolated in the lateral clamp. The clamp recurves and comes into contact with the uninucleate cell, and the walls between are dissolved. The nucleus from the clamp migrates into the hyphal cell, reconstituting the dikaryotic condition ($a''b''$).

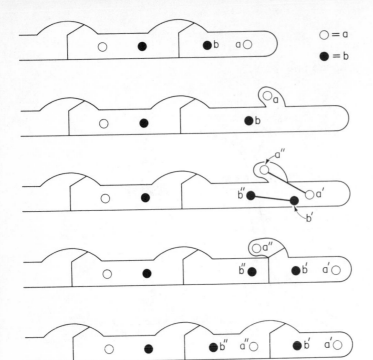

○ = a
● = b

Fig. 5-2. (left) Steps in the formation of a clamp connection. See text for the explanation.

Fig. 5-3. (below) Clamp connection of *Coprinus cinereus*. Note the continuity between the clamp and the main body of the hypha, and also the two dolipore septa. × 25,100. [Courtesy D. J. McLaughlin].

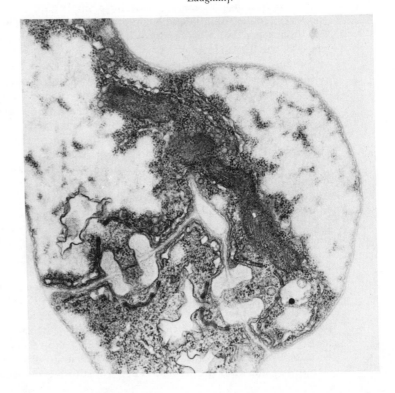

The clamp remains as a permanent part of the hypha, and it is often used as morphological evidence that the hypha is dikaryotic (Figs. 5-2, 5-3).

The dikaryotic mycelium may serve as the assimilative mycelium which is concealed deep within the substratum. When conditions are favorable for reproduction, some of the dikaryotic mycelium undergoes a complex morphogenesis to form a basidiocarp (such as the visible mushroom). All the mycelium within the basidiocarp is of the same type, in contrast to some ascomycetes in which two distinct mycelial systems take part in the formation of an ascocarp. Within the basidiocarp, some of the cells in sterile tissues may remain binucleate while others become multinucleate, depending upon the species.

Some of the cells in a specialized fertile region of the basidiocarp are the young basidia (Fig. 5-4). The basidia are at first dikaryotic and have two haploid nuclei (Fig. 5-5). The nuclei fuse to form a diploid nucleus, which then undergoes meiosis. Usually four uninucleate basidiospores are produced on each basidium. Although basidiospores may be borne directly upon the surface of the basidium, they are usually produced upon elongate projections, the *sterigmata*. Each sterigma bears a basidiospore at its apex. If a basidium is to form its basidiospores upon sterigmata (the usual situation), the sterigmata originate as projections from the apex of the basidium. The wall at the tip of the sterigma is at first thin and swells into a more or less globose apex; the swelling then elongates and enlarges asymmetrically. Meanwhile additional wall layers are produced around the basidiospore initial and then gradually thicken. Cytoplasmic

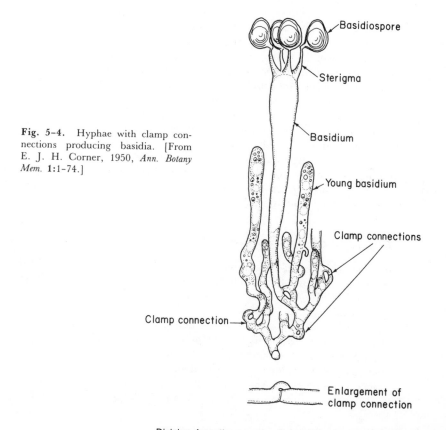

Fig. 5-4. Hyphae with clamp connections producing basidia. [From E. J. H. Corner, 1950, *Ann. Botany Mem.* **1**:1-74.]

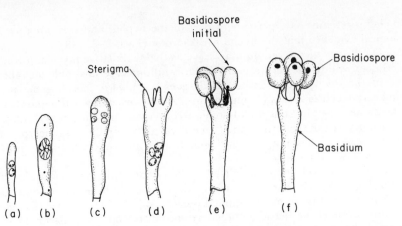

Fig. 5-5. Development of a typical basidium; (a) Young dikaryotic basidium; (b) Diploid basidium; (c) Basidium with four nuclei resulting from meiosis; (d) Basidium after development of sterigmata; (c) Migration of nuclei into basidiospore initials; (d) Mature basidium with basidiospores. [Adapted from A. H. Smith, 1934, *Mycologia* **26**:305-331.]

changes accompany maturation of the basidiospore initial: these may include the presence of many vesicles during the early stages and the appearance of mitochondria and other organelles during the later stages. A nucleus migrates into each basidiospore initial. In *Schizophyllum commune,* the nuclei and cytoplasm are apparently pushed through the sterigmata by the increasing vacuolation of the basidium (Wells, 1965), while in *Coprinus cinereus,* cytoplasmic microtubules may be responsible for nuclear migration as well as for the migration of other organelles (McLaughlin, 1977). Finally the cytoplasm within the basidiospore is separated from that of the sterigma by the formation of a plug at their juncture. At that point of juncture is the *hilar appendix* which consists primarily of a kneelike bulge enclosed by a wall continuous with that of the sterigma and containing an electron-light zone. The bulging is initiated after the nucleus migrates into the basidiospore initial, and the hilar appendix becomes increasingly distended while the basidiospores are maturing. The hilar appendix is extremely important in effecting basidiospore discharge and is discussed further in Chapter 10. In some species, a specialized electron-dense body is associated with the inner wall of the hilar appendix in its early developmental stages and may possibly contribute to its thickening (McLaughlin, 1977).

Variations of the General Pattern

The vast majority of the members of the Basidiomycotina have the characteristics outlined in the general life cycle on pages 188–192. There are, however, some that do not comply with one or more aspects of the general pattern. We will examine some of these deviations.

Most basidiomycete species are heterothallic, requiring fusion of two genetically different but compatible hyphae to produce a fertile cross resulting in a basidiocarp (heterothallism is discussed further in Chapter 9). Some basidiomycete species are homothallic and do not require plasmogamy between genetically different hyphae to produce a basidiocarp. An example is the edible straw mushroom, *Volvariella volvacea,* considered a delicacy in China. Although hyphal fusion may occur between monokaryotic hyphae and nuclear migration may occur, the nuclei are randomly distributed in the multinucleate mycelium and there is neither a tendency for the nuclei to occur in pairs nor clamp connections. Both karyogamy and meiosis occur in the basidium (Chang and Ling, 1970).

Clamp connections may also be lacking in heterothallic basidiomycete species. This absence of clamp connections may be sporadic in some species but occurs regularly in others. Some species that normally form clamps may become clampless if there is inadequate aeration. Unlike the multinucleate cells in the homothallic species, the cells of the mycelium are typically binucleate and dikaryotic. The dividing nuclei are found in the terminal cell of the hypha, which may be either binucleate or multinucleate. As in those cells forming a clamp connection, paired nuclei undergo simultaneous divisions. In this case, the spindles are parallel to each other, thus maintaining the close association of nuclear pairs (Kühner, 1977).

In heterothallic species, fusion of the primary mycelium to form the dikaryotic state may occur at various points in the life cycle. Fusion usually occurs during mycelial growth, but in some species it occurs at the time that the basidiocarp is initiated, or rarely, even in the subhymenium or hymenium. Binucleate, dikaryotic basidiospores may be regularly produced by some species. When these dikaryotic basidiospores germinate, they give rise directly to a dikaryotic mycelium that does not need to undergo plasmogamy to produce a basidiocarp.

The cells of the sterile tissues of basidiocarps may maintain the regularly binucleate condition or may become multinucleate, depending upon the species.

Although the vast majority of the species have only a meiotic division in the basidium and subsequently produce four uninucleate basidiospores, there are deviations from this pattern. In some species, a mitotic division will follow meiosis, resulting in an eight-nucleate basidium; or alternatively, an additional nuclear division occurs in the sterigma or young basidiospore. Other possibilities are that some of the nuclei within the basidium may not be incorporated into basidiospores but instead degenerate, or that more than one nucleus will be incorporated in a basidiospore. Various combinations of these possible nuclear events sometimes result in basidia that form up to eight basidia (e.g., two, three, or five) or have basidiospores that are binucleate or contain up to eight nuclei (Kühner, 1977).

Variations in Basidia

We have just seen that two important events take place in the basidium: these are karyogamy and meiosis. There is some variation in both basidial structure (discussed further on pages 210-211) and the site at which karyogamy and meiosis occur. We may describe that cell in which karyogamy occurs as the *probasidium* and that cell in which meiosis occurs as the *metabasidium* (Talbot, 1973-a)*

The most common situation is that in which both karyogamy and meiosis occur in essentially the same location, and are separated only by time. This is the type used to illustrate the general life cycle on pages 188-192. Functionally, the young basidium at first is the probasidium and is the site of karyogamy, but it is entirely transformed into the metabasidium when meiosis occurs. Morphologically, there are no remnants of the probasidial stage. At maturity, the basidium consists of a metabasidium, possibly with sterigmata (Figs. 5-4, 5-5, and 5-6).

---

* The terms probasidium and metabasidium may be used differently in other mycological works. See Talbot (1973-a) for clarification if required.

In the second type, karyogamy and meiosis are separated both in time and in space. In its earliest stages, the basidium consists entirely of a probasidium, which may be a thick-walled resting spore. Karyogamy occurs, and the probasidium produces a thin-walled tubular extension into which the diploid nucleus migrates. The thin-walled extension, together with any remnants of the probasidium, functions as a single cell, the metabasidium. Meiosis occurs within the metabasidium. In some species, the metabasidium becomes divided by transverse septa after meiosis. The mature basidium consists of the metabasidium (possibly including remnants of the probasidium), and sterigmata, if formed (Fig. 5–23).

(a)

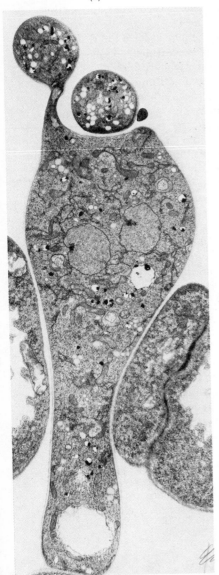

Fig. 5–6. Basidia with basidiospores. (a) Electronmicrograph of a longitudinal section of a basidium at interphase II of meiosis (two of the four nuclei are visible). One of the four basidiospores is attached to its sterigma while a second basidiospore lies above the basidium. × 5000; (b) Scanning electronmicrograph of a hymenium. Three of the four basidiospores are present on the basidium in the center. × 7,600. [(a) Courtesy D. J. McLaughlin; (b) Courtesy K. S. Yoon and D. J. McLaughlin.]

(b)

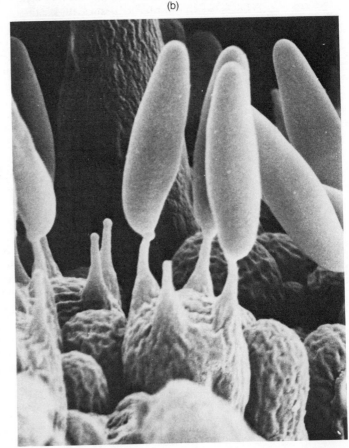

NONSEXUAL REPRODUCTION

Nonsexual reproduction occurs commonly in members of the Basidiomycotina, although much less frequently than in members of the Division Mastigomycota or the Subdivisions Zygomycotina and Ascomycotina. Nonsexual reproductive structures that are formed include buds (similar to those produced by yeasts), conidia, chlamydospores, and oidia.

Oidia commonly originate from a monokaryotic mycelium when a branch fragments into its component cells. The haploid oidia then germinate, yielding a haploid mycelium. Such oidial mycelia may unite with each other to form a dikaryotic mycelium, or an oidial hypha may unite with a haploid mycelium derived from a germinated basidiospore (Fig. 5-7). The most important role of oidia is apparently to increase the possibility that mating will take place. Nonsexual structures may be produced directly from the basidium, from the basidiospores, from the mycelium, and among the basidia in a hymenium. In a few cases, a separate sporocarp bearing the nonsexual spores may be produced.

Fig. 5-7. Nonsexual reproduction by means of oidia in *Coprinus lagopus:* (a) Modified hypha bearing oidia in a drop of liquid; (b) The hypha and oidia in (a) as it would appear if immersed in a film of water; (c) Fusion of oidial germ tubes of the oidia *a* and *c* of one mating type with the hyphae *b* and *d* of a mycelium derived from a basidiospore of the opposite mating type. [From H. J. Brodie, 1931, *Ann. Botany London* **45**:315–344.]

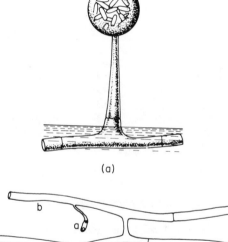

(a)

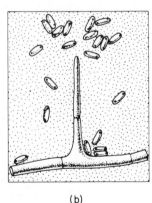

(b)

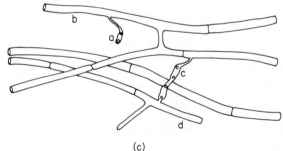

(c)

CLASSIFICATION

We shall divide the Subdivision Basidiomycotina into three classes, the Teliomycetes, Hymenomycetes, and Gasteromycetes (Ainsworth, 1973). The members of the Class Teliomycetes do not form a basidiocarp but produce resting spores that function as probasidia. Neither characteristic is found in the remaining two classes, as the majority of members produce a well-developed basidiocarp and the basidium does not originate from a resting spore. Members of the Class Hymenomycetes produce an exposed hymenium, while in the Class Gasteromycetes a hymenium is usually lacking but if present is formed in an enclosed basidiocarp. A further outline of the classification of this group follows:

Subdivision Basidiomycotina

Class Teliomycetes
    Order Uredinales
    Order Ustilaginales
Class Hymenomycetes
    Subclass Phragmobasidiomycetidae
        Order Tremellales
        Order Auriculariales
        Order Septobasidiales
    Subclass Holobasidiomycetidae
        Order Exobasidiales
        Order Brachybasidiales
        Order Dacrymycetales
        Order Tulasnellales
        Order Aphyllophorales
        Order Agaricales
Class Gasteromycetes
    Order Hymenogastrales
    Order Melanogastrales
    Order Gautieriales
    Order Phallales
    Order Tulostomatales
    Order Lycoperdales
    Order Sclerodermatales
    Order Nidulariales

Class Teliomycetes

Members of the Teliomycetes are plant parasites that do not form a basidiocarp. The probasidia are thick-walled resting spores that often overwinter before producing metabasidia. The resting spores are formed in a loose cluster, the *sorus*.

The Teliomycetes may be divided into two orders, the Uredinales and Ustilaginales (Durán, 1973; Laundon, 1973).

ORDER UREDINALES

The Uredinales are the ''rusts,'' a group of almost 5,000 specialized parasites occurring on ferns, gymnosperms, and angiosperms. The rust fungi are host-specific, and a given species or genetic strain of the fungus is known to occur only on certain genera, species, or genetic strains of the the host (for example, *Puccinia graminis* var. *tritici* occurs on wheat while *Puccinia graminis* var. *secalis* occurs on rye). These fungi are known to occur only as parasites in nature although they have been recently cultured in the laboratory (Laundon, 1973; Petersen, 1974). Rust diseases result in widespread physiological damage to the host, causing an increase in transpiration rate and a decrease in photosynthesis and respiration rates. Sometimes abnormal enlargements

(a)

Fig. 5-8. Sori in the Uredinales. (a) The uredinial stage of *Puccinia sorghi* on maize. Note that the mass of urediniospores in each sorus causes the host epidermis to rupture; (b) Scanning electronmicrograph of the telial stage of *Puccinia podophyllia.* The teliospores have ruptured the host epidermis. × 425. [Courtesy Plant Pathology Department, Cornell University; (b) Courtesy K. L. O'Donnell.]

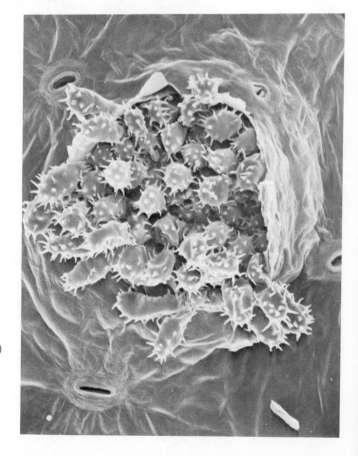

(b)

are induced to form. Rust diseases have been known since ancient times when it was thought that they were inflicted by a revengeful deity (see Chapter 12).

The mycelium occurs between cells of the host plant and penetrates them by forming haustoria. Clamp connections are not formed. The spores are usually borne in a loose cluster, the sorus, which often becomes exposed by rupture of the host epidermis (Fig. 5–8). Some sori formed by the rust fungi are orange or red giving the characteristic rustlike appearance to the diseased areas.

Although morphologically simple, some rusts have the most complex life cycle found in the fungi, in addition to possessing a high degree of host specificity. As many as five separate spore stages may be formed in succession in the life cycle of a rust. These spore stages may be defined according to their role in the life cycle (Hiratsuka, 1973). The spore stages are described below.

Stage 0. Pycnium bearing pycniospores. Pycnia are monokaryotic, haploid structures that produce *pycniospores.* The pycniospores are haploid, uninucleate structures that function as spermatia when transferred to another pycnium. The pycnia are hermaphroditic structures, containing the female receptive structures as well as the male pycniospores. The dikaryotic state is most often established by the transfer of pycniospores from one pycnium to the receptive hyphae in a second pycnium.

The pycnia vary in shape and are frequently flask-shaped but may also be flat or globose. They occur beneath or within the epidermis, beneath the cuticle, or within the cortex of the host plant (Fig. 5–9).

Stage I. Aecium bearing aeciospores. The *aecia* are usually associated with a pycnium that has functioned in plasmogamy to establish the dikaryotic state, and they usually bear the first dikaryotic spores, the *aeciospores,* in the life cycle. Upon germination, the aeciospores establish a dikaryotic mycelium.

A typical aecium occurs as an open cup, bearing the binucleate aeciospores in chains. The cupulate aecium is surrounded by a peridium, which encloses the aecium during its early stages and separates it from the host tissues in which the aecium is embedded. The peridium later ruptures and curves backward to form an irregular collar. The aeciospores are yellow to orange. A second common aecial form is that in which an elongated hornlike protrusion is formed and is surrounded by a peridium that opens by tearing into shreds (Fig. 5–10).

Stage II. Uredinium bearing urediniospores. The *urediniospores* are also binucleate and dikaryotic, but unlike the aeciospores, they represent the "repeating" stage. This means that a urediniospore may germinate, forming a dikaryotic mycelium that again gives rise to uredinia. The urediniospores are actually equivalent to an anamorph, as they are responsible for nonsexual reproduction.

Typically, the *uredinium* is a sorus that erupts through the host epidermis and contains one-celled urediniospores on short stalks. Germ pores are almost always visible. The urediniospore very often has spines protruding from its middle layer and penetrating the thin outer wall layer. The uredinia sometimes have a peridium surrounding the edge or contain scattered sterile hyphal tips among the urediniospores (Fig. 5–11).

Stage III. Telium bearing teliospores. The *telium* is a sorus bearing *teliospores.* The teliospores function as probasidia and are the site of karyogamy; they subsequently produce a metabasidium within which meiosis occurs. Ultimately the teliospores bear basidiospores. Teliospores are usually dark, thick-walled spores with two or more cells. They may be sessile or borne on a short stalk. The

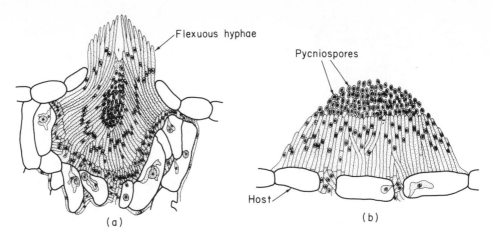

Fig. 5–9. Pycnia in the Uredinales. (a) Flask-shaped pycnium of *Puccinia graminis.* × 510: (b) Cushionlike superficial pycnium of *Puccinia fusca.* × 850. [From P. Sappin-Trouffy, 1896, *Botaniste* **5**:59–244.]

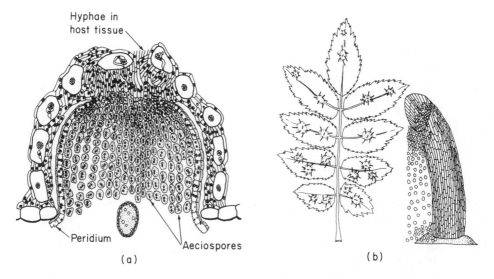

Fig. 5–10. Aecia in the Uredinales. (a) Aecium of *Puccinia graminis* showing typical cupulate form. × 510. An enlarged aeciospore is also shown; (b) A hornlike aecium of *Gymnosporangium juniperinum* (× 80) and its appearance on the host plant. [From P. Sappin-Trouffy, 1896, *Botaniste* **5**:59–244.]

teliospores usually occur in a sorus embedded in the host mesophyll or in the epidermal region. In some species, the teliospores occur on gelatinous strands or hornlike structures (the telial column) that extend from the host (Fig. 5–12). Other morphological forms of both teliospores and telia occur.

Stage IV. Basidiospores. The haploid basidiospores are produced on the metabasidium and are capable of reinfecting a host. Because this reinfection usually results in the formation of pycnia, the life cycle is initiated again.

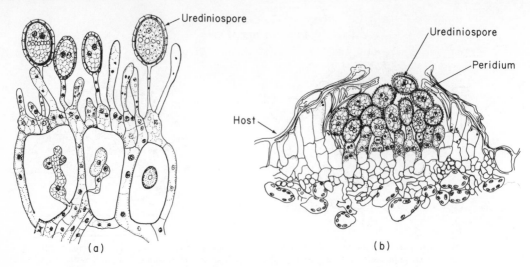

(a)

(b)

Fig. 5-11. Uredinia in the Uredinales. (a) A portion of an uredinium and urediniospores of *Puccinia graminis.* × 700; (b) A uredinium of *Cronartium ribicola* that has a peridium and is buried within the host tissue. × 500 [(a) from P. Sappin-Trouffy, 1896, *Botaniste* 5:59–244. (b) from R. H. Colley, 1918, *J. Agr. Res.* **15**:619–659.]

Fig. 5-12. Telia in the Uredinales. (a) Telium and teliospores of *Puccinia graminis.* × 450; (b) Telial horn formed by *Cronartium ribicola.* × 170. [(a) from P. Sappin-Trouffy, 1896, *Botaniste* 5:59–244; (b) from R. H. Colley, 1918, *J. Agr. Res.* **15**:619–659.]

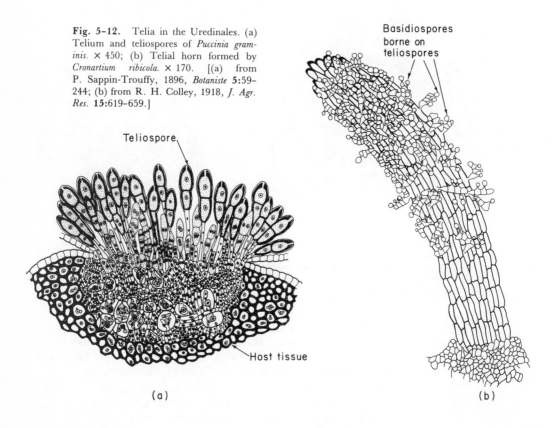

(a)

(b)

Some rusts must have all the above five spore stages in their life cycle, occurring in the order presented. Such a rust is termed *long-cycled,* in contrast to those that are *short-cycled* and have only some of the spore stages (see page 203). In addition, rusts differ in whether they require one or two hosts to complete their life cycle. An *autoecious* rust completes its entire life cycle on a single host plant. A *heteroecious* rust requires two hosts to complete its life cycle. The haploid phase (stages 0 and I) occur on one host species, while the dikaryotic and diploid stages (stages II, III, and IV) occur on a second host species. The first host is often termed the *alternate* host because it is not the host on which basidiospores are formed. An example of a heteroecious, long-cycled rust is *Cronartium ribicola,* the cause of white pine blister rust, which produces its pycnia and aecia on white pines (the alternate host) and its uredinia and telia on species of *Ribes* such as currants [Figs. 5–11(b), 5–12(b)]. A second example of a heteroecious, long-cycled rust is *Puccinia graminis,* which we discuss below.

Puccinia graminis. To clarify the life cycle of a typical long-cycled, heteroecious rust, we examine the life cycle of *Puccinia graminis,* the causal agent of the black stem rust of wheat and other cereals (Fig. 5–13).

In the spring, infection of the first host, the barberry, occurs when a basidiospore lands on the surface of the leaf. The basidiospore germinates, and a monokaryotic hypha invades the leaf. A flask-shaped pycnium which is immersed in the surface leaf tissue is formed. The pycnium has paraphyses and nonseptate flexuous hyphae. Pycniospores are produced within the pycnium and are exuded in a drop of nectar. These pycniospores function as spermatia and fuse with the flexuous hyphae (the female element) of a pycnium of another mating type (see Chapter 9 for additional information on mating types). The nucleus of the pycniospore migrates into the flexuous hyphae, and its progeny make their way down through the haploid hyphae to the fundaments of the aecium, thereby establishing the dikaryon. The aecial fundament, now dikaryotic, forms the cupulate aecium which bears binucleate, thin-walled aeciospores in chains. The aecia are formed on the underneath surface of the barberry plant.

The aeciospores are dispersed by air currents and may land on the second host, a cereal plant. The aeciospores germinate, and the dikaryotic hyphae penetrate the plant. The hyphae within the cereal plant eventually form pustules on the surface of the host. These pustules are uredinia and are reddish in appearance owing to the formation of thick-walled red urediniospores, each borne on a short stalk, within them. Each urediniospore is binucleate, thus preserving the dikaryotic condition. The urediniospores are capable of initiating new infections on the cereal plants, and this particular stage may repeat numerous times during the summer. It is this stage which is responsible for the huge buildup of spores which is a major factor in making this fungus a devastating parasite. As autumn approaches, a new type of sorus appears on the wheat plant. This is the telium which bears teliospores. The teliospores are composed of two binucleate cells and have thick, dark walls which make the telium appear black to the naked eye. The teliospore overwinters and, in the spring, each of its cells functions as a probasidium. Karyogamy takes place in each cell of the teliospore, and a metabasidium is produced from each cell. Meiosis takes place in the metabasidium, which is then divided by transverse septa into four cells, each containing a single nucleus. Each cell of the metabasidium forms a sterigma and the nucleus is included in the basidiospore which is then formed. This basidiospore is capable of initiating infection again on the barberry plant.

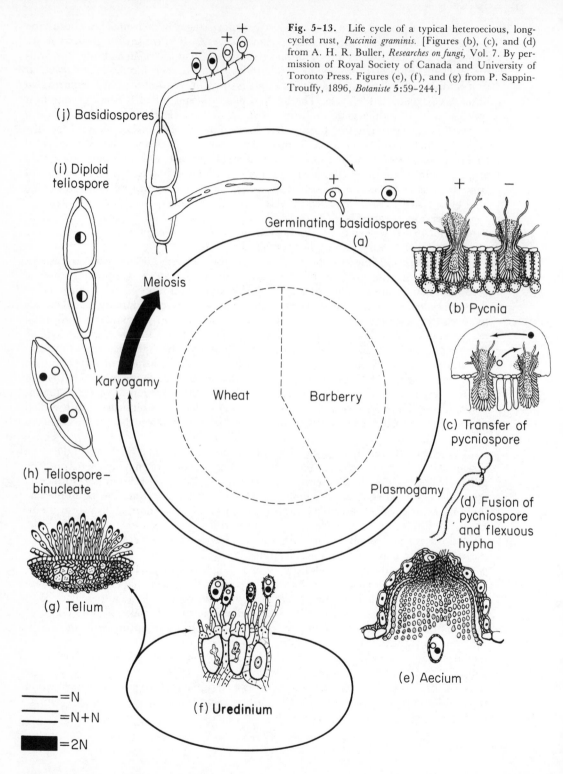

(j) Basidiospores

(i) Diploid teliospore

Meiosis

Karyogamy

(h) Teliospore– binucleate

(g) Telium

Wheat

Barberry

Germinating basidiospores
(a)

(b) Pycnia

(c) Transfer of pycniospore

(d) Fusion of pycniospore and flexuous hypha

Plasmogamy

(e) Aecium

(f) **Uredinium**

──── =N
════ =N+N
▐▐▐▐ =2N

Fig. 5–13. Life cycle of a typical heteroecious, long-cycled rust, *Puccinia graminis.* [Figures (b), (c), and (d) from A. H. R. Buller, *Researches on fungi,* Vol. 7. By permission of Royal Society of Canada and University of Toronto Press. Figures (e), (f), and (g) from P. Sappin-Trouffy, 1896, *Botaniste* **5**:59–244.]

Short-Cycled Rusts. In contrast to the long-cycled rusts, many rusts are short-cycled, as one or more of the spore stages does not occur in its life cycle. For example, a short-cycled rust may have all stages except one, possibly deleting the pycnial or the uredinial stages (that is, it would have spore stages I, II, III, and IV or 0, I, III, and IV, respectively). These rusts may be either heteroecious or autoecious. If the pycnial stage is lacking, plasmogamy frequently occurs by hyphal fusion within the host tissue, thus establishing the dikaryotic stage. Those rusts lacking the uredinial stage do not have a repeating stage.

Even shorter life cycles may be found, as more than one spore stage may be deleted. For example, both the aecial and uredinial stages may be absent, leaving only stages 0, III, and IV. One of the rusts with the simplest life cycle is *Puccinia malvacearum,* which has only stages III and IV and is autoecious, occurring on mallow and its allies. In *P. malvacearum,* monokaryotic mycelium is established in the host after the basidiospore germinates. A telium is formed, and plasmogamy occurs in some of the hyphae of the telium, thereby establishing the dikaryon. The dikaryotic hyphae produce the teliospores, which form the metabasidium and basidiospores without undergoing a rest period. The basidiospores become binucleate when their single haploid nucleus divides mitotically. Several cycles may occur in a single growing season.

In some life cycles, a particular spore stage may have the morphology of one spore stage but carry out the role of another (Hiratsuka, 1973). For example, let us examine *Endophyllum sempervivi,* in which the uninucleate basidiospore lands on the host and the mycelium overwinters in the host. In the spring, pycnia are formed but are nonfunctional. Plasmogamy occurs at the base of the telium, which is associated with the pycnium. Morphologically, this telium is indistinguishable from a typical aecium as it is cupulate, has a peridium, and produces binucleate spores in chains. Functionally, the sorus is a telium and its spores are teliospores. Karyogamy occurs in the teliospores, which then germinate, forming a metabasidium in which meiosis occurs, and finally, uninucleated basidiospores are produced. This life cycle may be designated as 0, III, IV.

ORDER USTILAGINALES

Unlike the Uredinales, the members of the Ustilaginales produce a yeastlike unicellular stage that may generally be cultured readily and is presumably capable of existing saprophytically in nature. A few of the genera include nonparasitic yeastlike fungi that complete their entire life cycles in the absence of a host (Lodder, 1970). One serious human disease, cryptococcus, is caused by the yeastlike nonsexual stage of *Filobasidiella neoformans* (Kwon-Chung, 1976). The vast majority of the members of the Ustilaginales are plant parasites and apparently require the plant for completion of their life cycles in nature. These fungi are the "smut" fungi.

The smut fungi number approximately 850 species. They are parasites on fungi and angiosperms, especially on grasses and sedges. The smut fungi derive their name from the sooty appearance of their sori, filled loosely with dark brown or black teliospores that form on the affected plant part (Fig. 5–14). They most commonly attack the reproductive structures of the plant, and depending upon the species, cause abnormalities of the ovaries or anthers, or they may invade the embryo in the developing seed. One particularly interesting abnormality is called induced "hermaphoditism" or alternatively, "parasitic castration." This phenomenon is caused by *Ustilago violacea,* which attacks members of the Carophyllaceae (e.g., the pinks and chickweed), plants that normally bear the male parts (stamens and anthers) in flowers

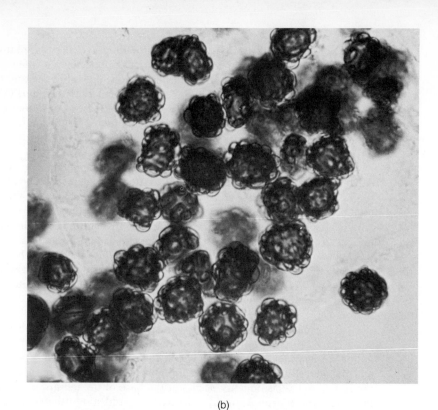

(b)

Fig. 5-14.   A smut fungus, *Urocystis cepulae,* on onion (a), and balls of teliospores from the onion (b). [Courtesy Plant Pathology Department, Cornell University.]

(a)

separate from those bearing the female parts (ovaries). Infection by *U. violacea* in the male plant results in replacement of the pollen grains in the anthers by teliospores (the only site where teliospores are produced) and may induce formation of ovaries. Conversely, infection in the female plant induces formation of sterile anthers. As a result of infection by *U. violaceae,* the appearance of the teliospore-laden anthers so closely resembles that of normal pollen-laden anthers that insects visit the anthers as usual and are probably responsible for transferring teliospores to other flowers. Other smut fungi may cause sterility, abnormal enlargement, or other modification of the reproductive structures. A common smut disease is "boil smut of corn," caused by *Ustilago maydis* and in which the most conspicuous symptom is the massive enlargement of the developing grain to form a large, fleshy, pearly gray gall consisting of host tissue and mycelium and filled with coal dust-like teliospores at maturity (Fig. 12–25). Smuts may also invade the stems, leaves, and occasionally the roots. An example of such a smut is *Tilletia contraversa,* cause of "dwarf bunt of wheat." Spores of *T. contraversa* in the soil invade seedlings and the resulting infection results in a marked shortening of the stem and mottling of the leaves. Some smuts produce their sori on vegetative organs (e.g., onion smut, Fig. 5–14).

The smut life cycle is considerably less complicated than that of the rusts, although these organisms share certain features. The smut mycelium that initiates the infection is usually dikaryotic, although in *Ustilago maydis* the infecting mycelium is haploid in its initial stages and the dikaryotic condition is soon established by mycelial fusion. The slender, septate mycelium develops intercellularly. Clamp connections and haustoria may be present. Eventually the mycelium in the sorus breaks apart into dikaryotic cells which differentiate into teliospores. The teliospores have dark, thick walls which may be smooth, spiny, or reticulated. In some instances, a number of

Fig. 5–15. Life cycle of a smut, *Tilletia eleusines.* The teliospores occur primarily in the ovaries of the host, replacing the seeds but occurring within the pericarp that normally surrounds the seeds. The binucleate mycelium formed by germination of the secondary spore is capable of invading the host. [Figures from N. C. Joshi, 1960, *Mycologia* **52**:829–836.]

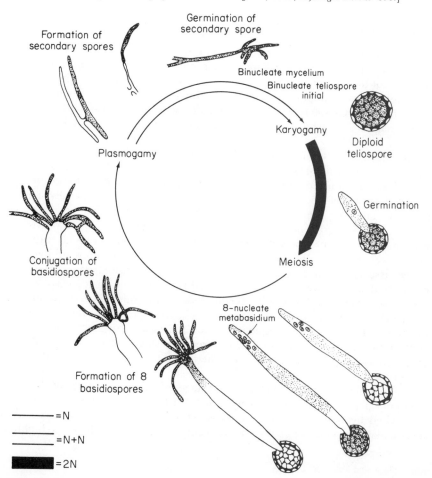

teliospores remain together in a cluster of definite pattern and are called *spore balls* [Fig. 5–14 (b)]. As the binucleate teliospore nears maturity, karyogamy occurs and the diploid teliospore gives rise to a metabasidium. Unlike its regularly triseptate counterpart in the rusts, the smut metabasidium may be nonseptate or, if septate, there may be any number of septa formed. The majority of the septate species form a three- or four-celled metabasidium. Meiosis occurs, and the metabasidium usually gives rise to uninucleate haploid basidiospores. Unlike the Uredinales, these basidiospores are not borne on sterigmata but are sessile on the metabasidium. If the metabasidium is septate, the basidiospores may be formed from each cell; and if the metabasidium is nonseptate, the basidiospores are typically formed at the apex of the metabasidium. An indefinite number of basidiospores are formed. They may give rise to additional spores and in culture will produce a yeastlike colony of spores. The haploid phase of the life cycle is very short, and plasmogamy typically follows with little delay. The majority of the smuts require different mating types to accomplish plasmogamy (see Chapter 9). Depending on the species, plasmogamy may occur between (1) metabasidia, which either may or may not have produced basidiospores; (2) basidiospores or spores produced from them; (3) haploid mycelium during the early infection stages; or (4) any combination of these structures. The dikaryotic hypha produced after plasmogamy is typically the structure which is capable of penetrating the plant. A typical life cycle of a smut is given in Fig. 5–15.

Class Hymenomycetes

The Class Hymenomycetes contains some of the most conspicuous and commonly known fungi, such as the jelly fungi, mushrooms, coral fungi, and bracket fungi. The majority of these fungi are saprophytes on wood or plant material in the soil, and the mycelium plays an important role in breaking down the plant remains and in producing humus (see Chapter 11). Many Hymenomycete species are partners in the mycorrhizal relationship where they are symbionts with the roots of higher plants (see Chapter 13). Many of these fungi are serious plant parasites, and one member, *Schizophyllum commune,* may cause a disease in humans. Species in the Hymenomycetes share the important characteristic of forming a hymenium that is exposed at maturity.

The Basidiocarp and the Hymenium. In almost all these fungi, the hymenium is borne on a basidiocarp (Fig. 5–16). The basidiocarp is generally a macroscopic structure and may be several centimeters in diameter or height. From species to species, there is a great deal of internal variation of the mature basidiocarp, both in arrangement and cellular morphology of the tissue zones. Generally at least three major anatomical zones may be distinguished: these are the hymenium, subhymenium, and sterile tissues. The hymenium is a layer of basidia, basidiospores, and often sterile elements that may cover the entire exposed surface of the basidiocarp or only part of the surface. The morphologically distinct tissue from which the hymenium arises is the subhymenium. Sterile tissue may separate adjacent hymenia or may comprise the bulk of the basidiocarp and is frequently called the *trama*. Portions of a basidiocarp not covered by a hymenium are usually covered by a skinlike layer, the *cuticle*. The cuticle may or may not be markedly differentiated from the trama tissue (we discuss the cuticle further on page 230).

Typically, the early developmental stages of either the whole basidiocarp or of an enlarging portion of the basidiocarp consist of thin-walled, tubular essentially undifferentiated hyphae that may be interwoven or cohere to form a tissue. Sometimes meristematic regions are active and are centers of growth in the more highly differen-

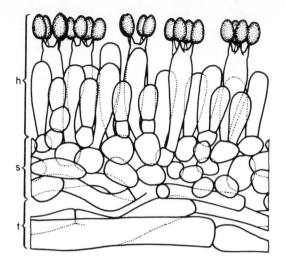

Fig. 5-16. Part of a cross section of a gill of *Agaricus campestris* showing a typical hymenium (*h*) with basidia in various stages of development. Some of the elements of intermediate length are probably basidioles. The subhymenium (*s*) gives rise to the cells in the hymenium, and the sterile tissue or trama (*t*) bears both the subhymenium and hymenium. × 400. [From A. H. R. Buller, 1922, *Researches on fungi,* Vol. 2, Longmans, Green & Co. Ltd., London.]

tiated basidiocarp (Reinjnders, 1977; Kennedy, 1972; Kennedy and Wong, 1978). These nonspecialized, thin-walled, and nonpigmented hyphae are called the *generative* hyphae. The generative hyphae grow by undergoing cell division, cell enlargement, extension at their tips, or a combination of these processes. When mature, the resulting tissue may consist entirely of generative hyphae, or alternatively the generative hyphae may give rise to other hyphal types or to discrete cells. When the generative hyphae produce other hyphal types, they may undergo a structural modification, become pigmented, or serve as depositories for metabolic by-products (Smith, 1966). Contents of these cells may be fats, glycogen, resins, tannins, and other materials (Lentz, 1954). Clamp connections may be either present or absent in the generative hyphae and in hyphae derived from them.

Examples of thick-walled hyphae include unbranched *skeletal* hyphae, which may make the basidiocarp texture somewhat tough, and the highly branched, relatively limited *binding* hyphae, which seem to weave other hyphal types together. The skeletal and binding hyphae are especially common in basidiocarps that are perennial or tough and leathery in texture. In these basidiocarps, generative hyphae may be intermixed with either skeletal or binding hyphae or with both types (Fig. 5-17). Other types of thick-walled hyphae may occur in fleshy basidiocarps. An example of a hyphal system accumulating a metabolic waste is provided by the *lactiferous* hyphae. The lactiferous hyphae form an undulating or anastomosing hyphal system filled with a milky fluid that appears as a turbid emulsion of droplets. A fungus with a lactiferous system will exude the milky fluid when broken, at first the fluid is usually white or clear but it may turn to other colors such as pink or yellow upon exposure to the air. Other hyphae which may be similar to the lactiferous hyphae in appearance may contain fats, aromatic oils, or a blood-colored fluid which may be exuded when the basidiocarp is damaged. The hyphae in some tissues will either break down to form a gel, yielding a soft watery tissue, or the hyphae will secrete a mucilaginous material, forming a sticky coating (Smith, 1966).

The generative hyphae may become divided by additional septa to form a number of separate cells. These cells sometimes remain tubular and elongated, but often they undergo inflation, becoming both wider and longer. This inflation contributes greatly to the overall increase in size of many basidiocarps. Cells in

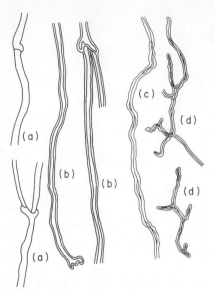

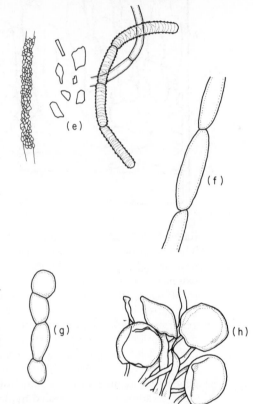

Fig. 5-17. Representative hyphae and cell types in the Hymenomycetes. (a) Thin-walled generative hyphae with clamps; (b) Thick-walled generative hyphae without clamps; (c) Skeletal hyphae; (d) Binding hyphae; (e) Two types of encrusted hyphae; (f) Hyphae with comparatively short, inflated cells; (g) Chains of subglobose to globose cells; and (h) Greatly inflated sphaerocysts. [(a)–(d) from S. Domanski, 1965, *Fungi,* U.S. Department of Agriculture Translation; (e, left) from F. L. Lombard and R. L. Gilbertson, 1966, *Mycologia* **58**:827–845: remaining figures adapted from D. Largent, D. Johnson, and R. Watling, 1977, *How to identify mushrooms to genus* III: *Microscopic features.* Mad River Press, Eureka, Calif. 148 pp.]

basidiocarps may become globose, barrel-shaped, sausage-shaped, or irregularly branched. Enlarged cells may comprise a tissue entirely, or they may be intermixed with elongated hyphae. The enlarged cells usually retain their thin walls, but in some cases the cell walls thicken. Soft, fleshy or brittle basidiocarps characteristically have an abundance of inflated, thin-walled cells in their tissues.

The hyphae often terminate as morphologically distinct entities. Most important are the basidia and *basidioles*, which terminate generative hyphae and occur in the hymenium. The basidiole may be a basidium which has not yet produced sterigmata, or it may be a cell morphologically identical to an immature basidium but which never produces basidiospores. Some of the basidioles become broad, forming a pavement of cells which helps support the basidia that produce basidiospores. *Cystidia* may occur in the hymenium as well as in the trama or sterile surfaces of the basidiocarp. A cystidium is morphologically distinct from an immature basidium but the morphology of the cystidia may vary greatly. Some of them are smooth and thin-walled like the basidia but have a different shape [Fig. 5–18(a)]. Others may have either colored or thickened walls and may be partially covered with incrustations (often calcium oxalate) [Fig. 5–18(b)]. Many cystidia contain distinctive contents that are oily, amorphous, and refractive or that stain deeply with specific stains [Fig. 5-18(c)]. Cystidia of this type often terminate hyphae having similar contents. The last cystidial type is that which is peculiarly branched [Fig. 5-18(d)].

The function of the cystidia apparently varies. In a few species, cystidia play a mechanical role in holding opposing hymenia apart (Fig. 5-34). The branched cystidia are thought to trap air and humidity in pockets around developing basidia, thereby maintaining favorable conditions for their further development. Many cystidia apparently provide an exit point for metabolic wastes accumulated in the hyphae. For example, some cystidia may secrete mucilage from a thin apex, and this secreted mucilage subsequently encrusts on the outer surface (Lentz, 1954; Smith, 1966). In an ultrastructural study of cystidia of *Suillus americanus,* Clémençon (1975) observed that a dark resin first accumulates within the endoplasmic reticulum but is then deposited between the plasmalemma and the cell wall. As the wall thickens, the dark material is first incorporated in patches in the cell wall but finally forms a granular coating on the surface of the cystidium and becomes interspersed with a slime that may be derived from the cell wall itself.

Fig. 5-18. Left: Cystidia arising from hypha that also produces basidia. Right: Types of cystidia that might occur in the hymenium. See text. × 1,000. [Left from E. J. H. Corner, 1950, *Ann. Botany Mem.* **1**:1-740; right from D. N. Pegler, 1966, *Persoonia* **4**:73-124.]

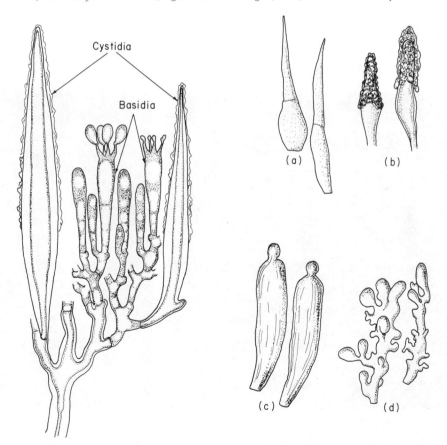

Frequently specific cells or tissues of the basidiocarp or basidiospores will react to chemical treatments and give characteristic color changes. In some species, entire structures will turn either green or blue when moistened with a reagent containing iodine. Moistening with KOH (potassium hydroxide) may induce a brown, yellow, red, or violet color of the tissue or spores of certain species (the particular color induced is constant for a given species). Lactiferous hyphae will sometimes give a positive reaction with the sulfovanillin test. Unfortunately, the chemical basis for many of these reactions is unknown (Watling, 1971). These and similar color reactions are of great value in making identifications of basidiocarps collected in the field (Baroni, 1978).

Vegetative structures produced by some species are *rhizomorphs* and sclerotia. Rhizomorphs are elongated, often hardened strands consisting essentially of bundles of parallel hyphae. Rhizomorphs may be divided into concentric tissue zones and elongate from the apex. They resemble roots and often may be found between the wood and bark of fallen trees or stumps [Fig. 5–19(a)]. Rhizomorphs are a multicellular link between the food supply and the basidiocarp, and they have a role in the conduction or storage of nutrients. Sclerotia are resistant structures which may give rise to mycelium or to a basidiocarp, depending on the species [Fig. 5–19(b)].

Basidium Types and Classification of the Hymenomycetes. At maturity, the morphology of the basidium may vary considerably from one genus to another.

Fig. 5–19. Vegetative structures produced by some basidiomycetes: (a) Rhizomorphs of *Armillaria mellea* on the surface of a log from which the bark has been removed. Approximately × ½; (b) Basidiocarp of *Typhula* arising from sclerotium. [Courtesy Plant Pathology Department, Cornell University.]

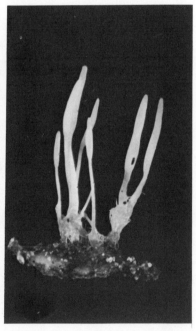

(a)

(b)

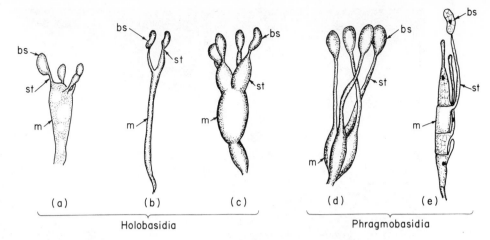

bs

bs
st

bs
st

bs
st

bs
st

st

m

m

m

m

m

m

(a)   (b)   (c)   (d)   (e)

Holobasidia   Phragmobasidia

Fig. 5-20. Different types of basidia in the Hymenomycetes. Mature basidia are represented; shown are the metabasidium (*m*), sterigmata (*st*), and basidiospores (*bs*). (a) A typical holobasidium found in the Agaricales that is not divided or cleft; (b) Deeply cleft holobasidium that bears only two sterigmata and basidiospores (*Dacrymyces*); (c) Holobasidium with swollen sterigmata (*Tulasnella*); (d) Phragmobasidium with longitudinal septa and with extremely long sterigmata (*Tremella*); (e) Phragmobasidium with transverse septa separating individual cells of the metabasidium (*Auricularia*).

The most commonly occurring basidium is a single-celled structure, the *holobasidium.* * The holobasidium is the type used to illustrate the general life cycle (pages 188-192). The holobasidium is nonseptate and generally clavate in shape, but it may also be deeply divided by a central cleft or may be somewhat spheroidal. Sterigmata may be present or absent. If present, the sterigmata may vary from comparatively short spicules as illustrated in Fig. 5-6 to elongate tubular or balloon-like structures (Fig. 5-20). A second major type of the basidium is a *phragmobasidium,* a basidium that is divided into separate cells by either transverse or longitudinal septa, which are formed after meiosis and separate the daughter nuclei from each other. The phragmobasidium is typical of the Subclass Phragmobasidiomycetidae, while the holobasidium is found in the Subclass Holobasidiomycetidae as well as in the Class Gasteromycetes.

SUBCLASS PHRAGMOBASIDIOMYCETIDAE

In this subclass, the basidiocarps may be waxy, fleshy, gelatinous, or dry and leatherlike in consistency. Those species which form a gelatinous basidiocarp are commonly known as ''jelly fungi'' (Fig. 5-21). The basidiocarp is not gelatinous early in its development, but later the hyphae in some tissues disintegrate and leave a gelatinous mass (Fig. 5-22) (Moore, 1965). These fungi are able to survive long periods of drought by drying to a horny texture. When moisture is again available, they absorb water and again become gelatinous. Spore formation and discharge can occur after the moisture is absorbed. Buller (1922) found that basidiocarps of a jelly fungus could revive and produce spores after being kept dry for 8 months. Many of these fungi are brightly colored and are some shade of orange, yellow, or red, while others are white or brown.

---

* This term is used differently by some authors. See Talbot (1973-a) for a discussion of its usage.

Fig. 5–21.   Basidiocarps of *Tremella fibulifera,* a jelly fungus. The larger basidiocarp is about 4 centimeters in diameter. [Courtesy J. S. Furtado.]

Fig. 5–22.   Gelatinous tissue from the center of the basidiocarp of a jelly fungus, *Pseudohydnum gelatinosum.* Note the apparent absence of hyphae. × 828. [From E. J. Moore, 1965, *Am. J. Botany* **52:**389–395.]

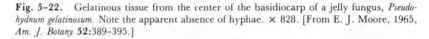

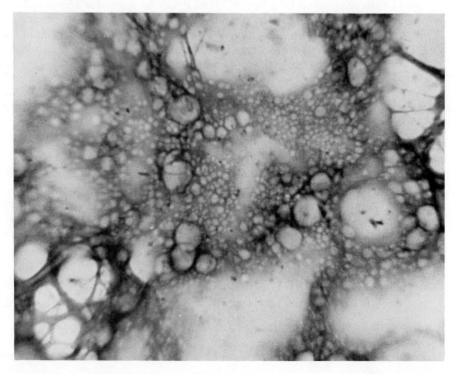

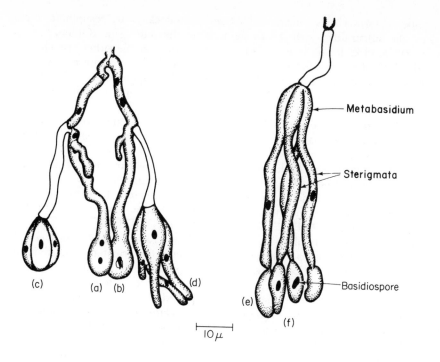

Metabasidium

Sterigmata

Basidiospore

(c) (a) (b) (d)

10 μ

(e)

(f)

Fig. 5-23. Basidia in Tremellales. Top: Developing phragmobasidia of *Exidia nucleata*. (a) Dikaryotic probasidium; (b) Diploid probasidium; (c) Segmented metabasidium after meiosis and the segmentation of the probasidium; (d) Sterigmata developing from the metabasidium; (e) Nuclear migration through the sterigma to the basidiospore initial; (f) Mature basidiospore borne on sterigma and metabasidium. Bottom: Basidia in various stages of development in hymenium within gel of *Tremella mesenterica*. Note that the basidiospores are formed above the surface of the gel. [Top series adapted from K. Wells, 1964, *Am. J. Botany* **51**:360–370. Bottom from P. A. Dangeard, 1895, *Botaniste* **4**:119–181.]

Basidiocarp form varies considerably. Some basidiocarps are resupinate, lying flattened on the substratum with the hymenium on the outer side. Others have a pustulate or cushionlike form, an erect hornlike form that may be either branched or unbranched, or the shape of a funnel or of a cap on a stalk. The hymenium may occur on either one or two sides and is exposed throughout development.

Phragmobasidia and possibly cystidia occur in the hymenium. The phragmobasidia may be buried within a gelatinous layer (Fig. 5–23). An important characteristic shared by these organisms is that basidiospores may frequently germinate by repetition, a process in which a sterigma and a new basidiospore form from the original basidiospore. Alternatively, basidiospores sometimes germinate by forming conidia. Nonsexual spores (oidia) are occasionally formed in some species in a separate sporocarp from that in which the basidiospores occur.

The Subclass Phragmobasidiomycetidae may be divided into three orders, which are distinguished primarily by the morphology of the basidium. These are the Orders Tremellales, Auriculariales, and Septobasidiales (McNabb, 1973).

Order Tremellales. Members of the Order Tremellales usually have *cruciately divided* metabasidia. That means that there are longitudinally oriented septa dividing the metabasidium into four cells. In a transverse section of a metabasidium, the septa appear in the form of a cross, as the septa are perpendicular to each other. One of the septa is formed immediately after meiosis is completed, and the second is formed shortly thereafter. As the majority of these fungi are gelatinous, the metabasidium is embedded within a layer of gel and produces elongate sterigmata that pass through this layer so that the basidiospores are borne on the surface.

Members of the Tremellales occur in diverse forms and are white, yellow, brown, or almost black in color. Most of these fungi are saprophytes on wood or the ground.

Order Auriculariales. These fungi form an elongate metabasidium that is divided by transverse septa, usually into four cells. The probasidial wall may or may not remain permanently distinct. Typically each cell of the metabasidium produces an elongate sterigma that bears a basidiospore. Some of these fungi are saprophytes on wood. An example is *Auricularia auricula,* which often forms brown, ear-shaped basidiocarps on trees and is therefore called "fungus ears." It is eaten by Oriental people. Many species are parasitic on other fungi, mosses, ferns, or higher plants. As in the Tremellales, a variety of basidiocarp forms are produced.

Order Septobasidiales. The phragmobasidium of the Order Septobasidiales is essentially similar to that of the Order Auriculariales. There is a persistent probasidial wall and an elongate metabasidium that is divided into as many as four cells. The basidiospores are not usually repetitive. These fungi are parasitic on insects, with which they establish a unique relationship favorable to some insects in the colony (Chapter 13). The basidiocarp is resupinate and crustlike and is somewhat leathery in texture. Only a single genus, *Septobasidium,* is represented in this order.

SUBCLASS HOLOBASIDIOMYCETIDAE

The members of the Subclass Holobasidiomycetidae include those Hymenomycetes that form a nonseptate basidium or a holobasidium. This group includes diverse types of fungi, including simple plant parasites, jelly fungi, and the mushrooms and bracket fungi with which you are undoubtedly familiar.

The simplest of the plant parasites are found in the Order Exobasidiales. They are parasites on stems, leaves, or flower buds of vascular plants and cause leaf spots or

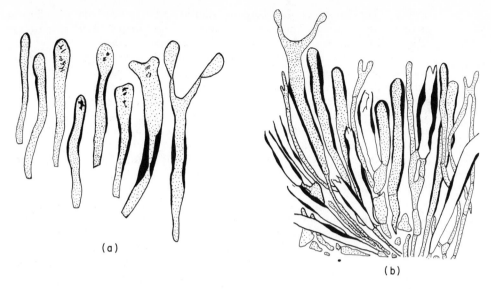

(a)

(b)

Fig. 5-24. A member of the Brachybasidiales, *Ceraceosorus bombacis.* × 1,125. (a) Developmental stages of the basidium, proceeding from the thick-walled probasidium (left) to the metabasidium (right). The metabasidium supports two sterigmata and basidiospores and retains the thickened wall that was present in the probasidium; (b) A portion of the hymenium, showing the persistent walls of the probasidia at various levels. [From J. L. Cunningham, B. K. Bakshi, P. L. Lentz, and M. S. Gilliam, 1976, *Mycologia* **68:**640–650.]

abnormal enlargements to occur. The mycelium occurs inside the leaf tissue and forms basidia which penetrate through epidermal cells or stomates of the host. The basidia are typical cylindrical to clavate holobasidia and may occur singly, in fascicles, or in a hymenium on the surface of the host. Basidiocarps are not formed. The few members of a second order, the Brachybasidiales, are somewhat similar plant parasites. These fungi form an extremely small pustule or disk-shaped basidiocarp which emerges through the epidermis or stomate and also have a persistent probasidial wall from which the undivided metabasidium emerges (Fig. 5-24).

The remaining orders in this subclass are the Dacrymycetales, Tulasnellales, Aphyllophorales, and Agaricales.

Orders Dacrymycetales and Tulasnellales. In many respects, these orders are similar to those in the Subclass Phragmobasidiomycetidae discussed on pages 211–214, and have been classified with them previously or by other authors. Some of these fungi form gelatinous basidiocarps and are also known as jelly fungi. The orders are separated primarily on the basis of basidial morphology (McNabb and Talbot, 1973).

In the Order Dacrymycetales, the typical basidium is a deeply divided form but without septa. The metabasidium is divided into two elongated extensions, each bearing a sterigma and basidiospore. This basidium type resembles a tuning fork, and is sometimes called a tuning-fork basidium. The basidiospores may be nonseptate but are usually septate. The septa often do not appear until after the basidiospores are discharged, when one to seven septa may be formed. The basidiocarps may range greatly in form, as may those of the Subclass Phragmobasidiomycetidae. They are usually gelatinous but may be waxy or dry and form a thin sheet or horny mass on drying. These fungi are often brightly colored and may be shades of yellow or orange as well as deep brown. They occur on wood as saprophytes. Oidia sometimes occur in separate sporocarps similar to those which bear the basidia.

In the Tulasnellales, the basidium bears balloonlike sterigmata. The sterigmata may be sporelike and deciduous or elongated and nondeciduous. The basidiocarp occurs as a flat layer which varies from a filmy, cobweblike covering to a crust. The consistency of the basidiocarp may also vary, both gelatinous and dry forms being known. The colors are soft hues of gray, pink, violet, or lilac. Members of the Tulasnellales are saprophytes on wood or old fungi or may occur as parasites on plants, especially plant parts near the soil. The anamorph of some species may form mycorrhizae with orchids (Chapter 13) (McNabb and Talbot, 1973). Also, some species trap nematodes.

Order Aphyllophorales. The Aphyllophorales is a large order containing about 2,000 known species. Many of these are the bracket and coral fungi. The vast majority of these fungi are saprophytic on plant debris. Many species are significant in decomposing plant remains, as they are able to digest cellulose or lignin that occurs in plant cell walls. Also, many members of the Aphyllophorales cause destruction of timber and wood products or decay wood in standing trees (Chapter 11).

Many of these fungi have been involved in everyday human affairs. A few were used medicinally by the Greeks and Romans as a remedy for many complaints, including colic, fractured limbs, and bruises. Later, in the fifth century, a bracket fungus was immersed in saltpeter and dried and then used to cauterize surgical wounds. Other bracket fungi have been used as curry combs for horses, as snuff, as razor strops, and as a source of dye for clothing (Rolfe and Rolfe, 1925). A few species are edible.

There is a great deal of morphological variation in this group. Those characters unifying the group are the possession of a holobasidium in an exposed hymenium borne on a dry or comparatively dry basidiocarp. The texture of the basidiocarp may be similar to that of cork, wood, leather, paper, or cartilage. Unlike the basidiocarps of the Order Agaricales, the basidiocarps of the Aphyllophorales are *not* fleshy and moist.

The simplest basidiocarp forms are crustlike and resupinate, lying appressed to the surface of the substratum and with a hymenium covering the entire exposed surface. Other basidiocarps may be similarly flat and appressed to the substratum but have a portion of the basidiocarp reflexed or turned away from the substratum (in these, the hymenium would occur on one surface only). Upright basidiocarps also occur and may assume various forms such as a funnel (with the hymenium on the outer surface) or upright clubs or stalks, which may be branched or unbranched (Fig. 5–27). Some of the basidiocarps form a definite cap, the *pileus*, which bears the hymenium on the lower surface. The pileus may be bilaterally symmetrical, resembling a shell or shelf, and may be attached directly to the substratum at its side (*sessile*) (Fig. 5–32) or it may be attached by a stalk (the *stipe*). A few of the species that form a pileus have one that is radially symmetrical, and is supported by a stipe attached to its center. Geotropism can be demonstrated in the upright or pileal forms (see Chapter 10).

We have seen that these basidiocarps range from somewhat fleshy, cartilaginous forms to forms that are hard and woody. The trama is generally comprised of elongate hyphae. Generative hyphae, sometimes with inflated segments or cells, predominate in the softer forms. The hard, woody forms often have skeletal or binding hyphae or both in addition to the generative hyphae. We use the term *monomitic* to refer to a tissue having only generative hyphae, *dimitic* for a tissue having generative hyphae and one additional hyphal type (either skeletal or binding hyphae), and *trimitic* for a tissue with all three hyphal types (Fig. 5–25).

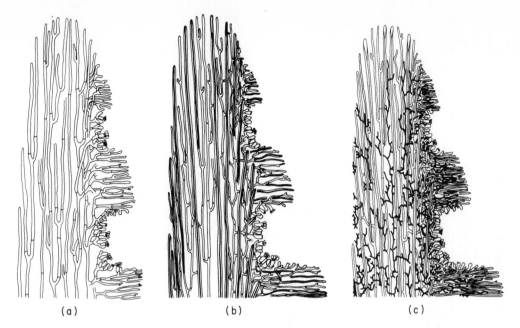

Fig. 5–25. (a) Monomitic tissue consisting of generative hyphae only; (b) Dimitic tissue consisting of generative hyphae (light) and skeletal hyphae (dark); (c) Trimitic tissue consisting of generative, skeletal, and binding hyphae. [From E. J. H. Corner, 1953, *Phytomorphology* **3**:152–167.]

The form of the hymenium is most often determined by that of the trama immediately beneath it. Because this trama bears the hymenium and determines its shape, it is called the *hymenophore*. The hymenophore may be smooth or in the form of warts or spines; it may line the interior of tubes or cover the surface of *gills* (leaflike lamellae that resemble the pages of a half-opened book). In terms of development of the hymenium, two distinct types may be distinguished, the *catahymenium* and *euhymenium*. The catahymenium consists at first of a layer of sterile hyphal elements; the developing basidia then push into the catahymenium and gradually extend through and beyond it. In contrast, the euhymenium lacks such a layer, but the first hymenial elements to appear are the basidia intermixed with a few cystidia. Sometimes the euhymenium may become progressively thicker as it ages. If the euhymenium thickens, it will produce successive layers of basidia, with the older basidia and cystidia buried beneath a layer of younger basidia. As each layer of basidia reaches maturity and then disintegrates, it is penetrated by hyphae that arise from the subhymenium and form new basidia on the surface. The euhymenium does not thicken in many species, and if it does, all basidia in it will mature simultaneously and will not be replaced by younger basidia.

Division of the members of the Aphyllophorales into genera was originally made on the basis of gross morphology of the basidiocarp and hymenium. In the more modern systems, a complex of characters is used. These include many microscopical and anatomical features, as well as the more conspicuous ones of gross morphology. Donk (1964) recognizes 22 families in this order, some of which are shown in Table 5-1. We shall briefly consider some representatives of this order.

Table 5-1: Some characteristics of selected families in the Aphyllophorales as delimited by Donk (1964)

Family	Basidiocarp	Tissue/Cell Types	Hymenophore/Hymenium	Representative Genera
Cantharellaceae	Tubular to funnel-shaped or with central stipe and pileus; fleshy to tough and membranous	Monomitic with thin-walled inflated hyphae; clamps ±; cystidia rare	Smooth or with shallow gills; hymenium thickening; 2–8 spores/basidium	*Cantharellus*
Clavariaceae	Erect, simple, or branched; usually fleshy, waxy, or gelatinous	Monomitic with inflating hyphae; dimitic with skeletal hyphae; clamps ±; cystidia rare	Covering all surfaces; smooth or becoming wrinkled; often thickening; 2–8 spores/basidium	*Clavaria*
Corticiaceae	Flat, usually weblike to membranous, sometimes waxy or gelatinous	Monomitic usually, rarely dimitic with skeletal hyphae; clamps ±; cystidia ±	Smooth, hymenium may or may not thicken; 2–4 spores/basidium	*Corticium*
Hericiaceae	Pileate, upright and branching, or flat; fleshy to membranous	Monomitic, sometimes with inflating hyphae, or dimitic with skeletal hyphae; clamps present; cystidia with distinctive contents in hymenium	Smooth or toothed; nonthickening hymenium; 4 spores/basidium	*Hericium*
Hydnaceae	Stipe and pileus; fleshy	Monomitic, hyphae thin-walled and inflating; clamps present; cystidia absent	Teeth or pointed spines; nonthickening hymenium; 2–6 spores/basidium; basidia often aborting	*Hydnum*
Hymenochaetaceae	Highly variable, including flat, one-sided sessile and those with a stalk and pileus; fibrous to woody; annual or perennial	Monomitic or dimitic with skeletal hyphae; hyphae almost always colored, becoming red and then permanently dark in KOH; clamps absent; prominent pointed, dark cystidia almost always present	Smooth, tubular, or with teeth; nonthickening hymenium; 2–4 spores/basidium	*Hymenochaete*
Polyporaceae	Highly variable, including flat, one-sided sessile and those with a central stipe and pileus; texture variable but often woody and hard; annual or perennial	Mono-, di-, or trimitic with skeletal or binding hyphae; clamps ±; cystidia ±	Typically tubular, but also with gills or other forms; nonthickening hymenium; 2–4 spores/basidium	*Polyporus* *Fomes*
Thelephoraceae	Highly variable, includes flat, those with a stipe and pileus, and many other forms	Monomitic, sometimes with inflating hyphae; becomes greenish in KOH; clamps ±; Cystidia rare	Smooth, warted, toothed, poroid, or folded; hymenium thickening; 2–4 spores/basidium	*Thelephora*

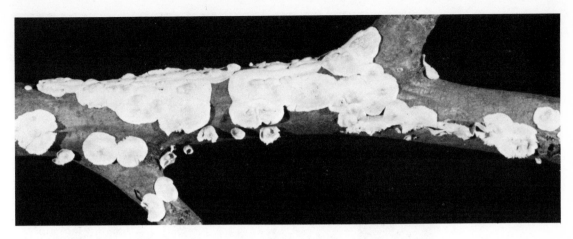

Fig. 5-26. Basidiocarps of *Stereum* having a smooth hymenophore.
[Courtesy Plant Pathology Department, Cornell University.]

Corticium. Probably the simplest basidiocarps are those formed by species of *Corticium*. These are thin, flattened resupinate basidiocarps that are completely appressed to the wood or bark on which they occur. The hymenium is flat and covers the entire surface of the basidiocarp. When fresh, the basidiocarp is somewhat membranous or crusty, but when dry it may crack in a manner similar to the old paint that it resembles. Basidiocarps of species of *Stereum* are similar except that they have a leathery texture and are not completely resupinate but have a recurved edge (Fig. 5-26).

Clavaria and Allies. Upright basidiocarps with a smooth hymenium occur in members of several genera, including *Typhula* and *Clavaria* (Figs. 5-27 and 5-28). *Typhula* species form a club-shaped unbranched basidiocarp arising from a sclerotium. The outer surface of this basidiocarp is covered by a smooth hymenium, and the texture is fleshy to waxy. Many species of *Clavaria* and allied genera also form unbranched basidiocarps, but this group of fungi is especially notable for producing repeatedly branched basidiocarps that often resemble antlers or coral, hence their common name, the "coral fungi." This branching is often quite regular, and may occur in a dichotomous or symmetrical radial pattern. These fungi are often colorful and may be yellow, red, purple, or violet as well as the more sedate shades of white, cream, gray, or brown. Large specimens may weigh several pounds. These fungi usually form their basidiocarps directly on the ground, usually in humus. The hymenium is smooth and covers virtually the entire basidiocarp. In members of some genera, basidia bear only two basidiospores each. The trama has generative hyphae that may be inflated but lacks binding hyphae. Clamp connections may be present or absent, depending upon the species. Cystidia are frequently present in the hymenium.

Cantharellus and Allies. These fungi form a fleshy basidiocarp which is usually trumpet- or funnel-shaped. They are most often some shade of yellow or orange but may occur in other colors and usually occur on the ground in autumn. The hymenium is borne on the lower surface of the pileus but not on the stipe. In members of the genus *Craterellus,* as well as in some species of *Cantharellus,* the hymenium is smooth (Fig. 5-28). In most species of *Cantharellus,* the hymenium is folded into shallow, rounded ridges which are longitudinally oriented or repeatedly merge and separate

Fig. 5-27. Basidiocarp of *Clavaria.* The smooth hymeno-phore covers the upright branches. The basidiocarp is about 10 centimeters tall. [From E. J. Moore and V. N. Rockcastle, 1963, *Fungi,* Cornell Science Leaflet.]

Fig. 5-28. Basidiocarps of *Craterellus taxophilus.* The smooth hymenophore lines the outer surface. These basidiocaps are about 5 to 10 centimeters in height. [Courtesy Plant Pathology Department, Cornell University.]

from each other in a reticulate pattern. According to Corner (1966), the ridges originate when numerous new basidia are formed among those already existing, thereby causing the hymenium to buckle and stretch the underlying hyphae of the trama. This stretching may continue until there is only a loose weft of hyphae remaining. Also there may be further thickening of the gill fold by the addition of new basidia. The trama consists of generative hyphae only, and cystidia are usually lacking.

Hydnum and Allies. In members of the genus *Hydnum* and its allies, the hymenium covers spines or teeth. There is great variety in the form of the basidiocarps. They may be resupinate and have the hymenium covering the entire surface or have a pileus. The pileus, which has the hymenium on the underside, may be either laterally attached or borne on a stipe (Fig. 5-29). The hedgehog fungus, *Hericium caput-ursi* (Fig. 5-30) forms a basidiocarp consisting principally of a fleshy mass of pendant teeth hanging down from a mass of fungal tissue. The size of these tooth fungi varies considerably, as well as their textures, which may be fleshy, leathery, woody, waxy, or membranous. They are most often drab in color and are usually some shade of brown. The majority occur as saprophytes on wood, while others occur on the ground in forests.

Polyporus and Allies (the Polypores). These fungi occur predominantly as saprophytes on dead woody plant parts such as tree trunks, stumps, and branches. A few occur on living trees, where they cause heartrot, and others occur on soil.

Fig. 5-29. The lower surface of a basidiocarp of *Echniodontium tinctorium* showing the toothed hymenophore. [From J. R. Weir and E. E. Hubert. 1918. U.S. Department of Agriculture Bull. 722, 39 pp. Courtesy U.S. Department of Agriculture.]

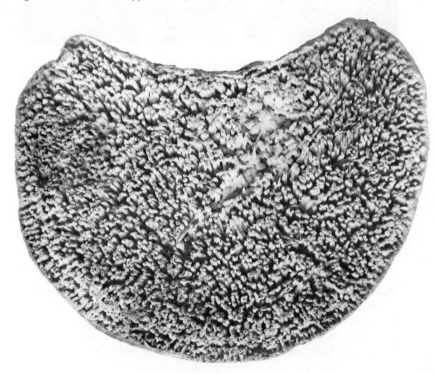

Fig. 5-30. Basidiocarps of the hedgehog fungus *Hericium caput-ursi*. The hymenophore covers the downward-hanging teeth. The basidiocarps are among the largest in the Aphyllophorales. [From E. J. Moore and V. N. Rockcastle, 1963, *Fungi,* Cornell Science Leaflet.]

Species of *Polyporus* and allied genera form tubes lined with the hymenium (Figs. 5-31, 5-32, and 5-33). If these tubes are examined from the end (or from the lower surface of the basidiocarp) one sees a surface punctuated by many pores. This may be easily visualized if one examines a handful of parallel drinking straws from the end. The generic name is derived from this poroid appearance (Greek: polys = many, poros = passage) and provides a common name for these fungi, the "polypores." Although the pores are usually circular (Fig. 5-33), they may be elongated and labyrinthine (Fig. 5-32), elongated and hexagonal, or elongated and almost gill-like.

The basidiocarps may be fleshy when they are young, but become dry and like wood, cork, or leather as they mature. The basidiocarps are often quite large, sometimes reaching 25 centimeters or more in diameter. Some of the basidiocarps are resupinate and have the hymenial tubes covering the surface (Fig. 5-31). Others form a pileus that is attached by a broad surface to the substratum and thus resemble a shelf. These shelflike fungi are often called "bracket fungi." Still other basidiocarps may have a stalk, extending from either the side or the center of the pileus. The majority of the fungi are annual and the mycelium produces new basidiocarps every year.

Fig. 5-31. A resupinate basidiocarp of *Poria europa*. Note the tiny pores in the hymenophore, which covers the entire surface. × 2. [Courtesy Plant Pathology Department, Cornell University.]

Fig. 5-32. Basidiocarps of the bracket fungus *Daedalea confragosa*. Note the irregularly shaped pores. Basidiocarps of this species range from 3 to 15 centimeters at their widest point. [From E. J. Moore and V. N. Rockcastle, 1963, *Fungi,* Cornell Science Leaflet.]

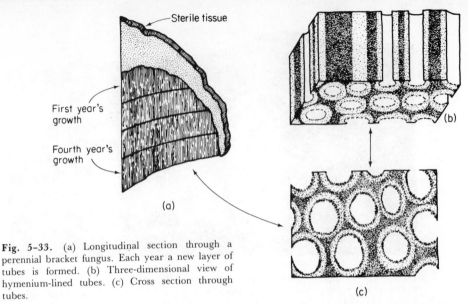

Fig. 5-33. (a) Longitudinal section through a perennial bracket fungus. Each year a new layer of tubes is formed. (b) Three-dimensional view of hymenium-lined tubes. (c) Cross section through tubes.

Species of *Fomes*, however, are perennial and the laterally attached basidiocarps are viable for many years. In these perennial polypores, a new layer of tubes develops each year (Fig. 5-33). The new layer of tubes is formed beneath the one that was produced the previous season, and only the most recently formed layer of tubes produces basidiospores. The basidiocarps are usually cream, gray, brown, or black in color but are sometimes shades of yellow, orange, or red. In those species forming a pileus, the color of the upper, sterile surface may be the same or different from that of the tubular region.

Order Agaricales The Agaricales include the familiar mushrooms and toadstools, well known for their edibility or deadly effects (Chapter 14). Popularly, "mushroom" is used to refer to edible members of this group while "toadstool" refers to poisonous members. The word toadstool is a corruption of the German name "Todestuhl," literally meaning "death's chair" (Bessey, 1950). The hallucinogenic fungi, used in many tribal religious rites, also belong to this group. Other members include the "boletes."

In nature, the members of the Agaricales occur predominantly as saprophytes on plant debris in the soil, on wood, or on dung. Some species are parasitic on plants, while others are involved in symbiotic mycorrhizal associations with vascular plants (Chapter 13). Members of the Agaricales may be found in diverse habitats such as pastures, lawns, meadows, old fields, bogs, on wood or trees, or in humus on the forest floor. They are often found in city environments, where they may be associated with stumps or roots left from a tree that had been removed. They are important in decomposing plant materials in nature (see Chapter 11).

More than 5,000 species are known in this order. The feature that unifies this group is their formation of a fleshy, moist basidiocarp bearing a hymenium that is exposed at maturity. The hymenium may be borne in tubes on the lower surface of a pileus. Those that bear tubes are members of the family Boletaceae and are often informally called "boletes." The majority of the Agaricales bear the hymenium on gills that hang downward from the pileus. The gill-bearing members of this order are classified in 13 families (Smith, 1973) and are often informally called "agarics." The characteristics of some families of the Agaricales are given in Table 5-2.

Table 5-2: Selected Characteristics of Representative Families of the Agaricales Which Are Recognized by Smith (1973)

Family	General Features	Accessory Structures	Basidiospores	Representative Genera
Boletaceae	Pileus on central stipe; fleshy; hymenium in tubes; associated with trees	Annulus and veil present or absent	Basidiospore deposit various colors	*Boletus*
Hygrophoraceae	Pileus on central stipe; gills thick, edge sharp, waxy in texture, decurrent; pileus brightly colored; on duff or ground	Absent	Spore print white; spores smooth	*Hygrophorus*
Tricholomataceae	Stipe present or absent; pileus may be seated eccentrically or centrally on stipe; gills attached but not decurrent; may occur on wood	Annulus present or absent	Spore print white or yellow to pink	*Marasmius* *Pleurotus* *Clitocybe* *Schizophyllum* *Armillaria*
Amanitaceae	Central stipe, separating cleanly from pileus; gills free; on ground in forests	Volva and annulus present; scales on pileus present or absent	Spore print white; spores smooth, thin-walled	*Amanita*
Lepiotaceae	Central stipe, separating cleanly from pileus; gills free; in grassy areas or humus, litter	Annulus and scales on pileus present	Spore print white to yellow or green	*Lepiota* *Chlorophyllum*
Agaricaceae	Central stipe, separating cleanly from pileus; gills gray to pink when young, becoming brown at maturity; cuticle of the pileus usually filamentous; on humus, grassy areas, duff	Annulus and veil present	Spore print chocolate brown; spores subglobose to elliptical; pore at apex	*Agaricus*
Strophariaceae	Stipe eccentric or central; stipe and pileus confluent; gills attached; pileus cuticle filamentous, often gelatinous; in grassy areas or plant debris such as mulch or wood chips	Annulus and veil present or absent	Spore print violet to brown; spores elliptical with pore at apex	*Stropharia* *Psilocybe*
Coprinaceae	Central stipe; basidiocarps thin and fragile; gills usually attached, may undergo autodigestion; pileus cuticle cellular; on dung or other organic waste	Annulus and vein present or absent	Spore print purple to brown or black; spores elliptical with pore at apex	*Coprinus* *Psathyrella*
Russulaceae	Central stipe; trama of pileus and stipe brittle, comprised of nests of large rounded cells bounded by connective hyphae; latex sometimes present; on ground associated with trees (mycorrhizal)	Absent	Spore print white to yellow; markings on spore in form of warts or ridges which become blue in iodine.	*Russula* *Lactarius*

Basidiocarp Structure and Development. The Agaricales possess a pileus that bears the hymenium on its lower surface on either a tube or gill-shaped hymenophore (Figs. 5–34, 5–35, and 5–36). The pileus may or may not be supported by a stipe. We may recognize the following anatomical divisions of the basidiocarp: the hymenium, subhymenium, trama of the hymenophore, trama of the pileus, trama of the stipe, and cuticle.

Fig. 5–34. (a) Longitudinal section through *Coprinus,* a member of the Agaricales with gills. (b) Cross section through pileus. (c) Three-dimensional view of a section of the gills. The hymenium covers the sheetlike surface. (d) Section through the gills. Note the cystidia that extend between the gills. [Figures (a), (c), and (d) adapted from A. H. R. Buller, 1924, *Researches on fungi,* Vol. 3, Longmans and Green & Co. Ltd., London.]

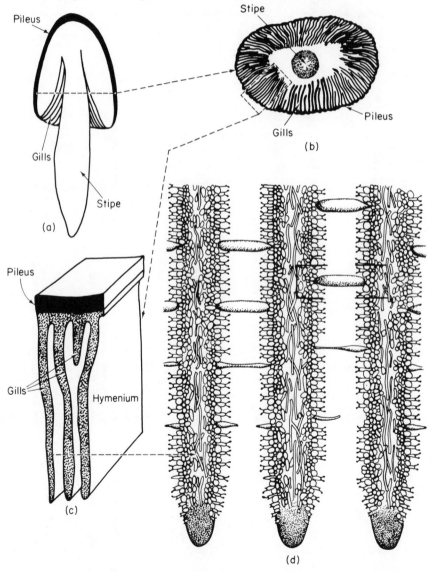

Fig. 5-35. Laterally attached basidiocarps of *Pleurotus,* a member of the Agaricales. Note the gills on the underneath surface. [Courtesy Plant Pathology Department, Cornell University.]

Fig. 5-36. A bolete, *Suillus granulatus.* About natural size. [Courtesy A. H. Smith.]

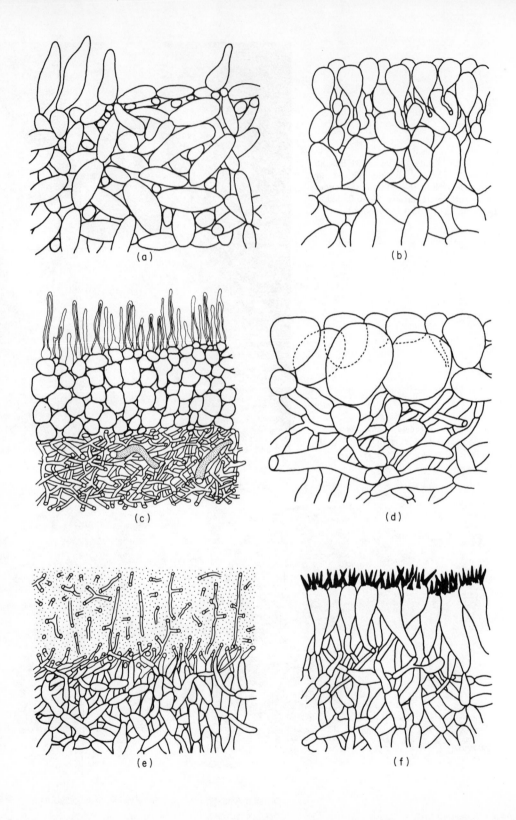

(a)

(b)

(c)

(d)

(e)

(f)

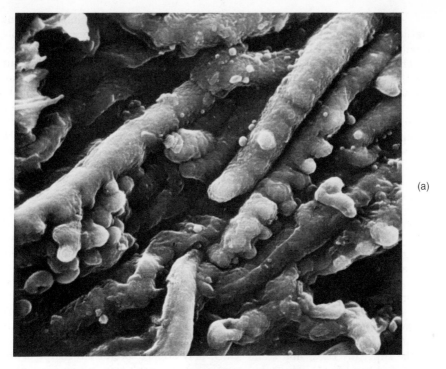

(a)

Fig. 5–38. Scanning electronmicrographs of (a) the pileus cuticle and (b) trama of the pileus in two members of the Agaricales. [(a), × 1,650, (b) × 475]. [Courtesy R. L. Homola and J. Kimball.]

(b)

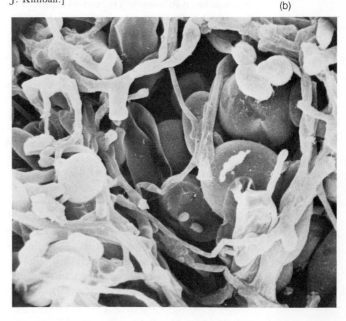

Fig. 5–37. (left) Some modifications of the cuticle found in the Agaricales. In each drawing, the cuticle is shown with some of the underlying trama. (a) Cuticle is structurally homogeneous with the underlying trama; (b) Cuticle is an upright palisade of cells; (c) Cuticle is of a pseudoparenchymatous layer and bears a layer of cystidia; (d) Cuticle is a single layer of greatly enlarged cells; (e) Cuticle has undergone gelatinization; (f) Cuticle is an upright palisade of cells, each with a hairlike projection. [(a)–(e) × 525, (f) × 1,500]. [From A. H. Smith, 1949, *Mushrooms in their natural habitat.* Hafner Press, New York, 626 pp. Copyright GAF Corporation 1949.]

The cuticle may consist of one to three layers and is a skinlike covering of the sterile portions of the basidiocarp which sometimes may be peeled off easily. Most often the cuticle consists of interwoven hyphae (Fig. 5-38), but in some species the outer layer may consist of separate cells. The cuticle may be poorly differentiated from the trama [(Fig. 5-37(a)]. In some cases, the outermost layer of the cuticle may form a palisade layer resembling the hymenium [Fig. 5-38(b)], pseudoparenchymatous cells [Fig. 5-38(c)], greatly inflated cells [Fig. 5-38(d)], or other modifications. The cuticle of some species may become viscous or sticky owing to the gelatinization of some of its components [Fig. 5-38(e)]. A cuticle that becomes gelatinous makes the surface feel slippery or sticky. In still other cases, some of the cells of the cuticle may separate from each other and form a powdery coating on the basidiocarp's surface.

The trama of the pileus and stipe generally consist primarily of generative hyphae or of thin-walled hyphal types or cells derived from the generative hyphae [Fig. 5-38(b)]. Some species have lactiferous or other specialized hyphal types. The structure of the pileus trama and stipe trama often differ in a given basidiocarp. The gill trama consists of hyphae projecting downward from the pileus trama. Within the gill, the tramal hyphae may lie parallel to each other, may be interwoven more or less at random, or may diverge when approaching the subhymenium (Fig. 5-39). The texture of the hymenophore is generally moist, but in *Hygrophorus* the gills have a distinct waxy texture.

In addition to the pileus, stipe, and hymenophore, other structures may be found on mature basidiocarps. These include a skirtlike ring around the stipe, the *annulus;* a veil on the margin of the pileus; a cuplike membrane surrounding the base of the stipe, the *volva;* and scales which may adhere to either the stipe or the pileus (Figs. 5-40 and 5-41). These accessory structures result from the presence of specialized *veils* during the early stages of development. The veils were torn through the expansion of the basidiocarp, resulting in the accessory structures. Any combination of accessory structures may occur, depending upon the particular developmental pattern that has been followed. Three developmental patterns may be distinguished, which are described below.

1. *Gymnocarpic development.* The basidiocarp always begins as a primordium of generative hyphae formed on the mycelium. This primordium becomes recognizable as the rounded ''button'' stage. In gymnocarpic development, the button continues to enlarge to form the basidiocarp, with the apical portion expanding to form the pileus. The hymenium differentiates from a region on the lower side of the developing pileus and is never covered by veils at any point in development. The mature basidiocarp typically lacks accessory structures (Fig. 5-42) (some gymnocarpic species have scales on the pileus but they do not originate from veils).

2. *Pseudoangiocarpic development.* Development of the primordium more or less proceeds as in the gymnocarpic type, but the enlarging pileus recurves and the edge of the young pileus becomes rejoined to the stipe by a loose weft of fundamental hyphae, the *partial veil.* A small cavity is enclosed, and the tubes or gills form within the cavity. As the pileus expands, the partial veil is torn and may be left as the collarlike annulus around the stipe, or remnants may remain on the edge of the pileus as a *marginal* veil, scales, or flaps of tissue. Sometimes the remaining veil is cobweblike and called a *cortina.* Therefore any basidiocarp which underwent this development may have the annulus alone; marginal veil, scales, or cortina alone; or both the annulus and the marginal veil, scales, or cortina (Fig. 5-42).

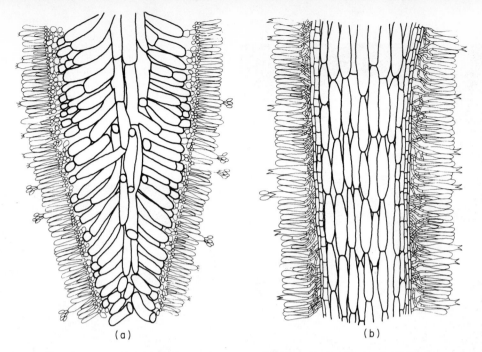

Fig. 5-39. Semidiagrammatic drawings of cross sections of gills showing the four important types of arrangement of the hyphae: (a) Divergent type; (b) Parallel type; (c) Convergent type; (d) Interwoven type. [From A. H. Smith, 1949, *Mushrooms in their natural habitat,* Hafner Press, New York, 626 pp. Copyright GAF Corporation 1949.]

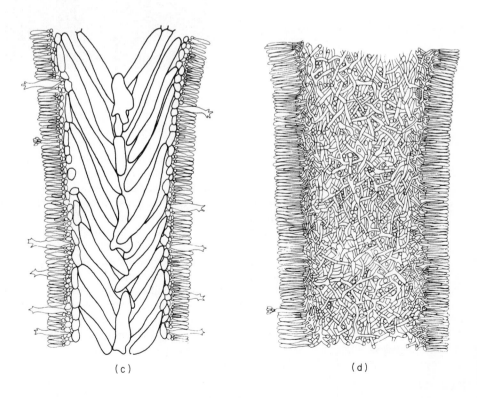

Fig. 5-40. An agaric which forms a universal veil. Left, a young basidiocarp is just beginning to rupture the universal veil. Right, the universal veil is left as a volva at the base of an older basidiocarp. Note that a cortina and annulus are lacking. [Courtesy Plant Pathology Department, Cornell University.]

Fig. 5-41. The field mushroom, *Agaricus campestris,* with a veil and annulus remaining after the recent rupture of the partial veil. [Courtesy Plant Pathology Department, Cornell University.]

3. *Hemiangiocarpic development.* Unlike the previous types, the hymenium is formed endogenously within gill cavities in the tissue of the young basidiocarp, and the hymenium is not exposed until late in the developmental sequence. The outer layer of the primordium differentiates into a distinct surrounding layer of tissue, the *universal veil.* A partial veil consisting of loose tissue joining the margin of the expanding pileus and stipe is also formed. As the basidiocarp enlarges, the universal veil ruptures. If the universal veil was poorly differentiated, it may remain behind in the substratum, gelatinize, or form scales on the stipe or pileus. If the universal veil was a well-defined layer, it leaves a cuplike volva at the base of the stipe. After rupture, the partial veil may remain as scales on the stipe or as an annulus (Fig. 5-42).

Any combination of structures may remain in the mature basidiocarp as a result of hemiangiocarpic development. For example, both the universal and partial veils may remain only as scales on the stipe and pileus. Many, including the cultivated mushroom *Agaricus brunnescens,* have only an annulus remaining from the partial veil. Some may have a volva with remnants of the partial veil remaining as patches on the stem. The deadly species of *Amanita* have a volva, annulus, and scales.

Morphological variation. As we have just read, there may be considerable variation in the presence or absence of accessory structures such as the volva, scales, or annulus. There are many other macroscopic variations that are important not only for recognizing species in the field but also for understanding relationships. We shall examine a small number of these variations.

Basidiocarps may range from those that are extremely small and perhaps also delicate to those that are large and perhaps also massive. The smallest species may have a pileus that is only about 2 millimeters in diameter, and there may be so few cells in the pileus that one may read print through the trama. An example of a small, delicate agaric is *Marasmius rotula,* often called the "horsehair fungus" because of the extremely long black stalks which support the delicate white pileus. This fungus occurs in dense populations, and a dried basidiocarp will revive when moisture becomes available. In contrast to such delicate agarics, some may have a pileus which is 30 centimeters or more in diameter. The most massive basidiocarps are often found among the boletes, which often have very thick, bowl-like pilei with a great deal of trama and the stout stipes required to support such a massive pileus.

Fig. 5-42. Developmental patterns in the Agaricales.

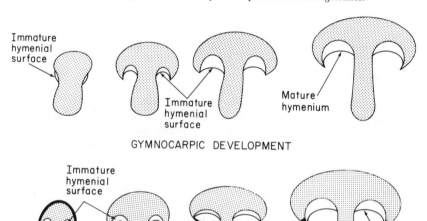

GYMNOCARPIC DEVELOPMENT

PSEUDOANGIOCARPIC DEVELOPMENT

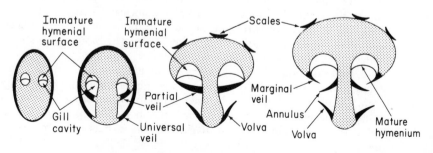

HEMIANGIOCARPIC DEVELOPMENT

The stipe may be either present or absent. If present, the stipe may have the same diameter throughout its length, or it may be somewhat enlarged and bulbous near the base. In the boletes, the stipe often has a reticulated pattern or glandular areas on its surface. The stipe may be attached to the pileus at its side, at its center, or in an intermediate position.

The pileus shapes include those which are asymmetrical and somewhat shell-shaped. These are typically associated with either the absence or the noncentral insertion of the stipe. In the majority of the species, the pileus shape is symmetrical and the stipe is centrally inserted. At maturity, the pileus shapes include those which are conical and more or less closed, as well as some which are bell-shaped, flattened, or even funnel-shaped. During maturation, the pileus usually changes from a more enclosed, somewhat globose or conical form to one that is more open and bell-like or flattened. A fully open pileus can easily discharge its basidiospores into open space, while a partially closed, conical form is not as well adapted for basidiospore discharge. Other variations of the pileus include the nature of the margin, which may be flared, curved, or of some other forms. Towards the middle of the pileus, there may be an indentation or a knob.

If we examine the underneath surface of the pileus, we might find that the gills may be distantly spaced from each other or so close that they almost touch. There is some variation in the shape of the gills as well as in their margin. At the point where the hymenophores approach the stipe, there may be a free space where these structures are not attached and only a short section where they are attached to the stipe, or they may be attached to the stipe throughout their entire length. In some species, the gills are decurrent, running down the stipe and gradually merging with it. In most species, the edges of the gills are wedge- or knife-shaped. Also in most species, the basidiospores will mature randomly over the entire surface of the hymenophore. In contrast, in species of *Coprinus* the basidiospores begin to mature at the bottom of the gill and the process then progresses upward. The basidiospores of *Coprinus* cannot fall out of the pileus readily because of the nontapering crowded gills, thickened gill margin, and the somewhat closed, conical shape of the pileus. Instead, the gills of some species undergo autolysis beginning at the base and progressing upwards as the basidiospores mature. The resulting black fluid drips away, carrying the basidiospores with it. For this reason, these fungi have been called the ''inky caps'' and the black fluid has been collected and used as ink (Fig. 5–43).

Basidiocarps may occur in virtually any color, ranging from somber shades such as white, cream, tan, gray, or brown to brightly colored shades of yellow, orange, red, violet, or green. The pigments responsible for these colors may occur within the cytoplasm of hyphae, in hyphal walls, or as encrustations such as crystals or spirals on the outer surface of hyphal walls. In some species, damaged tramal tissue will turn a distinctive color such as blue or red. The colors of the pileus, hymenophore, and stipe may be similar or dissimilar in a given basidiocarp. Further, the color of the gills or tubes may or may not correspond exactly to the color of the basidiospores that they bear. When viewed in quantity in a spore deposit, basidiospores may appear to be shades of white, pink, purple, ocher, or black.

Bioluminescence. Some agarics are bioluminescent, emitting light of a green, greenish white, or bluish green color. This light is sometimes quite bright and may be bright enough to read by or to be used as the sole light source in prolonged exposure of photographic plates. In the past, this phosphorescent material was known as ''fox fire'' and chunks were used to mark trails or on hats to illuminate the pathway at night.

(a)   (b)

Fig. 5-43. Basidiocarps of the inky cap mushroom, *Coprinus.* (a) Unopened basidiocarps with conical pileus; (b) Autolysis of the pileus results in the dripping of a black fluid containing basidiospores. [From E. J. Moore and V. N. Rockcastle, 1963, *Fungi,* Cornell Science Leaflet.]

The luminescent portion of the fungus varies: for some species of *Collybia*, it is the germinating sclerotium; for *Clitocybe illudens,* it is both the mycelium and basidiocarp; and for *Armillaria mellea,* it is both the mycelium and rhizomorphs. *Clitocybe illudens* is perhaps the most conspicuous of these fungi, as this species forms yellow-orange basidiocarps that occur in clusters containing up to 50 individuals. Each basidiocarp is extremely large: the diameter of the pileus and length of the stipe may be up to 20 centimeters. Basidiocarps appear around old stumps in the autumn and this species is called "jack-o'-lantern" fungus. Bioluminescence may be demonstrated by collecting the luminescent part of the fungus and then examining it in a dark room after allowing sufficient time for the eyes to become accustomed to the dark. Wood penetrated by the mycelium of *Armillaria mellea* is good material for this purpose. Such wood may be found by locating the basidiocarps of *A. mellea* or the long, dark rhizomorphs which grow beneath the bark of logs or stumps (Fig. 5-19).

Class Gasteromycetes

The Gasteromycetes include the puffballs, false truffles, stinkhorn fungi, bird's nest fungi, and earthstars. These fungi usually occur on the surface of the ground, but some members occur on logs or dung or beneath the surface of the soil. The Gasteromycetes are especially prevalent in the drier regions of the world and in sandy, dry habitats. The unifying characteristic of this class is that its members form basidiocarps which enclose the basidiospores until maturity is reached.

The basidiocarp is surrounded by a peridium, an outer covering consisting of one or more layers (Fig. 5-44). The peridium may be thick and leatherlike, papery, scaly, or membranous. Often the peridium will persist and remain intact throughout development, but in some species it will flake away or dehisce rather early. The peridium surrounds the fertile portions of the basidiocarp in which the basidiospores are borne. This internal, fertile tissue is the *gleba*. The gleba consists of sterile tissue, the trama, in combination with the basidia and basidiospores. The basidia may be scattered throughout the gleba or alternatively may occur in a hymenium lining elongated or branched chambers within the gleba. Most of the basidia are somewhat globose in shape and often bear sessile basidiospores seated directly on the basidium rather than on sterigmata. If sterigmata are present, the entire sterigma breaks off and remains attached to its basidiospore. Basidiospores are not forcibly ejected (see Chapter 10). At maturity, the gleba may undergo autodigestion to become slimy or may become powdery and dry. In addition to the gleba and peridium, the basidiocarp may have a columella as well as a stipe. The columella is a central column of sterile tissue which occurs within the gleba. It may occur only near the base of the basidiocarp or may transverse the entire gleba. Some species have an external stipe upon which the peridium and gleba are seated, and this stipe may or may not extend into the columella.

Development of many Gasteromycetes has been studied and various developmental patterns described. One is the *lacunar* type, in which the gleba is at first a fleshy mass of hyphae. During development, the tissues pull apart to form various lacunae (cavities) which later become lined by the hymenium or filled with basidia. In other developmental types, plates or branches of trama tissue originate from the columella and grow upward or originate in undifferentiated tissue near the apex of the basidiocarp and grow downward. To a certain extent, the course of development controls the final morphology of the gleba (Fig. 5-45).

Fig. 5-44. A Gasteromycete, *Trunocolumella citrina* (Hymenogastrales). (a) External view of a basidiocarp; (b) A median section through a basidiocarp; (c) A section of the peridium and gleba. [From S. M. Zeller, 1939, *Mycologia* **31**:1–32.]

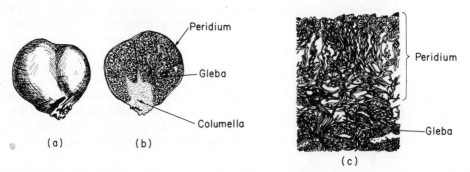

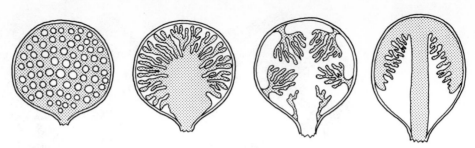

Fig. 5-45. Diagrams of longitudinal sections of basidiocarps in the Gasteromycetes showing different arrangements of the gleba derived from distinctive developmental patterns. The lacunar form is at the left. [Adapted from E. Fischer, 1933. In A. Engler and K. Prantl, *Die natürlichen Pflanzenfamilien*, Vol. 7A, W. Engelmann, Leipzig.]

The Gasteromycetes may reproduce nonsexually by producing conidia. The conidiophores and conidia may be borne in the hymenium with the basidia or in a separate sporocarp. If they are in a separate sporocarp, that sporocarp superficially resembles the basidiocarp but bears conidia rather than basidiospores (Zeller and Dodge, 1924).

Among the Gasteromycetes there are important differences, such as whether a hymenium is present or absent and differences in the texture of the gleba and the nature of the stipe. The orders of the Gasteromycetes, as recognized by Dring (1973), follow.

ORDER HYMENOGASTRALES

In this heterogenous order, the peridium is connected to plates of trama which divide the interior of the basidiocarp into chambers. These chambers are often elongated or irregular (Fig. 5-44) but may also be boxlike. A distinct hymenium lines the surface of the chambers. These chambers are usually empty early in development but later may become partially or completely filled with basidiospores. Two to eight basidiospores are borne on each basidium. The gleba is cartilaginous to fleshy and usually retains this structure until maturity. In some members, the gleba becomes powdery at maturity.

Some of the members of this order strikingly resemble members of the Agaricales and may be an evolutionary link between these two groups. These fungi are comparatively large and may occur below or above the ground. They have a stipe and a pileus (Fig. 5-46). The peridium forms the surface tissue of the pileus and extends into plates of trama which resemble the gills of the agarics or tubes of the boletes. The tramal plates may anastomose and crumple to form irregular cavities. The stipe extends as a columella through the fertile portion of the basidiocarp. At maturity, the peridium breaks away from the stipe or may crack in a longitudinal manner. In some cases, there may be further expansion of the basidiocarp. The morphology of these fungi has been compared with that of an agaric that failed to open during its development and had subsequent internal maturation of its basidiospores (Smith, 1973).

The remaining members of this group superficially resemble truffles and are sometimes called "false truffles" (Figs. 5-44, 5-47.). They often occur underground or may occur in the duff on the forest floor. The basidiocarp is usually somewhat globose or pear-shaped and may be lobed. It is often connected to fairly prominent rhizomorphs, and the size is typically quite small, the basidiocarps being often only a

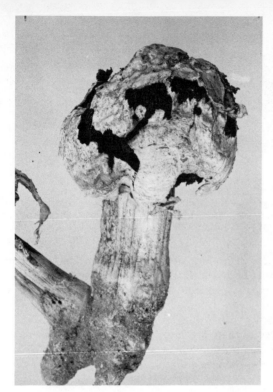 

Fig. 5-46. A member of the Hymenogastrales, *Gyrophragmium delilei,* which forms both a stipe and a pileus. Left: an intact basidiocarp. Note the rupturing of the peridium. Right: a longitudinal section through the basidiocarp showing the gill-like tramal plates in the gleba. [Courtesy U.S. Department of Agriculture.]

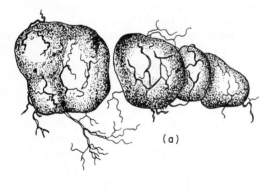

(a)

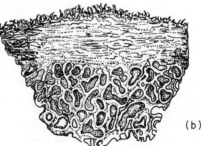

(b)

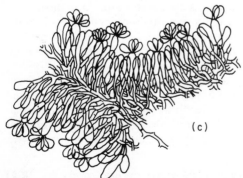

(c)

Fig. 5-47. A false truffle, *Rhizopogon luteolus.* (a) Mature basidiocarps with rhizomorphs (natural size); (b) A portion of the periphery of the basidiocarp showing the peridium (top) and gleba (bottom), × 450; (c) A portion of the gleba, × 450. [From C. Tulasne and H. Rehsteiner. *In:* E. A. Gäumann and C. W. Dodge, 1928, *Comparative morphology of fungi,* McGraw-Hill Book Company, New York, 701 pp.]

few millimeters or centimeters in diameter. These basidiocarps have a fleshy gleba that disintegrates along with the basidiocarp at maturity. The gleba often develops a strong odor at maturity, which helps attract rodents and insects that eat the basidiocarps.

An important genus of the false truffles is *Hymenogaster*. The development of several species of *Hymenogaster* has been studied, and although the developmental pattern is not necessarily the same, many of them have the lacunar type of development (Fig. 5–48). Generally, the peridium and trama tissues of the gleba are poorly differentiated from each other in the young stages and may undergo no further differentiation in some species. As development proceeds in other species, the outer hyphae become large and form a coarse outer layer that gradually merges into the gleba. The cavities develop by splitting and tearing in the upper portion of the basidiocarp, leaving a hemispherical to conical sterile base that may persist to maturity. When the cavities expand above, the lower ones next to the sterile base are stretched longitudinally so that they appear to radiate from the sterile base. This expansion causes the plates of trama to become gradually thinner (Dodge and Zeller, 1934). Other developmental patterns may occur within the Hymenogastrales (Hawker, 1954, 1955).

Fig. 5–48. Development of the basidiocarp of *Hymenogaster tener,* with all figures shown in longitudinal section. (a) Very young basidiocarp, showing a cavity formed by stretching and tearing of hyphae. Hymenium (shaded zone) is already forming at upper margin of cavity; (b, c) Slightly older basidiocarps in which cavity is clear of torn hyphae and wrinkling of hymenium is beginning; (d) Older basidiocarp in which hymenium is greatly wrinkled and infolded; (e) Mature basidiocarp showing the sterile base and elongated, more or less radially arranged chambers. [From L. E. Hawker, 1954, *Roy. Soc. London Phil. Trans.* **B. 237**:429–546.]

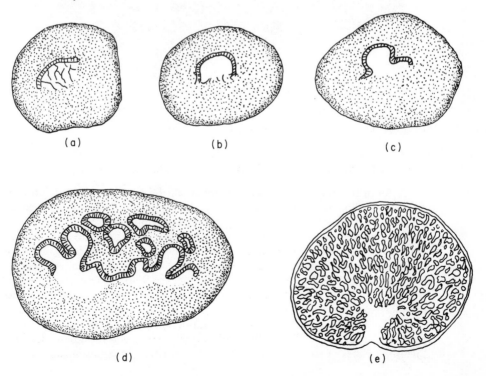

(a)   (b)   (c)

(d)   (e)

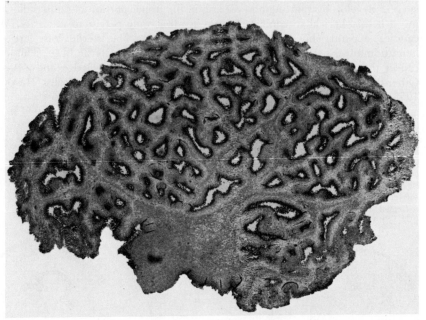

Fig. 5–49. *Gautieria* sp., a member of the Gautieriales. Top: Entire basidiocarps. Bottom: Longitudinal section showing the gleba. Note the absence of a peridium. [Courtesy Plant Pathology Department, Cornell University.]

ORDERS MELANOGASTRALES AND GAUTIERIALES

As in the Hymenogastrales, members of the Melanogastrales and Gautieriales are usually hypogeous. Their basidiocarps are similar in their overall form to those of the so-called "false truffles" described in the previous section.

In the Melanogastrales, the chambers within the hymenium do not become lined with a well-defined hymenium as in the Hymenogastrales. Instead, the basidia and basidiospores are formed more or less randomly within the chambers along with the loose "stuffing" hyphae that make the basidiocarp solid. At maturity, the hyphae and basidia undergo autolysis and fill the cavities with a gelatinous slime in which the basidiospores are distributed.

The basidiocarps in the Order Gautieriales are unique in that they lack a peridium at maturity (Fig. 5–49). The cartilaginous gleba terminates in a number of pileuslike plates of tissue which more or less surround the basidiocarp.

ORDER PHALLALES

The Order Phallales contains approximately 65 species which are especially widespread in the tropical regions of the world, although they are common in the woods of the temperate zone in the late summer and fall. In their early developmental stages, these fungi form globose or ovoid basidiocarps, which often superficially resemble an egg and occur beneath the surface of the soil. These range in size from approximately 0.5 to 5 centimeters in diameter. The young basidiocarp is covered by a peridium, and underneath the peridium is a well-defined gelatinous layer. The gleba, which is somewhat fleshy in this early stage, may be attached to the peridium at the base only, or it may be attached by platelike extensions which extend through the gelatinous layer. When mature, the gleba undergoes autolysis to form a slime.

Fig. 5-50. *Phallogaster saccatus,* a member of the Phallales. (a) A longitudinal section through a basidiocarp showing the thick, persistent peridium with a gelatinous layer beneath it. The gleba with labyrinthine, hymenium-lined chambers occupies most of the internal space; (b) A portion of the gleba is shown in greater detail; (c) A mature basidiocarp showing the rupture of the peridium, slimy contents, and conspicuous rhizomorphs. [Courtesy Plant Pathology Department, Cornell University.]

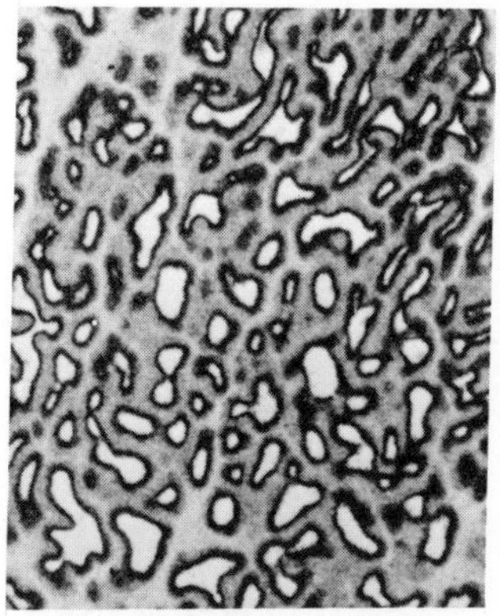

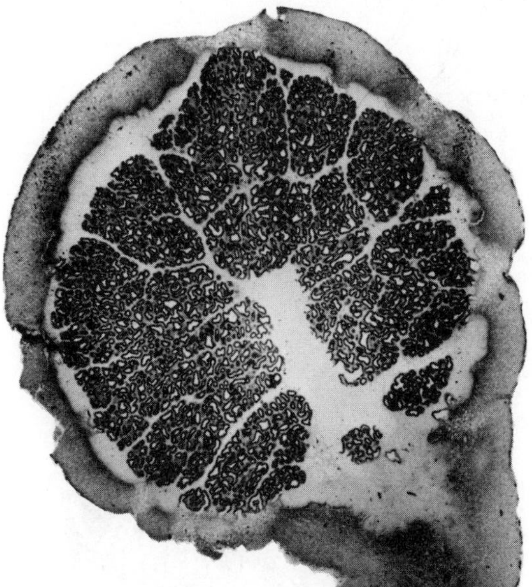

(a)

(b)

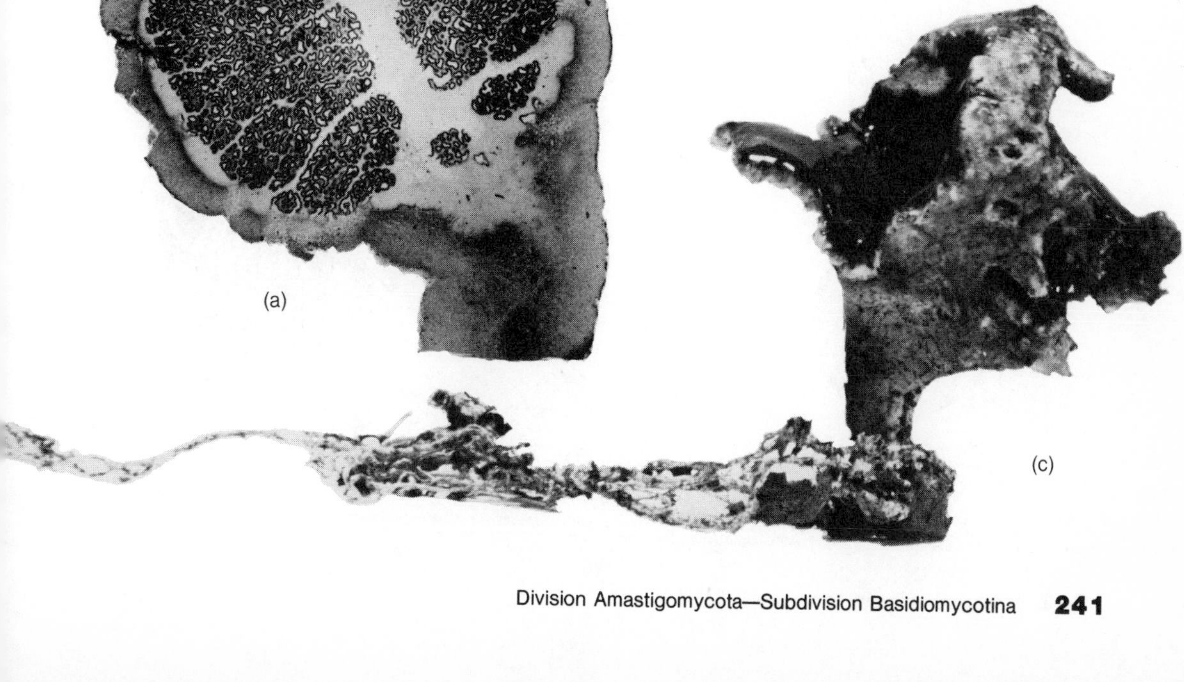

(c)

Some members such as *Phallogaster* (Fig. 5–50) have a persistent peridium and will remain egg- or pear-shaped permanently. These hypogeous fungi share some similarities with the Hymenogastrales (for example, the chambered gleba) and are apparently transitional forms between that order and the remainder of the Phallales (Dring, 1973).

In the remaining members (the majority), the internal portion of the young basidiocarp is occupied by some spongy, sterile tissue. This tissue is capable of undergoing rapid expansion to form a *receptacle* (the stem), which may fully expand in

Fig. 5–51. Representative members of the Phallales showing variations in the form of the receptacle. Note that a prominent volva is present at the base of the receptacle of each. (a) *Aseröe* sp. with receptacle extended into arms; (b) *Simblum* sp. with receptacle terminating as a pileuslike hollow network; (c) *Clathrus* sp. with receptacle as a sessile hollow network. [Figures originally from M. J. Berkeley, J. Gerard, and V. Fayod, respectively. In: E. A. Gäumann and C. W. Dodge, 1928, *Comparative morphology of fungi,* McGraw-Hill Book Company, New York. 701 pp.]

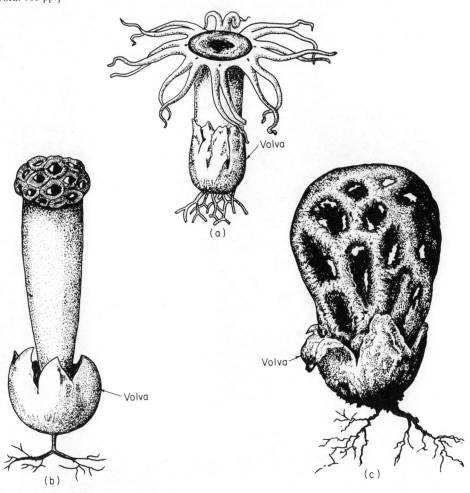

Fig. 5–52. A stinkhorn, *Mutinus caninus.* Early stages, still surrounded by the peridium, are on the right. Mature basidiocarps are on the left. Note the ruptured peridium (the volva) and the slimy apex on the mature basidiocarps. [Courtesy Plant Pathology Department, Cornell University.]

only a few hours. When the receptacle expands, it ruptures the peridium, which remains behind in the ground as the volva. The receptacle, which is now aboveground, may occur in various forms, such as an unbranched column, a hollow, netlike ball, or arms which may be united at the top or be spread in a raylike fashion (Fig. 5–51). The gleba is carried upward by the expansion of the receptacle, and meanwhile the gleba undergoes autolysis to form a foul-smelling slime in which the basidiospores are embedded. The slimy, putrid gleba is very attractive to flies and attracts them to the basidiocarps. The flies carry the spores to new locations.

North American Phallales are represented principally by the so-called "stinkhorn" fungi, members of the family Phallaceae. The volva is prominent in these fungi, and either cups around or sheaths the elongate, unbranched receptacle (Figs. 5–52 and 5–53). The hollow receptacle may be white, delicate pink, lilac, or orange-red and somewhat spongy in consistency. The receptacle may either have a bell-shaped pileus or may terminate bluntly without such a cap. The terminal portion of the pileus bears the foul-smelling olive-colored slime containing the basidiospores. A skirtlike veil may be present beneath the pileus (Fig. 5–53).

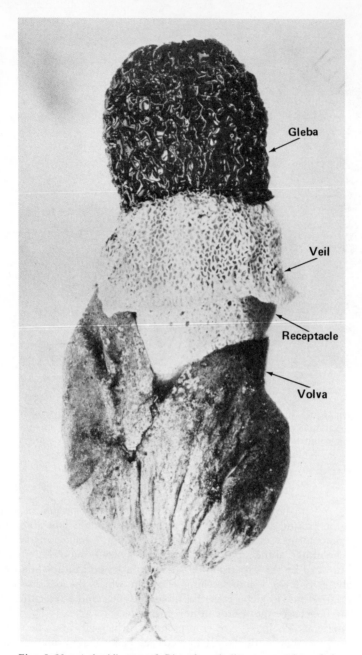

Fig. 5-53. A basidiocarp of *Dictyophora duplicata,* a member of the Phallales. This stinkhorn has a lacy, skirtlike veil. [Courtesy U.S. Department of Agriculture.]

These fungi form basidiocarps that have dry contents at maturity. The gleba is fleshy when young but the generative hyphae undergo autodigestion, resulting in breakdown of the gleba. This autodigestion frees a mass of hyphae, the *capillitium*, and basidiospores. At first, the disintegrated tissue is watery but it eventually dries. Most of the capillitial hyphae are thick-walled, brown, and without septa and are derived from skeletal or binding hyphae. However, some of them are hyaline, thin-walled, and septate and will collapse. These two types of capillitium may occur in the same or different basidiocarps. Further, the capillitium may be branched or unbranched and may sometimes have external markings such as spines. The capillitium is a tangled mass that traps the basidiospores and may also break into even shorter fragments. At maturity, the basidiocarp is filled with a powdery material consisting of fragments of the capillitium and basidiospores. Both the capillitium and basidiospores may escape through an ostiole or fissures in the persistent peridium, or the peridium may scale away during development. If the peridium is struck lightly (perhaps by raindrops) or is broken suddenly (perhaps by being stepped upon by an animal), a powdery cloud of basidiospores is suddenly emitted. For this reason, these fungi are often called "puffballs." Usually, however, the basidiospores are released over a period of time, as they are blown away by breezes but are prevented from being blown away all at once because they are trapped within the tangled capillitium.

Fig. 5-54. A member of the Tulostomatales, *Tylostoma lysocephalum* (natural size). Note the prominent pore in the peridium. [From W. H. Long, 1944, *Mycologia* **36**:318–339. Courtesy U.S. Department of Agriculture.]

Members of the Tulostomatales have a prominent stalk consisting mostly of hyphae arranged parallel to the long axis. The stalk supports a rounded peridium (Fig. 5-54). In most other respects, the Tulostomatales are similar to the members of the Lycoperdales, which lack such a well-developed stalk (Fig. 5-55, 5-56).

Fig. 5-55. Basidiocarps of a puffball, *Lycoperdon.* [From E. J. Moore and V. N. Rockcastle, 1963, *Fungi,* Cornell Science Leaflet.]

Members of the Lycoperdales are almost always epigeous, varying in size from a few millimeters to over a meter in diameter. They form a sessile basidiocarp or have a short, somewhat spongy stalk which consists of modified glebal tissue. A columella may also be present. The basidiocarps are often globose, subglobose, or pyriform in shape. The peridium is usually subdivided into two or more layers, ranging from those that are extremely thick to those that are comparatively thin and flake away at maturity. The surface of the peridium may be smooth or may have ornamentations such as spines or warts. The color of the peridium is usually white or some shade of tan or yellow. The color, thickness, and markings of the peridial layers of a basidiocarp may differ among species. During the early developmental stages, glebal chambers are formed that are lined by a well-defined hymenium. Various types of development have been described, including the upward or downward growth of trama tissue as well as the lacunar type. At first, the gleba is light in color (e.g., white or pale yellow) but the contents darken as the basidiocarp matures because of the maturation of the dark-colored basidiospores. The capillitium and intermixed basidiospores often become deep shades of brown or olive. In addition to being dark, the basidiospores are usually globose and warted or spiny.

Fig. 5-56. Basidiocarps of the giant puffball, *Calvatia gigantea*. Compare their size with the 36-inch ruler. [Courtesy Plant Pathology Department, Cornell University.]

Some members of the Lycoperdales are known as "earthstars" (Fig. 5-57). Earthstars possess a prominent, thickened outer peridium of two or three layers and a thin inner membrane. As maturity is reached, the outer peridium separates from the inner peridium and dehisces in a stellate manner. Segments of the outer peridium recurve and superficially resemble petals of a flower while the inner peridium retains its rounded shape. Some earthstars have been used as hygrometers because of their response to moisture (the peridium closes when the weather is dry and opens when the weather is damp).

Fig. 5-57. An earthstar, *Geastrum*. Note that the peridium of the basidiocarp at the left has not recurved and also note the ostiole in the thin inner peridium of the central basidiocarp. The peridium of the basidiocarp at the right has collapsed after discharge of the capillitium and basidiospores. Approximately × ½. [From E. J. Moore and V. N. Rockcastle, 1963, *Fungi*, Cornell Science Leaflet.]

Most fungi in these orders have fewer peridial layers and do not have outer peridial layers that recurve as do those of the earthstars. This group is generally called puffballs because of the cloud of capillitial fragments and basidiospores that emerges. Common puffball genera include *Lycoperdon* and *Calvatia,* both in the Lycoperdales. Species of *Lycoperdon* form pear-shaped or subglobose basidiocarps that are about 3 to 5 centimeters in diameter. The peridium has two layers and the outer layer forms warts or spines. An ostiole is formed that allows the basidiospores to escape at maturity (Fig. 5-55). *Calvatia gigantea,* the giant puffball, is the best known member of the genus *Calvatia* (Fig. 5-56). *C. gigantea* forms creamy white, globose, or subglobose basidiocarps up to 120 centimeters in diameter. These basidiocarps have a smooth, leathery texture. *C. gigantea* is edible when young while the gleba is still firm and white, and it is often sold in farmers' markets. At maturity, the thin peridium flakes away to expose the capillitium. Large quantities of soft capillitium have been used to dress surgical wounds and to stop bleeding from injuries and was kept on hand in rural households for such purposes (Rolfe and Rolfe, 1925).

ORDER SCLERODERMATALES

This order shares many similarities with the Lycoperdales. These fungi occur in similar habitats and the overall appearance of the basidiocarp is generally similar to that of the sessile puffballs or earthstars.

Unlike the gleba in members of the Lycoperdales, the gleba in this order is divided into numerous chambers by thick or thin plates of trama, a characteristic that this group shares with the Hymenogastrales. Development is of the lacunar type. Basidia may be scattered within the glebal tissue or may occur in poorly defined clusters on the lining of the chambers (unlike the well-defined hymenium in members of the Lycoperdales). At maturity, the peridium contains a dry, powdery material. Most often, this consists only of the disintegrated gleba that is mixed with basidiospores, but in a few species a scant amount of capillitium is formed.

A common genus is *Scleroderma.* Species of *Scleroderma* occur above the ground and form sessile, globose basidiocarps with a thick, tough peridium. At maturity, the outer layer of the peridium breaks into scales. Usually the peridium is some shade of yellow or tan while the gleba is a dark purple-brown at maturity.

ORDER NIDULARIALES

Most of the members of the Nidulariales are called "bird's nest fungi," as their basidiocarps resemble miniature bird's nests filled with eggs. The basidiocarps range from several millimeters to a centimeter in diameter. These fungi are frequently found in large numbers in lawns, fields, and gardens (especially where manure has been distributed). In addition, they occur commonly on decaying wood in forests and are especially common in the autumn. The "nest" is actually a funnel- or goblet-shaped peridium with a large open mouth (Fig. 5-58). It is usually cream, tan, or ocher, sometimes has a hairy exterior, and is typically leathery and tough in texture. In some species, the base is immersed within a hyphal cushion (Fig. 5-59). The peridium may have one to three distinct layers. The peridium contains egg-shaped peridioles loosely piled on top of each other. Each peridiole is a membrane-bounded segment of the gleba with a central cavity containing basidiospores. In species of the common genera *Crucibulum* and *Cyathus,* the peridiole is attached to the peridium by a tubular structure that contains a coiled cord consisting of parallel hyphae, the *funiculus* (Fig. 5-59). A funiculus is lacking in some bird's nest fungi as well as in *Sphaerobolus,* another

Fig. 5-58. A bird's nest fungus, *Cyathus striatus,* growing on dead twigs. [Courtesy H. J. Brodie.]

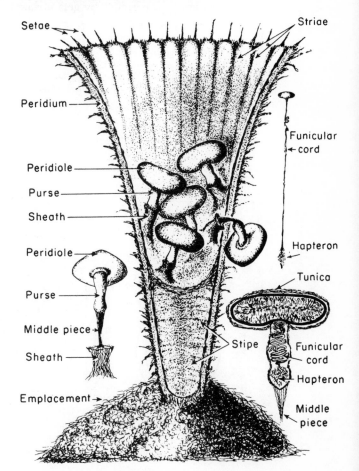

Fig. 5-59. Longitudinal section through a basidiocarp of *Cyathus striatus* showing the attachment of the peridioles to the peridium. When a peridiole is ejected by a raindrop, the purse is torn at its lower end, freeing the funicular cord (funiculus) which bears an adhesive organ (hapteron) that aids in attachment. [Figure reproduced by permission of the National Research Council of Canada from *Can. J. Botany,* 1951, **29:**224–234 and courtesy H. J. Brodie.]

member of this group which forms only a single spherical peridiole in a globose peridium. The bird's nest fungi and *Sphaerobolus* have interesting adaptations for spore discharge which are discussed in Chapter 10.

When the basidiocarps of the bird's nest fungi are first evident, they consist of undifferentiated, hyaline hyphae and are more or less globose in shape. The outer layers of the peridium gradually differentiate, becoming dark or hairy depending upon the species. At first the differentiating peridium surrounds an undifferentiated, more or less homogenous ground tissue (the gleba). As the basidiocarp elongates and expands, the peridium gradually becomes somewhat funnel-shaped [Fig. 5-60(d)] and opens at the apex. In a few species, the peridium tears to form the opening of the funnel, but in most species the peridium has a central pore that may contain a drumlike membrane concealing the gleba. At a fairly early stage in the above sequence, the glebal tissue becomes distinct from that of the peridium because of the gelatinization of a thin layer of tissue [Fig. 5-60(a)]. Both the funiculus and peridiole are initiated when some hyphae grow inward from the peridium, penetrating the previously homogenous ground tissue and converging with other hyphae that originate in the ground tissue. The convergent hyphae delimit a peridiole initial [Figs. 5-60(b) and 5-61] that gradually becomes more distinct as it enlarges and layers of tissue differentiate and mature [Fig. 5-60(c)]. Eventually each peridiole is covered by three distinct layers of tissue. Depending on the species, the peridioles mature simultaneously or differentiate successively, beginning from the top and progressing downward. A cavity forms within each peridiole by gelatinization of ground tissue, and both the peridiole

Fig. 5-60. Progressive stages in development of a basidiocarp of the Nidulariales. (a) Young stage with a homogenous gleba; (b) Peridiole initials are forming within the gleba; (c) Further enlargement of the basidiocarp and differentiation of the periodoles; (d) Nearly mature basidiocarp just prior to opening. [Adapted from L. B. Walker, 1920, *Botan. Gaz.* **70**:1–24.]

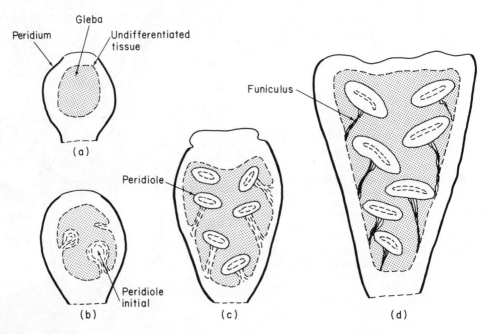

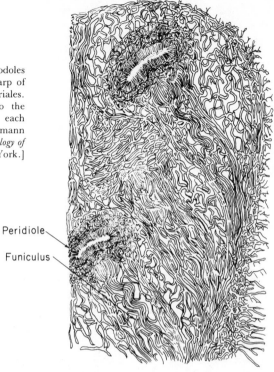

Fig. 5-61. Differentiation of the periodoles within the gleba of an immature basidiocarp of *Crucibulum vulgare,* a member of the Nidulariales. Note that the peridioles are attached to the peridium by the parallel hyphae within each funiculus. [From J. Sachs. *In:* E. A. Gäumann and C. W. Dodge, 1928, *Comparative morphology of fungi,* McGraw-Hill Book Company, New York.]

Peridiole

Funiculus

and the enclosed cavity enlarge further [Fig. 5-60(d)]. The cavity is lined by basidia when it is only about half its final size, but by the time the peridiole is mature, the cavity is filled with basidiospores that have been released by the collapse and gelatinization of the basidia. While the peridioles are maturing, the ground tissue which surrounds them undergoes gelatinization. By the time that the peridium expands fully and the apical membrane ruptures, the peridioles are fully formed and are embedded in a pool of gelatinous material that eventually dries. Because the peridioles are no longer supported by glebal tissue, they often settle to the bottom of the cup-shaped peridium.

One member of this group, *Cyathus stercoreus,* is tetraploid, having 12 chromosomes which represent the 4N condition. Evidence to support this conclusion is that the haploid nucleus contains similar pairs of chromosomes and that quadrivalent configurations are formed in meiosis. Apparently, *C. stercoreus* is the first basidiomycete in which polyploidy has been demonstrated (Lu, 1964).

ORIGIN AND EVOLUTION OF THE BASIDIOMYCOTINA

Numerous theories of the origin and evolution of the Subdivision Basidiomycotina have been proposed, and there is no universally accepted scheme. A tenable theory has been presented by Savile (1955) and is considered here.

The phylogenetic age of the host plant may be used as an index of the approximate age of the parasite (Fig. 5-62). The most primitive existent basidiomycete genus

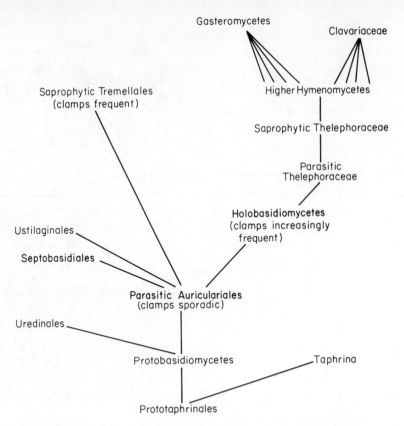

Fig. 5-62. A phylogeny of the Subdivision Basidiomycotina. [Reproduced by permission of the National Research Council from D. B. O. Savile, 1955, *Can. J. Botany* **33**:60–104.]

is assumed to be *Uredinopsis,* a rust genus with species parasitic on ferns in the genus *Osmunda,* the phylogenetically oldest genus with members that are parasitized by rust fungi. Presumably, the hypothetical ancestral Protobasidiomycete was a rust that shared some resemblance to *Uredinopsis.* This rust had no pycnia, aecia, or uredinia, and it had simple teliospores that may have merely been rounded hyphal cells. The mycelium inhabited fern fronds and was self-sterile, clampless, and dikaryotic. The teliospores each produced a basidium bearing four spherical basidiospores upon sterigmata. Presumably the spores germinated upon the fern frond, and the hyphae from two compatible spores fused either immediately upon or after infection. This hypothetical rust ancestor is remarkably similar to the ascomycete genus *Taphrina,* which also forms a dikaryotic mycelium in its host and has rounded cells from which the ascus forms. Under abnormal conditions, *Taphrina* may bud externally, producing subspherical spores upon sterigmalike projections. Both the rusts and members of *Taphrina* parasitize plants of ancient lineage, the ferns and woody dicots, indicating that they existed during the same geological age span; of greatest importance is the fact that a fungus similar to *Taphrina* would be unspecialized and plastic enough to be ancestral to phylogenetic lines that must adapt to different environments. A *Taphrina*-like fungus, or a Prototaphrina ancestor, is assumed to have given rise to both the present-day *Taphrina* and the basidiomycetes.

Both the Prototaphrina and Protobasidiomycete ancestors are extinct. The Protobasidiomycetes gave rise to the Uredinales, beginning with the *Uredinopsis*-like forms. Life cycles of the rust fungi lengthened with the introduction of more spore forms. The first heteroecious rusts originated when part of the life cycle transferred from the telial to the aecial host.

The Protobasidiomycete ancestor also gave rise to a primitive parasitic fungus in the Phragmobasidiomycetidae. The unstable and highly variable heterobasidium is assumed to be primitive and not likely to be derived from the undivided basidium that has become a stable feature in the Holobasidiomycetidae. This primitive fungus was of the type found in the family Auriculariaceae in the Auriculariales, which has a transversely septate basidium (Fig. 5–20). This basidium type is considered to be primitive because it resembles a mycelium and because the fungi in which it appears are morphologically simple parasitic forms on phylogenetically primitive hosts and, most importantly, because clamps appear sporadically in this group. This primitive, *Auricularia*-type ancestor gave rise to (1) the Septobasidiales (parasitic and with no clamps), (2) the Ustilaginales (parasitic with clamps), (3) the Tremellales (saprophytic with clamps), and (4) the Holobasidiomycetidae.

The early members of the Holobasidiomycetidae line were parasitic fungi like some Aphyllophorales (the family Thelephoraceae) that form a flat hymenium on a resupinate basidiocarp, either with or without clamps. Through evolutionary pressure, the hymenophore developed a greater spore-producing surface by forming ridges, teeth, gills, and pores. In addition, there was a tendency for the hymenophore to project downward, as this increased the efficiency of spore discharge. This evolutionary line gave rise to the more advanced Hymenomycetes. The Gasteromycetes represent a polyphyletic group in which convergent evolution under the pressures of the same habitat produced superficial similarities. Both the Gasteromycetes and the Clavariaceae (containing *Clavaria*) represent polyphyletic groups originating from the Hymenomycetes.

References

Ainsworth, G. C. 1973. Introduction and keys to higher taxa. *In:* Ainsworth, G. C., F. K. Sparrow, and A. S. Sussman, Eds., The fungi—an advanced treatise. Academic Press, New York. **4A**:1–7.

Arthur, J. C. 1929. The plant rusts (Uredinales). John Wiley & Sons, Inc., New York. 446 pp.

Atkinson, G. F. 1914-a. The development of *Agaricus arvensis* and *A. comtulus*. Am. J. Botany **1**:3–22 (Agaricales).

———, 1914-b. The development of *Lepiota clypeolaria*. Ann. Mycol. **12**:346–360 (Agaricales).

Baroni, T. J. 1978. Chemical spot-test reactions—boletes. Mycologia **70**:1064–1076 (Agaricales).

Bessey, E. A. 1950. Morphology and taxonomy of fungi. Hafner Press, New York. 791 pp.

Blizzard, A. W. 1917. The development of some species of agarics. Am. J. Botany **4**:221–240 (Agaricales).

Brodie, H. J. 1936. The occurrence and function of oidia in the Hymenomycetes. Am. J. Botany **23**:309–327.

———, 1975. The bird's nest fungi. University of Toronto Press, Toronto. 199 pp. (Nidulariales).

Buller, A. H. R. 1922. Researches on fungi. Vol. 2. Longmans Green and Co., London. 492 pp.

———, 1941. The diploid cell and the diploidisation process in plants and animals with special reference to the higher fungi. Botan. Rev. 7:335–431.

Burt, E. A. 1926. The Thelephoraceae of North America. XV. *Corticium*. Ann. Missouri Botan. Gardens **13**:173–354 (Aphyllophorales).

Chang, S. T., and K. Y. Ling. 1970. Nuclear behavior in the basidiomycete, *Volvariella volvaceae*. Am. J. Botany **57**:165–171 (Agaricales).

——, and C. K. Yau. 1971. *Volvariella volvacea* and its life history. Am. J. Botany **58**:552–561 (Agaricales).

Clémençon, H. 1975. Ultrastructure of hymenial cells in two boletes. Nova Hedwigia Beih. **51**:93–98 (Agaricales).

Coker, W. C., and A. H. Beers. 1943. The Boletaceae of North Carolina. University of North Carolina Press, Chapel Hill, N.C. 96 pp. (Agaricales).

——, and ——. 1951. The stipitate *Hydnums* of the eastern United States. University of North Carolina Press, Chapel Hill, N.C. 86 pp. (Aphyllophorales).

——, and J. N. Couch. 1928. The Gasteromycetes of the eastern United States and Canada. University of North Carolina Press. Chapel Hill, N.C. 195 pp.

Corner, E. J. H. 1950. A monograph of *Clavaria* and allied genera. Oxford University Press, London (Ann. Botany Mem. 1). 740 pp. (Aphyllophorales).

——, 1966. A monograph of Cantharelloid fungi. Oxford University Press, London (Ann. Botany Mem. 2). 255 pp. (Aphyllophorales).

——, 1970. Supplement to "A monograph of *Clavaria* and allied genera. Nova Hedwigia Beih. **33**:1–299 (Aphyllophorales).

Couch, J. N. 1933. Basidia of *Septobasidium (Gelenospora) curtisii.* J. Elisha Mitchell Sci. Soc. **49**:156–162 (Septobasidiales).

——, 1938. The genus *Septobasidium.* University of North Carolina Press, Chapel Hill, N. C. 480 pp. (Septobasidiales).

Craigie, J. H. 1942. Heterothallism in the rust fungi and its significance. Trans. Roy. Soc. Can. Ser. 3, Sec. 5, **36**:19–40 (Uredinales).

——, and G. J. Green. 1962. Nuclear behavior leading to conjugate association in haploid infections of *Puccinia graminis.* Can. J. Botany **40**:163–178 (Uredinales).

Dodge, C. W., and S. M. Zeller. 1934. *Hymenogaster* and related genera. Ann. Missouri Botan. Gardens **21**: 625–708 (Hymenogastrales).

Donk, M. A. 1964. A conspectus of the families of Aphyllophorales. Persoonia **3**:199–324.

Douglas, G. E. 1916. A study of development in the genus *Cortinarius.* Am. J. Botany **3**:319–335 (Agaricales).

——, 1918. The development of some exogenous species of agarics. Am. J. Botany **5**:36–54 (Agaricales).

Drechsler, C. 1969. A *Tulasnella* parasitic on *Amoeba terricola.* Am. J. Botany **56**:1217–1220. (Tulasnellales)

Dring, D. M. 1973. Gasteromycetes. *In:* Ainsworth, G. C., F. K. Sparrow, and A. S. Sussman, Eds., The fungi—an advanced treatise. Academic Press, New York. **4B**:451–478.

Durán, R. 1973. Ustilaginales. *In:* Ainsworth, G. C., F. K. Sparrow, and A. S. Sussman, Eds., The fungi—an advanced treatise. Academic Press, New York. **4B**:281–300.

Elrod, R. P., and W. H. Snell. 1940. Development of the carpophores of certain Boletaceae. Mycologia **32**:493–504 (Agaricales).

Fischer, G. W., and C. S. Holton. 1957. Biology and control of the smut fungi. Ronald Press Company, New York. 622 pp. (Ustilaginales).

——, 1965. The romance of the smut fungi. Mycologia **57**:331–342 (Ustilaginales).

Fitzpatrick, H. M. 1913. A comparative study of the development of the fruit body in *Phallogaster, Hysterangium,* and *Gautieria.* Ann. Mycol. **11**:119–149 (Gautieriales, Phallales).

Flegler, S. L., and G. R. Hooper. 1978. Ultrastructure of *Cyathus stercoreus.* Mycologia **70**:1181–1190 (Nidulariales).

Halisky, P. M. 1965. Physiologic specialization and genetics of the smut fungi. III. Botan. Rev. **31**:114–150 (Ustilaginales).

Harrison, K. A. 1973. Aphyllophorales III. Hydnaceae and Echinodontiaceae. *In:* Ainsworth, G. C., F. K. Sparrow, and A. S. Sussman, Eds., The fungi—an advanced treatise. Academic Press, Inc. New York **4B**:351–368.

Hawker, L. E. 1954. British hypogeous fungi. Phil. Trans. (Roy. Soc. London) **B 237**:429–546.

——, 1955. Hypogeous fungi. Biol. Rev. **30**:127–158.

Hiratsuka, Y. 1973. The nuclear cycle and the terminology of spore states in Uredinales. Mycologia **65**:432–443.

——, and G. B. Cummins. 1963. Morphology of the spermagonia of the rust fungi. Mycologia **55**:487–507 (Uredinales).

Kennedy, L. L. 1972. Basidiocarp development in *Calocera cornea*. Can. J. Botany **50**:413-417 (Dacrymycetales).

——, and R. J. Larcade. 1971. Basidiocarp development in *Polyporus adustus*. Mycologia **63**:69-78 (Aphyllophorales).

——, and W. W. Wong. 1978. *Fomitopsis cajanderi:* development in nature and in culture. Can. J. Botany **56**:2319-2327 (Aphyllophorales).

Kühner, R. 1977. Variation of nuclear behaviour in the Homobasidiomycetes. Trans. Brit. Mycol. Soc. **68**:1-16.

Kwon-Chung, K. J. 1976. Morphogenesis of *Filobasidiella neoformans,* the sexual state of *Cryptococcus neoformans.* Mycologia **68**: 821-833 (Ustilaginales).

Laundon, G. F. 1973. Uridinales (sic). *In:* G. C. Ainsworth, F. K. Sparrow, and A. S. Sussman, Eds., The fungi—an advanced treatise. Academic Press, New York. **4B**:247-279 (Uredinales).

Lentz, P. L. 1954. Modified hyphae of Hymenomycetes. Bot. Rev. **20**:135-199.

Lodder, J. 1970. The yeasts, Second ed. North Holland Publishing Co., Amsterdam. 1485 pp.

Lowe, J. L., and R. L. Gilbertson. 1961. Synopsis of the Polyporaceae of the western United States and Canada. Mycologia **53**:474-511 (Aphyllophorales).

Lowy, B. 1971. Tremellales. Flora Neotropica, Monogr. 6, pp. 1-153.

Lu, B. C. 1964. Polyploidy in the Basidiomycete *Cyathus stercoreus.* Am. J. Botany **51**:343-347 (Nidulariales).

Martin, G. W. 1952. Revision of the North Central Tremellales. State Univ. Iowa Studies Nat. Hist. **19** (3).1-122.

McLaughlin, D. J. 1977. Basidiospore initiation and early development in *Corprinus cinereus.* Am. J. Botany **64**:1-16 (Agaricales).

McNabb, R. F. R. 1973. Phragmobasidiomycetidae: Tremellales, Auriculariales, Septobasidiales. *In:* Ainsworth, G. C., F. K. Sparrow, and A. S. Sussman, Eds., The fungi—an advanced treatise. Academic Press, New York. **4B**:303-325.

——. 1973. Taxonomic studies in the Dacrymycetaceae VIII. *Dacrymyces* Nees ex Fries. New Zealand. J. Botany **11**:461-524 (Dacrymycetales).

——, and P. H. B. Talbot. 1973. Holobasidiomycetidae: Exobasidiales, Brachybasidiales, Dacrymycetales, Tulasnellales. *In:* Ainsworth, G. C., F. K. Sparrow, and A. S. Sussman, Eds., The fungi—an advanced treatise. Academic Press, New York. **4B**: 317-325.

Miller, O. K., Jr. 1972. Mushrooms of North America. E. P. Dutton. New York. 360 pp.

Moore, E. J. 1965. Fungal gel tissue ontogenesis. Am. J. Botany **52**: 389-395 (Tremellales).

Nobels M. K. 1937. Production of conidia by *Corticium incrustans.* Mycologia **29**:557-566 (Aphyllophorales).

Olive, L. S. 1953. The structure and behavior of fungus nuclei. Botan. Rev. **19**: 439-586.

Overholts, L. O. 1953. The Polyporaceae of the United States, Alaska, and Canada. University of Michigan Press, Ann Arbor, Mich. 466 pp. (Aphyllophorales).

Pegler, D. N. 1973. Aphyllophorales IV: Poroid families. *In:* Ainsworth, G. C., F. K. Sparrow, and A. S.Sussman, Eds., The fungi—an advanced treatise. Academic Press, New York. **4B**: 397-420.

Petersen, R. H. 1973. Aphyllophorales II: the clavarioid and cantharelloid basidiomycetes. *In:* Ainsworth, G. C., F. K. Sparrow, and A. S. Sussman, Eds., The fungi—an advanced treatise. Academic Press, New York **4B**: 351-368.

——, 1974. The rust fungus life cycle. Botan. Rev. **40**: 453-513 (Uredinales).

Reinjnders, A. F. M. 1977. The histogenesis of bulb and trama tissue of the higher basidiomycetes and its phylogenetic implications. Persoonia **9**:329-362.

Rolfe, R. T., and F. W. Rolfe. 1974. The romance of the fungus world. Dover Publications, Inc., New York. 301 pp. (Reprint of 1925 Edition published by Chapman & Hall.)

Savile, D. B. O. 1955. A phylogeny of the basidiomycetes. Can. J. Botany **33**:60-104.

——. 1968. Possible interrelationships between fungal groups. *In:* Ainsworth, G. C., and A. S. Sussman, Eds., the fungi—an advanced treatise. Academic Press, New York. **3**: 649-675.

Shaffer, R. L. 1975. The major groups of basidiomycetes. Mycologia **67**:1-18.

Singer, R. 1962. The agaricales in modern taxonomy. J. Cramer, Weinheim. 915 pp.

——, and E. E. Both. 1977. A new species of *Gastroboletus* and its phylogenetic significance. Mycologia **69**:59-72.

Smith, A. H. 1962. Notes on astrogastraceous fungi. Mycologia **54**: 626-639.

——, 1966. The hyphal structure of the basidiocarp. *In:* Ainsworth, G. C., and A. S. Sussman, Eds., The fungi—an advanced treatise. Academic Press, New York. **2:**151-177.

——. 1968. Speciation in higher fungi in relation to modern generic concepts. Myocologia **60:**742-755.

——, 1973. Agaricales and related secotioid Gasteromycetes. *In:* Ainsworth, G. C., F. K. Sparrow, and A. S. Sussman, Eds., The fungi—an advanced treatise. Academic Press, New York. **4B:** 421-450.

——, and H. D. Thiers. 1964. A contribution toward a monograph of North American species of *Suillus.* Private Publication Ann Arbor, Mich. 116 pp. (Agaricales).

——, and ——. 1971. The boletes of Michigan. University of Michigan Press, Ann Arbor Mich. 428 pp.

Talbot, P. H. B. 1973-a. Towards uniformity in basidial terminology. Trans. Brit. Mycol. Soc. **61:**497-512.

——. 1973-b. Aphyllophorales I: General characteristics: thelephoroid and cupuloid families. *In:* Ainsworth, G. C., F. K. Sparrow, and A. S. Sussman, Eds., The fungi—an advanced treatise. Academic Press, New York **4B:**327-349.

Teixeira, A. R. 1962. The taxonomy of the Polyporaceae. Biol. Rev. **37:**51-81 (Aphyllophorales).

Townsend, B. B. 1954. Morphology and development of fungal rhizomorphs. Trans. Brit. Mycol. Soc. **37:**222-233.

Trappe, J. M. 1951. A revision of the genus *Alpova* with notes on *Rhizopogon* and the Melanogastraceae. Nova Hedwigia Beih. **51:**279-309 (Melanogastrales and allies).

Walker, L. B. 1920. Development of *Cyathus fascicularia, C. striatus,* and *Crucibulum vulgare.* Botan. Gaz. **70:**1-24 (Nidulariales).

——, 1940. Development of *Gasterella lutophila.* Mycologia **32:**31-42 (Hymenogastrales).

Watling, R. 1971. Chemical tests in agaricology. *In:* C. Booth, Ed., Methods in microbiology. Academic Press, New York. **4:**567-597 (Agaricales).

Wells, K. 1964. The basidia of *Exidia nucleata.* I. Ultrastructure. Mycologia **56:** 327-341 (Tremellales).

——, 1965. Ultrastructural features of developing and mature basidia and basidiospores of *Schizophyllum commune.* Mycologia **57:**236-261 (Aphyllophorales).

——, and A. H. Smith. 1964. The genus *Calvatia* in North America. Lloydia **27:**148-186 (Lycoperdales).

Yoon, K. S., and D. J. McLaughlin. 1979. Formation of the hilar appendix in basidiospores of *Boletus rubinellus.* Am. J. Botany **66:**870-873. (Agaricales).

Zeller, S. M. 1939. New and noteworthy Gasteromycetes. Mycologia **31:**1-32.

——, and C. W. Dodge. 1924. *Leucogaster* and *Leucophlebs* in North America. Ann. Missouri Botan. Gardens **11:**389-410 (Hymenogastrales).

SIX

Division Amastigomycota— Form-Subdivision Deuteromycotina

Members of the Form-Subdivision Deuteromycotina are often known as "deuteromycetes," and also were formerly called the "imperfect fungi," or Fungi Imperfecti, terms that you may see in other literature. This is a group of about 15,000 species. They are combined in a single group because they apparently lack a sexual stage and therefore have no convenient place in the remainder of the fungus classification system, which is based principally on the mode of sexual reproduction. It is possible either that many species never had sexual reproduction in their life cycles or that once-existent sexual reproduction was deleted. It is generally believed that the majority of the deuteromycetes are the nonsexual stages, or anamorphs, of sexually reproducing fungi that belong to the Ascomycotina or Basidiomycotina, with the largest number occurring in the Ascomycotina. Frequently the sexual stage is discovered later, and then both stages of the fungus, telomorph and anamorph, are taxonomically transferred to the group to which the telomorph belongs. This transfer is consistent with the concept that the anamorph and telomorph together constitute the complete life cycle, the holomorph. As an example, *Curvularia intermedia* was classified in the Form-Subdivision Deuteromycotina and placed in the genus *Curvularia* because it possesses the enlarged central cell of the conidium that is typical for this genus [Fig. 6–1(c)]. Its telomorph was later discovered and found to be an ascomycete in the genus *Cochliobolus*. This fungus has since then been considered to belong to the Subdivision Ascomycotina in the genus *Cochliobolus* and is identified as *Cochliobolus intermedius* with

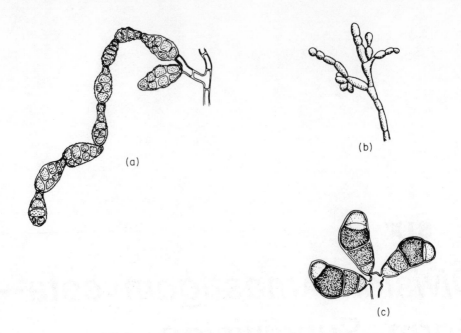

(a)

(b)

(c)

Fig. 6-1. Some common members of the deuteromycetes: (a) *Alternaria;* (b) *Clado-sporium;* (c) *Curvularia;* (d) *Fusarium;* (e) *Gonatobotrys;* (f) *Helicosporium;*

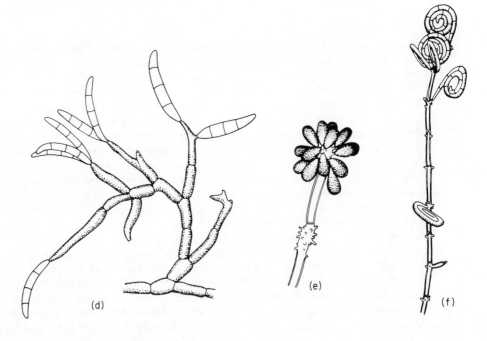

(d)

(e)

(f)

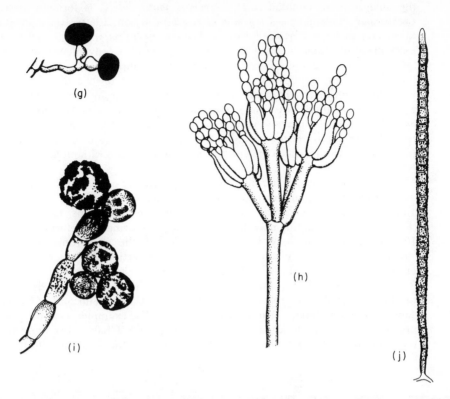

Fig. 6–1. *(continued)* (g) *Nigrospora;* (h) *Penicillium;* (i) *Periconia;* (j) *Septonema;* (k) *Thielaviopsis;* (l) *Trichoderma;* (m) *Verticillium.* [From drawings by H. H. Swart.]

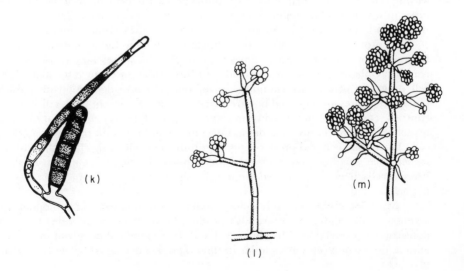

the anamorph or conidial stage *Curvularia intermedia.* It is formally designated as *Cochliobolus intermedius (= Curvularia intermedia)* (Nelson, 1960). It is evident from this discussion that the Form-Subdivision Deuteromycotina represents a ''wastebasket'' assemblage of fungi classified together for the sake of convenience and, as a group, lacks any common phylogenetic origin or relationship. This taxon is designated as a form-subdivision rather than as a subdivision because of its artificial composition.

Although members of the deuteromycetes as a group are unrelated, it may be assumed that there are some natural phylogenetic groups within the overall assemblage and also that speciation may occur even in the absence of sexual reproduction. The required genetic variation for speciation occurs through mutation and the parasexual cycle, in which genetic recombination occurs without sexual reproduction (Chapter 9).

Members of the deuteromycetes invade virtually every possible habitat and type of substratum. Large numbers are saprophytic inhabitants of soil, and many are plant and animal parasites. Some occur in swiftly running streams, forming tetraradiate (four-armed) conidia that may be collected in foam formed on a turbulent stream (Fig. 10–17). Some commonly known members of the deuteromycetes include the ''molds,'' which usually may be easily cultured on artificial media and yield a prolific growth of mycelium. The molds have wide application as research tools for genetic (Chapter 9), nutritional, and biochemical studies (Chapters 7 and 8). Some species produce metabolic by-products that are commercially useful and have, in essence, supplied the foundation for industrial mycology. For example, these fungi are used to produce cheese, antibiotics, and various organic acids (Chapter 14). Some members of the deuteromycetes that are important elsewhere in this book are illustrated in Fig. 6–1.

SPORE TYPES AND MORPHOLOGY

Some members of the deuteromycetes are single-celled, yeastlike organisms that reproduce by budding. The vast majority are mycelial. The mycelium may be of the ascomycete type, in which septa with simple pores are formed, or of the basidiomycete type, in which dolipore septa and clamp connections are formed. The structure of the mycelium is useful in determining the natural affinities of these fungi.

Reproduction may occur by cellular division or by germination of mycelial fragments or spores of various types (Fig. 6–2). The spores may be *chlamydospores* or conidia. Chlamydospores are formed by transformation of an intercalary vegetative cell or group of cells by rounding up of the cells and deposition of a thick wall. Chlamydospores are not shed but are released when the parent hypha disintegrates. In contrast, conidia are produced from a specialized cell that is not part of the vegetative mycelium; when mature, conidia are deciduous and separate readily from the parent cell. The parent cell that produces the conidium is usually a specialized, fertile cell, the *conidiogenous cell,* but may sometimes be another conidium. One to several conidiogenous cells may be produced from and supported by a *conidiophore,* a specialized hypha.

Most deuteromycetes rely upon conidia for reproduction. Conidiophores, conidiogenous cells, and conidia are also produced by many ascomycetes to accomplish nonsexual reproduction. Morphological and developmental details of these structures among the ascomycetes are similar to those features described below for the deuteromycetes.

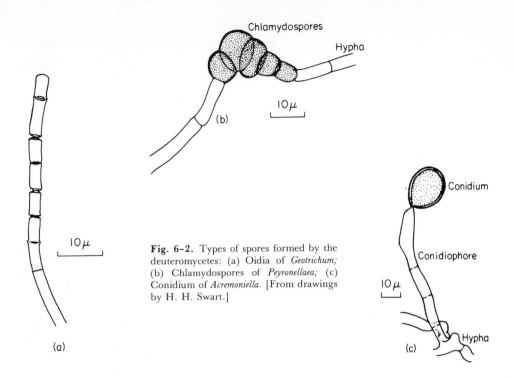

Fig. 6-2. Types of spores formed by the deuteromycetes: (a) Oidia of *Geotrichum;* (b) Chlamydospores of *Peyronellaea;* (c) Conidium of *Acremoniella.* [From drawings by H. H. Swart.]

The Conidiophores and Conidiogenous Cells

Conidiophores are usually morphologically distinct from the vegetative hyphae. They may be light or dark, unbranched or branched in a complex manner, and of uniform thickness throughout or with enlarged parts, most often the apices. Conidiophores bear conidiogenous cells that demonstrate various arrangements (e.g., Fig. 6-3). In turn, the conidiogenous cells differ in the manner of conidial production, including formation of the conidia singly, in clusters, or in succession. If the conidia are produced in succession, they may occur in chains, either with the oldest conidium located at the apex of the chain and pushed upward by younger ones formed at the base, or conversely with the youngest conidium at the apex.

Some important morphological differences may be distinguished among conidiogenous cells that are related to the way that they produce their conidia (Kendrick and Carmichael, 1973). In some species, the conidiogenous cells may resemble vegetative hyphae and are not specialized for the production of large numbers of conidia. In others, conidia are produced in succession from the tip of determinant, nonelongating conidiogenous cells known as *phialides* (Fig. 6-4). Phialides are usually flask or bottle-shaped fertile cells that terminate in a narrow neck. The neck is sometimes surrounded by a flaring collarette, which is the ruptured outer wall that remains at the phialide apex after differentiation of the first conidium. The *annellidic* conidiogenous cell, unlike that of the phialides, produces a succession of conidia, but the tip of the conidiogenous cell continues to elongate. Growth of the conidiogenous cell tip occurs through a ringlike tear (annellation) remaining after a conidium is shed. A series of these rings is left behind on the tip of the conidiogenous cell (Fig. 6-10). Other variations of conidiogenous cell development also occur in the deuteromycetes.

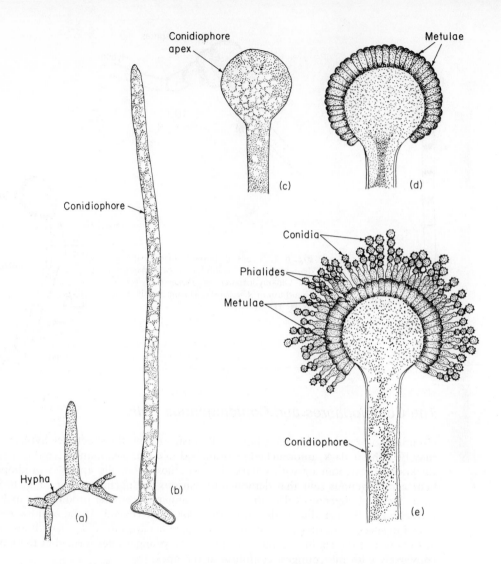

Fig. 6-3. Development of the conidial apparatus in *Aspergillus niger:* (a) Foot cell originating from hypha and bearing young conidiophore as a vertical branch; (b) Developing conidiophore; (c) Swelling of the terminal portion of the conidiophore; (d) Development of metulae from the conidiophore apex; (e) Young sporulating apex showing phialides bearing chains of conidia. (a) and (b) × 172; (c)-(e), × 265. [Figures from C. Thom and K. B. Raper, 1945, *A manual of the Aspergilli,* The Williams & Wilkins Company, Baltimore.]

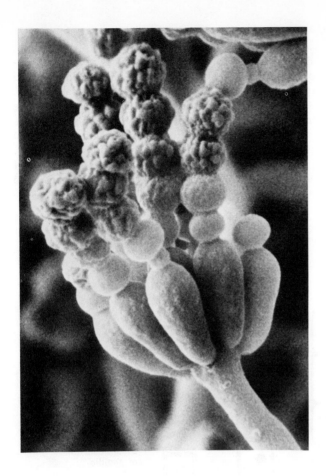

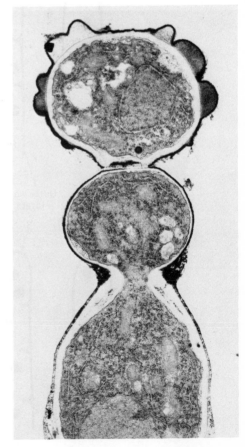

Fig. 6-4. Phialides and their conidia in *Memnoniella echinata*. Left: A scanning electronmicrograph showing a surface view of a cluster of phialides and their chains of conidia. Note that the basal conidia are still incompletely developed, and have not yet formed their rough markings. Right: A transmission electronmicrograph showing a section through the phialide tip and two conidia. [Courtesy R. Campbell, Bristol University; from 1975, *Mycologia,* **67**:760–769].

The Conidia: Ontogeny

The actual mode by which a conidiogenous cell forms its conidia is of great importance. Various systems based on light microscopical observations have been proposed to describe developmental types (Hughes, 1953; Subramanian, 1962; Tubaki, 1966). Precise observations of the conidiogenous cells and conidial development are sometimes difficult with light microscopy because of the extremely small size of some conidia and conidiogenous cells and the consequent difficulty in distinguishing critical features such as the nature of wall differentiation and conidial initiation. Conversely, observations with the electron microscope have assisted greatly in resolving some of these finer details of development. Recently a descriptive system has been proposed

Fig. 6-5. Top: Holothallic conidium development. Bottom: Enterothallic conidium development. Successive stages are shown. In each type, a random number of nuclei are included in each conidium.

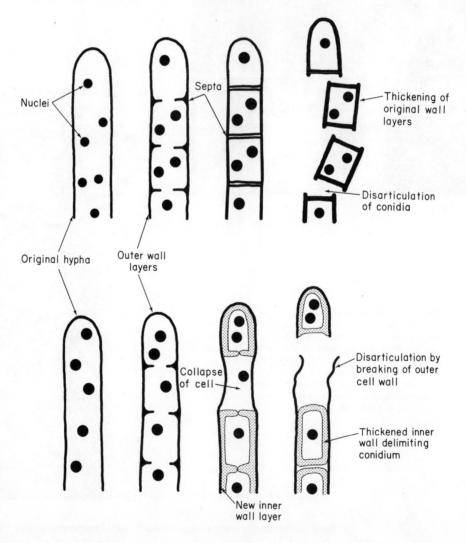

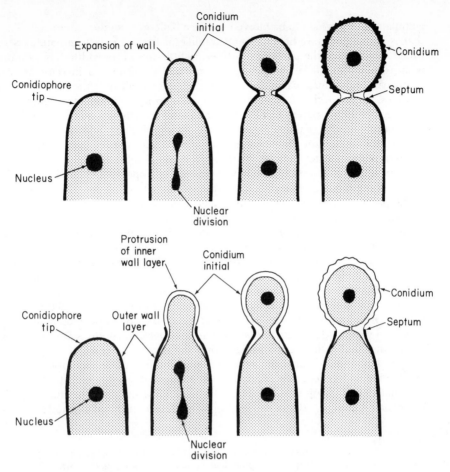

Fig. 6–6. Top: Holoblastic conidium development. Bottom: enteroblastic conidium development. Successive stages are shown in each series.

that incorporates evidence from ultrastructural studies. According to this system, two major types of development may be distinguished (Cole and Samson, 1979; Kendrick, 1971). These are the *thallic* and *blastic* types.

Thallic development occurs in the formation of the chlamydospores described in the previous section as well as in some conidia (Fig. 6-5). The spore initials are delimited from the remainder of the fertile hypha or conidiogenous cell by septa. The septa are formed only after apical growth of the fertile hypha has ceased. If any enlargement of the spore initial occurs, it takes place only *after* the septa are formed. Spores are released when the parent hypha decays (as in chlamydospores) or when the parent hypha undergoes disarticulation. Conidia that undergo this type of development are often given special names such as *oidia* or *arthrospores*.

In blastic development, conidia are not derived from a preexisting vegetative cell but rather result from the formation of new cells arising from the conidiogenous cell or conidium (Fig. 6-6). Unlike the thallic type, the conidial initial may enlarge considerably before being cut off by a septum. Generally, in this type of development the conidial initial appears to be blown out like a balloon from the parent cell. The portion that is expanding may be small and budlike at first or may originate as a relatively broad extension of the parent cell, differing little from it in width. The conidial initial is surrounded at all stages by a cell wall and continues to increase in size. Both cytoplasm and a nucleus migrate from the parent cell into the conidial initial as it enlarges. The conidial initial undergoes substantial enlargement before it is finally separated from the parent cell by the centripetal growth of a septum. Further changes may occur in the septum, such as the thickening or addition of more layers. Sometimes the septum develops in such a way that a special abscission layer forms where separation later occurs, resulting in conidial release.

In both thallic and blastic development, an important subdivision can be made based on the mode of wall differentiation during conidium formation. All wall layers of the parent hypha may extend into or be incorporated within all wall layers of the developing conidium, which makes formation of the conidium strictly exogenous, i.e., of the *holo*-type (we designate exogenous thallic development as *holothallic* and exogenous blastic development as *holoblastic*). The vast majority of conidia are apparently either holothallic or holoblastic (Figs. 6–5, 6–6). Alternatively, some conidia develop endogenously; the outermost wall layer of the parent cell is not included in the conidial cell wall. The new wall of the endogenous conidium may be a modification or continuation of the innermost wall layers of the parent structure or it may be newly synthesized inside the wall of the parent cell. Such development is the *entero*-type, and we may designate the *enterothallic* or *enteroblastic* subdivisions, respectively, for those fungi that form their conidia within the parent structure.

Enterothallic conidia are relatively uncommon, while enteroblastic conidia are much more widely represented. This latter mode of development is best exemplified in the phialides (Figs. 6–4, 6–7). Phialides may form the first conidium by either the holoblastic (Hanlin, 1976) or enteroblastic (Hammill, 1974) modes. If this first

Fig. 6–7. Development of conidia from a phialide. (a) Slightly elongated phialide showing a single wall; (b) The phialide apex has ruptured, and a new wall is protruding through it. This wall will become the apical wall of the newly developing conidium; (c) The new conidium has been formed, has been delimited by a septum, and has undergone enlargement; (d) The process is being repeated to form a second conidium beneath the older one. [Adapted from G. T. Cole and W. B. Kendrick, 1969, *Can. J. Botany* **47**:779–789.]

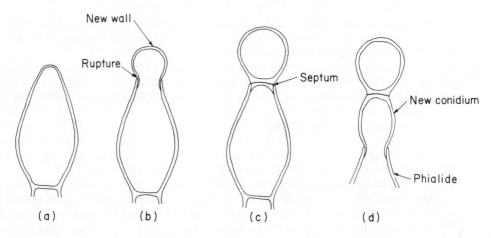

conidium was formed by the enteroblastic method, it ruptures the phialide wall as it emerges. The ruptured wall may then flare outward as a minute collarette. All subsequent conidium development is enteroblastic, each conidium originating when the innermost layer(s) of the phialide wall are blown out through the mouth of the phialide. These wall layers eventually become the outermost layers of the conidium as new wall material is laid down inside the cell. A septum is formed within the phialide itself, separating the exogenously maturing conidium from the endogenous conidium initial. The delimited, maturing conidium is extruded from the phialide by pressure exerted by development of the next conidium (e.g., Hammill, 1972-a, 1974; Hanlin, 1976; Trinci et al., 1968). Fungi that form ringlike annellations on the conidiogenous cells also produce conidia by enteroblastic development (Hammill, 1971, 1974, 1977-c).

The Conidia: Morphology

Mature conidia may be smooth or bear ornate markings such as warts. Conidia are often round or ellipsoid but may also be sickle-shaped, curved, coiled, or of other elaborate shapes. Usually conidia consist of a single cell, but they may have two to several cells. They may be either light or dark, and when seen in a mass may occur in a variety of colors such as white, blue, green, yellow, brown, or black. Some of this variation may be noted in Fig. 6-1.

CLASSIFICATION

Because the Form-Subdivision Deuteromycotina is an artificial assemblage of fungi, any further division into taxa is not intended to group closely related organisms together. Each taxon within this form-subdivision is designated as a *form* taxon to indicate that it is based on empirically selected criteria. Classification of the deuteromycetes into form-orders, form-families, and form-genera has been largely looked upon as a convenient way to group these fungi for identification and not as a means to organize related fungi together in such a way that phylogenetic trends may be recognized. These taxa are based primarily on convenient, easily distinguishable morphological characteristics. The criteria that are typically used are the color, shape, and septation of the conidia and whether or not the conidiogenous cells and conidiophores are borne loose or in some organized manner, perhaps in a sporocarp.

The effect of this system has been to lump unrelated fungi together in the same genus and to distribute fungi that are probably related in different genera. For example, many fungi have been placed in the genus *Helminthosporium* because they form dark, ellipsoid conidia with three or more cells on a dark mycelium. All members of *Helminthosporium* are not closely related, however, and when sexual stages have been discovered, they have belonged to the ascomycete genera *Trichometasphaeria* and *Cochliobolus*. To increase this problem, the conidial genus *Curvularia* has also been found to have a telomorph in the ascomycete genus *Cochliobolus*. The assumption may be made that the telomorph may be used as an index of natural relationships. Therefore this evidence points to the conclusions: (1) that at least two natural but phylogenetically distant or unrelated lines are included in *Helminthosporium,* and (2) that a close phylogenetic relationship exists between some members of *Helminthosporium* and *Curvularia* (Nelson, 1964).

At present, a satisfactory classification system for the deuteromycetes does not exist. However, efforts are being made to devise a classification system for them that will reflect the natural relationships in the group. In order to do this, it is necessary to find criteria that will indicate such relationships. Attention is being given to the mode of development of conidiogenous cells and conidia, the sequence of conidial maturation in multicellular forms, the type of scars left on the conidiogenous cell after loss of the conidium, the mode of germination of the conidium, and anamorph-telomorph interrelationships.

The following artificial taxa are currently utilized:

Form-Subdivision Deuteromycotina
    Form-Class Blastomycetes
    Form-Class Hyphomycetes
    Form-Class Coelomycetes
    Form-Class Mycelia Sterilia

Form-Class Blastomycetes

Members of this form-class include yeastlike fungi. These may be unicellular or form a limited mycelium and reproduce by budding (essentially similar to holoblastic spore formation). Many seem to be similar to the yeasts in the Ascomycotina but fail to produce ascospores. Others appear to have affinities with the Basidiomycotina, as they produce a limited mycelium with clamp connections or produce sterigmata and spores which may be subsequently ejected in a manner similar to that of the basidiospores. These yeasts may be isolated from a variety of habitats, including water, the soil, or plants. Many are pathogenic on animals or humans. Pathogenic nonsexual yeasts include some species of *Cryptococcus* and *Candida,* which we discuss further in Chapter 12 (Kreger Van-Rij, 1973).

Sometimes a sexual stage may be discovered for a yeastlike organism which was previously thought to be nonsexual only. These sexual stages are usually found to belong to the Hemiascomycetes or to the basidiomycete orders Tremellales or Ustilaginales (Kreger Van-Rij, 1973).

Form-Class Hyphomycetes

The Hyphomycetes include those deuteromycetes that form a mycelium but lack a sporocarp. These fungi are often commonly called "molds" or "mildews." The conidiophores are often scattered singly and at random on the mycelium. A culture of a Hyphomycete can be easily recognized by the powdery or fluffy appearance of the colony. The colony may be uniformly colored owing to the mass of colored conidia overlying the mycelium. The mycelium may be light or dark, and the color of the young colony or nonsporulating edges of the colony is the color of the mycelium and not that of the conidia. Some Hyphomycetes may release pigments into the culture medium, making the reverse side of a culture a different color from the mycelium or spores.

In some species, the conidiophores may be tightly clustered together to form a pulvinate mass, the *sporodochium,* which may be formed upon a stroma. Another type of arrangement is a *synnema,* a cluster of conidiophores that form a fascicle and may adhere to each other for part of their length, forming an elongate, bristlelike structure.

Synnemata may be fleshy, brittle, or very hard in consistency. The unattached apical portions of the conidiophores radiate outward, and frequently they produce a mucoid slime in which the conidia are borne (Fig. 6-8).

The Hyphomycetes may be divided in form-families on the basis of whether or not sporodochia or synnemata are formed, as well as on the basis of the morphology of the conidiophores, conidiogenous cells, and conidia. Characteristics of the conidia which are considered are whether or not they are light or dark, their shape and size, and their possible division into more than one cell by septa.

Although the Hyphomycetes are morphologically simple, they are both ecologically and industrially important. They include numerous fungi that are active in soil (Chapter 11) or are plant or human pathogens (Chapter 12), as well as some of the predacious fungi that trap nematodes (Chapter 12). Many industrially important products are derived from genera such as *Penicillium* and *Aspergillus* (Chapter 14).

Fig. 6-8. (a) A sporodochium formed by *Epicoccum andropogonis* on its host with details of the conidia: (b) A synnema of *Calostilbella* and a detail of its conidiogenous cell and conidium. [(a) From drawings by H. H. Swart. (b) From E. F. Morris, 1963, *Western Illinois Univ. Ser. Biol. Sci.* **3**:1–143.]

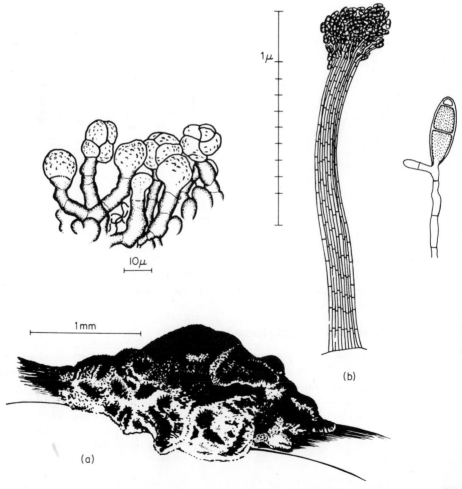

(b)

(a)

Form-Class Coelomycetes

In this form-class, the conidiophores are borne on or within fungal structures which resemble ascocarps. The following two form-orders are based upon the morphology of these multicellular structures (Sutton, 1973).

FORM-ORDER SPHAEROPSIDALES

The members of the Form-Order Sphaeropsidales are characterized by the production of *pycnidia* (Fig. 6-9). A pycnidium superficially resembles a perithecium as it is a closed ascocarp bearing its conidiophores, conidiogenous cells, and conidia within a cavity. The pycnidia may be discoid, globose, or flasklike. Some are originally globose but flatten to become cupulate. Pycnidia may be entirely closed or may open to the outside by an ostiole, slit, or tear. Walls vary from dark and tough, leathery, or carbonaceous to brightly colored and fleshy types. Pycnidia may be immersed within stromata or host tissues or may form on the surface of the substratum. Stromata may enclose one or more pycnidia and may resemble stromata of the Pyrenomycetes or Loculoascomycetes in their variations.

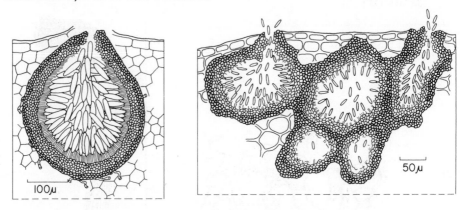

Fig. 6-9. Longitudinal sections through pycnidia of two species of *Macrophoma*. [From G. Morgan-Jones, 1971, *Can. J. Botany* **49**:1921-1929.]

Members of this form-order often occur as saprophytes on decaying plant debris or as parasites on living plants. These fungi are often responsible for a variety of leaf spot diseases in plants; for example, species of *Septoria* may invade celery, azaleas, gladiolus, and wheat.

FORM-ORDER MELANCONIALES

A second type of conidial sporocarp is the *acervulus* which characterizes the members of the Form-Order Melanconiales (Fig. 6-10). An acervulus is an open mass of closely packed conidiophores and conidiogenous cells which may form a flat discoid cushion of conidia or may become cupulate or slightly pulvinate. Acervuli are usually formed on a plant host and are often erumpent from the epidermis of the host, pushing aside flaps of host tissue as they emerge. Superficially, an acervulus resembles an apothecium. Long, dark bristles may occur among the conidiogenous cells in the acervulus.

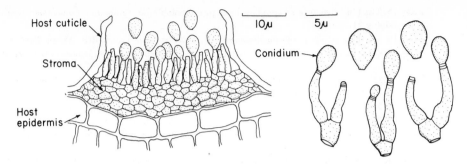

Fig. 6-10. Left: Longitudinal section through an acervulus of *Cryptocline betularum.* Right: Details of the conidia and conidiogenous cells. The conidiogenous cells are of the annellidic type and have ringlike markings (annellations). [From G. Morgan-Jones, 1971, *Can. J. Botany* **49:**1921-1929.]

These fungi are primarily plant pathogens, and include many fungi which cause important diseases. An example is *Colletotrichum,* which causes the anthracnose diseases of citrus, banana, cucumber, and other crops.

Form-Class Mycelia Sterilia

Some members of the deuteromycetes lack spores entirely and reproduce by undergoing mycelial fragmentation. Examples include the common genus *Rhizoctonia,* which may enter into a symbiotic relationship with orchids (Chapter 13) or may cause plant diseases or parasitize other fungi (Chapter 12).

References

Abe, S. 1957. The *Penicillia*—atlas of microorganisms. Kanehara Shippan, Tokyo. 319 pp.

Barnett, H. L. 1955-1960. Illustrated genera of imperfect fungi, Second ed. Burgess Publishing Co., Minneapolis. 225 pp.

Campbell, W. P., and D. A. Griffiths. 1974. Development of endoconidial chlamydospores in *Fusarium culmorum.* Trans. Brit. Mycol. Soc. **63:**221-228.

Carroll, F. E., and G. C. Carroll. 1974. The fine structure of conidium initiation in *Ulocladium atrum.* Can. J. Botany **52:**443-446.

Cole, G. T. 1975. The thallic mode of conidiogenesis in the Fungi Imperfecti. Can. J. Botany **53:**2983-3001.

——, and R. A. Samson. 1979. Patterns of development in conidial fungi. Pitman. San Francisco. 190 pp.

Crane, J. L., and J. D. Schoknecht. 1973. Conidiogenesis in *Ceratocystis ulmi, Ceratocystic piceae,* and *Graphium penicillioides.* Am. J. Botany **60:**346-354.

Gilman, J. C. 1957. A manual of soil fungi, Second ed. Iowa State University Press, Ames, Iowa. 450 pp.

Goos, R. D. 1956. Classification of the Fungi Imperfecti. Proc. Iowa Acad. Sci. **63:**311-320.

Hammill, T. M. 1971. Fine structure of annellophores. I. *Scopulariopsis brevicaulis* and *S. koningii.* Am. J. Botany **58:**88-97.

——. 1972-a. Electron microscopy of conidiogenesis in *Chloridium chlamydosporis.* Mycologia **64:**1054-1065.

——. 1972-b. Fine structure of annellophores II. *Doratomyces nanus.* Trans. Brit. Mycol. Soc. **59:**249-253.

——. 1972-c. Fine structure of annellophores. V. *Stegonosporium pyriforme.* Mycologia **64:**654-657.

——. 1972-d. Electron microscopy of phialoconidiogenesis in *Metarrhizium anisopliae.* Am. J. Botany **59:**317-326.

———. 1973-a. Fine structure of conidiogenesis in the holoblastic, sympodial *Tritirachium roseum*. Can. J. Botany **51**:2033-2036.

———. 1973-b. Fine structure of annellophores. IV. *Spilocaea pomi*. Trans. Brit. Mycol. Soc. **60**:65-68.

———. 1974. Electron microscopy of phialides and conidiogenesis in *Trichoderma saturnisporum*. Am. J. Botany **61**:15-24.

———. 1977-a. Additional electron microscopy of phialoconidiogenesis in *Metarrhizium anisopliae*: microtubules in phialidic necks. Mycologia **64**:1058-1061.

———. 1977-b. Karyology during conidiogenesis in *Gliomastix murorum*: light microscopy. Am. J. Botany **64**:1140-1151.

———. 1977-c. Transmission electron microscopy of annellides and conidiogenesis in the synnematal hyphomycete *Trichurus spiralis*. Can. J. Botany **55**:233-244.

Hanlin, R. T. 1976. Phialide and conidium development in *Aspergillus clavatus*. Am. J. Botany **63**:144-155.

Hartmann, G. C. 1966. The cytology of *Alternaria tenuis*. Mycologia **58**:694-701.

Hughes, S. J. 1953. Conidiophores, conidia, and classification. Can. J. Botany **31**:577-659.

———. 1971. Annellophores. *In:* Kendrick, B., Ed., Taxonomy of Fungi Imperfecti. University of Toronto Press, Toronto. pp. 132-140.

Kendrick, B. 1971. Conclusions and recommendations. *In:* Kendrick, B., Ed., Taxonomy of Fungi Imperfecti. University of Toronto Press, Toronto. pp. 253-262.

———, and J. W. Carmichael. 1973. Hyphomycetes. *In:* Ainsworth, G. C., F. K. Sparrow, and A. S. Sussman, Eds., The fungi— an advanced treatise. Academic Press, New York. **4A**:323-509.

Kreger-Van Rij, N. J. W. 1973. Endomycetales, basidiomycetous yeasts, and related fungi. *In:* Ainsworth, G. C., F. K. Sparrow, and A. S. Sussman, Eds., The fungi—an advanced treatise. Academic Press. New York. **4A**:11-32.

Luttrell, E. S. 1963. Taxonomic criteria in *Helminthosporium*. Mycologia **55**:643-674.

———, 1964. Systematics of *Helminthosporium* and related genera. Mycologia **56**:119-132.

Morris, E. F. 1963. The synnematous genera of the Fungi Imperfecti. Western Illinois Univ. Ser. Biol. Sci. **3**:1-143.

Nelson, R. R. 1960. *Cochliobolus intermedius*, the perfect stage of *Curvularia intermedia*. Mycologia **52**:775-778.

———, 1964. The perfect stage of *Helminthosporium spiciforum*. Mycologia **56**:196-201.

Pirozynski, K. A. 1971. Characters of conidiophores as taxonomic criteria. *In:* Kendrick, B., Ed., Taxonomy of Fungi Imperfecti. University of Toronto Press, Toronto. pp. 37-49.

Raper, K. B., and D. I. Fennel. 1965. The genus *Aspergillus*. The Williams & Wilkins Company, Baltimore. 686 pp.

Rogers, J. D. 1965. The conidial stage of *Coniochaeta ligniaria*: morphology and cytology. Mycologia **57**:368-378.

Subramanian, C. V. 1962. A classification of the Hyphomycetes. Current Sci. **31**:409-411.

———. 1971. The phialide. *In:* Kendrick, B., Ed., Taxonomy of Fungi Imperfecti. University of Toronto Press, Toronto. pp. 92-119.

Sutton, B. C. 1973. Coelomycetes. *In:* Ainsworth, G. C., F. K. Sparrow, and A. S. Sussman, Eds., The fungi—an advanced treatise. Academic Press, New York. **4A**:513-582.

Tanaka, K., and T. Yanagita. 1963. Electron microscopy on ultrathin sections of *Aspergillus niger* II. Fine structure of conidia-bearing apparatus. J. Gen. Appl. Microbiol. **9**:189-200.

Trinci, A. P. J., A. Peat, and G. H. Banbury. 1968. Fine structure of phialide and conidiospore development in *Aspergillus giganteus* "Wehmer." Ann. Botany **32**:241-249.

Tubaki, K. 1966. Sporulating structures in Fungi Imperfecti. *In:* Ainsworth, G. C., and A. S. Sussman, Eds., The fungi—an advanced treatise. Academic Press, New York. **2**:113-131.

Part Two

PHYSIOLOGY
AND   REPRODUCTION

SEVEN

Growth

All organisms have the potential to increase in mass by cell division, cell enlargement, or both. Such an increase in mass is termed *growth*. These processes play roles of varying importance in different organisms. In comparatively simple, unicellular organisms (including yeasts), growth occurs predominantly by cell division, which increases the population and is therefore a reproductive process. Cell enlargement, accompanied by nuclear division and synthesis of cytoplasm, is primarily responsible for growth of fungi with coenocytic, nonmycelial thalli such as members of the Blastocladiales. Mycelial fungi, like other complex multicellular organisms, grow through a combination of cell division and enlargement.

MECHANICS OF GROWTH IN MULTICELLULAR FUNGI

A fungal hypha consists primarily of a rigid cell wall surrounding a vacuolated cytoplasm and a hyphal tip (with a plastic wall, a disproportionately large number of nuclei, and a nonvacuolated cytoplasm). This hyphal tip constitutes approximately the last 50 to 100 microns of the hypha. As described in Chapter 1, growth in mycelial fungi occurs by extension of the hyphal tip, while the older portions of the hyphae are incapable of growth. Although the older hypha is not capable of growth, it has an important role in supporting growth of the tip as new protoplasm is formed throughout the hypha and transported to the tip by active cytoplasmic streaming. (The use of radioactive precursors indicates that the rates of protein and ribonucleic acid synthesis do not decrease substantially in hyphal regions quite distant from the hyphal tip). It is this active translocation of synthesized components that makes possible the rapid growth rate of the hyphal tip (Zalokar, 1959), which for *Neurospora* in culture is about 3.6 to 3.8 millimeters/hour (Ryan, et al., 1943).

The developing hypha usually forms branches in acropetal succession (proceeding towards the tip) behind the hyphal tip. Branch formation is preceded by a localized softening of the previously hardened hyphal wall, and a branch develops from this site. This primary branch in turn usually gives rise to secondary and tertiary branching systems. When a branch is initiated, a thickened cell wall must be differentiated and a new growing apex established. This process is extremely rapid, requiring only about 40 to 60 seconds (Burnett, 1976). Apical dominance is typically apparent; that is, a hypha that is producing branches continues to grow at a more rapid rate and is usually longer than the branches it is producing. The typical growth pattern, therefore, is that of a long central axis from which shorter axes branch; each central axis is longer than the branches that it bears, the whole strongly resembling a two-

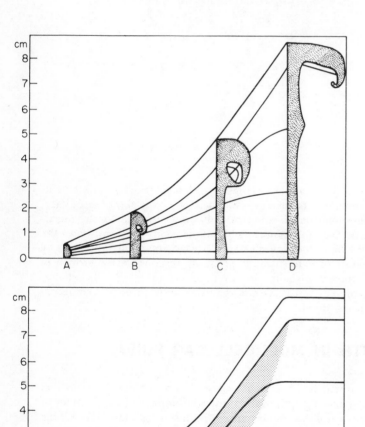

Fig. 7–1. Above: A diagrammatic growth curve of *Agaricus campestris* showing four different stages. The lines are drawn through homologous points. The horizontal axis is only a very rough approximation of time. Below: This is the same graph as above, but those regions of the stalk that are in the process of elongation (i.e., the growth zone) lie within the shaded area. In this way it is possible at each stage of development to see the extent of the growth zone. [Figures and legend from J. T. Bonner et al., 1956, *Mycologia* **48:** 13–19.]

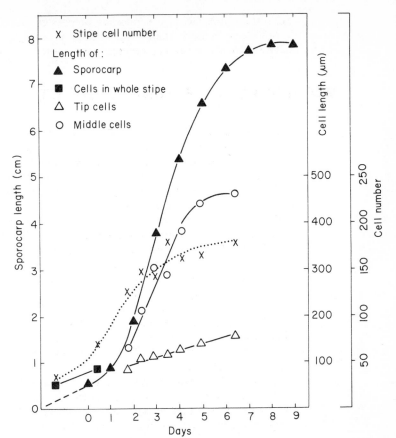

Fig. 7-2. Growth of a sporocarp is determined by the growth rate and number of all individual cells, as illustrated in the agaric *Flammulina velutipes*. [Adapted from W. M. Wong and H. E. Gruen, 1977, *Mycologia* **69**:899–913].

dimensional replica of a pine tree. This growth pattern may break down in culture, where the mycelium shows a strong tendency to occupy all medium within the circumference of the developing colony. In this case, apical dominance of the lead hyphae may be lost and branches from the lead hyphae may grow vigorously into unoccupied space. The effect of this type of growth is to form a circular colony on a solid medium on which there are no obstructions or to form a spherical colony within a liquid medium. Mechanisms controlling branching and apical dominance are not known.

Complex sporocarps and sclerotia are composed of mycelium that is modified to form distinctive cells and tissue. Sporocarps are characterized by zones of tissues which elongate, expand, and change during development to give the sporocarp its final mature form (Fig. 7-1). Some of the developing tissue zones (usually consisting of tightly packed parallel hyphae) appear to be actively growing from their apices, which have the dense cytoplasm and numerous nuclei characteristic of a growing hyphal tip. Unlike hyphae, the cells in an ascocarp or basidiocarp may undergo intercalary enlargement, elongation, or septation after their formation (Fig. 7-2). In an ascocarp, additional hyphae may push their way up through partially developed tissue, thereby increasing the mass of the sterile tissue of the ascocarp and the hymenium.

Sclerotia which lack the somewhat precise growing points found in ascocarps and basidiocarps achieve their final size by prolific hyphal elongation and branching, followed by a marked enlargement of the cells.

MEASUREMENT OF FUNGAL GROWTH

The simplest method of assessing fungal growth is by a linear measurement. The change in the radius of a developing colony on agar may be observed over a period of time. Alternatively, a horizontal growth tube of about 40 centimeters in length containing a layer of agar may be employed by measuring from the point of inoculation to the advancing edge of the mycelium at intervals or by the microscopic measurement of individual hyphae. These methods have the virtue of being extremely simple, are of value in making rough estimates of growth, and are nondestructive; this allows repeated observations of the same mycelium. They fail to account for differences in aerial mycelium or mycelium submerged in the agar and also fail to distinguish between mycelial spread over the medium and the actual total production of mycelium.

A more accurate method of assaying growth of a fungus is assessment of dry-weight gain. A liquid medium is inoculated with the fungus and, after the growth period, the resulting mycelium is removed by filtration or centrifugation, is ashed, and then is weighed. Unlike the linear techniques, this method is a destructive one and large numbers of cultures must be used successively in order to obtain data about growth rates over a period of time, and it may measure accumulated storage products rather than new protoplasm.

Growth of single-celled fungi, such as the yeasts, may be assessed by bacteriological methods, such as changes in the light absorption or light scattering properties of the liquid culture medium. These techniques are simple to carry out and allow repeated observations on a single culture.

KINETICS OF FUNGAL GROWTH

If the growth rate of single-celled organisms such as bacteria or yeasts is followed in a liquid culture from the time of inoculation until decline, a typical pattern evolves which may be divided into stages (Fig. 7–3). Stage I (the *lag* phase) is a period occurring after inoculation and before any cell division takes place. Stage II (the *acceleration* phase) is initiated when cell division begins and marks the transition between dormancy and active growth. Stage III (the *exponential* phase) is attained when cell division reaches and maintains a uniform rate. It is during stage III that the greatest increase in cell numbers takes place; this is an exponential increase in numbers. Stage IV (the *deceleration* phase) is characterized by a decrease in cell division until Stage V (the *stationary* phase) is reached where either there is no additional growth or production of new cells is exactly balanced by death of old cells. Eventually senescence occurs and is marked by a decrease in the number of viable cells (Stage VI, the *decline* phase). The actual time required for each phase varies with the particular organism and with the specific conditions under which it is cultured.

Although stages can be rather precisely defined in the case of unicellular organisms in a liquid medium, mycelial fungi do not lend themselves to such a precise quantitative analysis. Division of each cell into two cells resulting in exponential

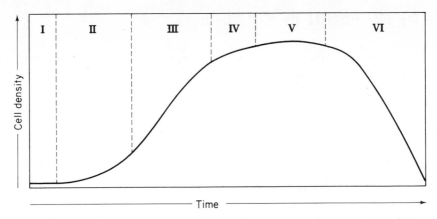

Fig. 7-3. An idealized growth curve for a single-celled organism such as a yeast. See text for a discussion of the stages indicated.

increase in unicellular organisms is lacking and there is differential activity of the cells. Growth is restricted to the hyphal tip only, which grows more or less at a constant linear rate and therefore extends the margin of the colony at a linear rate. Although each hypha extends at a linear rate, growth of *Chaetomium globosum* occurs exponentially during the early stages of colony development if the entire branching mycelium is considered. During the later stages of colony development, however, growth in the center of the colony declines while the marginal hyphae maintain their linear rate of growth (Plomely, 1959). In general, mycelial growth may be qualitatively divided into the following: Stage I (lag phase) with no apparent growth, Stage II (*linear* phase) with rapid and approximately linear growth, and Stage III (decline phase) during which there is no growth or there is a decline in dry weight owing to autolysis (Fig. 7-4).

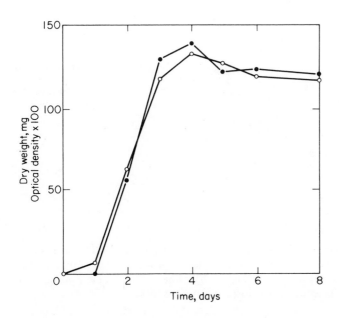

Fig. 7-4. The growth of the mycelium of *Fusarium solani* in an aerated medium. Open circles indicate dry weight; closed circles indicate optical density. [From V. W. Cochrane, 1958, *Physiology of fungi,* John Wiley & Sons, Inc., New York.]

GROWTH OF FUNGI IN CULTURE

Fungi grow normally as saprophytes or parasites on naturally occurring animal or plant products; these are extremely complex materials that provide a number of unknown nutrients for growth of the fungus. It is usually impossible or at least extremely inconvenient for mycologists to study physiology of fungi in their natural habitats, and it is desirable to grow fungi in the laboratory under conditions where nutritional and environmental factors can be controlled and duplicated. Aside from physiology, many other aspects of mycology require that fungi be grown and readily available in the laboratory (such as routine identification of fungi isolated from natural substrata, genetic or developmental studies, and teaching). A requisite, therefore, is to create in the laboratory nutritional and physical conditions that will allow fungi to grow and develop as "normally" as possible.

A primary requisite for laboratory cultivation of fungi is selection of a suitable substratum. The simplest of these are the materials on which the fungi would ordinarily occur in nature (dung, grains, wood, etc.). These are autoclaved and used directly, or fungi may also be grown on water in which these materials were soaked and heated. Media of this type are extremely cheap and easy to prepare and will often support fungi that will not grow on other media. A disadvantage of natural media is that they can never be precisely duplicated for physiological studies as they are of an unknown composition. Alternatively, a completely synthetic medium consisting of chemicals (such as glucose, asparagine, magnesium sulfate, or biotin) which are of both known composition and concentration may be used. A synthetic medium is the only type of medium which may be precisely duplicated and is suitable for meticulous physiological studies, but unfortunately some fungi may grow poorly or not at all on such media.

Basal semisynthetic medium

Glucose _ _ _ _ _ _ _ _ _ _ _ _ _ _ _ _ _ _ 10 g.

Asparagine _ _ _ _ _ _ _ _ _ _ _ _ _ _ _ _ . 2 g.

KH_2PO_4 _ _ _ _ _ _ _ _ _ _ _ _ _ _ _ _ _ 1 g.

$MgSO_4 \cdot 7H_2O$ _ _ _ _ _ _ _ _ _ _ _ _ _ 0.5 g.

Fe^{+++} _ _ _ _ _ _ _ _ _ _ _ _ _ _ _ _ _ 0.2 mg.

Zn^{++} _ _ _ _ _ _ _ _ _ _ _ _ _ _ _ _ _ _ 0.2 mg.

Mn^{++} _ _ _ _ _ _ _ _ _ _ _ _ _ _ _ _ _ 0.1 mg.

Biotin _ _ _ _ _ _ _ _ _ _ _ _ _ _ _ _ 5 μg

Thiamine _ _ _ _ _ _ _ _ _ _ _ _ _ _ _ 100 μg

Distilled water to make _ _ _ _ _ _ _ 1 liter

Agar (for solid media) _ _ _ _ _ _ _ _ 20 g.

Fig. 7-5. A semisynthetic medium which supports growth of a wide variety of fungi. [From V. G. Lilly and H. L. Barnett, 1951, *Physiology of the fungi.* Used with permission of McGraw-Hill Book Company, New York.]

Such fungi may grow quite well on a semisynthetic medium, one which is basically similar to a synthetic medium but to which some natural products such as malt extract or yeast hydrolysate have been added (Fig. 7-5). The natural materials provide unknown nutrients which stimulate growth of the fungus.

A semisynthetic or synthetic growth medium may be used as a liquid or as a liquid solidified with agar. Agar is a gelatinous polysaccharide prepared from red algae. Agar is mixed with the nutrient liquid, heated to melt the agar, and then allowed to cool. When the preparation reaches about 45°C, the agar solidifies to a gel in which the nutritional medium is entrapped. The preparation may be placed in flasks or tubes and then sterilized by steaming, filtration, chemical treatment, or, most commonly, autoclaving; this kills contaminating microorganisms. The medium is then inoculated with spores or mycelium of the fungus.

NUTRITIONAL REQUIREMENTS FOR GROWTH

Nutrient Uptake

DIGESTIVE ENZYMES

Fungi are in direct contact with their nutrients in the environment. Smaller molecules (such as simple sugars and amino acids) in solution in the watery film surrounding the hyphae can be directly absorbed by the hyphae. Larger insoluble polymers such as cellulose, starch, and proteins must undergo a preliminary digestion before they can be used.

Molecules that are too large to be absorbed by the fungus are attacked by extracellular enzymes. Like all digestive enzymes, the digestive enzymes of fungi control hydrolysis reactions which cleave the large molecules into simpler components. The digestive enzymes are highly specific and are able to control hydrolysis of particular molecules only. Complete digestion of a large polymer is a stepwise process involving different enzymes, until finally a simple, soluble molecule is released. It is this simple molecule which is taken up by the fungus. Once it is absorbed into the cell, this small molecule is further acted upon by intracellular enzymes (Fig. 7-6).

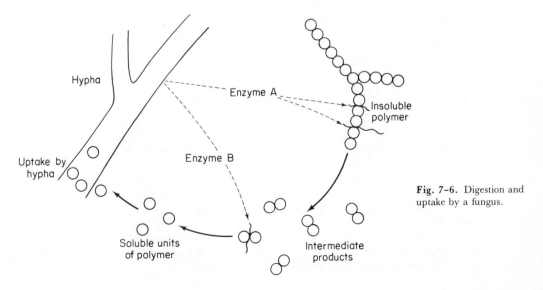

Fig. 7-6. Digestion and uptake by a fungus.

The ability to utilize large molecules ultimately depends on the ability of the fungus to digest them, which in turn depends on the enzymes with which the fungus is equipped. Fungi typically have a large number of enzymes but, for the most part, many of them lie idle until the fungus comes into contact with a substrate on which particular enzymes can act. Growth of the fungus occurs equally well on a medium containing either complex or simple nutrients. This is not always the case, however, and the necessary enzymes may be entirely lacking and the fungus may be unable to grow on a medium which contains an undigestible substrate.

In some instances, *adaptive enzymes* are formed. A fungus may not have the enzymes necessary to digest starch, for example; but if it is transferred to a medium containing starch, it may eventually produce the necessary enzymes and utilize starch. Probably a single mutant nucleus among the many nuclei in the inoculum or those arising during the incubation period survives. Selection pressures are such that the mutant nucleus survives and gives rise to nuclei that allow the genetic machinery to produce the enzymes which were previously nonexistent.

UPTAKE

All ions and molecules entering the fungal cell must pass through both the cell wall and the plasmalemma, a unit membrane consisting primarily of lipid and protein. The wall itself is somewhat porous, allowing ions and molecules to pass through it. The exact nature of this porosity is unknown, but it has been postulated that minute pores or channels may exist (Burnett, 1976). A nonelectrolyte in solution may pass through the plasmalemma in response to a concentration gradient, moving from a greater to a lesser solute concentration. Ions in solution will likewise pass across the membrane from a greater to a lesser solute concentration, but there the driving force is a difference in electrochemical potential rather than the difference in concentration. In many cases, nonelectrolytes or ions are accumulated by the cell, but this accumulation cannot be accounted for by existing gradients. The mechanism is known as *active transport* and requires that cells expend metabolic energy to accumulate these materials. Active transport may be explained by assuming that a *carrier system* operates: that is, the solute combines with a specific substance (the carrier) which is capable of moving across the membrane and then dissociating from the solute which has thus been transferred from one side of the membrane to another, thereby gaining entry to the cell. The necessary metabolic energy to drive the carrier system may be provided by ATP or by an electrochemical gradient.

Uptake of solutes has been studied extensively in yeast cells. The membrane of the yeast cell has a low permeability to ions, allowing few to enter or leave in response to an electrochemical gradient. This has an ecological advantage when one considers that the normal environment of the yeast cell is a dilute medium that would draw ions from the cytoplasm. Instead, the yeast cell has a highly efficient transport capacity for ions of physiological importance, especially potassium, magnesium, and phosphate. The size of the yeast cell, like that of other walled cells, is regulated by the cell wall and not by the osmotic concentration. There is no need for the cell to control osmotic concentration, and the yeast cell can therefore accumulate and store large quantities of ions. These stored ions may be used in growth and cell division when nitrogen becomes available.

Of the nonelectrolytes, the yeast cell accumulates sugars, amino acids, and nucleotides by specialized transport systems and organic acids both by a specialized transport system and by free diffusion.

Experimental work on the uptake of electrolytes and nonelectrolytes has not been as extensive in the filamentous fungi as in the yeasts. The mycorrhizal fungi (Chapter 13) have been studied to a greater extent than other filamentous fungi. Generally the data derived from studies on filamentous fungi correspond to those obtained from studies of yeast. Most nutrients enter fungal cells by specialized transport systems; the exceptions are ammonia and possibly nitrate ions, which enter through passive diffusion. Many factors such as the pH, temperature, particular combination of electrolytes, or physiological condition of the fungus may affect uptake of a given material (Burnett, 1976).

Little is known about the manner in which water enters the fungal cell, but it is generally assumed to enter by osmosis.

Once the nutrients and water have gained entrance to the cell, they are often translocated from one site to another. Translocation is especially important in fungal hyphae, where growth at the tips must be supported, and when macroscopic structures such as sporocarps or sclerotia are formed. Cytoplasmic streaming is generally believed to play a role in intracellular transport. Rates of cytoplasmic streaming vary greatly. They may be as high as 90 centimeters/hour (Burnett, 1976) but are usually in the range of 2 to 20 centimeters/hour. Cytoplasmic streaming occurs in all types of hyphae and is not affected by the presence or absence of pores. This streaming is also bidirectional.

Essential Elements

Some elements are essential for the growth of fungi, and if an essential element is lacking from the medium or substratum, the fungus will not survive no matter how abundant other elements are.

Studies to determine which elements are essential may be of two basic types. First, mycelia and spores may be analyzed to determine what elements are present. If certain elements (such as carbon, nitrogen, or potassium) are always present regardless of the composition of the medium on which the fungus was grown, one may assume that these elements are essential for growth of the fungus. The second, and more commonly used method, is that of conducting nutritional studies using synthetic media from which the elements in question are eliminated one at a time. If a fungus is unable to grow when sulfur, for example, is eliminated, it is assumed that sulfur is essential. Nutritional studies require that sources of chemical contamination be eliminated as nearly as possible; for this reason, chemicals must be of the greatest purity, water must be glass-distilled, glassware must be chemically clean and preferably of quartz, and agar either must not be used or must be highly purified (agar normally contains several trace elements and vitamins that exert a favorable effect on the growth of fungi).

ESSENTIAL MACROELEMENTS

Almost the entire fungus mycelium is composed of the nonmetallic elements carbon, nitrogen, hydrogen, oxygen, sulfur, and phosphorus and the metallic elements potassium and magnesium. These elements are the *macroelements* which are required in comparatively large quantities by fungi. Carbon, nitrogen, hydrogen, and oxygen are used in the formation of fungal walls, and all these elements take a functional role in the constantly occurring metabolic events in the protoplasm. Hydrogen is obtained from water or when organic compounds are metabolized. Oxygen is obtained from the atmosphere during respiration (Chapter 8). We shall concentrate our attention on carbon, nitrogen, and the remaining macroelements.

Carbon. About half of the dry weight of fungus cells consists of carbon, which gives an indication of the important role of carbon compounds within the cell. Organic compounds are used as structural materials and also provide energy to the cell upon oxidation. Carbon is required in greater quantities than any other essential element by the fungus, and carbon nutrition is of paramount importance to the fungus.

Fungi may use a wide variety of organic compounds or carbon dioxide as a source of carbon. Even though atmospheric carbon dioxide can be utilized by some fungi, it cannot supply enough carbon to serve as the only source. Organic compounds which may be used by the fungi include carbohydrates (mono-, di-, oligo-, and polysaccharides) and organic acids. Carbohydrates constitute the most important carbon sources. Fungi differ widely in their abilities to use different carbon sources, and their ability to utilize a particular carbon source may be altered by the combination of nutrients present or by other cultural conditions, such as pH. If a fungus is provided with a mixture of carbon sources, it will sometimes use one carbon source preferentially over the others present. Also a mixture of carbon sources may support more growth than a single carbon source (e.g., galactose and glucose in combination may give much more growth than galactose or glucose alone) (Burnett, 1976).

Carbohydrates. MONOSACCHARIDES AND THEIR DERIVATIVES. Monosaccharides are simple sugars which have either five carbon atoms (pentoses) or six carbon atoms (hexoses). Each sugar molecule has an aldehyde (—CHO) and ketone (= CO) group. These groups may be reduced to form an alcohol derivative of the sugar or oxidized to form an acid derivative of the sugar. The sugars may differ in the spatial configuration of the molecule; one series consists of the D and L forms which differ in the position of the hydroxyl group nearest the aldehyde grouping. Only some of the D and L forms occur naturally.

The sugar supporting growth of almost all fungi is D-glucose, a naturally occurring hexose (Fig. 7–7). Many fungi can make equally as good growth on D-fructose and D-mannose as on glucose (exceptions to this general rule are found among the Chytridiales, Blastocladiales, and Saprolegniales). D-Galactose is used by the majority of the fungi, but few of them grow as well on D-galactose as on D-glucose. Fungi which at first may make no growth on D-fructose, D-mannose, or D-galactose may initiate growth after a period of time due to adaptive enzyme formation. Although a sugar other than glucose may give maximum growth of a fungus, it appears that the more closely the configuration of a sugar resembles glucose, the more fungi use it. This is perhaps because the ability of a fungus to use a particular sugar depends on how easily it can be converted to a phosphorylated derivative of glucose which can enter the respiratory pathways.

Of the naturally occurring pentoses, D-xylose (Fig. 7–8) supports growth of the largest number of fungi. Growth of a particular fungus on D-xylose may be equal to or better than that on D-glucose, although some fungi may not grow at all on D-xylose. L-Arabinose may support satisfactory growth of some fungi but the amount of growth is usually less than on either D-glucose or D-xylose. The remaining pentoses may be utilized by some fungi but in general appear to be poor sources of carbon.

Many sugar alcohols, such as sorbitol, glycerol, and mannitol, occur in nature. These sugar alcohols may be utilized by fungi as a carbon source, but usually somewhat less satisfactorily than the corresponding simple sugar. An exception is mannitol, derived from the reduction of D-fructose or D-mannose, which may give growth equivalent to that obtained on glucose for some fungi. However, many fungi which have wide substrate ranges will not grow at all on mannitol.

Acids derived from simple sugars and glycosides may serve as a carbon source for certain fungi, but data are rare and it is not possible to make generalizations about the utilization of these compounds by fungi.

DISACCHARIDES AND POLYSACCHARIDES. Simple sugars or their derivatives may combine to form complex, chainlike polymers. Either like or unlike sugar units (monomers) may combine. Repeating units of the same sugar may form two types of polymers which differ only in the spatial configuration (alpha or beta) of the glycoside linkage. If the polymer consists of only two monomers, it is a disaccharide; longer polymers are polysaccharides.

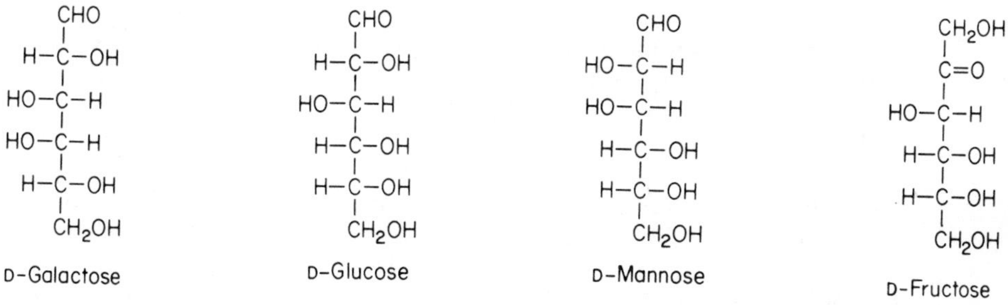

$$
\begin{array}{cccc}
\text{CHO} & \text{CHO} & \text{CHO} & \text{CH}_2\text{OH} \\
\text{H}-\text{C}-\text{OH} & \text{H}-\text{C}-\text{OH} & \text{HO}-\text{C}-\text{H} & \text{C}=\text{O} \\
\text{HO}-\text{C}-\text{H} & \text{HO}-\text{C}-\text{H} & \text{HO}-\text{C}-\text{H} & \text{HO}-\text{C}-\text{H} \\
\text{HO}-\text{C}-\text{H} & \text{H}-\text{C}-\text{OH} & \text{H}-\text{C}-\text{OH} & \text{H}-\text{C}-\text{OH} \\
\text{H}-\text{C}-\text{OH} & \text{H}-\text{C}-\text{OH} & \text{H}-\text{C}-\text{OH} & \text{H}-\text{C}-\text{OH} \\
\text{CH}_2\text{OH} & \text{CH}_2\text{OH} & \text{CH}_2\text{OH} & \text{CH}_2\text{OH} \\
\text{D-Galactose} & \text{D-Glucose} & \text{D-Mannose} & \text{D-Fructose}
\end{array}
$$

Fig. 7–7. Naturally occurring hexoses that are widely used by fungi.

$$
\begin{array}{cc}
\text{CHO} & \text{CHO} \\
\text{H}-\text{C}-\text{OH} & \text{H}-\text{C}-\text{OH} \\
\text{HO}-\text{C}-\text{H} & \text{H}-\text{C}-\text{OH} \\
\text{HO}-\text{C}-\text{H} & \text{H}-\text{C}-\text{OH} \\
\text{CH}_2\text{OH} & \text{CH}_2\text{OH} \\
\text{L-Arabinose} & \text{D-Ribose}
\end{array}
$$

$$
\begin{array}{c}
\text{CHO} \\
\text{H}-\text{C}-\text{OH} \\
\text{HO}-\text{C}-\text{H} \\
\text{H}-\text{C}-\text{OH} \\
\text{CH}_2\text{OH} \\
\text{D-Xylose}
\end{array}
$$

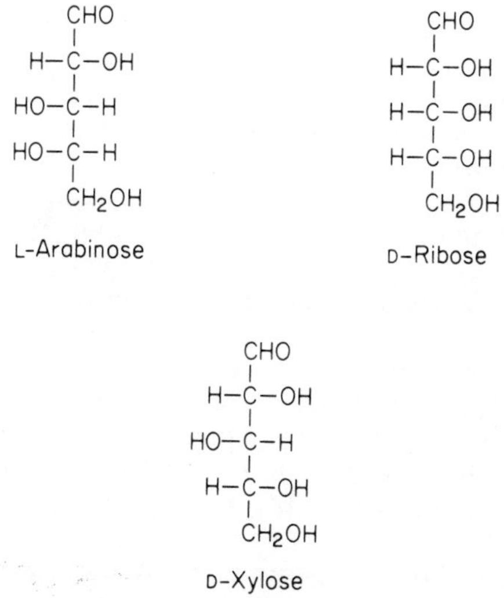

Fig. 7–8. Naturally occurring pentoses that are widely used by fungi.

Both disaccharides and polysaccharides are important sources of carbon in nature. High-molecular-weight polysaccharides constitute the bulk of the carbohydrates produced by animals and plants. Disaccharides are abundantly produced and also may be freed from polysaccharides upon their breakdown. In order to utilize either di- or polysaccharides, the fungi must produce extracellular digestive enzymes which will cleave the glycoside linkages between the monomers. After the sugars or their derivatives are freed by digestion, the fungus can usually absorb and utilize the simple sugar in the same fashion that it would utilize a freely occurring sugar. It is evident that the ability of the fungus to utilize these large compounds depends on its ability (1) to digest them and (2) to absorb the component simple sugars. Usually any inability to utilize a complex carbohydrate is due to the failure to hydrolyze it and not to the inability to absorb the simple sugars. A fungus capable of hydrolyzing a given polymer is usually able to grow on any of the component sugars if they are provided in their simple form (any fungus which can grow on a glucose polymer can grow on free glucose).

The most commonly occurring disaccharides are maltose, cellobiose, lactose, and sucrose. Maltose, a hydrolysis product of starch, consists of glucose molecules joined by an alpha-glycoside linkage. Cellobiose, like maltose, also consists of glucose molecules but differs by having a beta-glycoside linkage. Cellobiose is a hydrolysis product of cellulose. Lactose, a component of milk, is composed of glucose and a molecule of galactose. Sucrose, which occurs in plants, contains one molecule of glucose and one of fructose. These disaccharides are digested to their component molecules by the enzymes maltase, cellobiase, lactase, and sucrase, respectively. Maltose and cellobiose are widely available to fungi; fewer fungi use sucrose; and even fewer use lactose.

Polysaccharides widely available in nature include the pentosans (polymers of pentose); glycogen, starch, and cellulose (all polymers of D-glucose but differing in degree of branching or configuration of glycoside linkage); pectins (polymers of D-galacturonic acid, an acid derivative of D-galactose); and hemicelluloses, lignins, and gums (all of a complex and poorly understood composition). All of these polysaccharides are utilized by at least some fungi. Starch and cellulose are widely utilized as carbon sources by fungi. The breakdown of starch, a storage product of plants, is diagrammed in Fig. 7–9. Digestion of cellulose, hemicelluloses, and lignins and the role of these in nature is discussed more fully in Chapter 11.

ORGANIC ACIDS. An organic acid possesses one or more carboxyl groups. Organic acids include the monocarboxylic acids (fatty acids that form a fat when esterified with glycerol), dicarboxylic acids (the four-carbon acids of the citric acid cycle), and both monocarboxylic and dicarboxylic amino acids that may combine to form a protein. Both fatty acids and amino acids may be derived from these larger molecules by digestion. Most fungi apparently produce the enzyme lipase, which hydrolyzes fats to release fatty acids. Similarly, peptones may be hydrolyzed to release their component amino acids.

The organic acids may serve as a sole carbon source for some fungi but, in general, fungi grow poorly or not at all on them. Aside from the intrinsic incapability of many cells to metabolize organic acids, the general unavailability of these acids may be partially a result of impermeability of the cell to them and of their tendency to interfere with growth by causing an increase in pH of the culture medium if used in the neutralized form (Cochrane, 1958).

Nitrogen. Nitrogen is required by all organisms to synthesize amino acids and,

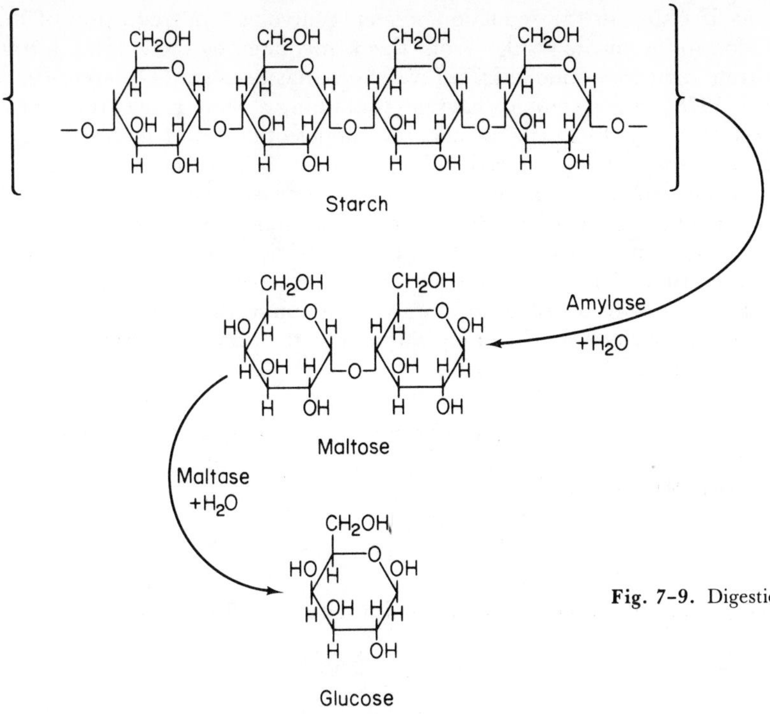

Starch

Maltose

Glucose

Amylase
+H₂O

Maltase
+H₂O

Fig. 7–9. Digestion of starch.

from these, proteins, which are required to build protoplasm. Without protein, growth could not occur. Nitrogen is also a component of the nucleic acids and some vitamins.

Fungi may use inorganic nitrogen in the form of nitrates, nitrites, or ammonia or organic nitrogen in the form of amino acids. It is questionable whether fixation of atmospheric nitrogen, which is so well-known among the bacteria, occurs in the fungi. Not all fungi use nitrogen sources with equal facility, and a fungus may have a requirement for nitrogen in a specific form.

Nitrates. Numerous fungi utilize nitrates as a form of nitrogen, but inabilities to use nitrates are common among the Blastocladiales, Saprolegniaceae, yeasts, and the higher basidiomycetes. The nitrate ion may be incorporated into the medium as ammonium nitrate, potassium nitrate, sodium nitrate, and calcium nitrate (Cochrane, 1958).

The nitrate ion, which is taken up by active transport, must be reduced to the oxidation level of ammonia (NH_3) before the nitrogen can be assimilated into organic compounds. Overall, this stepwise reduction may be summarized:

$$\text{Nitrate} \longrightarrow \text{nitrite} \longrightarrow \genfrac{}{}{0pt}{}{\text{Unknown}}{\text{intermediate}} \longrightarrow \text{hydroxylamine} \longrightarrow \text{ammonia}$$

Oxidation states:

$$+5 \qquad +3 \qquad +1 \qquad -1 \qquad -3$$

The first step in the reduction sequence involves the reduction of the nitrate (NO_3^-) ion to the nitrite (NO_2^-) ion: this is mediated by the enzyme nitrate reductase. Nitrate reductase obtained from *Neurospora* has been studied extensively (Garrett and Amy, 1978). It is a complex enzyme containing a heme moiety (cytochrome b_{557}), sulfhydryl group(s), FAD (flavin adenine dinucleotide), and molybdenum (Garrett and Nason, 1969). When the reduction of nitrate occurs, the enzyme moieties participate in the sequential transfer of electrons. This sequence begins when NADPH (reduced nicotinamide adenine dinucleotide phosphate) serves as the electron donor and the FAD, heme (cytochrome b_{557}), and molybdenum of the enzyme function as sequential carriers. Ultimately, the nitrate ion receives electrons and is reduced to nitrite. This pathway operates in close association with cytochrome c, which may receive part of the electrons. We may summarize the flow of electrons in *Neurospora* as follows (Garrett and Nason, 1969):

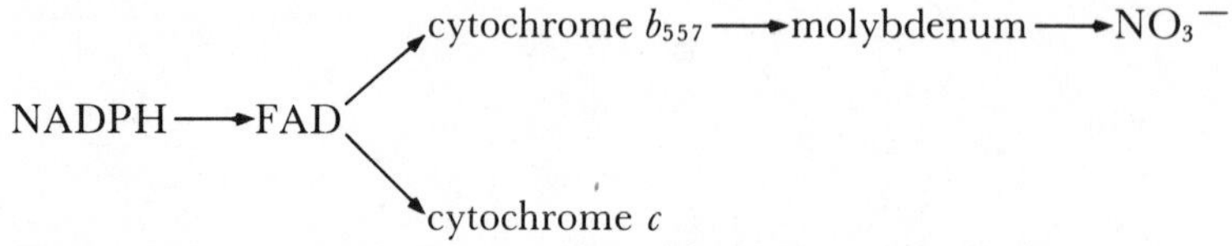

$$NADPH \longrightarrow FAD \nearrow \text{cytochrome } b_{557} \longrightarrow \text{molybdenum} \longrightarrow NO_3^- \\ \searrow \text{cytochrome } c$$

The final portion of the above scheme culminating in nitrate reduction is:

$$\begin{array}{c} Mo^{5+} \\ Mo^{6+} \end{array} + 2e \begin{array}{c} NO_3^- \\ NO_2^- + H_2O \end{array}$$

Subsequent reduction of the nitrite ion (NO_2^-) to ammonia is not well understood, and it is not known if the intermediates are all inorganic or are partially organic. If it is assumed that the pathway is entirely inorganic, and that two electron changes are made for each step, then the following may be postulated (Nason, 1962; Nason and Takahashi, 1958):

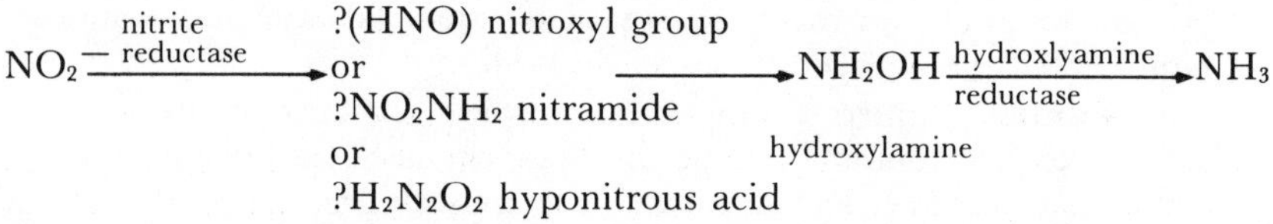

$$NO_2^- \xrightarrow[\text{reductase}]{\text{nitrite}} \begin{array}{l} ?(HNO) \text{ nitroxyl group} \\ \text{or} \\ ?NO_2NH_2 \text{ nitramide} \\ \text{or} \\ ?H_2N_2O_2 \text{ hyponitrous acid} \end{array} \xrightarrow{} NH_2OH \xrightarrow[\text{reductase}]{\text{hydroxylamine}} NH_3$$

hydroxylamine

The identity of the $+1$ oxidation state intermediate is unknown, and it may be either an inorganic or organic compound. Nitrite reductase, which reduces nitrite, and hydroxylamine reductase, which reduces hydroxylamine, have been detected in extracts from *Neurospora*. Both are pyridine-nucleotide enzymes conjugated with an unidentified metal. Like nitrate reductase, nitrite reductase has a heme moiety (Garrett and Amy, 1978).

In general, any fungus that can reduce and assimilate the nitrate ion can utilize any form of nitrogen with an oxidation level equivalent to that of ammonia (such as the ammonium ion or organic nitrogen). There are, however, some fungi that utilize nitrate exclusively.

Nitrites. Nitrites may serve as the sole source of nitrogen for some fungi, and some nitrate fungi may be able to utilize nitrites. Nitrites tend to be toxic to most species of fungi, however, especially if they accumulate in the medium and cannot be used to satisfy the nitrogen requirement. For this reason, nitrites are not used in the preparation of routine laboratory media.

Ammonium. Numerous fungi are unable to utilize nitrates as a source of nitrogen but require a more reduced form of nitrogen, presumably because they cannot reduce the nitrate ion. These fungi may use nitrogen in the form of the ammonium ion or in the form of organic nitrogen, which has the same oxidation level as the ammonium ion. The ammonium ion may be supplied as an ammonium salt. Ammonium ions of different salts are equivalent physiologically, but the anions are not. Ammonium ions of organic salts do not cause a drop of the pH of the medium due to the buffering capacity of the organic ion. If ammonium nitrate is added to the medium, preferential utilization of the ammonium ion causes a decrease in the pH of the medium, perhaps to the point where it limits growth. When ammonium nitrate is added to the medium, preferential utilization of the ammonium ion is the rule, although the nitrate ion is present. Often the nitrate ion will not be utilized until the ammonium ion is exhausted. This preferential utilization is probably the result of a feedback repression, whereby the end product represses the activity of an enzyme(s) producing it. Ammonia will repress nitrate reductase activity, thereby repressing conversion of nitrate to nitrite. It is possible that preferential utilization of ammonia over nitrate is of some ecological value for soil fungi. In the soil, both nitrates and ammonia usually occur together, but it requires less energy expenditure by the fungus to use the reduced form of nitrogen.

Organic Nitrogen. Essentially all fungi utilize nitrogen supplied to them in the organic form, although only a comparatively small number must obtain their nitrogen in this form. The amino acid-requiring fungi include some of Oomycetes, (*Leptomitus lacteus* and *Apodachlya brachynema*), and species of *Blastocladia*. Fungi as a group actively decompose proteins in nature to their component amino acids, which can be assimilated. In culture, organic nitrogen may be supplied as amino acids, peptides, or peptones, the last two yielding amino acids upon hydrolysis. The fungus species vary greatly in this response to the different amino acids and use some of them much more readily than others. Although the response of different species to any single amino acid varies widely, asparagine most often gives good growth. Good growth is generally obtained also with glycine, glutamic acid, and aspartic acid. That amino acid most often giving poor growth is leucine. Growth is usually better with a mixture of amino acids than with a single acid.

Sulfur. Fungi have a requirement for sulfur, which is generally satisfied in culture by the incorporation of the sulfate ion, $SO_4{}^{2-}$ (usually as magnesium sulfate), in the medium. Fungi reduce the sulfate ion by an unknown mechanism and metabolize needed sulfur compounds from the reduced sulfur. Sulfur compounds detected in fungi include amino acids (cysteine, cystine, and methionine), the tripeptide glutathione, vitamins (thiamine and biotin), antibiotics (penicillin and gliotoxin), and other miscellaneous compounds.

Some fungi are unable to reduce the sulfate ion and have a requirement for reduced sulfur. These fungi include several members of the Saprolegniales, all members of the Blastocladiales, and some induced mutants of higher fungi. Some fungi may be able to utilize organic sulfur although they do not require it. Reduced sulfur may be supplied as cysteine, cystine, glutathione, and methionine, although the fungus may utilize one form better than another.

Phosphorus, Potassium, Magnesium. Approximately 0.001 M to 0.004 M concentrations of phosphorus, potassium and magnesium are sufficient to support normal fungus growth. The inorganic form is utilized. Phosphorus metabolism is closely interrelated with respiration and carbohydrate metabolism, and a high concentration of phosphorus in the medium increases utilization of carbohydrates. Magnesium is involved with activation of enzyme systems, particularly those utilized in aerobic and anaerobic respiration. Although fungi will not grow in the absence of potassium, its role in the cell is not known.

ESSENTIAL MICROELEMENTS

Metallic elements required in trace quantities by fungi are iron, zinc, copper, manganese, molybdenum, and either calcium or strontium. Some fungi may have a requirement for a microelement which is not generally shared by other fungi. *Aspergillus niger,* for example, requires gallium (Steinberg, 1938) and scandium (if glycerol is used as a carbon source) (Steinberg, 1939), while other fungi may require cobalt. Boron, an essential microelement for green plants, is needed in only trace quantities, if at all, by fungi.

Microelements are usually required in concentrations ranging from 0.0001 ppm to 0.5 ppm (parts per million). Molybdenum is needed in such small quantities that it is difficult to establish the actual amount needed, and estimates range between 0.1 ppb and 10 ppb (parts per billion). Quantities of microelements greatly in excess of the required amount are generally toxic. When media are prepared, these microelements are not usually directly added, since chance impurities present on glassware, in the distilled water, or from chemicals usually provide them in sufficient amounts. Some of them may even be introduced on dust particles that fall into the medium during preparation. All the essential microelements, in addition to 15 other minor elements, occur in yeast extract, often used in media (Grant and Pramer, 1962).

Microelements play diverse roles in the cell but are chiefly associated with enzymes. An enzyme may be activated by the microelement or may contain the microelement as part of its structure. Molybdenum, for example, participates as an electron carrier in the enzymatic reduction of nitrates in some fungi (see also p. 288). The microelements may also be structural components of vitamins and other metabolites and are therefore required for their synthesis. Iron, for example, is contained within the enzyme catalase, the cytochromes involved in electron transport, and other metabolites, including growth factors and pigments. Microelements are required for normal growth and sporulation, and deficiencies may have various effects on the fungus (Fig. 7-10). The effect of a deficiency for a particular microelement is largely determined by the function of that microelement; for example, a deficiency of zinc or manganese, which normally activate enzymes of the citric acid cycle, may impair the normal operation of this cycle. Manganese may also decrease the rate of sporulation generally, while a deficiency of copper may result in the reduction of pigmentation in the spores of some fungi (Cochrane, 1958).

Fungi may also take up metallic elements that are not known to be essential. These include mercury, nickel, lead, uranium, and possibly others. High concentrations of any of these elements may be toxic, and as mentioned above, excessive quantities of the essential microelements may also be toxic. Mercury and the essential microelement copper are widely used as major constituents of fungicides. In toxic concentrations, much of the copper binds to the plasmalemma and interferes with transport of other materials into the cell. Copper which has gained entrance to the cell may cause genetic mutations (Ross, 1975).

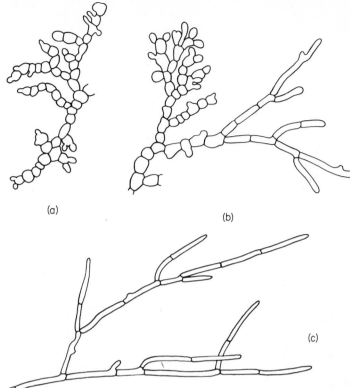

(a)

(b)

(c)

Fig. 7-10. (a) and (b) Severe manganese deficiency in the Pyrenomycete *Chaetomium globosum.* Symptoms are a decrease in elongation of the hypha, increased branching, and formation of globose cells. (c) Normal hyphae growing in medium with 0.5 ppm manganese. All × 750. [From H. L. Barnett and V. G. Lilly, 1966, *Mycologia* **58**:585–591.]

Vitamins

Vitamins are organic compounds that function as coenzymes or as constituent parts of coenzymes which catalyze specific reactions. They are effective in very small quantities and are not used in the manufacture of structural parts of the cell. All organisms require at least some vitamins, but the synthetic capacities vary. Green plants manufacture their own vitamins, but animals must obtain their vitamins from an outside source.

Some fungi, if supplied with a sugar, nitrogen source, and minerals, are able to synthesize all the vitamins that they require. They may even synthesize vitamins in such quantity that they are excreted into the medium. Other fungi may have a partial vitamin deficiency and make limited growth on a vitamin-free medium but much better growth if an exogenous vitamin source is provided (Fig. 7-11). Other fungi have a complete deficiency for a given vitamin and must rely totally on an exogenous supply. Fungi may have only a single vitamin deficiency, or they may have multiple deficiencies and require more than one vitamin. The vitamin deficiency may be absolute and unvarying, or it may be a conditioned deficiency which disappears as the mycelium ages or when external conditions are altered. Vitamin deficiencies that can be conditioned by external factors are partial deficiencies; complete deficiencies are virtually

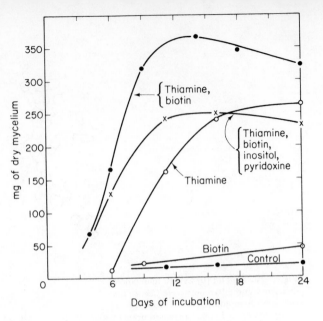

Fig. 7-11. Growth of *Lambertella pruni* in 25 milliliters of liquid glucose-casein hydrolysate medium containing various vitamins. Partial deficiencies for both thiamine and biotin are evident, being greater for thiamine. Note that the addition of inositol and pyridoxine to media containing thiamine and biotin depressed growth. [From V. G. Lilly and H. L. Barnett, 1951, *Physiology of the fungi.* Used with permission of McGraw-Hill Book Company, New York.]

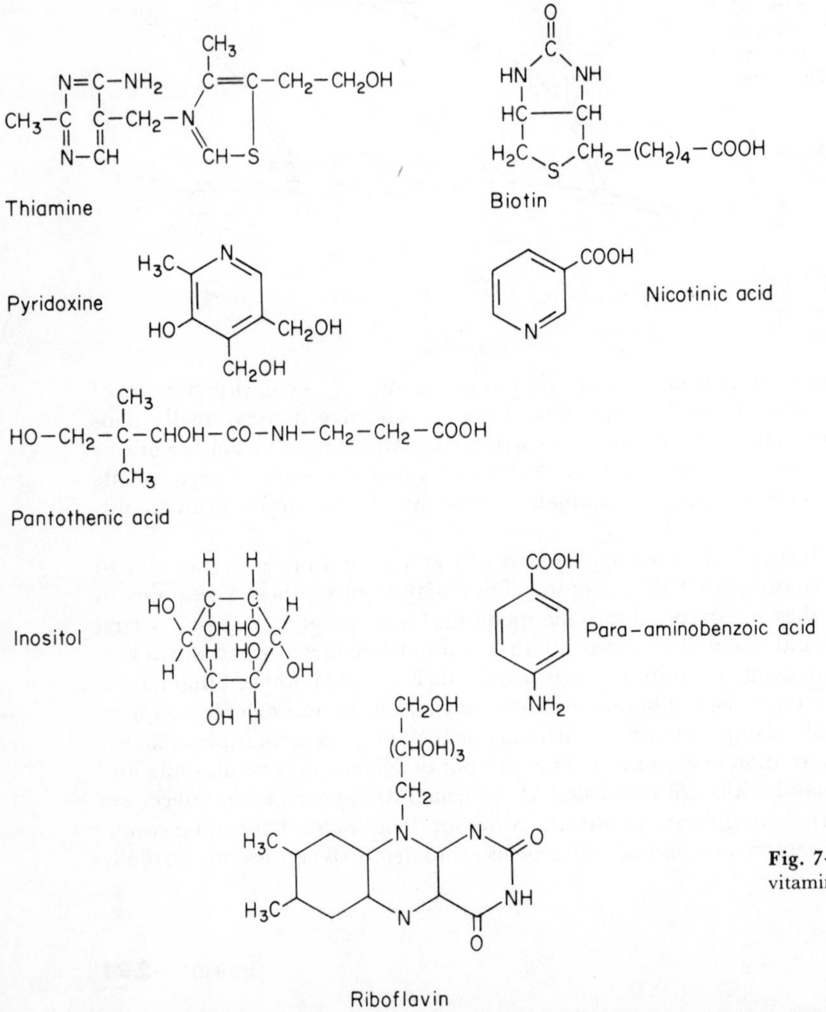

Thiamine

Biotin

Pyridoxine

Nicotinic acid

Pantothenic acid

Inositol

Para-aminobenzoic acid

Riboflavin

Fig. 7-12. Structural formulas of some vitamins important to fungi.

unalterable. A partial requirement becomes more evident as the upper temperature limits allowing growth are reached and sometimes when the composition of the medium is changed. As an example of the latter, *Pythium butleri* becomes thiamine-requiring when the mineral salt concentration exceeds a certain level (Robbins and Kavanagh, 1938). *Pellicularia koleroga* requires both thiamine and biotin when growing on sucrose; when growing on glucose, it requires thiamine alone (Mathew, 1952).

Fungi apparently do not have a need for vitamins A, D, and E, which are required by animals, and synthesis of these vitamins has not been detected in fungi. Fungi do synthesize and require the water-soluble B vitamins and vitamin H.

Fungi often respond with increased growth to additions of natural materials such as tomato or coconut juice, dung or wood extracts, extracts of the host plant, or ash. In many cases, an equivalent growth cannot be obtained by the addition of specific purified vitamins or minerals, and the increase in growth is attributed to unknown *growth factors* present in the natural materials. Those materials giving increased growth may eventually prove to be vitamins, specific minerals, amino acids, or other nutrients. In many cases the growth factors prove to be a chemical which does not ordinarily serve as a nutrient; for example, choline, hemin, or sterols may serve as growth factors for some fungi. Many growth factors have not been identified.

Vitamins manufactured by fungi are thiamine (vitamin B_1), biotin (vitamin H), pyridoxine (vitamin B_6), nicotinic acid, pantothenic acid, riboflavin (vitamin B_2), inositol, and *p*-aminobenzoic acid (Fig. 7–12).

THIAMINE

The greatest number of deficiencies are for thiamine, and any filamentous fungus which has a single vitamin deficiency is most likely to be thiamine-deficient (Fig. 7–13). All species of *Phytophthora* are thiamine-deficient, but only a few basidiomycetes lack this vitamin. Next to biotin, yeasts are most often deficient for thiamine. The thiamine molecule contains two ring structures, a substituted pyrimidine and a substituted thiazole:

thiamine pyrimidine   thiamine thiazole

Thiamine-deficient fungi vary in their capacity to utilize either the pyrimidine or thiazole moiety or the entire thiamine molecule. Many thiamine-deficient fungi must be supplied with the entire preformed thiamine molecule and are apparently unable to synthesize any part of it. Numerous fungi can make the thiazole part of the molecule and will synthesize the thiamine molecule if provided with the pyrimidine moiety. Some fungi have the reverse synthetic capacity and can manufacture the pyrimidine ring only. *Phycomyces* can combine the pyrimidine and thiazole rings to make thiamine, but it is unable to synthesize either intermediate.

Thiamine has a role in the regulation of carbohydrate metabolism. The active form of thiamine is thiamine pyrophosphate (a pyrophosphoric ester). This is also

Fig. 7-13. Effect of thiamine on growth of *Phycomyces blakesleeanus*. Growth is absent in the thiamine-deficient medium on the left, and is abundant in the thiamine-supplemented medium on the right. [Courtesy Plant Pathology Department, Cornell University.]

known as cocarboxylase, the coenzyme of carboxylase. Pyruvic acid, a key intermediate in carbohydrate metabolism, is transformed into acetaldehyde and carbon dioxide by the enzyme carboxylase. Pyruvic acid tends to accumulate in cultures of thiamine-deficient fungi.

BIOTIN

Numerous fungi are deficient for biotin, although this deficiency is not nearly so common as that for thiamine. Most yeasts are biotin-deficient. The biotin molecule is synthesized according to the following scheme:

$$\text{Pimelic acid} \longrightarrow \text{desthiobiotin} \longrightarrow \text{biotin}$$

Some biotin-deficient fungi can grow if provided with the precursor desthiobiotin and are apparently unable to convert pimelic acid to desthiobiotin. Such fungi include the Pyrenomycetes *Ceratostomella ips, Neurospora crassa,* and some yeasts. Other fungi cannot grow if provided with desthiobiotin but require biotin to satisfy their requirements. These fungi, which include the Pyrenomycetes *Ceratostomella pini* and *Sordaria fimicola,* have a metabolic block which makes it impossible for them to convert desthiobiotin to biotin.

The role of biotin is only vaguely understood. It is apparently involved in the synthesis of aspartic acid, which may partially replace a biotin requirement; in the synthesis of fatty acids as oleic acid, it may partially replace biotin for a *Neurospora crassa* mutant and for *Ceratostomella pini* (Cochrane, 1958).

THE REMAINING VITAMINS.

Pyridoxine (vitamin B_6) is required by fewer fungi than either thiamine or biotin. Partial or complete deficiencies for pyridoxine are known among some yeasts and are especially common in species of *Ceratostomella.* The coenzyme form of pyridoxine serves as a coenzyme for various enzymes involved in amino acid metabolism.

A deficiency for pantothenic acid is fairly common among the yeasts, especially among species of *Saccharomyces,* but it is virtually unknown among the filamentous fungi. Pantothenic acid is a constituent of coenzyme A which mediates in acyl transfers and related reactions. Pantothenic acid favors the accumulation of glycogen and increases respiratory activity in yeasts.

Deficiencies for inositol, riboflavin, and *p*-aminobenzoic acid are rare. Riboflavin provides the prosthetic group for the flavin enzymes, required in respiration, and for various enzymes involved in nitrogen metabolism. The roles of inositol and *p*-aminobenzoic acid are unknown.

Fungi as Test Organisms

In our complex technological and scientific age, it is frequently necessary to establish the composition of complex materials and mixtures such as foodstuffs for humans, animals, and microbes; metabolites produced by organisms; and soils. The amounts and types of vitamins, amino acids, and elements are often determined by using microorganisms that have a deficiency for the substance under investigation. In a similar manner, microorganisms may be used to detect substances that are antagonistic or poisonous to them. Use of microorganisms for such analyses is termed *bioassay.*

Fungi widely used for vitamin bioassays include *Phycomyces blakesleeanus* for thiamine (Fig. 7–13), *Saccharomyces carlsbergensis* for pyridoxine and pantothenic acid, mutants of *Neurospora crassa* for *p*-aminobenzoic acid and inositol, and *Saccharomyces cerevisiae* for biotin. Few fungi isolated from nature are deficient for amino acids, so that the majority of amino acid bioassays are done with bacteria, although induced mutants of *Neurospora* may also be used. *Aspergillus niger* is widely used to determine available copper, magnesium, potassium, and molybdenum in the soil.

BIOASSAY METHODS

The first requisite for conducting a bioassay is the selection of an appropriate organism deficient for the test vitamin, amino acid, or element or is sensitive to an antagonistic substance. The reaction of the organism should be specific; that is, it should respond to a vitamin but not to its moieties or precursors, for example. The organism should be sensitive enough to the presence of the substance to show a growth response to dosage increases which would plot in a linear fashion over some concentration range. Species differ greatly in their sensitivity to the test substance.

A synthetic medium is prepared that lacks the test substance but contains all other necessary nutrients. This medium is usually purified to remove traces of biologically active substances, the method depending on the substances which must be removed (filtration by activated charcoal is used to rid the medium of residual vitamins, for example). The test compound typically contains its vitamins or amino acids in a bound form unavailable to the test organism, so these must be extracted and made water-soluble by acid or enzymatic hydrolysis. The extracted compound is then added to the basal medium in increasing concentrations which, ideally, should result in an approximately linear growth versus dosage response.

The inoculum is of great importance. Prior to its use as inoculum, the fungus should be grown on a medium which provides adequate but not excessive amounts of all factors needed, including the test substance. If the organism is a single-celled yeast, the inoculum consists of drops of cells suspended in sterile water. If a filamentous fungus is used, the inoculum may consist of ungerminated or germinated spores or fragmented mycelium. The amounts of inoculum used must be carefully standardized to give repeatable results.

The conditions (light, temperature, etc.) under which the culture is incubated and the length of the incubation period must be rigidly controlled. At the end of the incubation period, the amount of growth is determined by either turbimetric procedures or a dry-weight analysis. The data achieved in this way must then be compared with a standard to determine the actual amounts of the test substance present.

The principles involved in establishing a standard for comparison are quite simple. Chemically pure samples of the vitamin, amino acid, element, or potential antagonist are added to the purified basal medium in a series of increasing concentrations. These are inoculated with the test fungus and the growth response to the pure substance recorded. From the data, a *standard curve* is drawn (Fig. 7–14). By comparing the growth of the fungus on the unknown test substance with the standard curve, the amount of the vitamin, amino acid, or other compound may be determined.

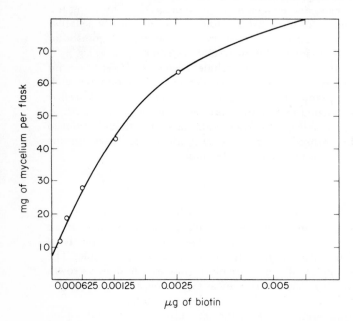

Fig. 7-14. A standard growth curve for *Neurospora sitophila* on biotin. [From L. H. Leonian and V. G. Lilly, 1945, *West Virginia Agr. Expt. Sta. Bull.* 319:1-35].

There are many advantages to using microorganisms for bioassay. In general, the methods are less complicated than those involved in chemical analysis. It is not necessary to obtain the substance in a chemically pure form, and minute quantities of the substance may suffice. Bioassays can be conducted in a short time with a minimum of equipment. In some cases, bioassays may be more accurate than a chemical analysis; for example, a bioassay of the mineral content of the soil measures the minerals which are available to plants and not the absolute amounts present.

PHYSICAL REQUIREMENTS FOR GROWTH

In addition to having nutrients, fungi must be exposed to favorable temperature, moisture, pH, and light conditions for growth to take place. For each of these physical environmental factors, there is a range in which growth will occur, delimited by a minimum point below which no growth will occur and a maximum point above which no growth will occur. There is usually an optimum over some small part of the range, indicating that the greatest growth rate occurs there (Fig. 7-15). Growth curves which are drawn as a function of temperature or other physical factors are characterized by the minimum, optimum, and maximum points. These cardinal points are a useful index to the behavior of a given fungus and as a point of comparison between fungi, but these are not absolute values because they may change with an alteration in cultural conditions, the age of the mycelium, or the particular genetic strain of the fungus.

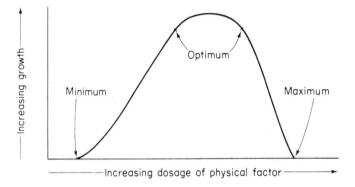

Fig. 7-15. A theoretical growth curve of a fungus in response to a physical factor, showing minimum, optimum, and maximum points.

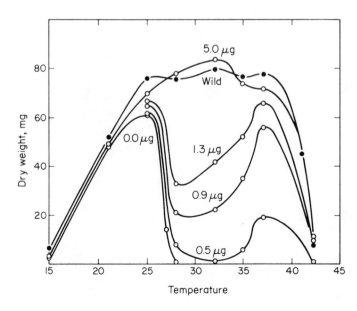

Fig. 7-16. Growth of a wild type *Neurospora* and a temperature-sensitive riboflavin-deficient mutant in response to temperature and to different concentrations of riboflavin (indicated on curves in micrograms per 20 milliliters of medium). Below 25°C, no riboflavin is required for growth, while no growth occurs above 28°C without an exogenous source riboflavin. [From H. K. Mitchell and M. B. Houlahan, 1946, *Am. J. Botany* **33**:31–35.]

Virtually all reactions within the cell are affected by physical factors, especially temperature and pH, and each reaction tends to have its own minimum, optimum, and maximum. Cell processes have widely different ranges in which they can operate; a given temperature, for example, favors some reactions and not others. The optimum value for growth, then, is that point at which a given physical factor is optimum, or at least favorable, for the greatest number of metabolic processes.

A change in the physical environment may alter the response of the fungus to other factors influencing growth. Conditional vitamin requirers, for example, may require an exogenous source of a vitamin at certain temperatures and pH values but not at others (Fig. 7-16).

Temperature

Temperature is extremely important in determining the amount and rate of growth of an organism. Increasing temperature has the general effect of increasing enzyme activity and chemical activity. Many chemical reactions increase tenfold for each 10°C rise in temperature, but enzyme activity usually increases only twofold for each 10°C

Fig. 7-17. Differential effect of temperature on the growth of *Phoma apiicola* in culture. [Courtesy Plant Pathology Department, Cornell University.]

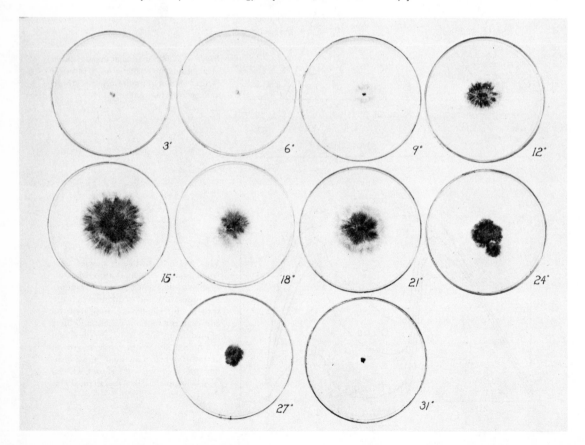

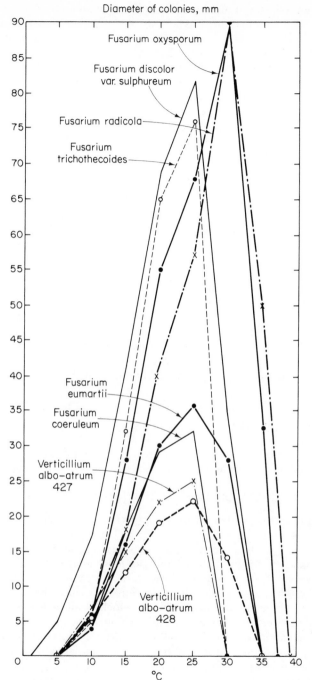

Fig. 7–18. Temperature-growth curves of eight fungi which cause potato-rot and wilt diseases. *(Fusarium coeruleum, F. discolor var. sulphureum, F. eumartii, F. oxysporum, F. radicola, F. trichothecoides,* and two isolates of *Verticillium alboatrum).* [From H. A. Edson and M. Shapovalov, 1920, *J. Agri. Res.* **18:**511–524.]

increase. Enzymes eventually become inactivated at high temperature levels; some enzymes are inactivated at temperatures as low as 30°C. In addition to having a general effect on the metabolic rate, temperature may directly affect the synthesis of vitamins, amino acids, or other metabolites—failure of some fungi to grow at high temperatures is directly due to an inability to synthesize such a needed component.

Beginning with the minimum temperature permitting growth, the amount of growth of the fungus increases with the increase in temperature until the optimum is reached. Throughout the optimum range, the growth rate becomes more or less steady. When the temperature increases above the optimum, there is a decline in growth rate until the maximum point permitting growth is reached (Fig. 7–17). If several cultures are grown at different temperatures and the amount of growth plotted, one can derive a typical growth curve illustrating this response.

In general, fungi have similar temperature requirements (Fig. 7–18), although there are exceptions. While some strains of food-spoiling fungi of the species *Cladosporium* and *Sporotrichum* can grow at temperatures as low as −5° to −8°C, most fungi have a minimum temperature of 0° to 5°C. In most cases, the optimum temperature range lies somewhere between 15° and 30°C, causing the temperature growth curve to be skewed to the right. A strain of *Coprinus fimetarius* has an unusually high optimum at 40°C and is still able to grow at 44°C [growth at 44°C is considerably increased by the addition of methionine to the medium (Fries, 1953)]. Few fungi are able to grow above the maximum temperatures of 35° to 40°C, but the point at which death results may be much higher. The thermal death point for many fungi lies approximately between 50° and 60°C in moist heat, but temperatures as high as 105°C may be required for 12 hours to kill all the wood-rotting fungi in dry blocks of wood (Cartwright and Findlay, 1934; Snell, 1923).

Temperature minima, optima, and maxima differ for growth, reproduction, and spore germination.

Hydrogen Ion Concentration

All fungi are in contact with aqueous solutions in nature or in the laboratory, and the hydrogen ion concentration (pH) of these solutions exerts control over fungal growth.

EFFECTS OF HYDROGEN ION CONCENTRATION

One effect of pH is on the availability of certain metallic ions. Metallic ions may form complexes that become insoluble at certain pH ranges. Magnesium and phosphate ions may coexist in their free form at a low pH, but at a higher pH they form an insoluble complex, reducing the availability of these ions to the fungus. Iron deficiency may result in an alkaline medium as the ferric ion complex becomes insoluble at a high pH. The effect is similar with the calcium and zinc ions.

A second effect of pH is on cell permeability, which is altered with different degrees of acidity or alkalinity. The effect is particularly noticeable on compounds that ionize. A possible explanation is that at a lower pH the protoplasmic membrane becomes saturated with hydrogen ions so that passage of essential cations is limited, while at a higher pH the membrane becomes saturated with hydroxyl ions and thereby limits the entrance of essential anions. *p*-Aminobenzoic acid exists predominantly as a free acid at a very low pH, and uptake of this vitamin is greatest at a low pH. About eight times as much *p*-aminobenzoic acid is required at pH 6.0 as at pH 4.0 to support the same amount of growth.

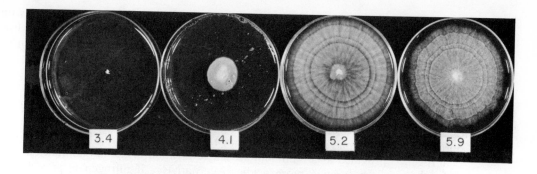

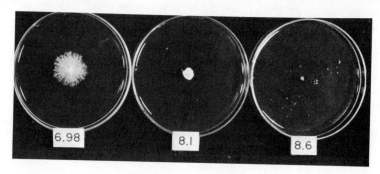

Fig. 7-19. Differential effect of pH on the growth of *Ceratocystis ulmi* in culture. [Courtesy Plant Pathology Department, Cornell University.]

External hydrogen ion concentration also affects pH within the cell, which in turn affects enzyme activity. Enzymes are inactivated at either pH extreme, but they have different optimum pH levels for activity. Some are more active in a weakly acidic solution, while others are more active in a weakly alkalinic solution. Optima for most enzymes lie between pH 4 and 8 (Fig. 7-19). An unfavorable pH may alter normal synthetic ability of the cell. *Sordaria fimicola,* for example, is usually able to synthesize its own thiamine; but when it is cultured on a medium with an initial pH of 3.6 to 3.8, it is unable to grow unless exogenous thiamine is provided (Lilly and Barnett, 1947).

GROWTH AND HYDROGEN ION CONCENTRATION

Fungi have a comparatively broad pH range over which they can grow and the optimum pH for most fungi is on the acidic side of the scale, under pH 7. Any pH-growth curve is a summation of all the effects of pH on the numerous factors which control growth and does not represent a unitary effect. Growth at low pH may be the direct result of greater iron availability, while a growth at higher pH may be due to the increased activity of enzymes with high pH optima. It is not surprising that many pH-growth curves show two optimal hydrogen ion concentrations. Such a case are species of *Coprinus,* for which the minima represent pH-dependent unavailability of different ions. If iron, zinc, and calcium are provided in available form, the apparent double optima merge into a single, broad optimum range (Fries, 1956).

pH-Growth curves are not constant but may be modified by an alteration of any number of environmental factors, including temperature, age of the mycelium at the time of harvest, calcium and magnesium levels, and nitrogen source. The ease with which minima, optima, and maxima in pH-growth relationships may be shifted is further testimony to the intricate involvement of hydrogen ion concentration with a large number of growth processes.

HYDROGEN ION CHANGES EFFECTED BY THE FUNGUS

Fungi almost invariably alter the pH of the medium in which they grow. An uptake of either anions or cations by the fungus can drive the pH in the opposite direction (Fig. 7-20). A good example of this may be found with different nitrogen sources. If nitrogen is provided as an ammonium salt, utilization of the ammonium ion will make the medium become more acidic. Alternatively, if nitrogen is provided as sodium nitrate ($NaNO_3$), the medium becomes more alkaline as the nitrate ion is removed.

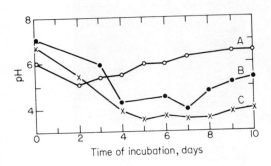

Fig. 7-20. Changes in pH of *Penicillium notatum* cultures using glucose and varying sources of nitrogen: (a) $NaNO_3$; (b) Mixture of $NaNO_3$ and the amino acids tryptophan, asparagine, and cystine; (c) Tryptophan, asparagine, and cystine. [From A. E. Dimond and G. L. Peltier, 1945, *Am. J. Botany* **32**:46–50.]

Conversion of a neutral component to an acidic or alkaline metabolite may also be responsible for a shift in pH of the medium. A common cause of a decrease in pH is the accumulation of organic acids (especially gluconic, pyruvic, citric, and succinic acids) formed from the metabolism of sugars. Carbon dioxide, a byproduct of carbohydrate metabolism, combines with water to form carbonic acid. Carbonic acid may drive the pH down, but it is dissociated in the presence of stronger acids and forms bicarbonates under alkaline conditions. In a similar fashion, the release of ammonium ions from the deamination of amino acids and protein may cause the pH to rise. The ultimate pH of the culture medium depends on the relative rate and occurrence of these pH-changing processes and the extent to which the medium is buffered.

Moisture

Fungi require relatively high moisture levels, although many of the higher fungi are able to grow in the absence of free water. A high relative humidity is also required. Maximum growth for most fungi occurs at a relative humidity of 95% to 100%, and growth declines or ceases in humidity of 80% to 85%. A few fungi will grow at a relative humidity as low as 65%.

Conditions for adequate moisture are satisfied in routine cultural techniques. Agar media contain water bound within a gel, and the atmosphere within a petri dish or flask has a relative humidity near 100%. In nature, the ability to grow on a given substrate is partially determined by the moisture content of the substrate. Wood, for example, is not decayed by fungi if the moisture content is below 20%. Soil fungi grow better in moderate than in high moistures, but this is because soil aeration (and therefore oxygen supply) is limited when the moisture content is high.

Fungi grow either within or in contact with watery solutions normally having at least some dissolved sugars or salts. As you probably know, the concentration of solutes may increase the osmotic concentration of the solvent in which they are dissolved. Most fungi grow best with a relatively low osmotic concentration (0.5 M or less). Many can tolerate higher osmotic concentrations, especially if they are allowed to adapt gradually to increasing levels of higher osmotic concentrations. The degree of tolerance differs from species to species. Most fungi cease growth or are inhibited when the concentration of soluble sugar exceeds about 2.0 M (Burnett, 1976). Fungi tolerant of the higher osmotic concentrations are often found growing on concentrated sugary materials such as jams, nectar, honey, and fruit juices. Such fungi may even require this high osmotic concentration; an example is the yeast ally *Eremascus albus* (Paugh and Gray, 1969). *E. albus* will not grow in a medium with a low sucrose concentration (less than 5%) but will grow directly on sucrose crystals. Optimum growth occurs on a medium with a sucrose concentration of approximately 1.4 M, but it will still grow well on 2.33 M sucrose. Optimum growth of *E. albus* is not dependent upon the availability of sucrose as a nutrient (an inert material producing a high osmotic concentration will give similar results). Other fungi are adapted to the high concentrations of salts that may be found in brine or in seawater. Marine fungi typically yield optimum growth when provided with a medium consisting of approximately 50% to 100% seawater, but similar salt concentrations inhibit growth or reproduction of terrestrial fungi. It has been suggested that this difference in degree of tolerance to salinity may be the principal factor maintaining the marine fungi as an ecological group distinct from terrestrial fungi (Jones and Byrne, 1976).

Light

Light is a natural component of any environment. The vast majority of fungi are exposed to alternating cycles of daylight and darkness, while other fungi occurring deep within the soil or host tissues are likely to be in total darkness. Light intensities vary in nature: a fungus growing beneath a leaf in the forest is not exposed to as great a light intensity as a fungus growing above the ground or on the surface of a plant in an unshaded clearing.

Although the growth rate and/or the synthetic capacity of some fungi may be controlled by light, the growth of most fungi is apparently not sensitive to light. The most widespread effect of light on fungal growth is that of inhibition in strong light. This inhibition may sometimes be overcome by adding natural materials to the medium, which suggests that strong light may destroy required vitamins (e.g., Barnett, 1968; Robbins and Hervey, 1960). However, the vegetative growth of some aquatic fungi may be stimulated by light. Stimulation of growth has been observed in the marine ascomycete *Buergenerula spartinae* (Gessner, 1976) and in *Blastocladiella emersonii* (a member of the Blastocladiales). The dry weight of *B. emersonii* grown in light may be as much as 141% of the dry weight of cultures grown in the dark (Cantino and Horenstein, 1956). Light may exert both stimulatory and inhibitory effects on the vegetative structures of the same fungus. An example is *Candida albicans,* which has growth inhibition when exposed to light but also has enhanced carbohydrate synthesis (Saltarelli and Coppola, 1979).

Blue light depresses synthesis of the pigment melanin and of tyrosinase in a mutant of *Neurospora crassa,* while it has an opposite effect in *Aureobasidium pullulans* and *Cladosporium mansonii* (Page, 1965).

Light may also affect the formation of reproductive structures or may control oriented phototropic movements of reproductive structures. Generally, many more instances are known of light affecting reproduction than vegetative growth. Additional information about the effects of light and the mode of its interaction with fungi is presented in Chapter 9.

Aeration

Two components of the air are of special importance to the fungi: these are oxygen and carbon dioxide. Oxygen is vital to cellular respiration, in which an energy source is oxidized to carbon dioxide and water and energy is made available to the cell. Carbon dioxide may accumulate as a result of cellular respiration by either the fungi or other organisms. Respiration is discussed further in Chapter 8. In addition, as mentioned on page 284, some fungi are able to fix atmospheric carbon dioxide and use this as a source of carbon. This may be experimentally confirmed by providing radioactive carbon dioxide to the fungus in the air and later demonstrating that the radioactive carbon has been incorporated into organic compounds within the fungus.

It has been widely recognized that most fungi are *obligate aerobes* and require at least some free molecular oxygen in the atmosphere. Fungi are generally found growing on or near the surface of their substrata (e.g., foods such as meats or cheese, wood of living trees, or soil) but are occasionally isolated from substrata far removed from the air. Many fungi will fail to grow in cultures if molecular oxygen is not freely available. Oxygen pressure in the atmosphere is normally measured as 160 millimeters Hg, but some fungi grow well in extremely small amounts of oxygen (as low as 10 to 40 millimeters Hg). This apparently widespread ability to adjust to wide ranges of available oxygen probably indicates that oxygen variation in nature is ordinarily not critical for fungal growth (Tabak and Cooke, 1968-a). Some fungi are *facultative anaerobes* and not only can survive in the absence of measurable oxygen but also can grow and sporulate under such conditions. Facultative anaerobes can use combined oxygen in addition to free molecular oxygen in the atmosphere. Fungi which are facultative anaerobes are able to survive in many environments having extremely limited oxygen levels; these include sewage sludge and polluted streams (Tabak and Cooke, 1968-b). Although many bacteria are *obligate anaerobes* which cannot grow in the presence of free molecular oxygen, we do not know of any fungi which are obligate anaerobes (Tabak and Cooke, 1968-a).

Effective utilization of carbon or nitrogen compounds as sources of nutrients may be affected by the amount of oxygen available in the atmosphere. For example, in an aerobic atmosphere, *Mucor rouxii* is able to utilize a large variety of carbon sources and amino acids for growth, but under anaerobic conditions it can utilize only hexoses as a carbon source and poorly utilizes amino acids as a nitrogen source (nitrate or ammonium ions serve as a better source of nitrogen under anaerobic conditions). Insufficient oxygen also increases nutritional demands in *M. rouxii* and decreases growth (Bartnicki-Garcia and Nickerson, 1962). Similarly, the growth of many fungi under anaerobic conditions may be improved by the addition of biotin and thiamine (Tabak and Cooke, 1968-b) or yeast extract to the medium (Deploey and Fergus, 1975).

Air usually contains about 0.3% carbon dioxide, but high concentrations of carbon dioxide resulting from cellular respiration may accumulate in tightly closed culture containers or in nature. Natural environments where high carbon dioxide levels occur often result from cellular respiration by various organisms in combination with poor aeration. These environments include stagnant waters and water-logged soils where bacteria may thrive and the woody tissues of a host plant where carbon dioxide produced by the host plant may accumulate. Fungi are usually inhibited by concentrations of carbon dioxide greater than 10% to 15% but the amount of carbon dioxide tolerated varies among the fungi. *Fusarium oxysporum* and *F. eumartii,* which grow under normal atmospheric conditions, can tolerate a carbon dioxide concentration up to 75.3% (Hollis, 1948). Some fungi that inhabit stagnant water can tolerate a higher carbon dioxide concentration. An example of such a fungus is *Aqualinderella fermentans,* a member of the Oomycetes. *A. fermentans* grows poorly under normal atmospheric conditions, but its growth is actually improved as the carbon dioxide concentration increases to 20%, and it can tolerate a concentration up to 99% (Emerson and Held, 1969). *A. fermentans* is unique because it actually requires carbon dioxide for optimum growth. In addition, *A. fermentans* can grow in the absence of aerobic oxygen.

Growth of fungi may sometimes be inhibited by the total absence of carbon dioxide or by concentrations of oxygen much greater than those that normally occur in the atmosphere (Tabak and Cooke, 1968-a).

Another property of the air is that it may contain volatile organic compounds. Fungi are known to produce a wide variety of volatile metabolites, including alcohols, aldehydes, and other organic compounds such as coumarin. Volatile metabolites are produced in nature by higher plants, bacteria, and actinomycetes. Volatiles may accumulate either in closed culture containers or in poorly aerated habitats. Volatiles may exert either inhibitory or stimulatory effects upon the growth and reproduction of the fungi: the differential effects depend upon the type and concentration of the volatiles as well as upon the fungal species. Deformation of the hyphae and abnormally thick walls have been observed in some fungi exposed to volatiles produced by bacteria (Moore-Landecker and Stotzky, 1973). Probably most of these volatiles exert their effects upon fungi by regulating internal metabolism and not by serving as a nutrient (Fries, 1973).

References

Barnett, H. L. 1968. The effects of light, pyridoxine, and biotin on the development of the mycoparasite, *Gonatobotryum fuscum.* Mycologia **60**:244–251.

Bartnicki-Garcia, S., and W. J. Nickerson. 1962. Nutrition, growth, and morphogenesis of *Mucor rouxii.* J. Bacteriol. **84**:841–858.

Bonner, R. D., and C. L. Fergus. 1960. The influence of temperature and relative humidity on growth and survival of silage fungi. Mycologia **52**:642–647.

Burkholder, P. R. 1943. Vitamin deficiencies in yeasts. Am. J. Botany **30**:206–211.

———, and D. Moyer. 1943. Vitamin deficiencies of fifty yeasts and molds. Bull. Torrey Botan. Club **70**:372–377.

Burnett, J. H. 1976. Fundamentals of mycology, Second ed. Crane Russak & Company, Inc. 673 pp.

Cantino, E. C., and E. A. Horenstein. 1956. The stimulatory effect of light upon growth and CO_2 fixation in *Blastocladiella.* I. The S. K. I. cycle. Mycologia **48**:777–799.

Cartwright, K. St. G., and W. P. K. Findlay. 1934. Studies in the physiology of wood-destroying fungi. II. Temperature and rate of growth. Ann. Botany London **48**:481–495.

Cochrane, V. W. 1958. Physiology of fungi. John Wiley & Sons, Inc., New York. 524 pp.

Cuppett, V. M., and V. G. Lilly. 1973. Ferrous iron and the growth of twenty isolates of *Phytophthora infestans* in synthetic media. Mycologia **65**:67–77.

Deploey, J. J. and C. L. Fergus. 1975. Growth and sporulation of thermophilic fungi and actinomycetes in O_2–N_2 atmospheres. Mycologia **67**:780–797.

Emerson, R., and A. A. Held. 1969. *Aqualinderella fermentens* gen. et sp. n., a phycomycete adapted to stagnant waters. II. Isolation, cultural characteristics, and gas relations. Am. J. Botany **56**:1103–1120.

Fries, L. 1956. Studies in the physiology of *Coprinus*. II. Influence of pH, metal factors, and temperature. Svensk Botan. Tidsskr. **50**:47–96.

——, 1973. Effects of volatile organic compounds on the growth and development of fungi. Trans. Brit. Mycol. Soc. **60**:1–21.

Fries, N. 1973. Effects of volatile organic compounds on the growth and development of fungi. Trans. Brit. Mycol. Soc. **60**:1–21.

Garrett, R. H., and N. K. Amy. 1978. Nitrate assimilation in fungi. Adv. Microbial Physiol. **18**:1–65.

——, and A. Nason. 1969. Further purification and properties of *Neurospora* nitrate reductase. J. Biol. Chem. **244**:2870–2882.

Gessner, R. V. 1976. In vitro growth and nutrition of *Buergenerula spartinae*, a fungus associated with *Spartina alterniflora*. Mycologia **68**:583–599.

Grant, C. L., and D. Pramer. 1962. Minor element composition of yeast extract. J. Bacteriol. **84**:869–870.

Harris, J. L., and W. A. Taber. 1970. Influence of certain nutrients and light on growth and morphogenesis of the synnema of *Ceratocystis ulmi*. Mycologia **62**:152–170.

Hasija, S. K., and H. C. Agarwal. 1978. Nutritional physiology of *Trichothecium roseum*. Mycologia **70**:47–60.

Hill, E. P. 1965. Uptake and translocation 2. Translocation. *In:* Ainsworth, G. C., and A. S. Sussman, Eds., The fungi—an advanced treatise. Academic Press, New York. **1**:457–463.

Hollis, J. P. 1948. Oxygen and carbon dioxide relations of *Fusarium oxysporum* Schlecht. and *Fusarium eumartii* Carp. Phytopathology **38**:761–775.

Jennings, D. H. 1974. Sugar transport into fungi: an essay. Trans. Brit. Mycol. Soc. **62**:1–24.

Johri, B. N., and H. J. Brodie. 1972. Nutritional study of *Cyathus helenae* and related species. Mycologia **64**:298–303.

Johnson, G. T., and F. McHan. 1975. Some effects of zinc on the utilization of carbon sources by *Monascus purpureus*. Mycologia **67**:806–816.

Jones, E. B., and P. J. Byrne. 1976. Physiology of the higher marine fungi. *In:* Jones, E. B. G., Ed., Recent advances in aquatic mycology. John Wiley & Sons, Inc. New York. pp. 135–175.

——, and J. L. Harrison. 1976. Physiology of marine Phycomycetes. *In:* Jones, E. B. G., Ed., Recent advances in aquatic mycology. John Wiley & Sons, Inc. New York. pp. 261–278.

Kinsky, S. C. 1961. Induction and repression of nitrate reductase in *Neurospora crassa*. J. Bacteriol. **82**:898–904.

Leonian, L. H., and V. G. Lilly. 1945. The comparative value of different test organisms in the microbiological assay of B vitamins. West Virginia Agr. Expt. Sta. Bull. 319. 35 pp.

Lewis, H. L., and G. T. Johnson. 1967. Growth and oxygen-uptake responses of *Cunninghamella echinulata* on even-chain fatty-acids. Mycologia **59**:878–887.

Lilly, V. G. 1965. The chemical environment for fungal growth I. Media, macro- and micronutrients. *In:* Ainsworth, G. C., and A. S. Sussman, Eds., The fungi—an advanced treatise. Academic Press, New York. **1**:465–478.

——, and H. L. Barnett. 1947. The influence of pH and certain growth factors on mycelial growth and perithecial formation by *Sordaria fimicola*. Am. J. Botany **34**:131–138.

——, and ——. 1951. Physiology of the fungi. McGraw-Hill Book Company, New York. 464 pp.

Mandels, G. R. 1965. Kinetics of fungal growth. *In:* Ainsworth, G. C., and A. S. Sussman, Eds., The fungi—an advanced treatise. Academic Press, New York. **1**:599–612.

Mathew, K. T. 1952. Growth-factor requirements of *Pellicularia koleroga* Cooke in pure culture. Nature **170**:889–890.

Moore-Landecker, E., and G. Stotzky. 1973. Morphological abnormalities of fungi induced by volatile microbial metabolites. Mycologia **65**:519–530.

Nason, A. 1962. Symposium on metabolism of inorganic compounds II. Enzymatic pathways of nitrate, nitrite, and hydroxylamine metabolism. Bacteriol. Rev. **26**:16–41.

——, and H. Takahashi. 1958. Inorganic nitrogen metabolism. Ann. Rev. Microbiol. **12**:203–246.

Nicholas, D. J. D. 1965. Utilization of inorganic nitrogen compounds and amino acids by fungi. *In:* Ainsworth, G. C., and A. S. Sussman, Eds., The fungi—an advanced treatise. Academic Press, New York. **1**:349–376.

——, and A. Nason. 1954. Mechanisms of action of nitrate reductase from *Neurospora*. J. Biol. Chem. **211**:183–197.

Nolan, R. A. 1976. Physiological studies on an isolate of *Saprolegnia ferax* from the larval gut of the blackfly *Simulium vittatum*. Mycologia **68**:523–540.

Page, R. M. 1965. The physical environment for fungal growth 3. Light. *In:* Ainsworth, G. C., and A. S. Sussman, Eds., The fungi—an advanced treatise. Academic Press, New York. **1**:559–574.

Paugh, R. L., and W. D. Gray. 1969. Studies on the growth of the osmiophilic fungus *Eremascus albus*. Mycologia **61**:281–288.

Plomley, N. J. B. 1959. Formation of the colony in the fungus *Chaetomium*. Australian J. Biol. Sci. **12**:53–64.

Pugliese, F. A., and J. P. White. 1973. Zinc stimulation of nitrogen assimilation by *Helminthosporium cynodontis*. Mycologia **65**:295–309.

Robbins, W. J. 1939. Growth substances in agar. Am. J. Botany **26**:772–778.

——, and A. Hervey. 1960. Light and the development of *Poria ambigua*. Mycologia **52**:231–247.

——, and F. Kavanagh. 1938. Thiamine and growth of *Pythium butleri*. Bull. Torrey Botan. Club **65**:453–461.

——, and ——. 1942. Vitamin deficiencies of the filamentous fungi. Botan. Rev. **8**:411–471.

Robertson, N. F. 1965. The mechanism of cellular extension and branching. *In:* Ainsworth, G. C., and A. S. Sussman, Eds., The fungi—an advanced treatise. Academic Press, New York. **1**:613–623.

Ross, I. S. 1975. Some effects of heavy metals on fungal cells. Trans. Brit. Mycol. Soc. **64**:175–193.

Rothstein, A. 1965. Uptake and translocation I. Uptake. *In:* Ainsworth, G. C. and A. S. Sussman, Eds., The fungi—an advanced treatise. Academic Press, New York. **1**:429–455.

Ryan, F. J., G. W. Beadle, and E. L. Tatum. 1943. The tube method of measuring the growth rate of *Neurospora*. Am. J. Botany **30**:784–799.

Saltarelli, C. G., and C. P. Coppola. 1979. Effect of light on growth and metabolite synthesis in *Candida albicans*. Mycologia **71**:773–785.

Snell, E. E. 1948. Use of microorganisms for assay of vitamins. Physiol. Rev. **28**:255–282.

Snell, W. H. 1923. The effect of heat upon the mycelium of certain structural-timber-destroying fungi. Am. J. Botany **10**:399–411.

Steinberg, R. A. 1938. The essentiality of gallium to growth and reproduction of *Aspergillus niger*. J. Agr. Res. **57**:569–574.

——, 1939. Relation of carbon nutrition to trace-element and accessory requirements of *Aspergillus niger*. J. Agr. Res. **59**:749–763.

——, 1950. Growth of fungi in synthetic nutrient solutions. II. Botan. Rev. **16**:208–228.

Suomalainen, H., and E. Oura. 1971. Yeast nutrition and solute uptake. *In:* A. R. Rose and J. S. Harrison, Eds., The yeasts. Academic Press, New York. **2**:3–74.

Tabak, H. H., and W. B. Cooke. 1968-a. The effects of gaseous environments on the growth and metabolism of fungi. Botan. Rev. **34**:126–252.

——, and ——, 1968-b. Growth and metabolism of fungi in an atmosphere of nitrogen. Mycologia **60**:115–140.

Zalokar, M. 1959. Growth and differentiation of *Neurospora* hyphae. Am. J. Botany **46**:602–610.

EIGHT

Metabolism

In the last chapter we considered nutrient uptake and those nutritional and environmental conditions that promote growth. The fate of nutrients is not ended with their preliminary digestion and entrance into the cell, because they will become involved in the metabolism of the cell. Metabolic pathways are of two general types: (1) catabolic pathways that degrade a substance to simpler forms (digestion, considered in the last chapter, comprises the first catabolic reactions) and (2) anabolic pathways through which syntheses of cell components take place. Synthetic processes include formation of carbohydrates, proteins, lipids, and nucleoproteins which are common to all forms of life and of some specialized fungal metabolites. It is beyond the scope of this treatment to consider all the known metabolic pathways which occur in fungi, and therefore attention is given to carbon, nitrogen, and lipid metabolism.

CARBON METABOLISM

Respiration

Organisms derive useful energy and intermediates for syntheses from oxidation of compounds in a process termed *cellular respiration*. Materials that may be oxidized by autotrophic bacteria include ammonia, nitrite, and sulfur. The substrate most commonly oxidized by fungi, plants, and animals is glucose. Since glucose is broken down in respiration, it is also said to be *dissimilated*.

BIOLOGICAL OXIDATIONS

Since cellular respiration is primarily an oxidation process, we must consider the mechanics of biological oxidation. Oxidation is the removal or loss of electrons from a substrate (Substrate A, for example). These electrons cannot remain free but must be gained by a second substrate (substrate B), which is reduced. An oxidation reaction must always be accompanied by a reduction reaction. Most organic compounds lose hydrogen ions upon the removal of electrons and, in these cases, oxidation is equivalent to dehydrogenation. Therefore substrate A, which is oxidized, is a *hydrogen donor* and the reduced substrate B is a *hydrogen acceptor*. A typical oxidation-reduction sequence may be outlined as follows: The oxidation of AH_2 (the hydrogen donor):

$$AH_2 \xrightarrow[\text{(dehydrogenation)}]{\text{oxidation}} A + 2H^2 + 2e^-$$

The reduction of B (the hydrogen acceptor):

$$B + 2H^+ + 2e^- \xrightarrow[\text{(hydrogenation)}]{\text{reduction}} BH_2$$

The coupling of these two reactions:

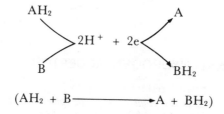

$$(AH_2 + B \longrightarrow A + BH_2)$$

Cellular respiration is not a single reaction as indicated above but rather a stepwise sequence of oxidation reactions in which hydrogen ions and electrons are passed from one substrate to another.

ENERGY RELEASE

A principal role of respiration is to release chemical energy from the substrate being oxidized and to make the energy available to drive other synthetic reactions within the cell. Oxidation is an exergonic chemical reaction, meaning that energy is released when oxidation takes place. Energy is released in a chemical form which may be used to allow endergonic reactions (those that require energy) to take place. Energy is released during respiration in the form of hydrogen ions and their electrons.

The hydrogen ions and electrons may be bound in a reduced compound. That most important in respiration is reduced NAD (nicotinamide adenine dinucleotide), or $NADH_2$. When $NADH_2$ is oxidized, energy is released and will drive an endergonic reaction:

$$NADH_2 \longrightarrow NAD$$

$$A \longrightarrow AH_2$$

Such a coupled reaction is an effective bridge for transfer of energy from respiration to energy-requiring reactions within the cell.

Alternatively, $NADH_2$ may not directly drive a synthetic reaction but may begin a sequence of oxidation-reduction reactions through the *cytochrome system* (also called the respiratory chain). The initial hydrogen and electron acceptor in the cytochrome system is a flavin, a yellow coenzyme. The next hydrogen and electron acceptors are cytochromes, enzymes with a heme prosthetic group, and the final acceptor is oxygen. Some exergonic reactions in this sequence make possible the energy-requiring union of ADP (adenosine diphosphate) with inorganic phosphate to form ATP (adenosine triphosphate). ATP is characterized by very high energy bonds joining two phosphate groups to the remainder of the molecule. These bonds represent a storehouse of energy, which may be used to drive an endergonic reaction. ATP may be thought of as an energy currency within the cell, much as money is a currency within our society. Energy from respiration is stored within the ATP molecule; it may be circulated within the cell and expended for an energy-requiring reaction. When energy is released from ATP, ATP undergoes an exergonic reaction that cleaves the high-energy inorganic phosphate group, leaving behind ADP. The high energy phosphate is then transferred to a new compound in an endergonic reaction thereby giving it additional energy with which it may drive other reactions. The reactions may be summarized as follows:

$$ATP\,(ADP \sim \textcircled{P}) \qquad ADP$$

$$A \diagdown \diagup A \sim \textcircled{P} \diagdown \diagup A$$
$$B \diagdown B \sim \textcircled{P}$$

THE FINAL HYDROGEN ACCEPTOR

The final substrate reduced by the hydrogen ions and terminating the oxidation-reduction sequence is the *final hydrogen acceptor*. The final hydrogen acceptor may be molecular oxygen which is reduced to water. If molecular oxygen is required as a final hydrogen acceptor, cellular respiration must take place in the presence of oxygen and is termed *aerobic respiration*. If some other molecule serves as the final hydrogen acceptor and oxygen is not required, *anaerobic respiration (fermentation)* takes place. Anaerobic respiration may take place in either the presence or absence of oxygen.

RESPIRATION—THE PROCESS

The most frequently occurring form of cellular respiration is aerobic oxidation of glucose, which may be expressed as a greatly simplified overall reaction:

$$C_6H_{12}O_6 + 6O_2 \longrightarrow 6CO_2 + 6H_2O + 38ATP$$

This equation represents two broad phases: (1) *glycolysis,* or the conversion of glucose to pyruvic acid, and (2) the citric acid cycle (CA cycle) which further oxidizes pyruvic acid to carbon dioxide and water, utilizing oxygen as the final hydrogen acceptor.

If adequate amounts of oxygen are lacking, or if anaerobic pathways are favored, the scheme above is altered. Glycolysis occurs, but intermediates of glycolysis act as the final hydrogen acceptor and are reduced to other end products. A typical overall reaction for anaerobic respiration is

$$C_6H_{12}O_6 \longrightarrow 2CH_3CH_2OH + 2CO_2 + 2ATP$$
$$\text{ethyl alcohol}$$

Note that the energy yield of fermentation is less than that of aerobic respiration because the end product is incompletely oxidized, still storing energy.

The discussion of respiration is divided into (1) glycolysis, (2) fermentative end products, and (3) the CA cycle.

Glycolysis. Glycolysis is the conversion of glucose to pyruvic acid by any biochemical pathway. The principal known glycolytic pathways are the Emden-Meyerhof (EM) and the hexose monophosphate (HMP) pathways, occurring in animals, plants, bacteria, and fungi. The EM pathway is especially common. A third glycolytic pathway is the Entner-Doudoroff (ED) pathway, known primarily from biochemical studies of bacteria. Details of these pathways are presented in Figs. 8-1, 8-2, 8-3, and 8-4).

Emden-Meyerhof Pathway. The EM pathway is the principal glycolytic pathway in fungi, as in most other organisms, and usually accounts for at least 50% of the glucose dissimilated in some fungi. Some of the glucose may be dissimilated by another concurrent pathway. In a few fungi, the EM pathway may be secondary in importance to either the HMP or ED pathway (Blumenthal, 1965) or totally lacking as in *Caldariomyces fumago* (Ramachandran and Gottlieb, 1963). The EM pathway yields only a small amount of ATP but is extremely important because it provides most of the pyruvic acid that enters the citric acid cycle and can provide large amounts of ATP.

Both the EM and HMP pathways can operate under either aerobic or anaerobic conditions, but the HMP pathway becomes much less active in the absence of free oxygen. Under anaerobic conditions, the EM pathway becomes increasingly important owing to impairment of the HMP pathway and may increase sufficiently to account for all the glucose dissimilated (Blumenthal, 1965). As indicated in the previous chapter, some fungi can grow under anaerobic conditions. Numerous other fungi that require aerobic conditions for growth can continue to metabolize under anaerobic conditions, although they may be incapable of further growth. Because the HMP pathway is impaired at low levels of oxygen, the EM pathway is probably chiefly responsible for meeting the energy requirements of fungi in these circumstances.

Most of the enzymes required for the EM pathway also operate in the HMP and ED pathways. One enzyme, phosphofructokinase, is unique for the EM pathway. Phosphofructokinase, as well as some other respiratory enzymes, requires zinc for its synthesis. A deficiency of zinc impairs the function of the EM pathway while favoring the HMP pathway (Blumenthal, 1965).

Hexose Monophosphate Pathway. The enzymes required to carry out the HMP cycle are widely distributed in fungi and may be universal. The presence of the enzymes, however, is not proof that the HMP pathway occurs. The HMP pathway occurs in a number of fungi, usually accounting for less than 40% of the glucose dissimilated; under anaerobic conditions, still less of the glucose is dissimilated by this pathway. Examples include the yeast *Candida utilis,* which dissimilates 41% of the glucose via the HMP pathway under aerobic conditions but only 4% under anaerobic conditions (Blumenthal, 1965). The relatively low level of the HMP reactions generally is perhaps due to limiting amounts of NADP. However, the HMP pathway may be the major route of glycolysis in some fungi. The yeast *Rhodotorula gracilis* metabolizes 60% to 80% of its glucose via the HMP cycle and also is unable to carry out metabolism under anaerobic conditions (Suomalainen and Oura, 1971).

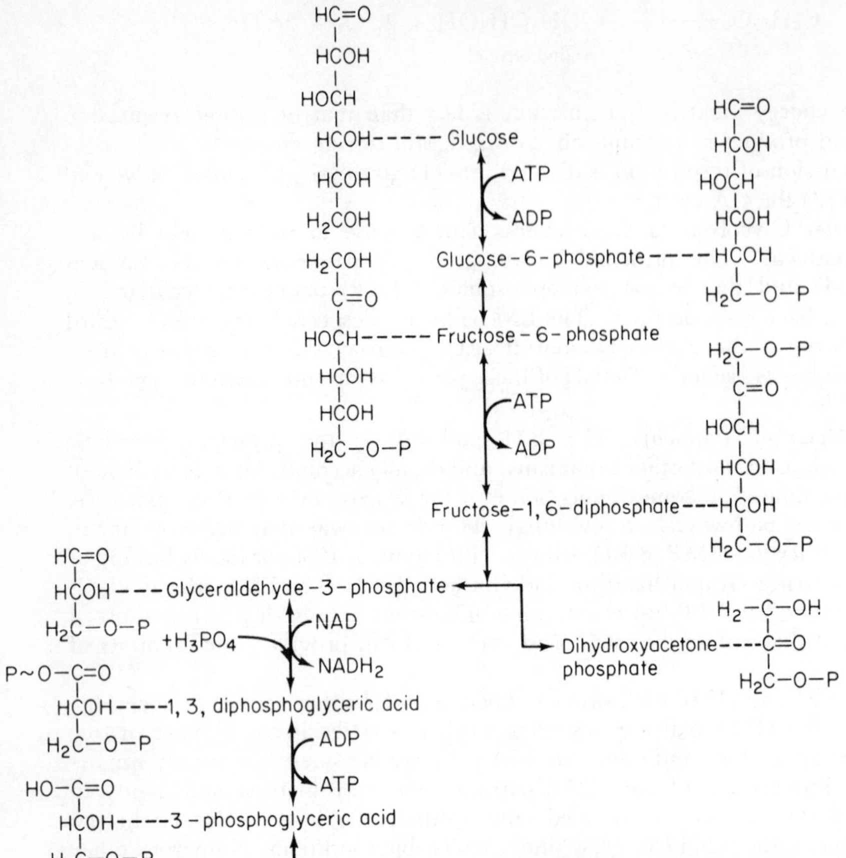

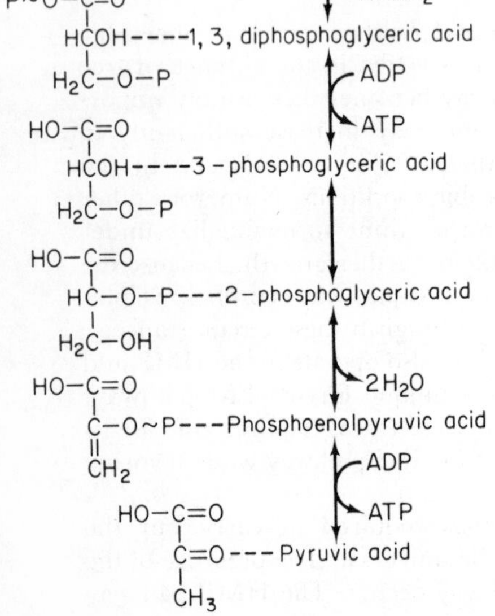

HC=O
|
HCOH
|
HOCH
|
HCOH ----- Glucose
|
HCOH
|
H_2COH

$\curvearrowright$ ATP
$\searrow$ ADP

Glucose – 6 – phosphate ----

HC=O
|
HCOH
|
HOCH
|
HCOH
|
H_2C–O–P

H_2COH
|
C=O
|
HOCH ----- Fructose – 6 – phosphate
|
HCOH
|
HCOH
|
H_2C–O–P

$\curvearrowright$ ATP
$\searrow$ ADP

H_2C–O–P
|
C=O
|
HOCH
|
HCOH
|
Fructose–1, 6–diphosphate --- HCOH
|
H_2C–O–P

HC=O
|
HCOH --- Glyceraldehyde – 3 – phosphate $\leftarrow$
|
H_2C–O–P   +H_3PO_4 $\curvearrowright$ NAD
$\searrow$ NADH$_2$

H_2–C–OH
|
$\rightarrow$ Dihydroxyacetone ---- C=O
phosphate |
H_2C–O–P

P~O–C=O
|
HCOH ----1, 3, diphosphoglyceric acid
|
H_2C–O–P

$\curvearrowright$ ADP
$\searrow$ ATP

HO–C=O
|
HCOH ----3 – phosphoglyceric acid
|
H_2C–O–P

HO–C=O
|
HC–O–P---2 – phosphoglyceric acid
|
H_2C–OH

HO–C=O
|
C–O~P---Phosphoenolpyruvic acid
||
CH_2

$\curvearrowright$ 2 H_2O

$\curvearrowright$ ADP
$\searrow$ ATP

HO–C=O
|
C=O --- Pyruvic acid
|
CH_3

Fig. 8–1. The Embden-Meyerhof (EM) pathway begins with the phosphorylation of glucose which is accomplished by the transfer of a phosphate group from ATP to glucose. The glucose-6-phosphate which is formed by this reaction undergoes a slight rearrangement to form fructose-6-phosphate, and a second phosphate group is added to form fructose-1,6-diphosphate. Fructose-1,6-diphosphate then splits to give a mixture of an aldehyde (glyceraldehyde-3-phosphate) and a ketone (dihydroxyacetone-phosphate), both 3-carbon molecules. The ketone may be transformed into the aldehyde, thereby giving two molecules of glyceraldehyde-3-phosphate from one fructose-1,6-di-phosphate molecule. The glyceraldehyde-3-phosphate molecules are oxidized, and two hydrogen ions and their electrons are removed by NAD, which is reduced to NADH$_2$. Accompanying this reaction is the addition of a second phosphate group from phosphoric acid (H_3PO_4). The compound formed is 1,3-diphosphoglyceric acid. An energy-rich group is lost from 1,3-diphosphoglyceric acid and a molecule each of ATP and 3-phosphoglyceric acid are formed. 3-phosphoglyceric acid is converted into 2-phosphoglyceric acid, and then dehydrated to form phosphoenolpyruvic acid. The phosphate group of phosphoenolpyruvic acid is then transferred to ADP to form ATP and a molecule of pyruvic acid, the end product of this sequence. Each of the above steps is controlled by specific enzymes. The net reaction is:

1 glucose + 2 ADP + 2 NAD + 2 H_3PO_4 $\longrightarrow$
2 pyruvic acid + 2 ATP + 2 NADH$_2$ + 2 H_2O

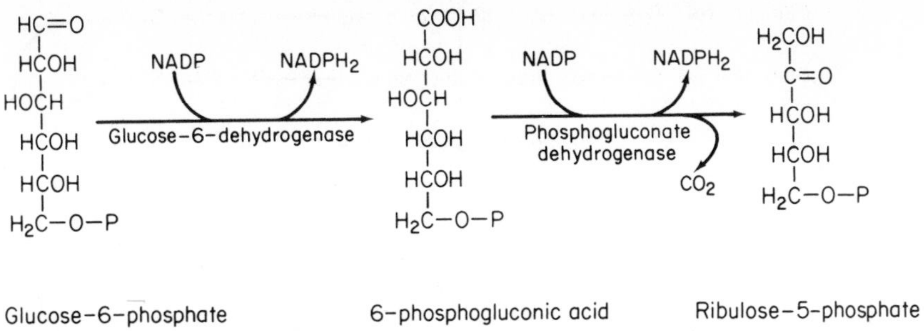

HC=O   COOH H_2COH

Figure structures:

Glucose-6-phosphate
$$HC=O$$
$$HCOH$$
$$HOCH$$
$$HCOH$$
$$HCOH$$
$$H_2C-O-P$$

NADP → NADPH$_2$, Glucose-6-dehydrogenase →

6-phosphogluconic acid
$$COOH$$
$$HCOH$$
$$HOCH$$
$$HCOH$$
$$HCOH$$
$$H_2C-O-P$$

NADP → NADPH$_2$, Phosphogluconate dehydrogenase →, CO_2

Ribulose-5-phosphate
$$H_2COH$$
$$C=O$$
$$HCOH$$
$$HCOH$$
$$H_2C-O-P$$

Fig. 8-2. The first set of reactions of the HMP pathway, which are oxidative. Like the EM pathway, the HMP pathway begins with glucose and the phosphorylation of glucose to form glucose-6-phosphate. Glucose-6-phosphate is usually oxidized to 6-phosphogluconic acid by the removal of electrons which are accepted by NADP (nicotinamide adenine dinucleotide phosphate). 6-Phosphogluconic acid is oxidized by a second NADP molecule, and the first carbon atom (in the C-1 position) is removed as carbon dioxide, leaving a 5-carbon intermediate, ribulose-5-phosphate. The sequence of reactions above completes the *oxidative* phase in which NADPH$_2$ is generated and made available as a hydrogen donor. The second phase of the HMP pathway is nonoxidative, consisting of the complex series of rearrangements shown in Fig. 8-3.

The HMP pathway is not simply an alternative route to pyruvic acid but fulfills certain important roles in the cell. The HMP pathway is the biosynthetic source of pentoses, required for synthesis of nucleic acids. Unlike NADH$_2$, which is generated from the EM pathway, the NADPH$_2$, formed from the HMP pathway apparently does not generate ATP through the cytochrome system. Therefore energy release through aerobic glycolysis occurs only from the EM cycle. The primary role of NADPH$_2$ is to function as a hydrogen donor, making possible the reduction of other compounds. These reactions include fatty acid, glycogen, and glutamic acid synthesis. In view of the role of NADPH$_2$ in syntheses, it is not surprising that the amount of glucose dissimilated by the HMP pathway is correlated with phases of growth. Some enzymes for the HMP pathway (and also for the citric acid cycle) are lacking in spores of the smut *Ustilago maydis* until 12 hours after their germination (Gottlieb and Caltrider, 1963). Typically the HMP cycle is more active in growing cells than in resting cells. For example, cells of *Candida utilis* dissimilate 40% of the glucose via the HMP pathway, and this value increases to 50% for actively growing cells. Similar increases have been noted for the Pyrenomycetes *Claviceps purpurea* and *Neurospora crassa* (Blumenthal, 1965).

Like the EM pathway, the HMP pathway produces glyceraldehyde-3-phosphate and fructose-6-phosphate. Either of these intermediates that are common to both pathways may be diverted into one or the other of the pathways, regardless of which one originally produced them.

Entner-Doudoroff Pathway in Fungi. The ED pathway is comparatively rare in nature and has been detected only in some gram-negative bacteria and in a few fungi. A modified ED pathway is reported to be the major pathway of glycolysis in *Caldariomyces fumago*, the HMP pathway being of minor importance. In *C. fumago*, the normal path from glucose through 6-phosphogluconic acid via a phosphorylated intermediate (glucose-6-phosphate) is not followed, but an alternate pathway from glucose to 2-ketogluconic acid and then to 6-phosphogluconic acid occurs (Ramchandran and Gottlieb, 1963). The sole glycolytic pathway in spores of the smut *Tilletia*

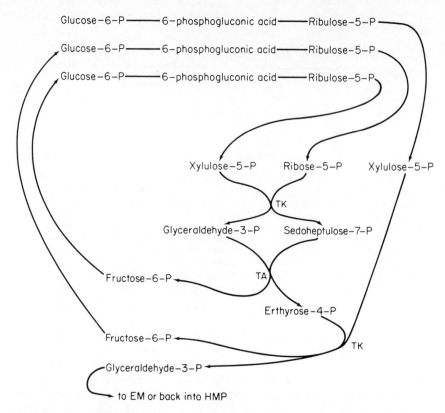

Fig. 8-3. The HMP pathway. The conversions of glucose-6-phosphate to ribulose-5-phosphate are oxidative (see text). The nonoxidative stage begins with the enzymatic conversion of ribulose-5-phosphate to ribose-5-phosphate and xylulose-5-phosphate. An enzyme,

transketolase, transfers an active glycoaldehyde $(CH_2OH-\overset{\overset{O}{\|}}{C}-)$ group from xylulose-5-phosphate to ribose-5-phosphate, thereby producing a 7-carbon intermediate (sedoheptulose-7-phosphate) and the 3-carbon glyceraldehyde-3-phosphate. In a similar reaction, a transaldolase enzyme cleaves an active dihydroxyacetone group $(H_2COH-CO-H_2COP)$ from the sedoheptulose-7-phosphate (leaving behind a hexose, erythrose-4-phosphate) and transferring it to glyceraldehyde-3-phosphate, producing a molecule of fructose-6-phosphate (a hexose). A second transketolase reaction converts xylulose-5-phosphate into glyceraldehyde-3-phosphate and erythrose-4-phosphate into fructose-6-phosphate. The fructose-6-phosphate molecules may be directly converted into glucose-6-phosphate and enter the cycle again, while the glyceraldehyde-3-phosphate may be diverted into the EM pathway or two glyceraldehyde-3-phosphate molecules may enter a transaldolase reaction to give another molecule of fructose-6-phosphate. The overall reaction is:

$$3 \text{ glucose-6-P} + 3 \text{ NADP} \longrightarrow 3 \text{ CO}_2 + 3 \text{ NADPH}_2 + 3 \text{ glyceraldehyde-3-P}$$

caries is the ED pathway, although it is replaced by the HMP and EM pathways in the mycelium (Newburgh and Cheldelin, 1958).

From these limited reports of the ED cycle in fungi, it is evident that this pathway is of little importance to the fungi generally, in terms both of its representation and of the quantity of glucose dissimilated via this route. Cochrane (1976) even questions whether or not the ED cycle occurs in the fungi, as he maintains that more experimental evidence is required.

Fermentation. An intermediate of glycolysis or the end product of glycolysis, pyruvic acid, may serve as the final hydrogen acceptor. This anaerobic reduction process is termed *fermentation* and culminates in the production of compounds such as ethyl alcohol, glycerol, and lactic acid. Fermentation may take place under aerobic conditions, occurring concurrently with pyruvic acid oxidation via the aerobic citric acid cycle, but it is usually favored by anaerobic conditions in which it does not compete with the aerobic citric acid cycle for pyruvic acid. (It is important to recall that almost all fungi are aerobic organisms and cannot grow under anaerobic conditions although preformed cells can survive in the absence of air.)

The EM pathway was long thought to be the only pathway culminating in fermentation products, but pyruvic acid which is reduced in fermentation, may also be derived from the HMP or ED pathways. The EM pathway is the major glycolytic pathway in most fungi and therefore accounts for the greatest quantity of fermentative end products formed.

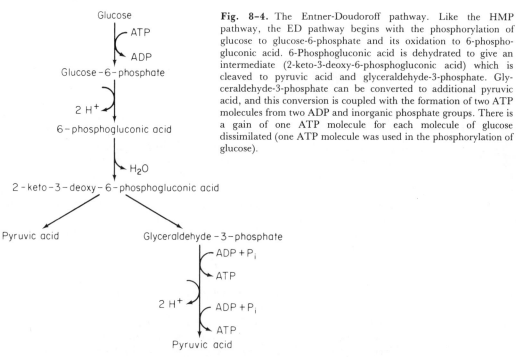

Fig. 8-4. The Entner-Doudoroff pathway. Like the HMP pathway, the ED pathway begins with the phosphorylation of glucose to glucose-6-phosphate and its oxidation to 6-phosphogluconic acid. 6-Phosphogluconic acid is dehydrated to give an intermediate (2-keto-3-deoxy-6-phosphogluconic acid) which is cleaved to pyruvic acid and glyceraldehyde-3-phosphate. Glyceraldehyde-3-phosphate can be converted to additional pyruvic acid, and this conversion is coupled with the formation of two ATP molecules from two ADP and inorganic phosphate groups. There is a gain of one ATP molecule for each molecule of glucose dissimilated (one ATP molecule was used in the phosphorylation of glucose).

Fermentation to Alcohols. Ethyl alcohol is the most important fermentation product commercially. Fermentation to alcohol was originally elucidated in the yeast *Saccharomyces cerevisiae,* and this organism continues to be the most important source of industrial alcohol (see Chapter 14). Several fungi produce ethyl alcohol, including the zygomycetes *Mucor* and *Rhizopus;* the ascomycetes *Ashbya gossypii* and *Neurospora crassa;* and the deuteromycetes *Aspergillus, Fusarium, Penicillium,* and others (Cochrane, 1958).

Ethyl alcohol formation (Fig. 8-5) is initiated by the union of pyruvic acid with a coenzyme, thiamine diphosphate (TPP), to form an *active pyruvate* complex on the surface of the enzyme pyruvate decarboxylase. This complex is decarboxylated to yield an active acetaldehyde complex, which breaks down to give acetaldehyde and TPP (Holzer and Beaucamp, 1961). The reactions may be summarized (Wilkinson and Rose, 1963):

$$TPP + CH_3CO\text{-}COOH \longrightarrow$$
$$\text{pyruvic acid}$$

$$CH_3CO\text{-}COOH\text{-}TPP \xrightarrow[\text{decarboxylase}]{\text{pyruvate}} \begin{array}{c} CO_2 \\ CH_3COH\text{-}TPP \end{array} \xrightarrow{} \begin{array}{c} TPP \\ CH_3CHO \end{array}$$
$$\text{acetaldehyde}$$

The acetaldehyde is reduced to ethyl alcohol, accepting hydrogen ions from $NADH_2$, by the enzyme alcohol dehydrogenase:

$$CH_3CHO + NADH_2 \xrightarrow[\text{dehydrogenase}]{\text{alcohol}} CH_3CH_2OH + NAD$$
$$\text{ethyl alcohol}$$

If the EM pathway was utilized to provide pyruvic acid, the $NADH_2$ required above may be provided by that pathway. An overall summary of these reactions is given in Fig. 8-5.

Fermentation to Glycerol and Other Polyhydric Alcohols. Under alkaline conditions, *Saccharomyces cerevisiae* forms glycerol. Part of the dihydroxyacetone phosphate which is produced in the EM pathway is reduced by $NADH_2$ to glycerol-3-phosphate, which is dephosphorylated to glycerol. This reduction results in a deficiency of $NADH_2$ available for the reduction of acetaldehyde. Rather than accumulating, acetaldehyde undergoes a conversion to a mixture of ethyl alcohol and acetic acid (Fig. 8-6).

A number of yeasts which are able to grow in the presence of high concentrations of carbohydrate carry out similar conversions in the absence of alkaline conditions. These yeasts form polyhydric alcohols such as arabitol, erythritol, and mannitol.

Fermentation to Lactic Acid. Fungi that form lactic acid are restricted to members of the Division Mastigomycota and Subdivision Deuteromycotina (Cochrane, 1958).

Lactic acid may be the sole fermentative end product or may appear with other compounds such as ethyl alcohol, acetaldehyde, or succinic acid. If lactic acid is the sole end product, its production is similar to that of ethyl alcohol. Pyruvic acid is reduced by $NADH_2$ or $NADPH_2$ and is catalyzed by lactic dehydrogenase (Fig. 8-7).

The Citric Acid (CA) Cycle. The CA cycle is an aerobic pathway resulting in the complete oxidation of pyruvic acid to carbon dioxide and water. Details of this pathway are given in Fig. 8-8.

For many years it has been known that the CA cycle occurs in animal and plant tissues, but techniques which successfully demonstrated the CA cycle in plants or animals failed to indicate its presence in fungi. The CA cycle was thought to be absent in the fungi, or at least of rare occurrence. After considerable effort involving refinement of existing techniques and development of new techniques (especially the use of isotopic tracers), enzymes for the CA cycle and a functional CA cycle have been found to be widespread among the fungi and to occur in representatives of every major group. The CA cycle has not been demonstrated to occur in several of the fungi

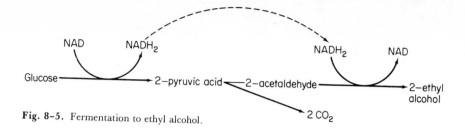

Fig. 8-5. Fermentation to ethyl alcohol.

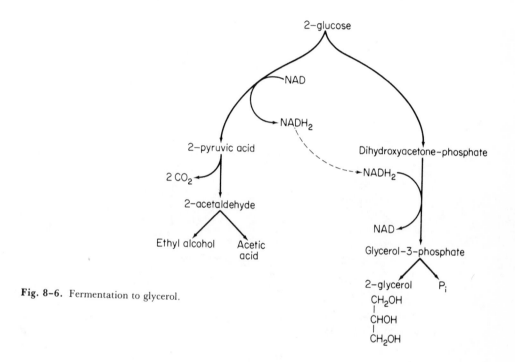

Fig. 8-6. Fermentation to glycerol.

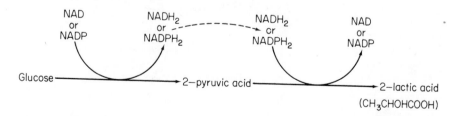

Fig. 8-7. Fermentation to lactic acid when lactic acid is the only end product.

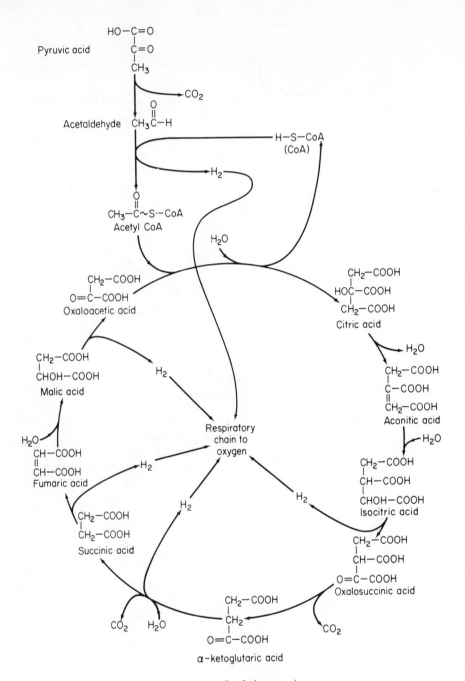

Fig. 8-8. (See facing page.)

Fig. 8–8. The conversion of pyruvic acid to acetyl-CoA and the citric acid (CA) cycle. Pyruvic acid is decarboxylated to leave an active 2-carbon acetaldehyde, which combines with coenzyme A to form acetyl-CoA. During the formation of acetyl-CoA, two hydrogen ions and two electrons are transferred first to NAD and then into the cytochrome system (the respiratory chain to oxygen). The acetate portion (2-carbon unit) of acetyl-CoA then joins with oxaloacetic acid (a 4-carbon acid) to form citric acid (a 6-carbon acid), the first compound in the CA cycle. Water is both removed from and added to citric acid, resulting in a rearrangement which produces isocitric acid. Isocitric acid is oxidized to form oxalosuccinic acid, which is decarboxylated to form α-ketoglutaric acid. A molecule of carbon dioxide and two hydrogen ions and two electrons are removed from α-ketoglutaric acid. The hydrogen ions and electrons are accepted by NAD. A molecule of water enters at this stage, and succinic acid is formed with the concurrent generation of one ATP molecule. Succinic acid is oxidized to fumaric acid by the loss of two additional hydrogen ions and two electrons. The oxidizing agent in this last reaction is flavin. Fumaric acid combines with a molecule of water to form malic acid. Malic acid is then oxidized, donating two hydrogen ions and two electrons to NAD and forming oxaloacetic acid. This oxaloacetic acid is then ready to combine with another active acetate group and to initiate the cycle again.

If no intermediates are drained from the CA cycle, the reactions may be summarized:

$$CH_3\text{-}CO\text{-}COOH + CoA\text{-}SH \longrightarrow CH_3\text{-}CO\text{-}SCoA + H_2 + CO_2$$
pyruvic acid   CoA   acetyl-CoA

$$CH_3\text{-}CO\text{-}SCoA + 3\ H_2O \longrightarrow CoA\text{-}SH + 4\ H_2 + 2\ CO_2$$
acetyl-CoA   CoA

Fig. 8–9. The glyoxylic bypass.

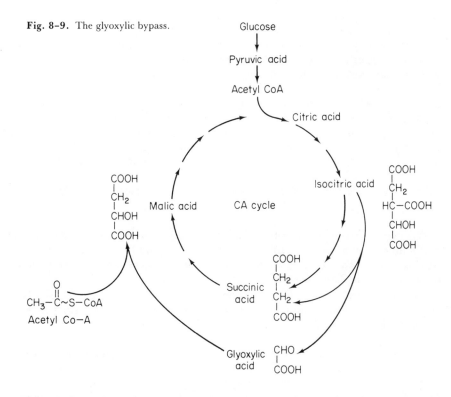

investigated, but this may be due to inadequate techniques and not to the absence of the cycle (Niederpruem, 1965). The possibility remains, however, that the citric acid cycle may not occur in all fungi.

A number of bacteria and fungi can utilize 2-carbon compounds (such as acetate and ethyl alcohol) as the sole source of carbon. Acetate may enter the CA cycle via the *glyoxylic bypass* (Fig. 8–9). Isocitric acid, which is formed in the CA cycle, is split by isocitrate lyase to succinic and glyoxylic acids. The succinic acid reenters the CA cycle, while the 2-carbon glyoxylic acid combines with the 2-carbon acetyl CoA, derived from the acetate, to form malic acid. This last condensation reaction is mediated by the enzyme malate synthetase. Malic acid then enters the CA cycle. It is possible that other 2-carbon compounds may also enter the CA cycle in this manner (Collins and Kornberg, 1960; Kornberg and Krebs, 1957; Kornberg and Madsen, 1957; Niederpruem, 1965). Additional enzymes, which have not been completely characterized, are required for the glyoxylic bypass (Casselton, 1976). As you may recall from Chapter 1, the enzymes required for the glyoxylic bypass are contained within specialized microbodies (glyoxysomes). The cellular location where the glyoxylic bypass occurs is unknown, although the CA cycle is associated with mitochondria.

Roles of the CA Cycle in Fungi. The CA cycle serves as a hub where many metabolic pathways converge. Some intermediates of the CA cycle may also serve as intermediates in either the catabolism or anabolism of compounds other than those in the CA cycle. Acetyl CoA, for example, is a degradation product of a fatty acid and may also serve in the synthesis of new fatty acids. α-Ketoglutaric acid has a similar role in amino acid metabolism. (Metabolism of amino acids and fatty acids is considered in greater detail on pp. 328 and 333.) The existence of intermediates which are common to catabolic pathways and to the CA cycle allow the degradation products of a number of compounds to enter the CA cycle, where they are further dissimilated. An extremely important role of the CA cycle is to make intermediates available that serve as precursors for biosynthesis of organic compounds.

Intermediates are drained from the CA cycle for synthesis of other compounds, but yet the cyclical nature of the CA cycle continues. Fixation of carbon dioxide makes it possible for the CA cycle to continue to operate because the necessary concentration of key intermediates is maintained. Although different mechanisms may operate, probably the most important in fungi is that in which pyruvic acid functions as an acceptor molecule that is carboxylated in the presence of pyruvate carboxylase:

$$\text{Pyruvic acid} + \text{ATP} + \text{HCO}_3^- \;\rightleftharpoons\; \text{oxalacetic} + \text{ADP} + \text{P}_i$$
$$\text{acid}$$

The enzyme pyruvate carboxylase contains biotin, and a monovalent ion, usually K^+, and a divalent ion such as Mn^{2+} or Mg^{2+} are required for the above reaction. The mechanism for carbon dioxide fixation in fungi is generally similar to that which occurs in mammalian tissues (Casselton, 1976; Utter et al., 1975).

Energy Production. The cytochrome system of fungi is generally similar to that occurring in mammals and consists of the overall linear sequence (or possibly a partially parallel sequence):

$$\text{flavin} \longrightarrow \underset{b}{\text{cytochrome}} \longrightarrow \underset{Q}{\text{coenzyme}} \longrightarrow \underset{c_1,\ c}{\text{cytochromes}} \longrightarrow \underset{a}{\text{cytochrome}} \longrightarrow \text{oxygen}$$

Each unit may be a complex of cytochromes; for example, cytochrome *a* may be subdivided into cytochromes a and a_3. The exact composition of the cytochrome system of fungi is highly variable. The cytochromes from different fungi often have different physical properties, and coenzyme Q differs in composition among the fungi. The cytochrome system may also vary among genetic strains or if a single species is grown under different environmental conditions—a cytoplasmic mutant ("petite colonie") of yeast and anaerobically grown normal yeast strains have the cytochromes a_1 and b_1, which are not normally present, but lack the normal cytochromes a, b, and c (Lindenmayer, 1965).

An important function of the CA cycle is to provide energy to the organism. This is primarily accomplished by the passage of hydrogen ions and their electrons along the cytochrome chain and the generation of three ATP molecules for each pair of hydrogens. Each mole of ATP stores about 7,000 calories of energy. The conversion of a molecule of pyruvic acid to acetyl CoA releases two hydrogen ions and two electrons, and the complete oxidation of one acetyl CoA via the citric acid cycle releases eight hydrogen ions and electrons. This accounts for the formation of 15 ATP moles, or about 105,000 calories, from each mole of pyruvic acid which is completely oxidized.

Role of the CA Cycle in Differentiation. We have just indicated that the CA cycle may provide important intermediates in metabolism. Further, from our consideration of morphology in Chapters 1 through 6, we know that the fungi often vary considerably in form during their normal life cycles. It is to be expected that morphological form is ultimately under metabolic and genetic control. In extensive studies of *Blastocladiella emersonii* Cantino (1966) established such a relationship and demonstrated that a modification of the CA cycle is involved.

Blastocladiella emersonii has light-stimulated carbon dioxide fixation that greatly enhances some aspects of metabolism. Carbon dioxide may be provided in the form of bicarbonate added to the culture medium. A culture that would ordinarily form only colorless zoosporangia may be induced to form resting spores by including $0.01\ M$ bicarbonate in the culture medium. A variety of biochemical differences exist between the colorless zoosporangium and the pigmented resting spore. For example, the resting spore wall (but not the wall of the zoosporangium) contains the dark pigment melanin and its precursor, tyrosine. The protoplast of the resting spore contains γ-carotene, which is absent in the zoosporangium. There are also quantitative differences in protein, polysaccharide, and lipid levels. Most of these differences seem to be ultimately triggered by alterations in respiration via the CA cycle (Cantino and Turian, 1959). As the resting spores differentiate, there is a marked decline in respiration, and ultimately the CA cycle is disrupted and its enzymes disappear (there is no similar decline during differentiation of the zoosporangium). Accompanying this decline of the CA cycle is a marked increase (about 6,500-fold) in the activity of isocitric dehydrogenase. The increase in isocitric dehydrogenase activity causes a reversal in an important step of the CA cycle (Fig. 8–10). In the normal CA cycle, isocitric acid is oxidized and decarboxylated to form α-ketoglutaric acid, but in this reversal, α-ketoglutaric acid is reduced and carboxylated to form isocitric acid. This effectively blocks the CA cycle and provides large quantities of isocitric acid. Rather than accumulating, the isocitric acid is shunted into a new metabolic pathway that involves its conversion to succinic and glyoxylic acids. The glyoxylic acid then becomes involved in transamination reactions to yield glycine (see also p. 328). The significance of this information relates not only to the manner in which the respiratory cycles may be altered during development but also to the control that this alteration may exert on normal development.

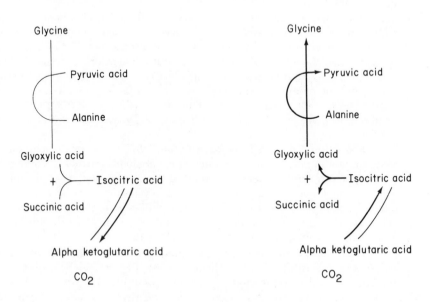

Fig. 8-10. The bicarbonate-induced shift in respiration associated with morphological changes in *Blastocladiella emersonii*. [Adapted from E. C. Cantino, 1961, *Symp. Soc. Gen. Microbiol.* **11**:243–271.]

Factors Influencing Respiration in Fungi. A number of environmental and intrinsic factors affect the course and rate of respiration.

External factors influencing respiration are pH, temperature, oxygen pressure, carbon dioxide, and nutrients. The pH and temperature affect respiration in a manner similar to their effect on growth, discussed in the last chapter.

Oxygen is a requirement for aerobic respiration to take place. The oxygen used in respiration is dissolved in liquid, and the amount of this dissolved oxygen is proportional to the amount of oxygen present in the gas phase. At very low concentrations of oxygen, the rate of respiration is directly correlated with the amount of oxygen available but, as the amount of oxygen increases, the rate of respiration becomes independent of the oxygen concentration. The point at which the respiration rate becomes independent of oxygen concentration is 0.022 millimoles/liter (about 16 millimeters Hg) for *Penicillium chrysogenum* and about 0.0046 millimoles/liter (about 4.0 millimeters Hg) for yeast cells (Finn, 1954).

Another extrinsic factor that may affect respiration is the sugar concentration. A high sugar concentration in the presence of adequate aeration may result in utilization of fermentative pathways in yeasts, as a glucose concentration over 5% totally inhibits synthesis of respiratory enzymes. As yeasts exhaust the sugar from their culture medium, there may be an increase in the activites of the enzymes of the CA cycle and also of the glyoxylic acid cycle (Suomalainen and Oura, 1971).

Intrinsic factors include the enzyme complement of the cell, cell permeability, phase of development, and age. The most important of these is the enzyme complement of a species. This will ultimately dictate which substrates can be oxidized and the biochemical pathways to be followed. Failure of a fungus to oxidize a substrate may be due to inability of the substrate to penetrate the cell membrane, but a cell-free preparation may oxidize the substrate readily, indicating that the cell has the necessary enzymes to accomplish the oxidation. Those substrates most often excluded from the cell are ionizable compounds such as organic acids. A fungus passes through many developmental stages (such as spore dormancy, spore germination, and mycelium development). Rates of respiration or the respiratory pathways may alter with the different stages or age. Respiration is greatest during periods of active growth and decreases with aging of the fungus and accompanying decrease in growth. Specific examples of the correlation of the respiratory pathway with development include *Tilletia caries* and *Blastocladiella emersonii,* mentioned on pp. 313 and 321, respectively, and the alterations that germinating spores undergo (Chapter 10). In *Allomyces,* the male gametangia have an impaired CA cycle and rely upon the glyoxylic bypass, while the female gametangia are not similarly impaired (Cantino, 1966).

Carbon Compounds Produced by Fungi

After carbon compounds are assimilated, a certain amount of the carbon is used by the fungi in the biosynthesis of cellular materials. These include wall constituents, storage carbohydrates, and a wide variety of metabolites that are of no apparent value to the fungi.

POLYSACCHARIDES

A large amount of the carbon assimilated by fungi is used in the biosynthesis of wall material. As indicated in Chapter 1, the fungal walls consist of complex mixtures of carbohydrates, proteins, and lipids, and the principal carbohydrate is usually chitin.

Biosynthesis of the wall polysaccharides and glycogen basically involves a repetition of the following basic reaction (Burnett, 1976):

$$\text{nucleoside diphosphate sugar} + \text{primer} \longrightarrow \text{(primer-sugar)} + \text{nucleoside diphosphate}$$

Chitin biosynthesis, which follows the above scheme, is generally similar in the fungi, crustaceans, and insects (McMurrough et al., 1971). Chitin biosynthesis begins with the conversion of glucose-1-phosphate into *N*-acetylglucosamine-1-phosphate, which requires a series of reactions (Fig. 8–11). The *N*-acetylglucosamine-1-phosphate then reacts with UTP (the nucleotide uridine triphosphate) to form UDP-*N*-acetylglucosamine, which is the nucleoside diphosphate sugar. Finally the UDP-*N*-acetylglucosamine transfers the *N*-acetylglucosamine moiety to the growing chitin chain, functioning as a primer, and becomes one of its subunits. The enzyme chitin synthetase and Mg^{2+} ions are required for this polymerization step. The completed chitin molecule is a long chain of sugar subunits which are joined by β-1, 4 links (Fig. 1–4).

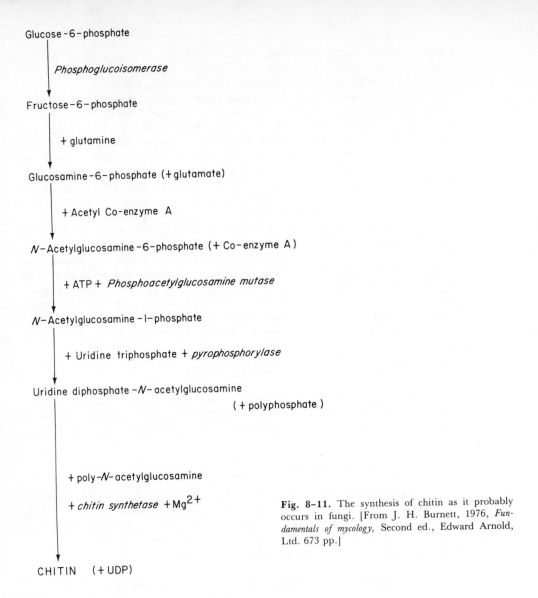

Glucose – 6 – phosphate

    Phosphoglucoisomerase

Fructose – 6 – phosphate

    + glutamine

Glucosamine – 6 – phosphate (+ glutamate)

    + Acetyl Co-enzyme A

N – Acetylglucosamine – 6 – phosphate (+ Co-enzyme A)

    + ATP + *Phosphoacetylglucosamine mutase*

N – Acetylglucosamine – 1 – phosphate

    + Uridine triphosphate + *pyrophosphorylase*

Uridine diphosphate – N – acetylglucosamine

                   (+ polyphosphate)

    + poly – N – acetylglucosamine

    + *chitin synthetase* + Mg^{2+}

Fig. 8–11. The synthesis of chitin as it probably occurs in fungi. [From J. H. Burnett, 1976, *Fundamentals of mycology*, Second ed., Edward Arnold, Ltd. 673 pp.]

CHITIN   (+ UDP)

Fungi form polysaccharides that serve as an energy reserve. The sugar alcohol mannitol is formed and utilized as a food reserve and is especially common among the ascomycetes and basidiomycetes. Other sugar alcohols formed by fungi (and presumably serving as storage products) are D-sorbitol, D-volemitol, D-arabitol, and L-erythritol. The most common of the storage products is glycogen (Fig. 8–12), which usually comprises about 5% of the dry weight of the mycelium and spores (Cochrane, 1958). Like the glycogen formed by animals, glycogen is a branched glucose polymer. In the main portion of the chain, the glucose units are joined by α-1, 4 linkages. In glycogen synthesis, UTP reacts with glucose-1-phosphate as follows (Burnett, 1976):

$$\text{UTP + glucose-1-phosphate} \rightleftharpoons \text{UDP-glucose + pyrophosphate}$$

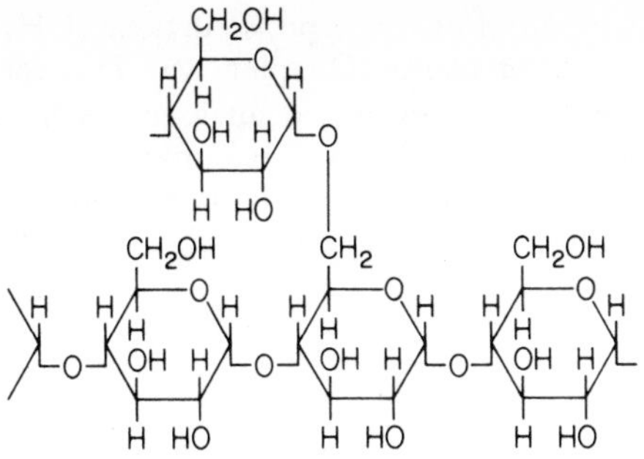

CH$_2$OH

Fig. 8-12. Portion of a glycogen molecule.

The UDP-glucose (the nucleoside diphosphate sugar) donates D-glucose in the presence of the enzyme glycogen synthetase, and the D-glucose is added to the growing glycogen molecule, which functions as an acceptor. These last steps may be summarized:

$$\text{UDP-glucose} + [\alpha\text{-1, 4- glucose}]_n \xrightarrow{\text{glycogen synthetase}}$$
$$[\alpha\text{-1, 4-glucose}]_{n + x} + x\,\text{UDP}$$

MISCELLANEOUS HYDROCARBON METABOLITES

Fungi are among the most active of chemists and are able to produce a large variety of organic compounds. Many of these compounds are of no apparent value to the fungus as they are not directly involved in supplying energy to the cell nor are they involved in structural components of the cell. Instead, the compounds often arise either directly or indirectly from a surplus of some intermediate of normal metabolism and represent ''overflow'' products. Alternatively, they may arise from an elaborate biosynthetic pathway, which in reality is an error in metabolism since it does not provide the fungus with energy or a needed synthetic product but instead requires energy and intermediates to take place. Some of these compounds are of commercial interest (see Chapter 14).

ALIPHATIC METABOLITES

A number of saturated aliphatic compounds are formed by fungi. These include alcohols, acids, and esters. Some fairly common examples include acids and various unsaturated aliphatics.

Acids from respiratory pathways may accumulate in such large quantities that they may not be further metabolized. These include gluconic, citric, oxaloacetic, malic, fumaric, succinic, lactic, and acetic acids. Some acids may be derived from respiratory intermediates. These include oxalic acid, a common metabolic product of fungi generally and especially of the Agaricales, derived from oxaloacetic acid; itaconic acid, formed by strains of *Aspergillus terreus,* derived from *cis*-aconitic acid; and acetonedicarboxylic acids, produced by *Aspergillus niger* from citric acid metabolism. Acids which may be formed directly from sugars are mannonic acid (from mannose), galactonic acid (from galactose), and glucuronic acid (from glucose) (Cochrane, 1958).

Unsaturated aliphatic compounds are also common, as an unsaturated double bond ($-CH = CH-$) and an unsaturated triple bond ($-C \equiv C-$) occur in a large number of fungal metabolites. The simplest compound of this type is ethylene ($CH_2 = CH_2$), produced by some fungi that induce wilts of plants (Chapter 12). The unsaturated double bond may also occur in chains, either with saturated single bonds or with other double bonds. Many of these compounds are colored, a common example being β-carotene, a yellow pigment (Fig. 8–13). The unsaturated triple bond is especially common in metabolites from the higher fungi and almost always occurs in chains with other triple bonds or with double bonds.

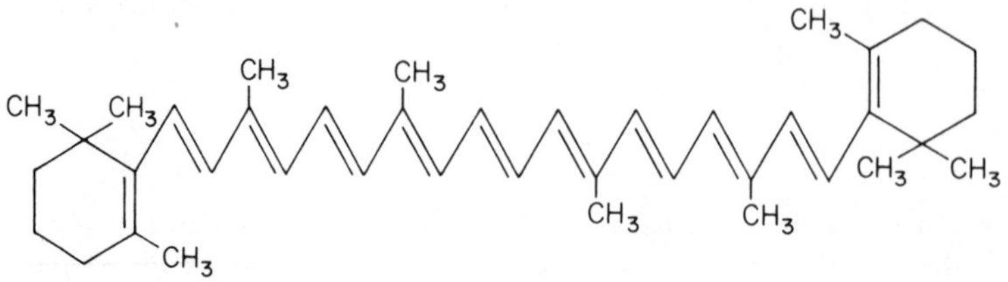

Fig. 8–13. β-Carotene.

AROMATIC METABOLITES

The aromatic nucleus is characterized by a ring of six carbon atoms and three unsaturated double bonds. There are two pathways by which the aromatic nucleus may be derived. These are the *shikimic acid* pathway and the *acetate* pathway. It is not known, however, whether all or even most aromatic compounds are derived from these pathways.

In the shikimic acid pathway, erythrose-4-phosphate (derived from the HMP pathway) condenses with phosphoenol pyruvic acid. The intermediate formed from this reaction undergoes successive rearrangements (including formation of a ring and a double bond) to form shikimic acid. Shikimic acid condenses with a 3-carbon unit and is converted to prephrenic acid, which has two double bonds. With the formation of a third double bond, prephrenic acid is converted to aromatic compounds (8–14).

The 2-carbon and 1-oxygen skeleton of acetic acid (acetate) has long been recognized as a potential building block for fungal metabolites. Various combinations of acetate units or modified acetate units could account for the structure of many metabolites. A pathway involving condensation of acetate molecules is that in which acetyl CoA combines with carbon dioxide to form malonyl CoA, a 3-carbon intermediate. Three malonyl CoA molecules undergo condensation to form a 6-carbon aromatic nucleus, meanwhile releasing three carbon dioxide molecules (Bu'Lock and Smalley, 1961). The intermediate reactions in this process are poorly understood.

There are large numbers of aromatic compounds in fungi, and they may occur in either monocyclic or polycyclic forms (Fig. 8–15). Most monocyclic aromatic metabolites are either aromatic alcohols or derivatives of these in which the alcoholic hydroxyl group is replaced by a methyl ($-CH_3$), methoxyl ($-OCH_3$), or carboxyl ($-CO_2H$) group or by an organic radical. The aromatic alcohols are phenol, pyrocatechol, resorcinol, quinol, hydroxyquinol, and pyrogallol; these differ in the number and position of their hydroxyl groups. Derivatives of phenol, resorcinol, and quinol are especially common.

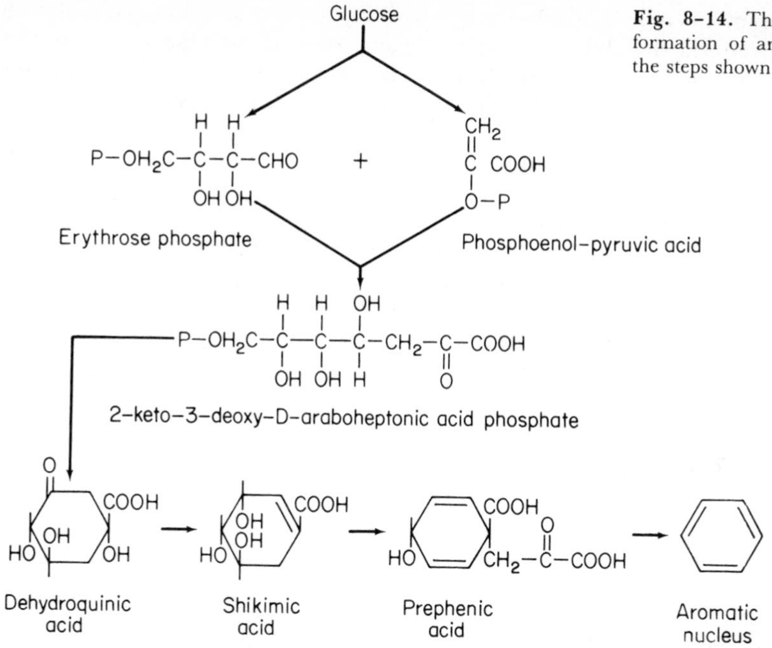

Glucose

Fig. 8-14. The shikimic acid pathway for formation of aromatic compounds. Some of the steps shown are multiple.

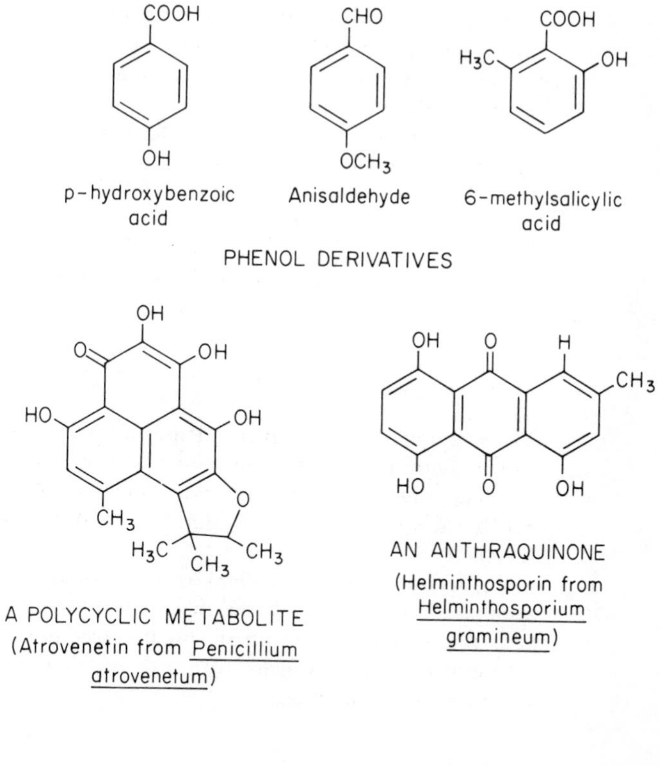

$P-OH_2C-\overset{H}{\underset{OH}{C}}-\overset{H}{\underset{OH}{C}}-CHO$ + $\overset{CH_2}{\underset{\overset{|}{O-P}}{\underset{\|}{C}}}COOH$

Erythrose phosphate   Phosphoenol-pyruvic acid

$P-OH_2C-\overset{H}{\underset{OH}{C}}-\overset{H}{\underset{OH}{C}}-\overset{OH}{\underset{H}{C}}-CH_2-\overset{}{\underset{O}{C}}-COOH$

2-keto-3-deoxy-D-araboheptonic acid phosphate

Dehydroquinic acid → Shikimic acid → Prephenic acid → Aromatic nucleus

PHENOL DERIVATIVES

p-hydroxybenzoic acid   Anisaldehyde   6-methylsalicylic acid

Fig. 8-15. Some aromatic metabolites.

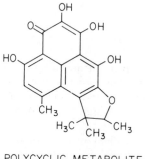

A POLYCYCLIC METABOLITE
(Atrovenetin from <u>Penicillium</u>
<u>atrovenetum</u>)

AN ANTHRAQUINONE
(Helminthosporin from
<u>Helminthosporium</u>
<u>gramineum</u>)

Examples of phenol derivatives are *p*-hydroxybenzoic acid and anisaldehyde which are derived via the shikimic acid pathway. Another member of this group, 6-methylsalicyclic acid, is an intermediate in the acetate pathway. The numerous resorcinol derivatives are presumably derived via the acetate pathway (Birkinshaw, 1965).

A quinol derivative may be oxidized by the fungus to form a *quinone*. There are several types of quinones, determined by the number of rings present and the manner in which they are joined. Of special interest are the *anthraquinones,* pigments that cause the mycelium to be brightly colored. The anthraquinones may account for as much as 30% of the dry weight of the mycelium. These compounds consist of three aromatic rings and may have from one to five hydroxyl groups, one of which may be methylated (Birkinshaw, 1965). Coenzyme Q in the cytochrome system is a quinone which is functional in the cell.

NITROGEN METABOLISM

Amino Acid Metabolism

DEAMINATION AND TRANSAMINATION REACTIONS

As noted in Chapter 7, proteins and peptides free amino acids upon hydrolysis, and these amino acids are a source of usable nitrogen. Before we can discuss utilization of amino acids by fungi, we must discuss two important types of reactions involving amino acids which are of general occurrence in organisms and are responsible for many of the transformations of amino acids. These are the oxidative deamination and transamination reactions.

A deamination reaction is one in which the amino group is removed from the amino acid, leaving a keto acid:

$$NH_2-\underset{\underset{H}{|}}{\overset{\overset{R}{|}}{C}}-COOH + \tfrac{1}{2}O_2 \longrightarrow O = \underset{}{\overset{\overset{R}{|}}{C}}-COOH + NH_3$$

amino acid   keto acid

The enzymes involved in the deamination reaction are D- or L-amino acid oxidases, which deaminate D- or L-amino acids, respectively. The amino acid oxidases of fungi are generally similar to those of animal systems, but they may differ in the rates with which they effect the deamination reaction and also in their degree of substrate specificity.

Transamination involves transfer of the amino group from one compound to another. It takes place between an amino acid and a keto acid, converting the keto acid into the corresponding amino acid (see next page). Transamination allows for the synthesis of an amino acid from any appropriate carbon skeleton, as long as a donor amino acid is available.

$$\underset{\substack{| \\ \text{H}}}{\overset{\substack{\text{R} \\ |}}{\text{NH}_2\text{—C —COOH}}} + \underset{}{\overset{\substack{\text{R}' \\ |}}{\text{O = C —COOH}}} \xrightarrow{\text{transaminase}}$$

amino acid A   keto acid B

$$\overset{\substack{\text{R} \\ |}}{\text{O = C —COOH}} + \underset{\substack{| \\ \text{H}}}{\overset{\substack{\text{R}' \\ |}}{\text{NH}_2\text{— C —COOH}}}$$

keto acid A   amino acid B

AMINO ACID CATABOLISM

Most of the amino acids assimilated by the fungus are used directly without being dissimilated. Transamination may convert amino acids into other required amino acids, which are then used directly in protein synthesis or in other syntheses. *Neurospora crassa* mutants, deficient for glutamic dehydrogenase and therefore unable to directly synthesize amino acids via α-glutaric acid (see next section), are able to utilize a wide variety of amino acids, indicating that an extensive transaminase system exists in that organism (Nicholas, 1965).

The principal route of amino acid catabolism in fungi is oxidative deamination, but a number of other mechanisms exist. For example, asparagine may be hydrolyzed to aspartic acid and ammonia in *Aspergillus niger* and other fungi; threonine is cleaved to glycine and acetaldehyde by *Neurospora crassa;* and arginine is degraded via the ornithine cycle (discussed in the next section) (Cochrane, 1958).

THE GLUTAMIC ACID CROSSROADS

Both the ammonia freed by the deamination reaction and inorganic nitrogen (nitrates, nitrites, and ammonia) which is directly taken up by the fungi enter into synthetic reactions. It is probable that all inorganic forms of nitrogen are reduced to ammonia and that ammonia enters amino acid biosynthetic pathways by combining with α-ketoglutaric acid, an intermediate in the CA cycle:

$$\text{HOOC–(CH}_2)_2\text{–CO–COOH} + \text{NADH} + \text{H}^+ + \text{NH}_3 \rightleftharpoons$$

α-ketoglutaric acid

$$\text{HOOC–(CH}_2)_2\text{–CH(NH}_2)\text{COOH} + \text{NAD}$$

L-glutamic acid

The enzyme which mediates this reaction, glutamic dehydrogenase, has been found in *Neurospora crassa* (Cochrane, 1958). The amino acid formed in this reaction, L-glutamic acid, is actively involved in transamination reactions and in specific synthetic pathways that give rise to other amino acids. The formation of glutamic acid from ammonia and its keto acid and the subsequent shunt of the amino group to other amino acids provides an important mechanism for introducing inorganic nitrogen into amino acid biosyntheses (Fig. 8–16).

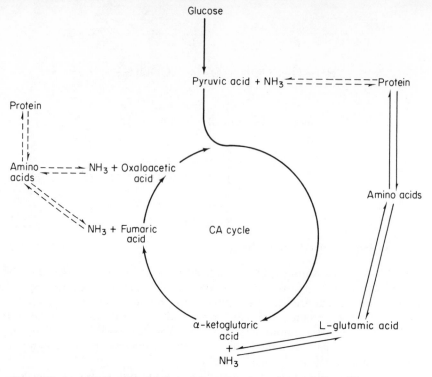

Glucose

Pyruvic acid + NH$_3$ = = = = = = = = → Protein

Protein

Amino acids — — — — → NH$_3$ + Oxaloacetic acid

NH$_3$ + Fumaric acid

CA cycle

Amino acids

α–ketoglutaric acid + NH$_3$

L–glutamic acid

Fig. 8-16. Interrelationship of carbohydrate and amino acid metabolism. Those conversions which are important in fungi are indicated by solid lines, while those in other organisms are indicated by a dashed line.

Just as glutamic acid provides a bridge leading to amino acid biosyntheses, it may function in reverse. The reaction above is reversible, and through glutamic acid the amino group may be lost as ammonia and the carbon skeleton of the amino acid introduced into the CA cycle as α-ketoglutaric acid, which is then oxidized.

A similar bridge involving aspartic acid and fumaric acid occurs in the bacteria, but it is apparently of minor importance in the fungi (Cochrane, 1958).

AMINO ACID BIOSYNTHESIS

Utilization of Nutrient-Deficient Mutants. The use of genetic mutants has contributed a great deal to our understanding of amino acid biosynthetic pathways and, in many cases, has provided the bulk of information about a pathway which is then supplemented by chemical and isotopic analyses. *Neurospora crassa* has been widely used in these studies.

For these studies, mutants are induced, usually by ultraviolet irradiation of the spores. The spores are inoculated onto a minimal medium (containing only the minimum amount of inorganic nutrients and vitamins that support growth of a normal strain). Any spore which fails to develop a germ tube is transferred to a medium which contains supplementary nutrients. If an arginine-deficient mutant is being sought, for example, the supplemented medium would contain arginine. If the spore then germinates and a mycelium develops on the supplemented medium, it indicates that the strain is deficient for that nutritional factor provided in the supplemented medium.

Not all mutants deficient for a single nutrient are genetically identical and may show quite different responses to chemically related nutrients that may or may not support growth. Each mutant has a genetic deficiency which does not allow it to convert one precursor to another compound in a biosynthetic sequence and this leads to the formation of the deficient nutrient. Any compound normally coming after the genetic block will support growth if it is provided in the medium, since the necessary end product can then be synthesized. Also if a genetic block occurs, that precursor occurring immediately before the block will not be utilized and will accumulate. The nutritional differences among the mutants are presumably due to genetic blocks at different points along the biosynthetic route.

We illustrate the elucidation of a biochemical pathway with methionine-requiring mutants (Horowitz, 1947). Both cysteine and homocysteine were recognized as probable precursors of methionine, and analysis of an accumulated substance in a group of mutants indicated that cystathionine might also be a precursor. The methionine-deficient mutants analyzed by Horowitz (1947) comprised four nutritional groups, each controlled by a single gene. These groups were

1. Strain me-1, which can only use methionine
2. Strain me-2, which can use homocysteine and methionine
3. Strain me-3, which can use cystathionine, homocysteine, and methionine
4. Strain me-4, which can use cysteine, cystathionine, homocysteine, and methionine

It is assumed that those mutants in group 4 have a genetically controlled block that does not permit the synthesis of the first precursor in this sequence. Cysteine does not support growth of group 3, although all other compounds will, and this indicates that the genetic block in group 3 interrupts the conversion of cysteine to the next precursor, which places cystine as the first precursor in this sequence. Similar reasoning shows us that cystathionine is the next precursor since group 2 cannot utilize either cysteine or cystathionine, both of which occur before its point of blockage. Since mutants exist which can use only methionine, it is apparently the last compound in the sequence. Homocysteine is then placed in the penultimate position. The sequence can then be constructed:

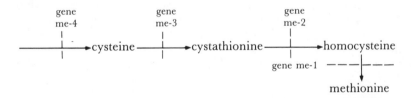

Biosynthesis of Other Amino Acids. The ornithine cycle was first established in mammalian liver, but it was later found to occur in *Neurospora crassa* through the use of single-gene nutritional mutants (Fincham and Boylen, 1957; Srb and Horowitz, 1944). In this cycle, glutamic acid is converted to ornithine by reduction and transamination (Fig. 8-17). Ornithine is converted to citrulline by the addition of a $O = C-NH_2$ group, and citrulline is converted to arginine (via arginosuccinic acid) by the replacement of this keto oxygen with $= NH$. Arginine may be degraded by the enzyme arginase in animals to give urea, an end product of metabolism, and ornithine, which can initiate the cycle again. Like animals, *Neurospora* also has arginase but the urea formed is broken down by an additional enzyme, urease, to form carbon

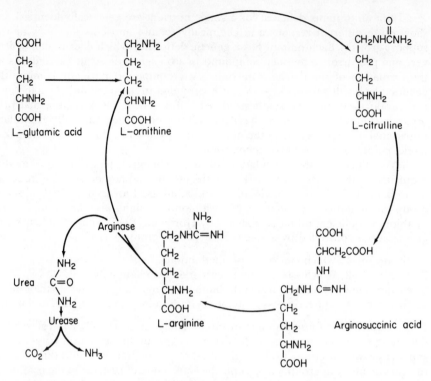

Fig. 8-17. The ornithine cycle in *Neurospora.*

dioxide and ammonia. Apparently the role of the ornithine cycle in *Neurospora* and other fungi is to supply arginine for protein synthesis.

We know from p. 326 that the shikimic acid pathway gives rise to aromatic intermediates that may be parent compounds for aromatic metabolites. The shikimic acid pathway also gives rise to the amino acids tryptophan, tyrosine, and phenyl-alanine, in addition to the vitamin *p*-aminobenzoic acid. All of these have an aromatic nucleus. A *Neurospora crassa* strain deficient for *p*-aminobenzoic acid, tryptophan, tyrosine, and phenylalanine will grow if shikimic acid is available (Tatum et al., 1954).

Nitrogen Compounds of Fungi

FUNCTIONAL NITROGEN COMPOUNDS

As in other organisms, nitrogen is an important component of the protoplasm of fungi. In a survey of cultivated fungi, the percentage of the dry weight attributable to nitrogen was found to vary between 2.27% in *Coprinus radicans* to about 5.13% in *Trichoderma lignorum* (Heck, 1929). The amount of nitrogen incorporated into the pro-toplasm varies with environmental conditions and with age of the mycelium. The nitrogen content is greatest when nitrogen is readily available in the medium, especially as ammonium salts. A young mycelium characteristically has a higher content than an older mycelium.

About 60% to 70% of the total nitrogen of the fungal cell is protein. The nitrogen which does not exist as protein occurs as nucleic acids, chitin, and nonessential metabolites.

MISCELLANEOUS NITROGEN METABOLITES

Fungi produce a large number of nitrogen metabolites that are of no apparent value to the fungus. We discuss some specific examples of these, which are pertinent to discussions elsewhere in this book and are indicative of the wide variety of these compounds which are formed by fungi.

Ammonia and its derivatives in which one or more of the hydrogen atoms are replaced by an alkyl group (the amines) are of frequent occurrence in the fungi. Many of the amines are apparently derived by decarboxylation of an amino acid. The simplest of these is methylamine (CH_3NH_2), which was found in 22 fungi in a survey of 105 species (as many as eight amines were found in a single species) (Birkinshaw, 1965). A hydroxylated derivative of an amine, ethanolamine (H_2NCH_2–CH_2OH), may be converted into an ionic compound by substitution with a fourth group. This ionic compound is choline:

$$(CH_3)_3N^+CH_2CH_2OH(Cl^-)$$

Choline is the parent compound for muscarine and muscaridine, poisons from *Amanita muscaria* (Chapter 14).

Other acyclic nitrogen metabolites are complex molecules consisting of many units. These include short peptide chains made up entirely of amino acids and molecules with mixtures of amino acids and nitrogen-free organic acids. Lycomarasmin, which induces wilting in plant diseases caused by *Fusarium*, is a molecule of this latter type (Chapter 12).

A number of nitrogen metabolites are heterocyclic compounds. These include five- and six-membered rings. Those most important to us are the metabolites which have the indole nucleus. Metabolites with the indole nucleus include the active components of hallucinogenic fungi, psilocin and psilocybin; the ergot alkaloids; and the poisons phalloidin and phalloin produced by *Amanita phalloides* (Chapter 14).

Many of the antibiotics are heterocyclic nitrogen compounds which include a sulfur or chlorine atom. Among these are penicillin and griseofulvin (see Chapters 11, 12, and 14).

LIPID METABOLISM

Lipids include a diverse variety of compounds that are soluble in nonpolar organic solvents. Many of the lipids contain glycerol, which may be esterified with one, two, or three fatty acid molecules to form mono-, di-, or tri-glycerides, respectively (these are called the neutral fats). The phospholipids also contain glycerol, as these compounds are derivatives of glycerol phosphate. Other lipids, including the sterols, may lack glycerol. Lipids may also be combined in nature with other compounds such as proteins, amino acids, or polysaccharides.

Lipid Catabolism

The lipids sometimes serve as a source of energy and carbon for fungi. Fungi do not usually require an exogenous source of lipids, as they are capable of synthesizing their own, and lipids are not ordinarily added to culture medium. Addition of lipids or fatty acids to the culture medium may enhance the growth of some fungi; for example, some boletes that are ordinarily difficult to grow in culture may grow quite well if vegetable oil is added (Schisler and Volkoff, 1977). The yeast *Pityrosporum ovale* actually requires an exogenous source of lipid (Hunter and Rose, 1971). Fungi produce triglycerides, which serve as an energy source when required (for example, upon spore germination). Triglycerides are also especially abundant in terms of quantity in natural substrata encountered by fungi.

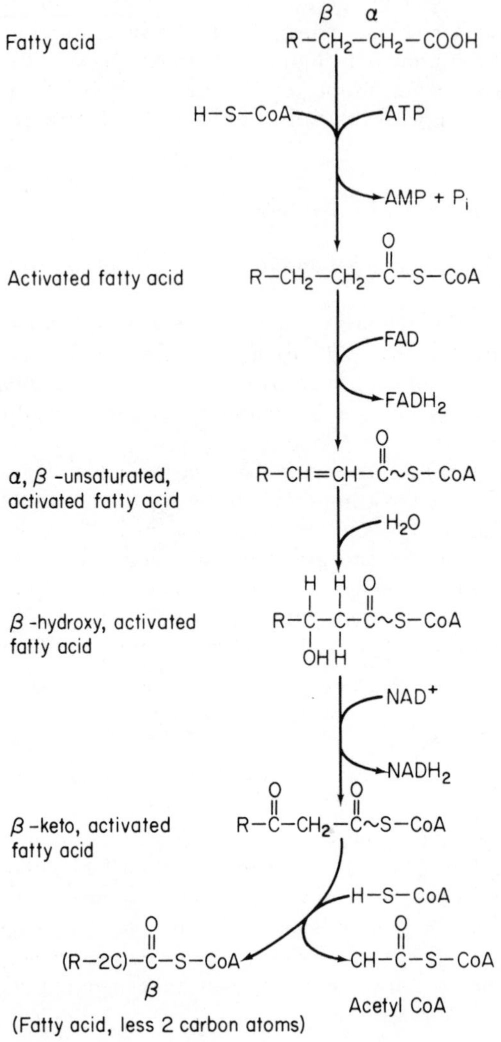

Fig. 8-18. β-Oxidation of a fatty acid as it occurs in animals. In fungi, the pathway occurs up to and including β-keto fatty acid. See text.

Triglycerides are hydrolyzed by lipases that release glycerol and the three fatty acids. Glycerol may enter glycolysis after its phosphorylation and oxidation to 3-phosphoglyceraldehyde and fatty acids. The fatty acids must undergo further degradation before they can be utilized.

The naturally occurring fatty acids have a straight chain with an even number of carbon atoms. That carbon atom next to the carboxyl group is the α-carbon, while the second one is the β-carbon. Fatty acid catabolism is primarily a sequence of oxidations of the β-carbon atoms (Fig. 8–18).

In animals, fatty acid catabolism begins with the activation of the fatty acid molecule by a linkage with a molecule of coenzyme A in the presence of ATP. This activated molecule is then dehydrogenated (with FAD as the hydrogen carrier) with one hydrogen atom being removed from both the α- and β-carbons, creating an unsaturated double bond ($-C = CH-$). The next step is a hydration, producing an alcoholic group on the β-carbon, which is then oxidized by a β-keto fatty acid (β oxidation). The hydrogen acceptor in this last reaction is NAD. The β-keto fatty acid then reacts with a second molecule of coenzyme A and yields an acetyl CoA and an activated fatty acid that has two fewer carbons than the original fatty acid. Successive β oxidations occur until only a 2-carbon fragment remains. The acetyl CoA molecules which are formed enter the CA cycle.

In fungi, the catabolism of fatty acid often proceeds in a similar fashion to the point where a β-keto fatty acid is formed, but the β-keto fatty acid may be decarboxylated to form a methyl ketone having one less carbon atom than the original fatty acid, i.e.,

$$\longrightarrow \text{R--CO--CH}_2\text{--COOH} \longrightarrow \text{R--CO--CH}_3 + \text{CO}_2$$

β-keto fatty acid   methyl ketone

These methyl ketones are formed in cheeses processed with fungi and in edible oils attacked by fungi, perhaps by a detoxification mechanism (Cochrane, 1958). If complete oxidation to acetyl fragments occurs, the fragments may be utilized in the glyoxylic acid cycle.

Enzymes for the β-oxidation of fatty acids, as well as those for the glyoxylic acid cycle, are contained in specialized microbodies, the glyoxysomes. This close association is functionally useful because the acetyl units derived from the oxidation to fatty acids are further metabolized via the glyoxylic acid cycle (Brennan and Lösel, 1978).

Not unexpectedly, there is sometimes a close association between particular phases of the life cycle in which a fungus is preferentially utilizing the glyoxylic acid cycle and in which it is metabolizing its lipid reserves. This has been observed in the motile zoospores of *Phytophthora erythroseptica* and upon spore germination (Brennan and Lösel, 1978).

Lipids Produced by Fungi

Like other organisms, fungi can synthesize a wide variety of lipids. These include the neutral fats, phospholipids, and sterols, as well as others. The lipid content of fungi fluctuates greatly in response to age, stage of development, and cultural conditions. The lipid content of yeasts usually ranges from approximately 7% to 15% of the dry weight, while that of filamentous fungi ranges from 6% to 9% when grown under favorable conditions (Brennan and Lösel, 1978). However, in fungi a lipid content of

20% of the dry weight is not uncommon, and the lipid content may be as great as 50% of the dry weight (Cochrane, 1958). Some yeasts which accumulate unusually high concentrations of lipid (up to 60% of their dry weight) may be commercially useful for the production of these fats (Hunter and Rose, 1971). Generally the triglycerides and their contained fatty acids comprise a rather large fraction of the lipids produced by fungi. Fatty acids may comprise as much as 88% of the lipid fraction, and among these, the 16- and 18-carbon acids are most abundant, the unsaturated fatty acids (especially oleic and linoleic acids) predominating (Cochrane, 1958). Sterols are probably produced by all fungi, but the amount of sterol is usually about 1% or less of the dry weight. The sterol usually present in the greatest quantity is ergosterol (Fig. 8–19), first detected in the sclerotia of *Claviceps purpurea* (the ergots) (Weete, 1973).

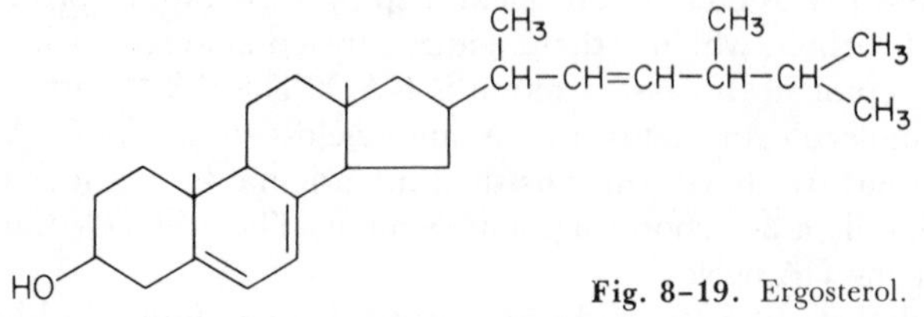

Fig. 8–19. Ergosterol.

Lipids serve diverse functions within the fungi. Lipids may serve as an energy reserve, and lipid droplets are often especially prominent in rapidly growing cells that are provided with abundant sugars and in various types of reproductive structures and their spores. The major component of these reserve lipids probably consists of triglycerides. Lipids are also an important component of the cellular membranes, which include the plasmalemma, endoplasmic reticulum, and organelle membranes. Lipids may comprise as much as 30% to 50% of these membranes (Weete, 1974). Most of the lipid within the membranes is phospholipid, but there are also sterols, sterol esters, and neutral lipids. Phospholipids are major constituents of cellular membranes, function in the transport of ions across membranes, and are involved in the functional organization of a number of complex systems within the membranes (e.g., the electron transport system within the mitochondrial membranes) (Brennan and Lösel, 1978). In experiments with various species of yeasts subjected to ultraviolet radiation, it was demonstrated that long-chain fatty acids or sterols would induce the recovery of some, but not all, species. This led to the suggestion that sterols that naturally occur in the membranes may induce recovery from ultraviolet radiation (Sarachek and Higgins, 1972). Lipids also occur in fungal cell walls and form a coating on spores; they may have a protective, waterproofing function. As outlined in the next chapter, some fungi produce sexual hormones. Three of the best known hormones (sirenin, antheridiol, and trisporic acid) are lipids. As the name implies, antheridiol is a sterol (Brennan and Lösel, 1978).

A number of lipid metabolites with no apparent function are produced by fungi. These include fatty acids that are not utilized in fat synthesis, the well-known phosphatides cephalin and lecithin (lipids containing nitrogen and phosphorus), cerebrins (the nitrogenous lipids), and some of the sterols. The cerebrins may constitute 0.2% to 0.3% of the dry weight and are end products of the autodigestion of some complex membranous lipids. The yellow pigments carotenes (Fig. 8–13), are often abundant in fungi.

References

Arnstein, H. R. V., and R. Bentley. 1953. The biosynthesis of kojic acid 1. Production from [1-^{14}C] and [3:4-^{14}C$_2$] glucose and [2-^{14}C]-1:3 dihydroxyacetone. Biochem. J. **54**:493–508.

Aronson, J. M. 1965. The cell wall. *In:* Ainsworth, G. C., and A. S. Sussman, Eds., The fungi—an advanced treatise. Academic Press, New York. **1**:49–76.

Birkinshaw, J. H. 1965. Chemical constituents of the fungal cell 2. Special chemical products. *In:* Ainsworth, G. C., and A. S. Sussman, Eds., The fungi—an advanced treatise. Academic Press, New York. **1**:179–228.

Blumenthal, H. J. 1965. Carbohydrate metabolism 1. Glycolysis. *In:* Ainsworth, G. C., and A. S. Sussman, Eds., The fungi—an advanced treatise. Academic Press, New York. **1**:229–268.

——, 1976. Reserve carbohydrates in fungi. *In:* Smith, J. E., and D. R. Berry, Eds., The filamentous fungi. Edward Arnold (Publishers) Ltd., London. **2**:292–307.

——, and S. Roseman. 1957. Quantitative estimation of chitin in fungi. J. Bacteriol. **74**:222–224.

Brennan, P. J., and D. M. Lösel. 1978. Physiology of fungal lipids: selected topics. Adv. in Microbial Physiol. **17**:47–179.

Bu'Lock, J. D., and H. M. Smalley. 1961. Biosynthesis of aromatic substances from acetyl- and malonyl-coenzyme A. Proc. Chem. Soc. **1961**:209–211.

Burnett, J. H. 1976. Fundamentals of mycology. Second ed. Crane Russak and Co., Inc., New York. 673 pp.

Cantino, E. C. 1966. Morphogenesis in aquatic fungi. *In:* Ainsworth, G. C., and A. S. Sussman, Eds., The fungi—an advanced treatise. Academic Press, New York. **2**:283–337.

——, and G. Turian. 1959. Physiology and development of lower fungi (Phycomycetes). Ann. Rev. Microbiol. **13**:97–124.

Casselton, P. J. 1976. Anaplerotic pathways. *In:* Smith, J. E., and D. R. Berry, Eds., The filamentous fungi. Edward Arnold (Publishers) Ltd., London. **2**:121–136.

Cochrane, V. W. 1958. Physiology of fungi. John Wiley & Sons, Inc., New York. 524 pp.

——, 1976. Glycolysis. *In:* Smith, J. E., and D. R. Berry, Eds., The filamentous fungi. Edward Arnold (Publishers) Ltd., London. **2**:65–91.

Collins, J. F., and H. L. Kornberg. 1960. The metabolism of C$_2$ compounds in micro-organisms 4. Synthesis of cell materials from acetate by *Aspergillus niger*. Biochem. J. **77**:430–438.

Dickens, F. 1958. Recent advances in knowledge of the hexose monophosphate shunt. Ann. N. Y. Acad. Sci. **75**:71–94.

Dowler, W. M., P. D. Shaw, and D. Gottlieb. 1963. Terminal oxidation in cell-free extracts of fungi. J. Bacteriol. **86**:9–17.

Elliott, C. G. 1977. Sterols in fungi: their functions in growth and reproduction. Adv. Microbial Physiol. **15**:121–173.

Entner, N., and M. Doudoroff. 1952. Glucose and gluconic acid oxidation of *Pseudomonas saccharophila*. J. Biol. Chem. **196**:853–862.

Fincham, J. R. S., and J. B. Boylen. 1957. *Neurospora crassa* mutants lacking arginosuccinase. J. Gen. Microbiol. **16**:438–448.

Finn, R. K. 1954. Agitation-aeration in the laboratory and in industry. Bacteriol. Rev. **18**:254–274.

Gooday, G. W. 1978. The enzymology of hyphal growth. *In:* Smith, J. E., and D. R. Berry, Eds., The filamentous fungi. Edward Arnold (Publishers) Ltd., London. **3**:51–77.

Goodwin, T. W. 1976. Carotenoids. *In:* Smith, J. E., and D. R. Berry, Eds., The filamentous fungi. Edward Arnold (Publishers) Ltd., London. **2**:423–444.

Gottlieb, D., and P. G. Caltrider. 1963. Synthesis of enzymes during the germination of fungus spores. Nature **197**:916–917.

Heck, F. L. 1929. A study of the nitrogenous compounds in fungous tissue and their decomposition in the soil. Soil Sci. **27**:1–45.

Holzer, H., and K. Beaucamp, 1961. Nachweis und Charakterisierung von α-Lactyl-thiaminpyrophosphat ("Aktives Pyruvat") und α-hydroxyäthyl-thiaminpyrophosphat ("Aktiver Acetaldehyde") als Zwischenprodukte der Decarboxylierung von Pyruvat mit Pyruvatdecarboxylase aus Bierhefe. Biochim. Biophys. Acta **46**:225–243.

Horowitz, N. H. 1947. Methionine synthesis in *Neurospora*. The isolation of cystathionine. J. Biol. Chem. **171**:255–264.

Hunter, K., and A. H. Rose. 1971. Yeast lipids and membranes. *In:* Rose, A. H., and J. S. Harrison, Eds., The yeasts. Academic Press, New York. **2**:211–270.

Kornberg, H. L., and H. A. Krebs. 1957. Synthesis of cell constituents from C$_2$-units by a modified tricarboxylic acid cycle. Nature **179**:988–991.

———, and N. B. Madsen. 1957. Synthesis of C$_4$-dicarboxylic acids from acetate by a "glyoxylate bypass" of the tricarboxylic acid cycle. Biochim. Biophys. Acta **24**:651–653.

Lindenmayer, A. 1965. Carbohydrate metabolism 3. Terminal oxidation and electron transport. *In:* Ainsworth, G. C., and A. S. Sussman, Eds., The fungi—an advanced treatise. Academic Press, New York. **1**:301–348.

MacGee, J., and M. Doudoroff. 1954. A new phosphorylated intermediate in glucose oxidation. J. Biol. Chem. **210**:617–626.

McCorkindale, N. J. 1976. The biosynthesis of terpenes and steroids. *In:* Smith, J. E., and D. R. Berry, Eds., The filamentous fungi. Edward Arnold (Publishers) Ltd., London. **2**:369–422.

McMurrough, I., A. Flores-Carreon, and S. Bartnicki-Garcia. 1971. Pathway of chitin synthesis and cellular localization of chitin synthetase in *Mucor rouxii*. J. Biol. Chem. **246**:3099–4007.

Newburgh, R. W., and V. H. Cheldelin. 1958. Glucose oxidation in mycelia and spores of the wheat smut fungus *Tilletia caries*. J. Bacteriol. **76**:308–311.

Nicholas, D. J. D. 1965. Utilization of inorganic nitrogen compounds and amino acids by fungi. *In:* Ainsworth, G. C., and A. S. Sussman, Eds., The fungi—an advanced treatise. Academic Press, New York. **1**:349–376.

Niederpruem, D. J. 1965. Carbohydrate metabolism 2. Tricarboxylic acid cycle. *In:* Ainsworth, G. C., and A. S. Sussman, Eds., The fungi—an advanced treatise. Academic Press, New York. **1**:269–300.

Pateman, J. A., and J. R. Kinghorn. 1976. Nitrogen metabolism. *In:* Smith, J. E., and D. R. Berry, Ed., The filamentous fungi. Edward Arnold (Publishers) Ltd., London. **2**:159–237.

Ramachandran, S., and D. Gottlieb. 1963. Pathways of glucose catabolism in *Caldariomyces fumago* (Ill.). Biochim. Biophys. Acta **69**:74–84.

Rattray, J. B. M., A. Schibeci, and D. K. Kidby. 1975. Lipids of yeasts. Bacteriol. Rev. **39**:197–231.

Sarachek, A., and N. P. Higgins. 1972. Effects of ergosterol, palmitic acid and related simple lipids on the recovery of *Candida albicans* from ultraviolet irradiation. Arch. Mikrobiol. **82**:38–54.

Schisler, L. C., and O. Volkoff. 1977. The effect of safflower oil on mycelial growth of Boletaceae in submerged liquid cultures. Mycologia **69**:118–125.

Sols, A., C. Gancedo, and G. DelaFuente. 1971. Energy-yielding metabolism in yeasts. *In:* Rose, A. H., and J. S. Harrison, Eds., The yeasts. Academic Press, New York. **2**:271–307.

Srb, A. M., and N. H. Horowitz. 1944. The ornithine cycle in *Neurospora* and its genetic control. J. Biol. Chem. **154**:129–139.

Suomalainen, H., and E. Oura. 1971. Yeast nutrition and solute uptake. *In:* A. H. Rose, and J. S. Harrison, Eds., The yeasts. Academic Press, New York. **2**:3–74.

Tatum, E. L., S. R. Gross, G. Ehrensvärd, and L. Garnjorbst. 1954. Synthesis of aromatic compounds by *Neurospora*. Proc. Natl. Acad. Sci. **40**:271–276.

Towers, G. H. N. 1976. Secondary metabolites derived through the shikimate-chorismate pathway. *In:* Smith, J. E., and D. R. Berry, Eds., The filamentous fungi. Edward Arnold (Publishers) Ltd., London. **2**:460–474.

Turner, W. B. 1976. Polyketides and related metabolites. *In:* Smith, J. E., and D. R. Berry, Eds., The filamentous fungi. Edward Arnold (Publishers) Ltd., London. **2**:445–459.

Utter, M. E., R. E. Barden, and B. L. Taylor. 1975. Pyruvate carboxylase: an evaluation of the relationships between structure and mechanism and between structure and catalytic activity. Advan. Enzymol. **42**:1–72.

Walker, P., and M. Woodbine. 1976. The biosynthesis of fatty acids. *In:* Smith, J. E., and D. R. Berry, Eds., The filamentous fungi. Edward Arnold (Publishers) Ltd., London. **2**:137–158.

Weete, J. D. 1973. Sterols of the fungi: distribution and biosynthesis. Phytochemistry **12**:1843–1864.

———, 1974. Fungal lipid biochemistry: distribution and metabolism. Plenum Press, New York. 393 pp.

Wilkinson, J. F., and A. H. Rose. 1963. Fermentation processes. *In:* Rainbow, C., and A. H. Rose, Eds., Biochemistry of industrial microorganisms. Academic Press, New York. pp. 379–414.

NINE

Reproduction

Reproduction is the procreation of new individuals by either sexual or nonsexual means. Sexual reproduction involves a cycle of plasmogamy (often accomplished by gametes or gametangia) followed by karyogamy and meiosis at regular points in the life cycle. This cycle is lacking in nonsexual reproduction. Reproductive structures in the fungi include gametes, gametangia, spores, and sporocarps. The formation of these structures and their behavior is controlled by genetic, hormonal, nutritional, and environmental factors.

At the cellular level, reproduction is accompanied by dramatic changes in the physiology of the fungus. Either internal or external factors cause vegetative growth to cease and the production of reproductive structures to be initiated. The precise biochemical mechanisms responsible for this shift are not generally known, but studies have been made of some alterations in physiology that accompany this shift. Major changes may occur in either the pathway or rate of respiration as well as in the activity of a variety of enzymes. Also, changes may occur in the levels of cellular components such as DNA, RNA, or protein. Examples of these internal shifts may be found on pages 321–322 and 357–358.

NONSEXUAL REPRODUCTION

In the majority of organisms, nonsexual reproduction takes place when a single parent forms progeny without a nuclear contribution from a second parent. Typically, no nuclear alterations take place, and the offspring is a genetic duplicate of the parent. In fact, agriculturalists take advantage of this by establishing large clones by nonsexual means from a plant with a favorable combination of genes. If progeny were derived by sexual means, the favorable combination of characters would almost assuredly be altered through meiotic recombination and outbreeding. The stability of nonsexual reproduction may at times be advantageous to the organism, but at other times it is not. Progeny that are genetically identical to their parents are usually very well adapted to a particular environment, as any poorly adapted individuals would not survive to leave progeny. As long as the environment remains static, there is an advantage to the genetic stability of nonsexual reproduction. But if the environmental conditions change, this stability does not allow for genetic adaptation to the changing conditions unless mutation takes place. In the majority of plants and animals, genetic variation occurs almost exclusively as a result of sexual reproduction.

The preceding discussion excludes the majority of fungi, as a unique mechanism for genetic variation is a regular feature of nonsexual reproduction in numerous fungi. This mechanism is *heterokaryosis*, which is often accompanied by the *parasexual cycle*.

Heterokaryosis

Heterokaryosis is the existence of different nuclear types in the same mycelium (the *heterokaryon*). Heterokaryons may originate from a hypha which has only a single nuclear type (a *monokaryon*). A monokaryon may give rise to a heterokaryon by (1) mutation of a nucleus within the mycelium, thus making that nucleus dissimilar to the other nuclei; (2) fusion of one hypha with another, followed by migration of nuclei and cytoplasm from one hypha to another; and (3) karyogamy of haploid nuclei to form diploid nuclei within the mycelium (*diploidization*) (Fig. 9-1). Both mutation and diploidization may occur spontaneously, and as we have seen from our discussions of sexual cycles, hyphal fusion occurs frequently in the fungi. These events may occur in any combination; for example, a heterokaryon which originated by hyphal fusion may have normal and mutated nuclei from both parents, and these nuclei may exist in the

Fig. 9.1. Heterokaryon formation by various mechanisms. (b) Heterokaryon breakdown occurring at sporulation when the nuclei are individually isolated into spores.

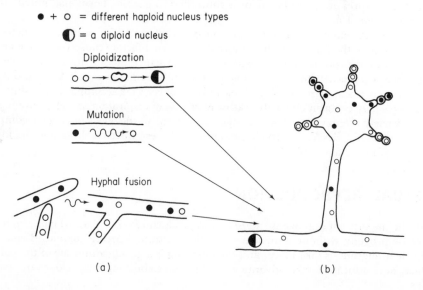

● + ○ = different haploid nucleus types

◐ = a diploid nucleus

Diploidization

Mutation

Hyphal fusion

(a)

(b)

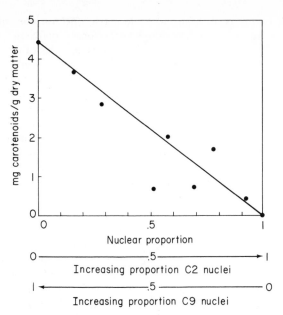

Fig. 9-2. Relationship of carotene production to nuclear ratios in heterokaryons of *Phycomyces blakesleeanus.* The white mutant strain C2 is unable to synthesize carotene, while the orange strain C9 produces lycopene, one of the carotenoid pigments. Both mutants have defective enzymes, but when the nuclei are brought together in a heterokaryon, complementation occurs and lycopene may be converted into other carotene pigments (γ-carotene and β-carotene). The wild type has large quantities of β-carotene. [Adapted from M. D. De La Guardia, C. M. G. Aragón, F. J. Murillo, and E. Cerdá-Olmedo. 1971. *Proc. Natl. Acad. Sci.* **68**:2012-2015.]

haploid state or form diploid nuclei by karyogamy between similar or dissimilar nuclei. The nuclei in the heterokaryon may divide mitotically for any number of nuclear generations, but these nuclei may finally be separated from each other when uninucleate, nonsexual spores are formed. Upon germination, these spores will give rise to a mycelium with a single nuclear type. Some fungi produce bi- or multinucleate spores that give rise to a heterokaryon if the nuclei in the spore are of different types.

Heterokaryon formation provides for a great deal of genetic variation of the mycelium. Phenotypically, a heterokaryotic mycelium may be quite different from its parental types, as it may contain cytoplasm from different sources as well as genetically different nuclei. In fungi, as in other organisms, the cytoplasm may carry genetic determinants and is sometimes responsible for influencing the phenotype. The phenotype of a heterokaryon is determined by the interaction of all nuclear types with the cytoplasm, and the phenotype is also influenced by the relative numbers of the various nuclei. It is important to realize that a single nuclear type can occur in any frequency, ranging from 0% to 100% of the total nuclei present within a mycelium. Therefore an allele carried by a particular nuclear type may occur in a wide range of dosages (1% to 99%) in a heterokaryon, and if this allele controls a particular phenotypic response, that response may also show a correspondingly flexible range (Fig. 9-2). This variation in relative nuclear numbers allows much more phenotypic diversity than is ordinarily encountered in typical diploid organisms, in which a particular allele is represented in dosages of 100%, 50%, or 0% in a diploid nucleus (consider the relative representation of *A* in *AA, Aa,* or *aa,* for example).

The interaction between nuclei becomes particularly important if the nuclei are biochemically deficient and are unable to direct the synthesis of a needed vitamin, amino acid, or other growth factor. Imagine, for example, that a mycelium which requires exogenous supplies of thiamine but not of the amino acid leucine fuses with a second mycelium which requires exogenous sources of leucine but can manufacture its own thiamine. The resulting heterokaryon, then, has nuclei that can control the biosynthesis of either thiamine or leucine, but not both. *Complementation* takes place, and one nuclear type directs the synthesis of thiamine while the second nucleus controls the synthesis of leucine. Each nuclear type has compensated for the deficiency of the other, and the resulting heterokaryon does not need exogenous supplies of either thiamine or leucine. A heterokaryon in which complementation occurs has a wide range of physiological flexibility. The fungus may "store" a nuclear type which could not survive alone in the prevailing environmental conditions but, with a change of conditions, this nuclear type may be necessary for the survival of the heterokaryon. The fungus may be allowed to take advantage of new food sources, for example, by utilizing the metabolic capacities of the stored nucleus. This is even more remarkable because it can be done without the delay of passing through sexual cycles and selection of appropriate nuclear types. If environmental changes occur, the chances of survival of a fungus are increased by having a large genetic reservoir from which to draw material for adaptations.

Parasexual Cycle

The parasexual cycle may take place within a heterokaryon. The parasexual cycle is a sequence involving (1) heterokaryon formation, (2) diploidization of nuclei, and (3) restoration of diploid nuclei to their haploid state (*haploidization*) (Fig. 9-3). Haploidization involves a series of atypical and irregularly occurring mitotic divisions of the diploid nuclei. Frequently, daughter nuclei resulting from mitosis have unequal numbers of chromosomes because sister chromatids failed to separate (nondisjunction) during anaphase. This yields a daughter nucleus with one chromosome too many $(2N+1)$, while the second daughter nucleus is lacking a chromosome $(2N—1)$. Both of these nuclei are now *aneuploids,* or they do not have even chromosome sets such as N or $2N$. The deficient aneuploid nucleus $(2N—1)$ may undergo additional loss of chromosomes in successive mitotic divisions of the same nature until it is reduced to the haploid condition. Pontecorvo (1958) estimates that only 1 out of every 1000 haploid nuclei in a heterokaryon of *Aspergillus niger* has undergone haploidization.

Mitotic crossing-over may accompany the parasexual cycle (Fig. 9-4). In mitotic crossing-over, segments of chromosomes are exchanged for exactly corresponding segments between homologous chromosomes, presumably occurring at the time when the chromosomes replicate. If dominant and recessive genes have been regrouped, subsequent haploidization may produce daughter nuclei which are genetically different from those that would be formed if this exchange had not taken place. For every 10 haploid nuclei derived through the parasexual cycle, perhaps only 1 nucleus is a recombinant derived from mitotic crossing-over (Pontecorvo, 1958).

The parasexual cycle mimics sexual reproduction. Unlike hyphae may fuse (plasmogamy) and their (gametic) nuclei occupy the same mycelium. These nuclei may fuse (karyogamy) to form a diploid (zygotic) nucleus. Genetic recombination may occur, and the diploid nucleus is reduced (but without meiosis) to the haploid condition. The recombinant, haploid nuclei are segregated into so-called nonsexual spores, which differ genetically from the parent mycelium.

Although the parasexual cycle is similar to sexual reproduction, it is less efficient in producing recombinant nuclei. Pontecorvo (1956) estimates that the initial frequency of mitotic crossing-over in nuclei is 500 to 1000 times lower than that of meiotic crossing-over. Although this initial number of nuclei produced by mitotic recombination increases through subsequent divisions before spore formation takes place, the possibility of any single nucleus being isolated in a single spore is slight. In sexual reproduction, virtually every spore produced as a result of meiosis contains a recombinant nucleus.

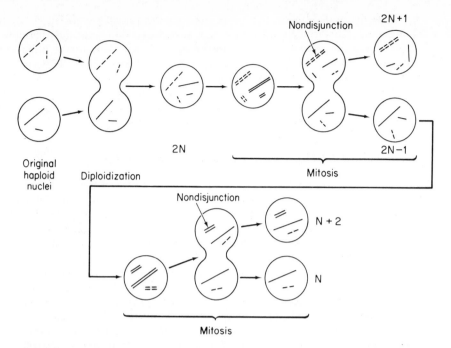

Fig. 9-3. Haploidization, beginning with genetically different haploid nuclei in the hetero-karyon, and terminating with the formation of a haploid nucleus which is genetically different from the original nuclei.

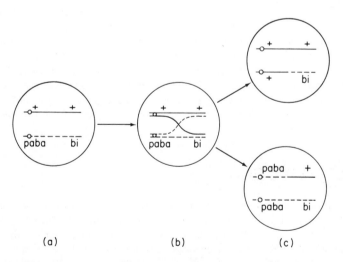

Fig. 9-4. Mitotic crossing-over between a single pair of homologous chromosomes. This may occur at some point during haploidization: (a) A diploid nucleus which is heterozygous for the genes *paba* (paminobenzoic acid–requiring) and *bi* (biotin-requiring); (b) Crossing-over at chromosome replication (four-strand stage); (c) Genetically different nuclei formed after chromosome segregation.

The parasexual cycle is a useful tool for the fungal geneticist. As discussed in Chapter 6, many fungi apparently lack sexual reproduction and therefore genetic studies based on sexual cycles cannot be carried out. Mutants or other genetically distinct strains may be induced to cross and recombinants may be derived from the parasexual cycle. By utilizing appropriate methods of genetic analysis, information may be gained about the location and activity of genes in these fungi. For example, some recent studies are focused upon the genetic control of aflatoxin production (aflatoxins are discussed further in Chapter 11) (Papa, 1977, 1978).

Distribution of Heterokaryosis and Parasexuality

Heterokaryosis and parasexuality occur in the members of the Deuteromycotina that have no known sexual reproduction and provide a significant source of genetic variation for these fungi, which are deprived of gametic fusion and meiosis. These fungi are omnipresent and many are highly specialized parasites, and generally speaking, they have survived quite well by utilizing the variation that heterokaryosis and parasexuality incorporate into their life cycles.

Heterokaryosis and the parasexual cycle are of probably universal occurrence among the filamentous ascomycetes and basidiomycetes but they are not known to occur in the Division Mastigomycota. These cycles provide for direct somatic variation in the vegetative phase of the life cycle, but additional and significant variation is incorporated through sexual reproduction. In certain instances, the ability to carry out sexual reproduction depends on the genetic ability to form heterokaryons (the basidiomycetes are fungi of this type). Heterokaryosis has been considered as a probable mechanism leading to the establishment of sex in fungi (Whitehouse, 1949).

A Concept of Nonsexual Reproduction

Few fungi reproduce ''without sex'' or without incorporating genetic variation in their life cycles through either ''cryptic'' or ''overt'' sexuality (Raper, 1959). The concept of nonsexual reproduction in fungi must be restricted to the actual formation of nonsexual reproductive spores and their associated structures which do not have meiosis as a prerequisite to their formation. These nonsexual spores include zoospores, sporangiospores, conidia, chlamydospores, and oidia. Associated structures would include zoosporangia, sporangia, conidiophores, pycnidia, and acervuli. Nonsexual spores are often produced as a regular part of the life cycle of a sexually reproducing fungus.

SEXUAL REPRODUCTION

Sexual reproduction allows for a great deal of genetic variability because it is a process in which genetically different gametes may unite and random assortment of alleles may occur at meiosis. This variability is especially important for those fungi that do not establish heterokaryons. The large number of genotypes within a population or species makes greater adaptation to changing environmental conditions possible than does the unvarying genotype of the typical nonsexually reproducing organism.

Further, there is greater genetic variation in a population in which the members are self-sterile but fertile with respect to other compatible individuals in the same or adjacent populations, which thus makes outbreeding or cross-breeding obligatory. Conversely, constant inbreeding (self-fertilization or mating with like individuals) tends to create a homogenous population lacking the greater adaptive qualities of an outbreeding population. Many highly evolved plants and animals have mechanisms that ensure cross-fertilization and exclude self-fertilization. For example, many hermaphroditic plants have male and female organs in the same flower, but cross-fertilization with other flowers is obligatory because the pollen and ovule mature at different times or the structure of the organs may be such that self-fertilization is impossible. In many plants and animals, the male and female organs are borne by different individuals, which thus gives rise to different "sexes."

Homothallism and Heterothallism

Among the fungi, we find that many are self-fertile while others are self-sterile. A fungus that is self-fertile is *homothallic* and will either fertilize itself or mate with a similar genetic strain. Some homothallic fungi bear differentiated male and female organs on the same thallus and mating will subsequently occur between these organs. A fungus of this type is *Pyronema domesticum*, which produces its antheridia and ascogonia from the same stalk cells (see Fig. 4–4). As indicated above, self-fertility and inbreeding tend to lead to genetic homogeneity. In contrast, some fungi are heterothallic and are self-sterile. The heterothallic fungus requires a compatible partner for reproduction to occur, and therefore outbreeding is obligatory.

Two fundamentally different types of heterothallism exist (Whitehouse, 1949). In *morphological heterothallism,* morphologically dissimilar sexual organs are produced by different thalli, and in order to mate the male gametangia or spermatia from one thallus must come into contact with the female gametangia produced by another thallus. Many members of the Division Mastigomycota have morphological heterothallism (these include some species of *Achlya,* discussed further on p. 350). *Physiological heterothallism,* which depends on genetic factors conferring compatibility or incompatibility, is entirely independent of morphological differences between male and female. A fungus having physiological heterothallism may lack differentiated sexual organs (such as the Hymenomycetes which mate by hyphal fusion, p. 189). To make the situation even more intricate, a fungus having physiological heterothallism may also have morphologically distinct gametangia. If this is the case, even if male and female organs occurred on the same thallus, the fungus would not be able to fertilize itself, as it has a basic requirement for genetically different nuclei. Fungi of this type include the Discomycete *Ascobolus stercorarius,* most species of the Pyrenomycete *Neurospora,* and the rust fungi. Fertilization in these fungi must occur between a male gametangium or spermatium borne on one thallus and a female gametangium borne on another thallus, and furthermore the thalli must be genetically compatible.

CONTROL OF PHYSIOLOGICAL HETEROTHALLISM

Physiological heterothallism is under genetic control. When the mating reactions of fungi are critically studied, it is usually found that genetic control is exerted either by two alleles at a single locus or by multiple-allelic series at either one or two loci. Occasionally a fungus is found that has physiological heterothallism but does not fit neatly into one of these categories and that may have some special features not generally shared by other fungi (for example, *Cochliobolus spicifer,* reported by Nelson et al., 1977).

Two-Allele Heterothallism. Two-allele heterothallism is determined by two alleles at one locus. Species possessing two-allele heterothallism are divided into two strains commonly referred to as *plus* and *minus* or *A* and *a* (Whitehouse, 1949). In this type of sexuality, a plus strain cannot mate with another plus strain but can mate with all minus strains, as illustrated in Figs. 9–5 and 9–6. Examples of fungi with two-allele heterothallism include members of the Mucorales such as *Mucor, Rhizopus,* and *Phycomyces*; most species of the Pyrenomycete *Neurospora;* the Discomycetes *Ascobolus magnificus* and *Sclerotinia gladioli;* the rust *Puccinia graminis;* and the smut fungus *Ustilago levis* (Whitehouse, 1949).

Bipolar Multiple-Allele Heterothallism. In bipolar multiple-allele heterothallism, there is a single locus that controls sexuality, but a multiple-allelic series occurs at that locus (Whitehouse, 1949). A cross is compatible if the alleles in the mating thalli are different. For example, if we designate the locus as *A,* there might be the multiple-alleles A_1, A_2, A_3, . . . A_n. A thallus carrying the gene A_1 cannot mate with a thallus also carrying A_1, but it will mate with a thallus containing any of the remaining alleles, as illustrated in Fig. 9–7. Bipolar multiple-allele heterothallism occurs in most smut fungi (Halisky, 1965; Raper, 1960), in a few Gasteromycetes (Burnett and Boulter, 1963; Fries, 1948), and in the mushroom *Coprinus comatus* (Whitehouse, 1949).

Tetrapolar Multiple-Allele Heterothallism. Tetrapolar multiple-allele heterothallism is essentially like the bipolar type, except that there are two loci, *A* and *B,* which control compatibility. These loci segregate independently at meiosis. Each locus is multiple-allelic, with at least 100 alleles at each locus (Raper, 1953). A compatible mating occurs if both the *A* and *B* alleles differ in both thalli. For example, a mycelium bearing the alleles A_1B_1 will not mate with A_1B_1 but will mate with A_2B_2 because the alleles at both loci are different (Fig 9–8). A cross between A_1B_1 is not fully compatible with A_1B_2 or A_2B_1 because either the alleles at the *A* locus are common or the alleles at the *B* locus are common.

In certain fungi, if crosses are made between hyphae bearing either common *A* alleles or common *B* alleles, heterokaryons may be established but sporocarps are rarely formed. If the heterokaryon has common *A*'s but dissimilar *B*'s the *flat reaction* occurs (Fig. 9–8). The growth of this heterokaryon type is sparse, depressed and with gnarled, irregularly branched hyphae in the older mycelium (Raper and San Antonio, 1954). The typical growth of a heterokaryon formed after mating of hyphae bearing common *B* alleles but dissimilar *A* alleles is also characteristic. The common *B* heterokaryon gives the *barrage* reaction (Fig. 9–8). The hyphae of the mating colonies grow toward each other but near the center growth ceases, leaving a strip (the barrage zone) with little growth separating the two colonies (Papazian, 1950; Raper, 1953). Common *AB* heterokaryons have also been found to occur (Middleton, 1964). In an *AB* heterokaryon, the original colonies overgrow each other.

In his research with heterokaryons of *Schizophyllum commune,* Parag (1965) demonstrated that common *A* alleles prevent the appearance of clamp connections, disrupt the regularity of nuclear distribution, and cause morphological and metabolic abnormalities in the mycelium. Common *B* alleles prevent nuclear migration, completion of the clamp connections, karyogamy, and meiosis. Apparently other tetrapolar fungi are controlled in the same manner.

Tetrapolarity occurs in the majority of Hymenomycetes, Gasteromycetes, and the smut fungus *Ustilago maydis.*

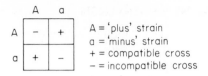

	A	a
A	−	+
a	+	−

A = 'plus' strain
a = 'minus' strain
+ = compatible cross
− = incompatible cross

Fig. 9-5. Mating reactions of possible isolates of a two-allele heterothallic fungus.

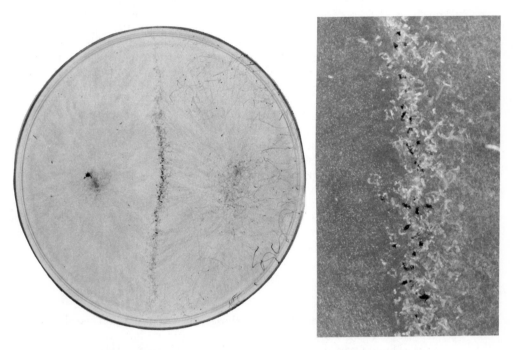

Fig. 9-6. Mating between plus and minus strains of a physiologically heterothallic organism, *Phycomyces blakesleeanus*. A line of zygospores forms where the strains meet. Right: An enlarged view of the line of gametangia (light) and zygospores (dark). [Courtesy Plant Pathology Department, Cornell University.]

+ = compatible cross
− = incompatible cross

	A₁	A₂	A₃	A₄
A₁	−	+	+	+
A₂	+	−	+	+
A₃	+	+	−	+
A₄	+	+	+	−

Fig. 9-7. Mating reactions of four possible isolates of a bipolar fungus.

FL = flat + = compatible cross
B = barrage − = incompatible cross

	A₁B₁	A₁B₂	A₂B₁	A₂B₂
A₁B₁	−	FL	B	+
A₁B₂	FL	−	+	B
A₂B₁	B	+	−	FL
A₂B₂	+	B	FL	−

Fig. 9.8. Mating reactions of four possible isolates of a tetrapolar fungus.

CONTROL OF MORPHOLOGICAL HETEROTHALLISM

Much less is known about the genetic determination of morphological sex than of physiological sex, and no generalized pattern of genetic control is known to exist. We consider a single example of genetic control of morphological sex.

The sexuality of the Pyrenomycete *Nectria haematococca* var. *cucurbitae* (formerly known as *Hypomyces solani* var. *cucurbitae)* has been extensively studied (Bistis and Georgopoulos, 1979; El-Ani, 1954; Hansen and Snyder, 1943; Snyder and Hansen, 1954). This fungus has two-allele heterothallism, with the compatibility factors *A* and *a* or plus and minus. In addition to this physiological heterothallism, morphological heterothallism under independent genetic control exists. The female sexual structures are protoperithecia which produce trichogynes. After plasmogamy occurs with a conidium (functioning as the male), the protoperithecia develop into perithecia. Some thalli bear only functional conidia (are designated "male") while other cultures bear only protoperithecia (are designated "female"). Other thalli are hermaphroditic and produce both the male and female structures. In addition to these fertile sexual forms, there is a reversion to a "neuter," sexually infertile mycelial form which bears neither conidia that function as spermatia nor protoperithecia. These four sexual types may be obtained from a single cross, approximately in a 1:1 ratio after mating a male with a female. It was determined (Hansen and Snyder, 1943) that these four sexual forms are controlled by the genes *M* and *m*, which determine the production or nonproduction of conidia, and by the genes *C* and *c* which determine the development or suppression of the protoperithecia. The cross above, with genotypes, may be designated

$$mC \text{ (female)} \times Mc \text{ (male)} \rightarrow mC \text{ (female)}:Mc \text{ (male)}:$$
$$mc\text{(neuter)}:MC\text{(hermaphrodite)}$$

Since these genes are nonallelic with those determining physiological compatibility, morphological sex and physiological sex will segregate independently in appropriate crosses (Hansen and Snyder, 1943; Snyder and Hansen, 1954):

$$+ , - = \text{physiological mating types}$$
$$N = \text{neuter} \qquad F = \text{female}$$
$$M = \text{male} \qquad H = \text{hermaphrodite}$$

I   H + × H −   ⟶ H + :H −
II   H + × M −   ⟶ H + :H − :M + :M −
III F + × H −   ⟶ F + :F − :H + :H −
IV   F + × M −   ⟶ F + :F − :M + :M − :H + :H − :N + :N −

Further discussion of *Nectria haematococca* var. *cucurbitae* is on page 353.

EVOLUTION OF HETEROTHALLISM

Homothallism is predominant in the Division Mastigomycota and Subdivision Zygomycotina, common in the Subdivision Ascomycotina, and apparently absent in the Subdivision Basidiomycotina, where heterothallism predominates. The majority of the fungi are homothallic. It may be that homothallism is the primitive condition and that heterothallism developed from homothallism (Raper, 1959, 1960). Heterothallism has selective advantages over homothallism as it increases outbreeding and variability. The actual evolution of heterothallism and of compatibility types is unknown, and various ideas have been presented.

To illustrate a possible evolution of physiological heterothallism, we explore the scheme devised by Whitehouse (1949). The primitive homothallic condition was replaced by a simple form of heterothallism in which the locus controlling compatibility had only two alleles. Hyphal fusion had previously occurred, but it was limited in significance as only nuclei of a single mating type were involved; but with the evolution of different nuclear types heterokaryons were formed which contained nuclei of different mating types. Sexual organs were no longer required to bring gametic nuclei together, and were either partially or wholly lost. With further addition of alleles and another locus, both bipolar and tetrapolar multiple-allelic heterothallism developed.

Development of Reproductive Structures

Morphogenesis of reproductive structures, including sexual organs, gametes, and sporocarps, is a precise process typically not under the same control as those factors which determine sexual compatibility. Control may be exerted by genes or hormones.

GENETIC CONTROL

As an example of genetic control of the development of reproductive structures, we consider *Glomerella cingulata,* a homothallic Pyrenomycete. A large number of mutants of *G. cingulata* were obtained which modify both the mating reaction and development of the perithecium (Wheeler, 1954; Wheeler and McGahen, 1952). Each mutant gene either partially or completely blocks events leading to the formation of a mature perithecium. The blocks occur at the following stages:

1. *Perithecial initiation.* Genes at the A locus control production of perithecia or conidia, and genes at the B locus control the arrangement (clumped or scattered) of the perithecia. Perithecial initials are not formed if the mutant gene A^1 is present, but their formation is only partially blocked by A^+ with B^+. This latter block is removed if either A^2 or B^1 is substituted at one of these loci (that is, if the genotype is either A^+B^1 or A^2B^+).
2. *Plasmogamy.* Most of the mutant genes block plasmogamy. The nonallelic genes arg^1, bi^1, and th^1, controlling requirements for the growth factors arginine, biotin, and thiamine, respectively, and the genes F^1 and st^1 interrupt plasmogamy.
3. *Karyogamy.* Fungi with the mutant gene B^1 develop normally up to and including the formation of croziers. Nuclear disintegration occurs prior to and during karyogamy. Sterile perithecia result.
4. *Meiosis.* The mutant dw_1 (dwarf) partially blocks meiosis and controls the production of dwarf ascospores. Approximately 70% to 80% of the asci abort as a result of nuclear disintegrations.

The following model demonstrates events leading to the formation of a mature perithecium and also the point at which each mutant exerts its effect:

$$
\begin{array}{c}
\text{mycelium} \xrightarrow[A^1]{A^+B^+} \text{perithecial initials} \xrightarrow[st]{F^1 \;\; \substack{arg^1\\bi^1\\th^1}} \text{plasmogamy} \xrightarrow{B^2} \text{karyogamy} \xrightarrow{dw^1}
\end{array}
$$

$$
\text{meiosis} \longrightarrow \text{mature perithecium}
$$

Matings leading to the production of normal perithecia require that at least one of each homologous pair of genes be of the wild type. After plasmogamy, the wild-type gene will mask the effect of its mutant allele through complementation in the heterokaryotic mycelium until karyogamy occurs. The wild-type genes presumably control synthesis of specific chemical substances which allow development to proceed normally (Wheeler, 1954).

HORMONAL CONTROL

Every organism, primitive or advanced, must have a certain degree of internal organization. Primary control of this organization is a result of the genes present, while secondary control is a result of the regulatory systems which operate. Regulatory systems themselves are under genetic control.

In animals, the regulatory systems partially operate through the secretion and actions of hormones. Animal hormones are secreted in minute quantities and have a specific effect on another organ, usually far-removed from that which secreted the hormone. There are also hormone-like substances, auxins, in higher plants. Like animal hormones, auxins are produced in small amounts and may have a physiological effect on a plant part far-removed from the original site of production. In contrast to animal hormones, however, the effects of auxins are nonlocalized, and the auxins serve as general growth factors which are not specific for any part of the plant, but they cause highly specific physiological changes in the cell.

Hormones controlling development of reproductive structures have been demonstrated to occur in certain fungi. These hormones, like those produced by animals, are produced in minute quantities, diffuse from their site of production, and finally exert a highly specific effect on a particular organ of the fungus. This description of hormones excludes substances such as vitamins, purines, and other growth factors which usually exert a more generalized effect on reproduction and growth.

Hormonal Control of Sexual Differentiation. Some species of the aquatic fungus, *Achlya* spp., are morphologically heterothallic and bear the male organs (antheridia) and female organs (oogonia) on separate thalli. Raper (1951) ingeniously demonstrated hormonal control of sexual differentation in *Achlya*. Potentially male and female thalli were alternated in sequential microaquaria. Water was then allowed to pass through the aquaria in one direction only. If growth factors or hormones were released into the medium, then theoretically those thalli occupying a stage in the sequence after hormonal release would respond. It was discovered through this arrangement that a series of highly specific hormones are formed that control the sexual process. The female vegetative plant produces hormones A and A^2 which induce the formation of the antheridial hyphae on the male plant. The male plant then produces hormone A^1 which passes back to the female plant, augmenting the continued production of hormones A and A^2. Eventually the male plant produces hormone A^3 which depresses the action of A and A^2; then the antheridial hyphae produce hormone B which passes to the female plant where it induces the formation of the oogonial initials. The oogonial initials then produce hormone C, which (1) induces the antheridial hyphae to grow along a concentration gradient toward the oogonium until physical contact is made between the organs and (2) induces the formation and delimitation of the antheridium in the male plant. The antheridia may then produce hormone D, which induces (1) the delimitation of the oogonium and (2) the differentiation of the oospheres. Therefore, as interpreted by Raper, sexual differentiation in *Achlya* is under the control of at least seven distinct hormones, four secreted by the male thallus and three secreted by the female thallus. The sequence is further clarified in Fig. 9–9.

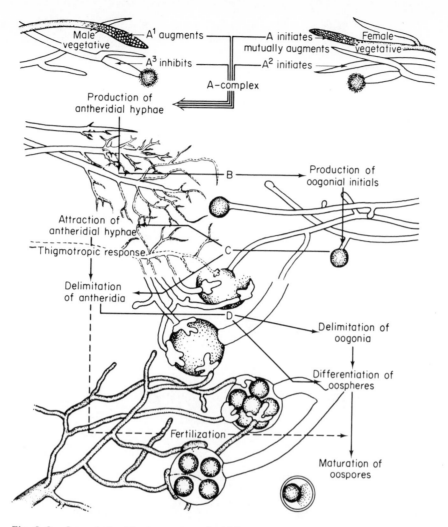

Fig. 9-9. Interrelationships between sequential hormonal secretions and morphology in heterothallic species of *Achlya*. [From J. R. Raper, 1955, *Biological specificity and growth,* E. G. Butler, ed., by Princeton University Press, Princeton, New Jersey.]

The hormones are effective in extremely minute quantities, and hormone *A* induced the formation of antheridial hyphae when the culture medium was tested in a dilution as low as 10^{-13} (Raper, 1951).

Both hormones *A* and *B* have been isolated and were found to be steroids (Fig. 9-10). Hormone *A* is now called *antheridiol* and the three steroids that give the effects of hormone *B* are called *oogoniol*. Because the hormones are effective in extremely minute dosages, large quantities of culture medium must be processed to obtain the hormones in pure form. Barksdale (1967) reported that 6,000 liters of the culture medium are required to yield only 20 milligrams of crystalline antheridiol. Action of the hormones is highly specific, and even steroids that are closely related in the structure typically fail to elicit a morphogenetic response when tested on cultures of *Achlya* (Barksdale et al.,

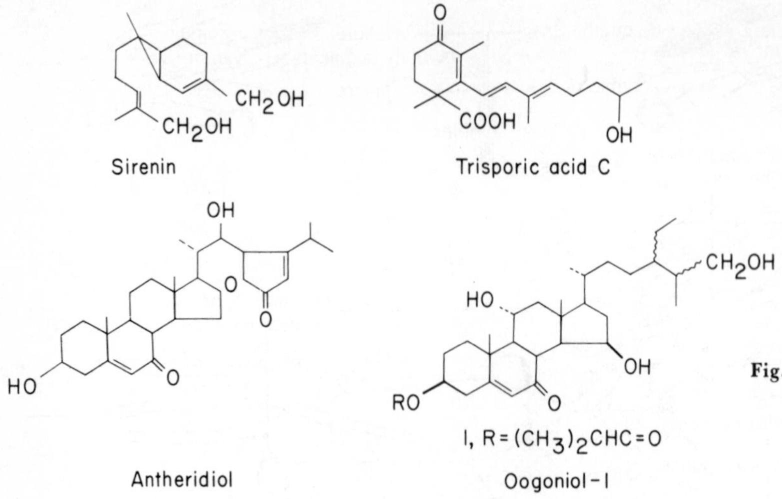

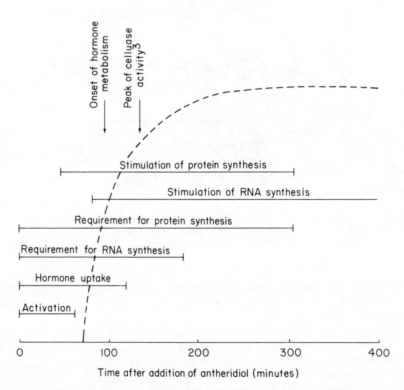

CH2OH
CH2OH
Sirenin
O
COOH
OH
Trisporic acid C
OH
O
O
HO
O
Antheridiol
CH2OH
HO
OH
RO
O
I, R=(CH3)2CHC=O
Oogoniol-I

Fig. 9-10. Fungal hormones.

Onset of hormone metabolism
Peak of cellulase activity3
Stimulation of protein synthesis
Stimulation of RNA synthesis
Requirement for protein synthesis
Requirement for RNA synthesis
Hormone uptake
Activation
0
100
200
300
400
Time after addition of antheridiol (minutes)

Fig. 9-11. Summary of events associated with antheridiol-induced differentiation in *Achlya.* The brackets indicate the time periods during development in which the process occurs. The dashed lines indicate relative morphological differentiation. [From W. E. Timberlake, 1976, *Develop. Biol.* **51**:202–214.]

1974). The mode of action of antheridiol has been investigated. When antheridiol is added to the culture medium in which a vegetative thallus is growing, the antheridiol is removed from the culture medium by the thallus and metabolized (Fig. 9-11). This uptake of antheridiol requires about 60 minutes and initiates an increase in the level of both messenger and ribosomal RNA as well as of the protein within the thallus. These metabolic changes may be detected in advance of the visible morphogenetic changes (initiation of the antheridial branches), which begin about 60 to 90 minutes after administration of the antheridiol but may require about 300 minutes for completion. Both the RNA and protein are required to support the growth of the antheridial branches. After about 180 minutes, no additional RNA synthesis is necessary, but protein synthesis must continue. There is also an increase in the enzyme cellulase, required to soften the parent walls to allow branching to take place (Timberlake, 1976), and an increase of catalase in microbodies (Choinski and Mullins, 1977).

A reserve glucan (a carbohydrate) is manufactured during vegetative growth and is present in the cytoplasm. If the thallus is transferred to a starvation medium lacking glucose, the glucan will be utilized to allow the gametangial initials to be formed (Faro, 1972). Nutritional conditions can influence the activity of antheridiol, as an adequate concentration of a nitrogen and carbon source must be present for the initiation of antheridial branches and also influences their number (Barksdale, 1970).

Hormonal Control of Gamete and Gametangia Attraction. One of the most important steps in a sexual cycle is that in which gametes or gametangia come into contact with each other so that plasmogamy may occur. Often the sexual structures are physically distant from each other and an attraction must be exerted to permit this contact. Hormones are sometimes responsible for this attraction.

Male gametes of *Allomyces,* a Chytridiomycete, are attracted to female gametangia before and during the release of female gametes. This attraction is due to the release of the hormone *sirenin* from the female gametes (Machlis, 1958-a). Once the males are attracted to the females, production of sirenin ceases. Sirenin has been obtained in pure form, and unlike the antheridiol and oogoniol, is not a steroid but rather a sesquiterpene (Nutting et al., 1968). Sirenin is of special historical interest as it was the first plant hormone to be chemically isolated and characterized (Machlis et al., 1968). Sirenin can attract male gametes in concentrations as low as 10^{-10} M, but at higher concentrations ($10^{-4}M$), it fails to attract them. Female gametes of *Allomyces* are not attracted to sirenin (Carlile and Machlis, 1965-b).

Nectria haematococca var. *cucurbitae,* discussed on p. 348, also has a hormonal system involved in gametic attraction. The trichogyne produces one or more substances that cause certain changes to occur in the conidium. These changes are redistribution of the cytoplasm and formation of papillae. The conidium then secretes a substance that causes the trichogyne to grow towards the conidium and to make contact with it (Bistis and Georgopolos, 1979).

Several members of the Mucorales are heterothallic, and as discussed in Chapter 3, gametangia are produced by hyphae of different mating types. Hyphae of each mating type produce a volatile compound that passes through the air and induces a response in hyphae of the opposite mating type. The active volatiles produced by the plus strain are 4-hydroxymethyltrisporates, while those produced by the minus strains are trisporins (Nieuwenhuis and van den Ende, 1975). The production of trisporic acids acts as a positive feedback mechanism and continues to stimulate production of the volatile precursors (Werkman and van den Ende, 1973). These active volatiles induce progametangia formation and synthesis of both carotene and trisporic acid. Mated cultures produce about 15 to 20 times as much carotene as separate cultures of each mating type and also produce trisporic acid, which can be chemically derived from β-carotene. Trisporic acid (Fig. 9–10) is formed by both the plus and minus strains, and if either partner is removed, production will cease. Purified trisporic acid in small concentrations ($10^{-8}M$) will induce formation of progametangia in unmated plus or minus cultures (Gooday, 1973; Kochert, 1978). Addition of trisporic acid to minus cultures of *Blakeslea trispora* will cause a massive (about 73-fold) increase in production of carotenoid pigments (Thomas and Goodwin, 1967). Although trisporic acids greatly enhance sexual reproduction,they apparently are not essential. Mutants of *Mucor mucedo* unable to respond to the addition of trisporic acids to the culture have been found capable of mating and producing progametangia (Wurtz and Jockusch, 1975).

Hormonal Control of Gametangium Response and Sporocarp Development. Bistis (1956, 1957) demonstrated that the control of apothecium development in the Discomycete *Ascobolus stercorarius* is under hormonal control. *A. stercorarius* is a physiologically heterothallic organism but hermaphroditic, bearing both male and female organs on the same thallus. Plasmogamy occurs by contact of an oidium and an ascogonium of different mating types. If an oidium is placed on agar near (but not in physical contact) to hyphae of the opposite mating type, the ascogonium-producing hyphae will differentiate. Evidently an ascogonium-inducing hormone is being produced by the oidium, as this effect occurs only within a radius of approximately 500 microns from that cell. There is no apparent control by the oidium during the initial stages of ascogonium development in which it is coiling. The trichogyne eventually appears, and that organ exhibits a strongly positive chemotropic response to the

Fig. 9-12. Hormonal attraction of the oidium for the trichogyne of *Ascobolus stercorarius:* (a) At 6:00 P.M., the oidium was placed in the vicinity of the ascogonium; (b) At 6:12 P.M., the ascogonial tip now shows a directional response to the oidium; (c) 6:25 P.M., there is now the development of a lateral, subterminal projection as a response to the relocation of the oidium, which was made at 6:16 P.M.; (d) 6:37 P.M., growth of the projection directly to the oidium and a change in direction of growth of the trichogyne tip; (e) 6:41 P.M., the oidium has been moved to a new position; (f) 6:48 P.M., both apices have now responded to the new location of the oidium. [Adapted from G. Bistis, 1957, *Am. J. Botany* **44**:436–443.]

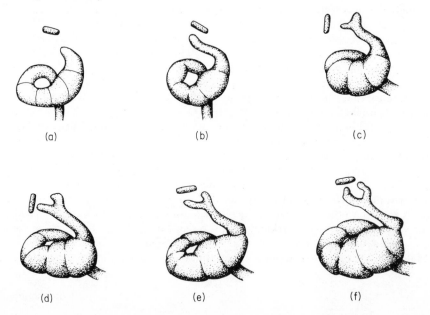

(a)   (b)   (c)

(d)   (e)   (f)

oidium. If the oidium is moved about on the agar surface, the trichogyne will change its direction of coiling and will always point toward the oidium (Fig. 9–12). If the oidium is removed at this point, growth of the trichogyne will cease, and in some cases the trichogyne will revert to the vegetative condition. Plasmogamy occurs, and development of the ascogonium is completed. The sheathing hyphae develop below the ascogonium, surround the ascogonium, and proliferate to form the sterile tissue of the apothecium. These ensheathing hyphae exhibit a directional growth which is undoubtedly a positive chemotropic response to hormone(s) produced by the ascogonium.

The principal steps in the hormonal mechanism include

1. Induction of ascogonial primordia by male hormone(s) produced by the oidium.
2. Directional growth of the trichogyne in response to male hormone(s) produced by the oidium
3. Attraction and proliferation of sheath hyphae in response to female hormone(s) produced by the ascogonium

It is also possible that maturation of the ascogonium and plasmogamy are controlled by hormones (Bistis, 1956).

Hormonal Control in Other Fungi. Hormonal control of specific phases of development has been demonstrated in other fungi. The particular fungi discussed in this chapter provide exceptionally good evidence of hormonal control of various developmental stages. Obtaining evidence for hormonal control is facilitated in a heterothallic fungus that allows the investigator to spatially separate the source of the hormone from the organ which it is affecting. However, hormonal control has been demonstrated in homothallic *Achlyas* (Raper, 1950). It is probable that many phases of development that were not cited are also under hormonal control, but proof is difficult if the organs and tissues are highly integrated in the same thallus.

NUTRITIONAL AND PHYSICAL FACTORS AFFECTING REPRODUCTION

Survival of a fungus species may depend on its ability to find and mate with a compatible partner, but it is equally dependent on the ability of the fungus to produce large quantities of viable spores under the prevailing environmental and nutritional conditions.

Fungi require a period of vegetative growth before formation of reproductive structures and spores will occur (Fig. 9–13). Materials produced in the mycelium are utilized in formation of reproductive structures, and there may even be a detectable decrease in the dry weight of the mycelium when sporulation takes place (Madelin, 1956–b). It is obvious that any factor affecting vegetative growth will indirectly influence reproduction.

Those factors that affect growth (discussed in Chapter 7) also affect reproduction. Environmental and nutritional requirements for sporulation (production of spores and accessory structures such as ascocarps and basidiocarps) are usually more precise, or more restricted, than those permitting vegetative growth. Vegetative growth will occur over a much wider range of conditions than reproduction will. The conditions allowing the formation of simple spore-bearing structures (usually nonsexual) conform more closely to conditions allowing vegetative growth than those permitting formation of complex sporocarps (usually sexual). There may be special

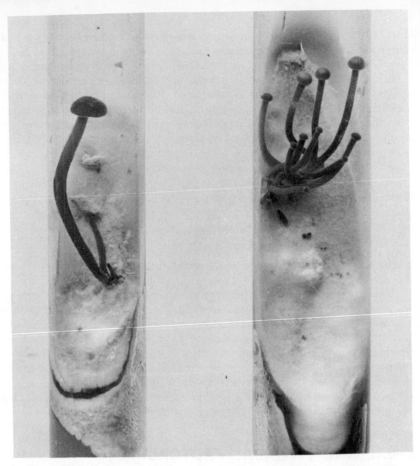

Fig. 9-13. Basidiocarps of *Collybia velutipes* arising from vegetative mycelium in culture. [Courtesy Plant Pathology Department, Cornell University.]

requirements for sporulation which are not required by the vegetative mycelium. An example of this is light.

There is no single set of optimum conditions for reproduction, as virtually each species responds uniquely to environmental and nutritional conditions. The majority of external effects on reproduction are quantitative, either decreasing or increasing the number of spores and other reproductive structures produced. Relatively few environmental or nutritional effects on reproduction are of a qualitative all-or-none nature or affect morphology.

Nutrition

CONCENTRATION OF MEDIUM

Media with a high nutrient concentration are usually unfavorable for reproduction. Klebs (1899) cultured *Saprolegnia mixta* in the vegetative state for $2\frac{1}{2}$ years by replenishing the enriched medium in which this organism was growing. When the fungus was transferred to water, sporangia and sporangiospore production began in a few days. From this, Klebs deduced that a fungus is most likely to sporulate after the vegetative mycelium exhausts the nutrient supply in the medium. This concept has been generally supported by experiments with other fungi. For example, Timnick and her coworkers (1952) leached nutrients out of the medium on which *Melanconium fuligineum* was growing by constantly dripping water through the medium and found that numerous spores were formed within 24 hours after inoculation rather than the 11 to 14 days normally required. In some cases, heavy growth of a fungus in culture will induce earlier sporulation because the nutrients are used more rapidly.

Sporulation of plant parasites in nature frequently accompanies a decrease in available nutrients; however, the reduction in nutrients may not be the direct stimulus for formation of reproductive structures. Leaf-inhabiting fungi such as *Venturia inaequalis* and *Erysiphe graminis* produce conidia in summer when the leaves are still alive and produce their sexual stages when the leaves are undergoing senescence or are dead (Hawker, 1957).

The reasons for inhibition of reproduction by high nutrient levels are not understood. It has often been suggested that cessation of vegetative growth may be the actual stimulus initiating reproduction and that the effect of high nutrient levels simply postpones starvation. Some insight into a physiological mechanism may be obtained by examining the yeast *Saccharomyces cerevisiae*, in which meiosis and sporulation may be induced by transfer from a nutrient-rich medium to a nutrient-poor medium (Croes, 1967–a, 1967–b). During vegetative growth, *S. cerevisiae* utilizes the glucose in its medium in fermentation pathways, which results in a decrease of glucose but an increase of ethanol in the medium (Fig. 9–14). Mitotic divisions occur in the presence of

Fig. 9–14. A comparison of the pathways involving carbohydrate and protein metabolism during active growth (left), the late phase of growth (center), and early sporogenesis (right) in *Saccharomyces cerevisiae*. Lines of different thickness mark the relative reaction rates of the processes. See the text for additional details. [From A. F. Croes, 1967, *Planta* **76**:227–237.]

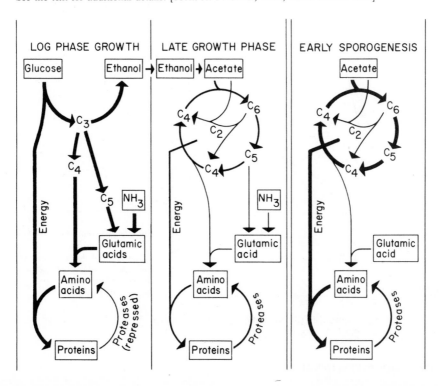

glucose. Near the end of the logarithmic phase of growth, a shift in the respiratory pathway occurs so that acetate and ethanol may be metabolized in an aerobic pathway (glyoxylic acid bypass of the CA cycle). The utilization of acetate or ethanol is essential to sporogenesis and appears to induce meiosis, as the addition of a small amount of glucose at this point will cause a reversion both to fermentation and to vegetative growth. Immediately after meiosis is induced, there is an initial sharp rise in aerobic respiration followed by its steady decline, reflecting the initial period of high metabolic activity followed by low metabolic activity. Metabolism of acetate makes the accumulation of glycogen and fats possible, which increases the dry weight. There is a steady decline in the RNA levels, which reflects the lowering of metabolic activity in later stages of sporogenesis. The level of protein is not altered greatly during sporulation, which indicates that the fungus is utilizing reserves that had been accumulated during growth. As we can see in the above account, the metabolic changes involved in reproduction in *S. cerevisiae* are directly related to glucose starvation (Croes, 1967–a, 1967–b).

Depletion of nutrients does not lead to sporulation in all cases, and certain fungi require a large quantity of nutrients to produce sporocarps. Madelin (1956–a) found that the size and dry weight of basidiocarps produced by *Coprinus lagopus* are proportional to the quantity of medium available. Only immature basidiocarps are produced at lower nutrient concentrations. Formation of basidiocarps accompanies a decrease in the dry weight of the vegetative mycelium (Madelin, 1956–b), thus leading to the hypothesis that materials are drawn from the mycelium to support basidiocarp formation.

Inhibition of reproduction at high nutrient concentrations may also be attributed to excretion of metabolites by the fungus that causes staling of the medium by accumulating or causing changes in pH. Presumably this staling would inhibit reproduc-

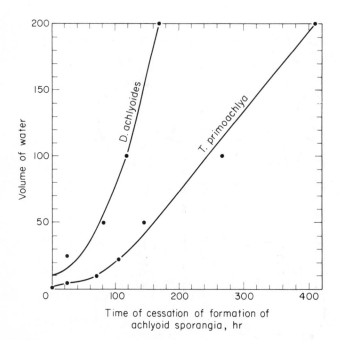

Fig. 9-15. Relation between the volume of water in culture, and the time at which formation of sporangia of *Dictyuchus achlyoides* and *Thraustotheca primoachlya* ceases. [From S. B. Salvin, 1942, *Am. J. Botany* **29**:97-104.]

tion without inhibiting vegetative growth. Much of the difficulty encountered in inducing sporocarp formation in the Ascomycotina and Basidiomycotina may be due to failure to provide sufficient nutrients without the deleterious effects that follow vigorous growth (Hawker, 1957). If it is staling of the medium which limits reproduction, then weak medium would support reproduction better because vegetative growth is reduced, which thus reduces the quantity of staling metabolites released into the medium (Fig. 9–15). Likewise, transfer of the mycelium to fresh, weaker medium leaves behind toxic metabolites and the mycelium would presumably produce less staling material in the weaker medium, thus encouraging sporulation. Brown (1925) noted that maximum sporulation of *Fusarium* occurred at that concentration which marks the dividing point between staling and nonstaling types of growth. Asthana and Hawker (1936) and Barksdale (1962) gave evidence to support the metabolite inhibition view by adding fresh medium to cultures in which sporulation had declined. The staling products were diluted without an accompanying reduction in nutrients, and the formation of reproductive structures resumed.

CARBON NUTRITION

Fungi differ in their ability to use different carbon sources for reproduction. The best carbon source for the formation of reproductive structures is not always that which yields maximum vegetative growth. Hawker (1939) found that the Pyrenomycete *Melanospora destruens* will grow vigorously on glucose but that reproduction is favored by sucrose. Polysaccharides such as starch were intermediate in effect between glucose and sucrose. Disaccharides and polysaccharides are sometimes more favorable for reproduction than the simple hexoses, although the hexoses may favor vegetative growth. Cellulose, which has limited availability for most fungi, may produce abundant sporulation in some cases.

The carbon source may influence the morphology of the reproductive structures. For example, zoosporangia of the chytrid *Chytriomyces hyalinus* were larger when grown on cellobiose than on fructose (starch supported the best vegetative growth) (Hasija and Miller, 1971).

The primary effect of different carbon sources is that of concentration. The fungus utilizes the complex molecules at a slower rate than the simple hexoses; therefore the amount of available carbohydrate is less and is effectively equivalent to lower carbohydrate concentrations. This explanation cannot be universally applied because glucose, a readily utilized hexose, is favored by some fungi both for formation of reproductive structures and vegetative growth.

Carbon concentration may affect the behavior or morphology of the reproductive structures. Hyphal fusions leading to dikaryon formation in *Schizophyllum commune* occur with a much greater frequency in the absence of sugar than in the presence of even small concentrations (0.5% or higher) (Ahmad and Miles, 1970). An increase in carbohydrate causes a decrease in the pycnidial size in some deuteromycetes and may also increase the length and curvature of the spore (Nitimargi, 1937). In species of *Fusarium*, an increase in carbohydrate produces degenerative changes (increase in fragmentation and a decrease in vacuoles) and decreases the number of septa formed (Horne and Mitter, 1927).

NITROGEN NUTRITION

Fungi differ in their ability to utilize different nitrogen sources for reproduction, and a nitrogen source that gives good vegetative growth may not favor reproduction.

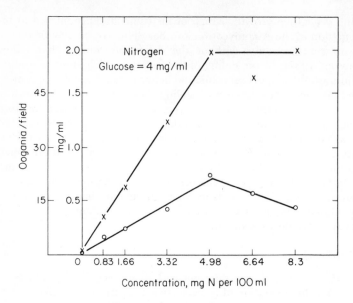

Fig. 9-16. Effect of the nitrogen concentration on vegetative growth (*x*) and reproduction (*o*) of *Achlya ambisexualis*. [From A. W. Barksdale, 1962, *Am. J. Botany* **49**:633–638.]

For example, asparagine is usually a good nitrogen source for vegetative growth but generally suppresses reproduction. Perhaps the inhibitory effect of asparagine stems from the accumulation of ammonia during the vegetative growth phase, which makes the medium alkaline and inhibits reproduction (Cochrane, 1958; Hawker, 1957), or alternatively the ammonia that is present in the culture as a gas may be toxic. In their study of species of *Phytophthora,* Leal et al. (1970) determined that the accumulation of ammonia as a gas is toxic and inhibits oogonium formation but that the increased pH and accumulation of ammonium ions in the medium apparently do not affect reproduction.

The minimum amount of nitrogen permitting sporulation is generally above that at which sparse vegetative growth can take place (Fig. 9–16), but at high nitrogen concentrations growth becomes vigorous and inhibits sporulation. Inhibition of sporulation at higher nitrogen concentrations may be due to exhaustion of other essential nutrients by the fungus or to the accumulation of toxic metabolites (Hawker, 1957).

Morphology of certain reproductive structures may be changed by varying the concentration of nitrogen. A high nitrogen concentration reduces the number of septa in the conidia of *Fusarium* (Brown and Horne, 1926). Pycnidia of *Phyllosticta antirrhini* are almost twice as large when grown on a medium containing an ammonium salt than when grown on a medium containing alanine (Maiello and Cappellini, 1976).

Nitrogen source may also affect some aspects of reproduction. Diploidization, required for the parasexual cycle, is greatly favored by nitrogen supplied as a nitrate but not by asparagine or ammonium in *Verticillium albo-atrum* (Ingle and Hastie, 1974). Ascospore ejection by *Leptosphaerulina briosiana* is favored by certain nitrogen sources more than by others (Pandey and Wilcoxson, 1967).

MINERALS AND VITAMINS

Essential mineral elements and vitamins were discussed in Chapter 7. Formation of reproductive structures typically requires a higher concentration of minerals or vitamins than those permitting vegetative growth. At that mineral concentration just permitting vegetative growth, sporulation may not occur (Fig. 9-17). For example, although the bird's nest fungus *Cyathus stercoreus* will grow vegetatively on a medium without calcium ions, basidiocarp formation will not take place in their absence (Lu, 1973). Similarly a fungus may be partially deficient for a vitamin, and this partial deficiency may be expressed as an inability to sporulate although vegetative growth is good. For example, *Sordaria fimicola* will grow vegetatively on a biotin-free medium but will not form perithecia unless biotin is added (Barnett and Lilly, 1947-a). Generally vitamins which favor mycelial growth also favor nonsexual reproduction, but the conditions required for sexual reproduction may be quite different (Hawker, 1957).

The optimum level of the deficient vitamin may be determined by the concentration of the other nutrients available or by the type of carbohydrate. The amount of thiamine required for perithecium development by *Ceratostomella fimbriata* increases as the available amount of carbohydrate increases (Barnett and Lilly, 1947-b). In a similar manner, the amount of available thiamine raises the optimum carbohydrate level for sporulation in some fungi, presumably because thiamine accelerates glucose utilization and brings about the onset of starvation (Hawker, 1950). Species of *Pythium* or *Phytophthora* as well as some other fungi require sterols for nonsexual and sexual reproduction but not for vegetative growth (Hendrix, 1970).

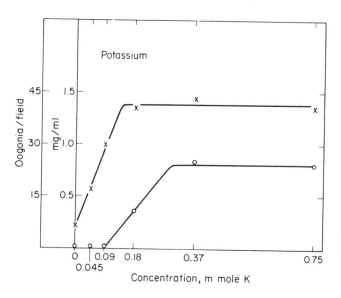

Fig. 9-17. Effect of the potassium concentration on vegetative growth (*x*) and oogonium formation (*o*) in *Achlya ambisexualis*. Note that the optimum concentration of potassium for vegetative growth is barely sufficient for reproduction. [From A. W. Barksdale, 1962, *Am. J. Botany* **49**:633–638.]

Physical Factors

We discussed physical factors that affect growth in Chapter 7. These include temperature, acidity, light, moisture, carbon dioxide, and oxygen. In general, the physical requirements for reproduction are more precise than those for vegetative growth. For example, there is usually a fairly broad range of temperatures permitting growth to take place but this range usually includes a low temperature range allowing vegetative growth to take place but too low for reproduction, a narrow temperature range where reproduction occurs, and a range too high to allow reproduction but still permitting vegetative growth (Fig. 9-18). The optima for vegetative growth and reproduction may or may not coincide. If they do not coincide, the optimum for

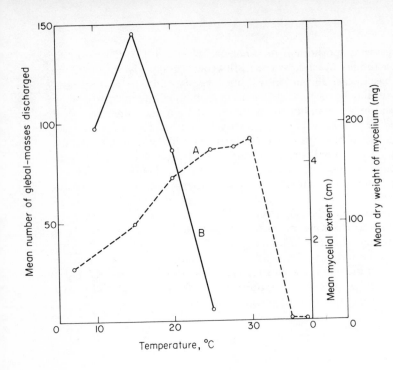

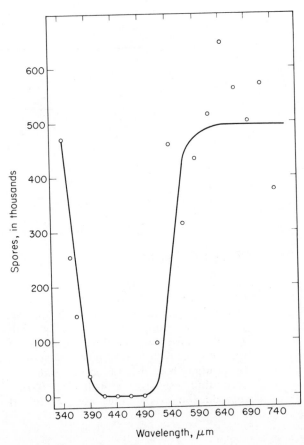

Fig. 9–18. Comparison of vegetative growth and basidiocarp production of an isolate of *Sphaerobolus* as affected by temperature. Key: *A*, vegetative growth, *B*, glebal masses discharged after 38 days. [Adapted from S. O. Alasoadura, 1963, *Ann. Botany London II.* **27**:123–145.]

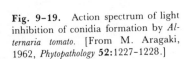

Fig. 9–19. Action spectrum of light inhibition of conidia formation by *Alternaria tomato*. [From M. Aragaki, 1962, *Phytopathology* **52**:1227–1228.]

reproduction may be either higher or lower than for vegetative growth. Unlike the growth of the vegetative mycelium, reproduction of some fungi may be favored by desiccation. For example, basidiocarps of *Gloeophyllum saepiarium* will form in culture only if the vegetative mycelium is exposed to periods of drying (States, 1975).

Aeration is often an extremely important parameter for reproduction. Fungi frequently will not sporulate in poorly aerated habitats or in tightly closed culture containers whereas, on the other hand, sporulation is favored by exposure to a moving air stream. Factors that may be important are decrease of oxygen, increase of respiratory carbon dioxide, alteration of the relative humidity, and possible accumulation of volatile metabolites.

LIGHT

Effects of light on reproduction in fungi are extremely complex. Closely related species or different isolates of the same species may differ in their response to light. Different developmental stages or spore forms of the same isolate may have varying light requirements. The intensity, duration, and quality of light all play a role in the overall effect of light on a fungus.

Generally, light may be either stimulatory (inductive) or inhibitory to formation of reproductive structures and spores in the fungi. The fungi may be divided into five groups on the basis of their response to light: (1) those which are apparently indifferent to light; (2) those in which sporulation is decreased or prevented on exposure to light (Fig. 9–19); (3) those which require alternating light and darkness to sporulate; (4) those which are able to produce viable spores in complete darkness but sporulate more abundantly with exposure to light; and (5) those which require light in order to produce reproductive structures and spores.

Mechanism of Photoresponse. Wavelengths of light that are either inductive or inhibitory are most often in the violet, blue, blue-green, near ultraviolet, or far ultraviolet regions of the spectrum. Occasionally yellow, red, or far red light may exert effects. A few cases are known in which the effects of one wavelength of light may be reversed by other wavelengths. For example, sporulation in *Botrytis cinerea* is inhibited by exposure to blue light, but this inhibition may be reversed and sporulation enhanced if the exposure to blue light is followed by exposure to far red or near ultraviolet light. The favorable effect of far red may be reversed by subsequent exposure to red or blue light. The last light to which *B. cinerea* is exposed is the one that controls the response (Tan, 1975).

To exert control on the fungus, the light must initially be received by a photoreceptor, a chemical compound within the fungus, and metabolic pathways must subsequently be altered. The biochemical effects of light on metabolism are complex and virtually unknown. To function as a photoreceptor, a chemical compound must be able to absorb the effective wavelengths of light. The identity of the photoreceptor(s) is unknown, but various compounds, including a porphyrin, a pteridine, carotenes, and flavins, have been detected in light-responsive fungi and the possibility has been considered that each may function as a photoreceptor. Both carotene pigments and flavins are yellow pigments and are capable of absorbing light in the blue range (the most frequently utilized light range). One particular flavin that may be a photoreceptor is riboflavin, which is capable of absorbing both visible and ultraviolet light. Current opinion most often favors the view that flavins function as photoreceptors. Light is thought to bring about the initial steps in photoresponse by (1) destroying riboflavin or (2) initiating activity in an electron-transfer system containing a flavoprotein component (Carlile, 1965). In the event of light induction, initiation of metabolic activity by path (2) above could account for stimulation. Lukens (1963) suggested that the mechanism responsible for photoinhibition is the destruction of a flavin required for sporulation. An alternative suggestion is that light may stimulate a flavoprotein electron-transfer path that depresses activity in alternate metabolic paths (Carlile, 1965).

Biochemical changes occurring as a result of light induction or inhibition are generally not well understood. However, light has been shown to alter the levels of some cellular constituents and therefore to alter metabolism. In *Blastocladiella emersonii,* exposure to white light causes the colorless zoosporangia to grow to a larger size than those cultured in the dark and also delays their maturation and zoospore discharge. (Cantino and Horenstein, 1959). The effects of light on the colorless zoosporangia of *B. emersonii* are to increase the rates of carbon dioxide fixation, glucose uptake, reproduction of nuclei, synthesis of DNA, protein synthesis, and polysaccharide synthesis (Goldstein and Cantino, 1962). The photoreceptor in *B. emersonii* may be a protein-bound porphyrin resembling a cytochrome (Cantino, 1966).

The response of fungi to light may be altered by their age or by other cultural conditions such as the medium composition or temperature.

Photoinduction. Light is required for reproduction by a numerous fungi of every major class. Although reproduction may require light, only a certain developmental phase may be light-dependent, or alternatively, different structures within the sporocarp may respond differently to light. Initials of the Discomycete *Sclerotinia sclerotiorum* will form in the dark, but light is necessary for the development of the pileate disks (Purdy, 1956). Light is required only for the pileus development in the basidiomycete *Lentinus tuber-regium* (Galleymore, 1949). Control by light is rather complex in apothecial development of the Discomycete *Pyronema domesticum* (Moore-Landecker, 1979-a, 1979-b). There is an absolute light requirement for formation of the gametangia, which require about 12 hours exposure to light for their full development. Following gametangial development, the ascogenous hyphae and asci can develop in the dark but their development is enhanced by additional exposure to light. The ascospores mature best if the light is alternating at the time of their development, and exposure to constant light inhibits ascospore maturation. Development of the sterile system, which occurs concurrently, is not light dependent after the apothecium is initiated although development may be altered by varying the length or intensity of light exposure.

The dosage of light required to induce reproduction is highly variable from species to species. Exposure to 0.1 foot-candle for 5 seconds increases basidiocarp formation by *Coprinus lagopus* (Madelin, 1956-b). In contrast, basidiocarp formation in *Cyathus stercoreus* requires 2.9 days of exposure at 240 foot-candles (Lu, 1965). The exposure time required for some fungi to sporulate will decrease as the light intensity is increased, indicating that the required dosage of light at a given wavelength remains constant. For example, *C. stercoreus* requires approximately 17,200 foot-candle-hours at 25°C to induce basidiocarp formation (Lu, 1965). In addition, the required dosage has been found to decrease as the wavelength of light becomes shorter.

Photoinduction in Combination with Photoinhibition. Alternation of light and dark is a constant environmental factor in nature. Certain fungi require both light and dark for complete formation of reproductive structures. Alternating light and dark may be required for sporulation, may regulate periodicity, or may produce zonation in cultures of fungi.

The deuteromycete, *Stemphylium botryosum* does not sporulate at all or forms only a few spores in constant darkness, and in constant light it forms sterile conidiophores.

Abundant sporulation is achieved only in alternating light and darkness. This fungus requires ultraviolet light for the production of condiophores, while the second developmental phase, during which the conidia are formed, requires darkness. Those wavelengths which are capable of inhibiting conidium formation are in the range of 240 to 650 nanometers and the particular length of the dark period required is temperature-dependent (Leach, 1968).

A second example is the zygomycete *Choanephora cucurbitarum,* which requires both light and dark for the formation of conidia, although sporangial formation is indifferent to light. Exposure of the cultures to continuous light (65 foot-candles), continuous darkness, or 2 days of darkness followed by light did not allow conidia to form. However, exposure of cultures either to light for 2 days followed by darkness or to continuous low intensity (less than 1 foot-candle) resulted in sporulation. Barnett and Lilly (1950) then made the assumption that light or darkness affects the metabolic sequences A and B. Presumably light is a requirement for the sequence A but inhibits sequence B, and the product(s) of A are required before B can occur. The completion of both sequences would be required before conidia would be formed. Therefore the cultures incubated in continuous light would only allow reaction A to occur. Those cultures incubated in either continuous darkness or darkness followed by light do not allow reaction A to occur; therefore, B cannot occur. Continuous light of low intensity allows both reactions to occur simultaneously. See Fig. 9–20.

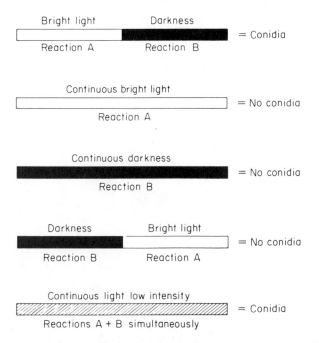

Fig. 9-20. The relationship of light to conidia formation in *Choanephora cucurbitarum.* See text for explanation. [From H. L. Barnett and V. G. Lilly, 1950, *Phytopathology* **40**:80–89.]

The natural processes of some fungi occur rhythmically at approximately 24–hour intervals under natural conditions. Such rhythms occur in many organisms other than the fungi and are called *circadian rhythms,* or 24–hour biological clocks. *Pilobolus* has a circadian rhythm. Periodicity of sporangium formation in *Pilobolus* occurs only in alternating light and dark, and the fungus will produce a daily crop of sporangiophores which are synchronized in their developmental stages. Ordinarily, sporangiophore maturation occurs in the morning (Fig. 9–21), but it can be made to take place at any hour of the day or night by manipulation of the 12–hour periods of light and darkness. Although alternation of light and darkness are required to regulate the biological clock, sporangiophores will form in total darkness, while both sporangiophore and sporangium production will occur in continuous light (McVickar, 1942).

Many fungi produce concentric rings of spores or sporocarps and vegetative mycelium when exposed to alternating light and dark. There are basically two types of response to light and darkness involved in zonation:

1. Reproductive structures are more abundant in those zones exposed to light, while vegetative mycelium is formed in the dark. These fungi either require light for the formation of reproductive structures or produce them more abundantly in light than in darkness. *Trichoderma lignorum* in Fig. 9–22 is of this type.
2. Reproductive structures are more abundant in those zones exposed to darkness, while vegetative mycelium is formed in the light.

Fluctuating temperatures may also cause zonation, and in some species zonation may be induced by either fluctuating light or fluctuating temperatures. Any variation from the optimum temperature (either higher or lower) is presumed to check vegetative growth, which is accompanied by an increase in reproduction (Ellis, 1931; Hafiz, 1951).

INJURY AND MECHANICAL BARRIERS

Injury to the mycelium, such as cutting or scraping, frequently stimulates the formation of reproductive structures. When cultures of *Alternaria solani* are scraped, conidiophores arise from the broken ends of the hyphae (Kunkel, 1918). Apothecia of

Fig. 9–21. Synchronous development of the sporangiophores of *Pilobolus*. Each sporangiophore bears a single black sporangium. [Courtesy Carolina Biological Supply.]

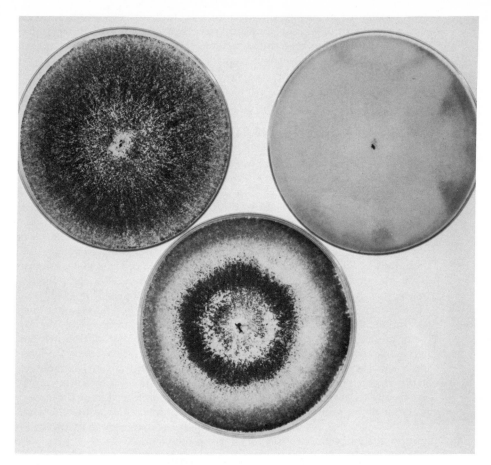

Fig. 9-22. The effect of light on sporulation of *Trichoderma lignorum*. The culture at the top left was grown in constant light, and is dark owing to the uniform distribution of spores. The culture at the top right was grown in constant darkness and has no spores. The bottom culture was grown in alternating light and darkness. That growth formed while exposed to light appears as dark rings, owing to spore production, while that growth formed in darkness is light, owing to the absence of spores. [Courtesy Plant Pathology Department, Cornell University.]

Pyronema will form along a cut through the culture or may be stimulated to form in the presence of crystals of toxic chemicals (Robinson, 1926). *Schizophyllum commune* typically requires formation of the dikaryon to form basidiocarps, but maceration or cutting of monokaryotic hyphae will lead to the production of strictly haploid basidiocarps. This response of *S. commune* is genetically controlled (Leonard and Dick, 1973). Although the exact mechanism controlling response to injury is not known, it has been suggested that the effect of limited injury may be to release reproduction-stimulating chemicals from the injured cells (Hawker, 1957). Supporting evidence comes from Lindegren and Hamilton's observation (1944) that ascus formation in yeast occurs only in portions of the colony where autolysis has taken place, which thus indicates that autolytic products favor sporulation in certain instances.

Reproduction **367**

Certain fungi such as *Pyronema* form reproductive structures upon reaching the petri dish edge or other mechanical barrier, such as a glass rod or partition in the dish. A similar response has been noted in nature. The hypogeous fungus *Hymenogaster luteus* tends to form its basidiocarps in contact with hard surfaces such as a layer of clay, stones, or tree roots (Hawker, 1954). The mechanism for the effect of the mechanical barrier is unknown.

REPRODUCTION IN THE NATURAL HABITAT

Even a casual observer of fungi in their natural habitats will note that fungi are often very specific in their occurrence on particular types of substrata; often they occur only on plants or plant debris of certain species and sometimes only on a specific part of that plant, on insects and also sometimes only on specific parts of certain insect species, on dung of certain species of animals, or on freshly burned material. Also fungi may occur only in a specific type of ecological niche, such as a bog, woodland, or open field. In addition, fungi fruit during limited parts of the growing season and may occur only sporadically. It is sometimes impossible to collect a given fungus species for years at a time because environmental conditions do not allow reproduction, but during a satisfactory season the fungus will suddenly reappear (Fig. 9-23).

Specificity of fungi for particular ecological niches or seasonal distribution is determined by any specificity for hosts that have a narrow ecological or seasonal distribution and by an ability to grow vegetatively or reproduce under the normally prevailing environmental conditions. A fungus in either nature or the laboratory will form sporocarps only if nutritional requirements and all environmental conditions (temperature, light, pH, moisture, and proper aeration of the substratum) are within the narrow tolerance ranges simultaneously. Therefore it is to be expected that a fungus will be somewhat restricted in its reproductive season and that its rather physiologically flexible mycelium will remain vegetative until the prevailing climate and conditions become satisfactory for reproduction.

Fig. 9-23. Formation of basidiocarps in nature. These mushrooms are forming at the edge of an underground mycelium, and are maturing simultaneously because all environmental conditions are suitable for reproduction. Such a ring of mushrooms is popularly called a "fairy ring." [From E. J. Moore and V. N. Rockcastle, 1963, *Fungi,* Cornell Science Leaflet.]

Those environmental factors most subject to daily and seasonal fluctuation are light, temperature, and moisture. Light-dependent fungi usually require only relatively small amounts of light to fruit, and it is doubtful that light intensity becomes too low to permit reproduction in nature. It has been determined (Wilkins and Patrick, 1940; Wilkins and Harris, 1946) that the seasonal and ecological distribution of fungi is determined by temperature and moisture content of the substratum through the season and that a certain period of suitable environmental conditions must prevail before a sporocarp is produced. This period of time, the lag phase, varies with the fungus species but is apparently proportional to the size of the fungus. Therefore a large sporocarp requires a longer period of suitable conditions than a very small one. Accordingly, the rapid rise of basidiocarps in the autumn may be partially a response to high maximum temperatures during the summer. Moisture is perhaps the most important single factor controlling reproduction, as all fungi require a certain amount of moisture but many fungi are apparently indifferent to temperature (Wilkins and Harris, 1946).

References

Ahmad, S. S., and P. G. Miles. 1970. Hyphal fusions in *Schizophyllum commune*. 2. Effects of environmental and chemical factors. Mycologia **62**:1008–1017.

Alasoadura, S. O. 1963. Fruiting in *Sphaerobolus* with special reference to light. Ann. Botany London II. **27**:123–145.

Aragaki, M. 1962. Quality of radiation inhibitory to sporulation of *Alternaria tomato*. Phytopathology **52**:1227–1228.

Aschan-Åberg, K. 1958. The production of fruit bodies in *Collybia velutipes* II. Further studies on the influence of different culture conditions. Physiol. Plant. **2**:312–328.

Asthana, R. P., and L. E. Hawker. 1936. The influence of certain fungi on the sporulation of *Melanospora destruens* Shear, and of some other Ascomycetes. Ann. Botany London **50**:325–344.

Barksdale, A. W. 1962. Effect of nutritional deficiency on growth and sexual reproduction of *Achlya ambisexualis*. Am. J. Botany **49**:633–638.

——, 1967. The sexual hormones of fungus *Achlya*. Ann. N.Y. Acad. Sci. **144**:313–319.

——, 1970. Nutrition and antheridiol-induced branching in *Achlya ambisexualis*. Mycologia **62**:411–420.

——, T. C. McMorris, R. Seshadri, T. Arunachalam, J. A. Edwards, J. Sundeen, and D. M. Green. 1974. Response of *Achlya ambisexualis* E 87 to the hormones antheridiol and certain other steriods. J. Gen. Microbiol. **82**:295–299.

Barnett, H. L., and V. G. Lilly. 1947-a. The effects of biotin upon the formation and development of perithecia, asci, and ascospores by *Sordaria fimicola* Ces. and DeNot. Am. J. Botany **34**:196–204.

——, and ——. 1947-b. The relationship of thiamin to the production of perithecia by *Ceratostomella fimbriata*. Mycologia **39**:699–708.

——, and ——. 1950. Influence of nutritional and environmental factors upon asexual reproduction of *Choanephora cucurbitarum* in culture. Phytopathology **40**:80–89.

Berquist, R. R., and J. W. Lorbeer. 1972. Apothecial production, compatibility and sex in *Botryotinia squamosa*. Mycologia **64**:1270–1281.

Bistis, G. N. 1956. Sexuality in *Ascobolus stercorarius*. I. Morphology of the ascogonium; plasmogamy; evidence for a sexual hormonal mechanism. Am. J. Botany **43**:389–394.

——, 1957. Sexuality in *Ascobolus sterocorarius*. II. Preliminary experiments on various aspects of the sexual process. Am. J. Botany **44**:436–443.

——, and S. G. Georgopoulos. 1979. Some aspects of sexual reproduction in *Nectria haematococca* var *cucurbitae*. Mycologia **71**:127–143.

Brown, W. 1925. Studies in the genus *Fusarium*. II. An analysis of factors which determine the growth-forms of certain species. Ann. Botany London **39**:373–408.

——. and A. S. Horne. 1926. Studies in the genus *Fusarium*. III. An analysis of factors which determine certain microscopic features of *Fusarium* strains. Ann. Botany London **40**:203-221.

Burnett, J. H. 1956. The mating systems of fungi I. New Phytol. **50**:50-90.

——, 1975. Mycogenetics—an introduction to the general genetics of fungi. John Wiley & Sons, Inc., New York. 375 pp.

——, and M. E. Boulter. 1963. The mating systems of fungi II. Mating systems of the Gasteromycetes *Mycocalia denudata* and *M. duriaena*. New Phytol. **62**:217-236.

Buxton, E. W. 1960. Heterokaryosis, saltation, and adaptation. III. Heterokaryosis as a factor in variability. *In:* Horsfall, J. G., and A. E. Dimond, Eds., Plant pathology—an advanced treatise. **2**: 365-380. Academic Press, New York.

Cantino, E. C, 1966. Morphogenesis in aquatic fungi. *In:* Ainsworth, G. C., and A. S. Sussman, Eds., The fungi—an advanced treatise. **2**:283-337.

——, and E. A. Horenstein. 1959. The stimulatory effect of light upon growth and carbon dioxide fixation in *Blastocladiella*. III. Further studies, *in vivo* and *in vitro*. Physiol. Plant. **12**:251-263.

Carlile, M. J. 1965. The photobiology of fungi. Ann. Rev. Plant Physiol. **16**:175-202.

——, and L. Machlis. 1965-a. The response of male gametes of *Allomyces* to the sexual hormone sirenin. Am. J. Botany **52**:478-483.

——, and ——. 1965-b. A comparative study of the chemotaxis of the motile phases of *Allomyces*. Am. J. Botany **52**:484-486.

Choinski, J. S., and J. T. Mullins. 1977. Ultrastructural and enzymatic evidence for the presence of microbodies in the fungus *Achlya*. Am. J. Botany **64**:593-599.

Cobb, F. W., C. L. Fergus, and W. J. Stambaugh. 1961. The effect of temperature on ascogonial and perithecial development in *Ceratocystis fagacearum*. Mycologia **53**:91-97.

Cochrane, V. W. 1958. Physiology of fungi. John Wiley & Sons, Inc., New York. 452 pp.

Croes, A. F. 1967-a. Induction of meiosis in yeast I. Timing of cytological and biochemical events. Planta **76**:209-226.

——. 1967-b. Induction of meiosis in yeast II. Metabolic factors leading to meiosis. Planta **76**:227-237.

Davis, R. H. 1966. Mechanisms of inheritance 2. Heterokaryosis. *In:* Ainsworth, G. C., and A. S. Sussman, Eds., The fungi—an advanced treatise. Academic Press, New York. **2**:567-588.

El-Ani, A. S. 1954. The genetics of sex in *Hypomyces solani* f. cucurbitae. Am. J. Botany **41**:110-113.

Ellis, M. 1931. Some experimental studies on *Pleospora herbarum* (Pers.) Rabenh. Trans. Brit. Mycol. Soc. **16**:102-114.

Esser, K. 1966. Incompatibility. *In:* Ainsworth, G. C., and A. S. Sussman, Eds., The fungi—an advanced treatise. Academic Press, New York. **2**:661-676.

Faro, S. 1972. The role of a cytoplasmic glucan during morphogenesis of sex organs in *Achlya*. Am. J. Botany **59**:919-923.

Fries, N. 1948. Heterothallism in some Gasteromycetes and Hymenomycetes. Svensk Botan. Tidsskr. **42**: 158-168.

Galleymore, H. B. 1949. The development of fructifications of *Lentinus tuber-regium* Fries in culture. Trans. Brit. Mycol. Soc. **32**:315-317.

Garnjobst, L. 1955. Further analysis of genetic control of heterokaryosis in *Neurospora crassa*. Am. J. Botany **42**:444-448.

Goldstein, A., and E. C. Cantino. 1962. Light-stimulated polysaccharide and protein synthesis by synchronized, single generations of *Blastocladiella emersonii*. J. Gen. Microbiol. **1962**:689-699.

Gooday, G. W. 1973. Differentiation in the Mucorales. Symp. Soc. Gen. Microbiol. **23**:269-294.

Hafiz, A. 1951. Cultural studies on *Ascochyta rabiei* with special reference to zonation. Trans. Brit. Mycol. Soc. **34**:259-269.

Halisky, P. M. 1965. Physiologic specialization and genetics of the smut fungi III. Botan. Rev. **31**: 114-150.

Hansen, H. N., and W. C. Snyder. 1943. The dual phenomenon and sex in *Hypomyces solani* f. cucurbitae. Am. J. Botany **30**:419-422.

Harnish, W. N. 1965. Effect of light on production of oospores and sporangia in species of *Phytophthora*. Mycologia **57**:85-90.

Hasija, S. K., and C. E. Miller. 1971. Nutrition of *Chytriomyces* and its influence on morphology. Am. J. Botany **58**:939-944.

Hawker, L. E. 1939. The influence of various sources of carbon on the formation of perithecia by *Melanospora destruens* Shear in the presence of accessory growth factors. Ann. Botany London II. **3**:455–468.

——. 1950. Physiology of fungi. University of London Press, London 360 pp.

——. 1954. British hypogeous fungi. Phil. Trans. (Roy. Soc. London) B**237**:429–546.

——. 1957. The physiology of reproduction in fungi. Cambridge University Press, New York. 128 pp.

——. 1966. Environmental influences on reproduction. *In:* Ainsworth, G. C., and A. S. Sussman, Eds., The fungi—an advanced treatise. Academic Press, New York. **2**:435–469.

Hendrix, J. W. 1970. Sterols in growth and reproduction of fungi. Ann. Rev. Phytopathol. **8**:111–130.

Horne, A. S., and J. H. Mitter. 1927. Studies in the genus *Fusarium*. V. Factors determining septation and other features in the section Discolor. Ann. Botany London **41**:519–547.

Ingle, M. R., and A. C. Hastie. 1974. Environmental factors affecting the formation of diploids in *Verticillium albo-atrum*. Trans. Brit. Mycol. Soc. **62**:313–321.

Klebs, G. 1899. Zur Physiologie der Fortpflanzung einiger Pilze. II. Jahrb Wiss. Botanik **33**:513–593.

Kochert, G. 1978. Sexual pheromones in algae and fungi. Ann. Rev. Plant Physiol. **29**:461–486.

Kunkel, L. O. 1918. A method of obtaining abundant sporulation in cultures of *Macrosporium solani* E. & M. Brooklyn Botan. Garden Mem. **1**:306–312.

Leach, C. M. 1963. The qualitative and quantitative relationship of monochromatic radiation to sexual and asexual reproduction of *Pleospora herbarum*. Mycologia **55**:151–163.

——. 1964. The relationship of visible and ultraviolet light to sporulation of *Alternaria chrysanthemi*. Trans. Brit. Mycol. Soc. **47**:153–159.

——. 1965. Ultraviolet-absorbing substances associated with light-induced sporulation in fungi. Can. J. Botany **43**:185–200.

——. 1968. An action spectrum for light inhibition of the "terminal phase" of photosporogenesis in the fungus *Stemphylium botryosum*. Mycologia **60**:532–546.

Leal, J. A., V. G. Lilly, and M. E. Gallegly. 1970. Some effects of ammonia on species of *Phytophthora*. Mycologia **62**:1041–1056.

Leonard, T. J., and S. Dick. 1973. Induction of haploid fruiting by mechanical injury in *Schizophyllum commune*. Mycologia **65**:809–822.

Lilly, V. G., and H. L. Barnett. 1951. Physiology of the fungi. McGraw-Hill Book Company, New York. 464 pp.

Lindegren, C. C., and E. Hamilton. 1944. Autolysis and sporulation in the yeast colony. Botan. Gaz. **105**:316–321.

Lu, B. C. 1965. The role of light in fructification of the Basidiomycete *Cyathus stercoreus*. Am. J. Botany **52**:432–437.

——, 1973. Effect of calcium on fruiting of *Cyathus stercoreus*. Mycologia **65**:329–334.

Lukens, R. J. 1963. Photo-inhibition of sporulation in *Alternaria solani*. Am. J. Botany **50**:720–724.

Machlis, L. 1958. A study of sirenin, the chemotactic sexual hormone from the watermold *Allomyces*. Physiol. Plant. **11**:845–854.

——, 1966. Sex hormones in fungi. *In*: Ainsworth, G. C., and A. S. Sussman, Eds., The fungi—an advanced treatise. Academic Press, New York. **2**:415–433.

——, 1972. The coming of age of sex hormones in plants. Mycologia **64**:235–247.

——, W. H. Nutting, and H. Rapoport. 1968. The structure of sirenin. J. Am. Chem. Soc. **90**:1674–1676.

Madelin, M. F. 1956-a. Studies on the nutrition of *Coprinus lagopus* Fr., especially as affecting fruiting. Ann. Botan. London. II. **20**:307–330.

——, 1956-b. The influence of light and temperature on fruiting of *Coprinus lagopus* Fr. in pure culture. Ann. Botany London. II. **20**:467–480.

Maiello, J. M., and R. A. Cappellini. 1976. The influence of carbon and nitrogen nutrition on pycnidium development in *Phyllosticta antirrhini*. Mycologia **68**:1174–1180.

Marsh, P. B., E. E. Taylor, and L. M. Bassler. 1959. A guide to the literature on certain effects of light on fungi: reproduction, morphology, pigmentation, and phototropic phenomena. Plant Disease Reptr. Suppl. **261**:251–312.

McMorris, T. C., R. Seshadri, G. R. Weihe, G. P. Arsenault, and A. W. Barksdale. 1975. Structures of oogoniol-1, -2, and -3, steroidal sex hormones of the water mold, *Achlya*. J. Am. Chem. Soc. **97**: 2544–2545.

McVickar, C. L. 1942. The light-controlled diurnal rhythm of asexual reproduction in *Pilobolus*. Am. J. Botany **29**:372–380.

Middleton, R. B. 1964. Evidences of common–AB heterokaryosis in *Schizophyllum commune*. Am. J. Botany **51**:379–387.

Moore-Landecker, E. 1979-a. Effect of light regimens and intensities on morphogenesis of the Discomycete *Pyronema domesticum*. Mycologia **71**:699–712.

——. 1979-b. Effect of cultural age and a single photoperiod on morphogenesis of the Discomycete *Pyronema domesticum*. Can. J. Botany **57**: 1541–1549.

Nelson, R. R., R. K. Webster, and D. R. MacKenzie. 1977. The occurrence of dual compatibility in *Cochliobolus spicifer*. Mycologia **69**:173–178.

Nieuwenhuis, M., and H. van den Ende. 1975. Sex specificity of hormone synthesis in *Mucor mucedo*. Arch. Microbiol. **102**:167–169.

Nitimargi, N. M. 1935. Studies in the genera *Cytosporina, Phomopsis,* and *Diaporthe*. VII. Chemical factors influencing sporing characteristics. Ann. Botany London **49**:19–40.

Nutting, W. H., H. Rapoport, and L. Machlis. 1968. The structure of sirenin. J. Am. Chem. Soc. **90**: 6434–6438.

Olive, L. S. 1958. On the evolution of heterothallism in fungi. Am. Nat. **92**:233–251.

Pandey, M. C., and R. D. Wilcoxson. 1967. Effect of carbon and nitrogen nutrition on reproduction in *Leptosphaerulina briosiane*. Am. J. Botany **54**:1170–1175.

Papa, K. E. 1977. Genetics of aflatoxin production in *Aspergillus flavus:* linkage between a gene for a high $B_1:B_2$ ratio and the histidine locus on Linkage group VIII. Mycologia **69**:1185–1190.

——, 1978. The parasexual cycle in *Aspergillus parasiticus*. Mycologia **70**:766–773.

——, W. A. Campbell, and F. F. Hendrix, Jr. 1967. Sexuality in *Pythium sylvaticum;* heterothallism. Mycologia **59**:589–595.

Papavizas, G. C. and W. A. Ayers. 1964. Effect of various carbon sources on growth and sexual reproduction of *Aphanomyces euteiches*. Mycologia **56**:816–830.

Papazian, H. P. 1950. Physiology of the incompatibility factors in *Schizophyllum commune*. Botan. Gaz. **112**: 143–163.

Parag, Y. 1965. Common–B heterokaryosis and fruiting in *Schizophyllum commune*. Mycologia **57**:543–561.

Pontecorvo, G. 1956. The parasexual cycle in fungi. Ann. Rev. Microbiol. **10**:393–400.

——, G. 1958. Trends in genetic analysis. Columbia University Press, New York. 145 pp.

Purdy, L. H. 1956. Factors affecting apothecial formation by *Sclerotinia sclerotiorum*. Phytopathology **46**: 409–410.

Raper, J. R. 1950. Sexual hormones in *Achlya*. VII. The hormonal mechanism in homothallic species. Botan. Gaz. **112**:1–24.

——, 1951. Sexual hormones in *Achlya*. Am. Scientist **39**:110–120, 130.

——, 1952. Chemical regulation of sexual processes in the thallophytes. Botan. Rev. **18**:447–545.

——, 1953. Tetrapolar sexuality. Quart. Rev. Biol. **28**:233–259.

——. 1959. Sexual versatility and evolutionary processes in fungi. Mycologia **51**:107–124.

——. 1960. The control of sex in fungi. Am. J. Botany **47**:794–808.

——. 1966-a. Life cycles, basic patterns of sexuality, and sexual mechanisms. *In:* Ainsworth, G. C., and A. S. Sussman, Eds., The fungi—an advanced treatise. Academic Press, New York. 2:473–511.

——. 1966-b. Genetics of sexuality in higher fungi. Ronald Press, New York. 283 pp.

——, and J. P. San Antonio. 1954. Heterokaryotic mutagenesis in 9 Hymenomycetes. I. Heterokaryosis in *Schizophyllum commune*. Am. J. Botany **41**:69–86.

Robinson, W. 1926. The conditions of growth and development of *Pyronema confluens*, Tul. [*P. omphalodes*, (Bull.) Fuckel]. Ann. Botany London **40**:245–272.

Roper, J. A. 1966. Mechanisms of inheritance 3. The parasexual cycle. *In:* Ainsworth, G. C., and A. S. Sussman, Eds., The fungi—an advanced treatise. Academic Press, New York. 2:589–617.

Sherwood, W. A. 1971. Some observations on the sexual behavior of the progeny of six isolates of *Dictyuchus monosporus*. Mycologia **63**:22–30.

Silver, J. C., and P. A. Horgen. 1974. Hormonal regulation of presumptive mRNA in the fungus *Achlya ambisexualis*. Nature **249**:252–254.

Smith, J. E., and J. C. Galbraith. 1971. Biochemical and physiological aspects of differentiation in the fungi. Adv. Microbial Physiol. **5**:45–134.

Snyder, W. C., and H. N. Hansen. 1954. Species concept, genetics, and pathogenicity in *Hypomyces solani.* Phytopathology **44**:338–342.

States, J. S. 1975. Normal basidiocarp development of *Gloeophyllum (Lenzites) saepiarium* in culture. Mycologia **67**:1166–1175.

Taber, W. A. 1966. Morphogenesis in basidiomycetes. *In:* Ainsworth, G. C., and A. S. Sussman, Eds., The fungi—an advanced treatise. Academic Press, New York. **2**:387–412.

Tan, K. K. 1975. Interaction of near-ultraviolet, blue, red, and far red light in sporulation of *Botrytis cinerea.* Trans. Brit. Mycol. Soc. **64**:215–222.

Thomas, D. M., and T. W. Goodwin. 1967. Studies on carotenogenesis in *Blakeslea trispora* I. General observations on synthesis in mated and unmated strains. Phytochemistry **6**:355–360.

Timberlake, W. E. 1976. Alterations in RNA and protein synthesis associated with steroid hormone-induced sexual morphogenesis in the water mold *Achlya.* Develop. Biol. **51**:202–214.

Timnick, M. B., H. L. Barnett, and V. G. Lilly. 1952. The effect of method of inoculation of media on sporulation of *Melanconium fuligineum.* Mycologia **44**:141–149.

———, V. G. Lilly, and H. L. Barnett. 1951. The effect of nutrition on the sporulation of *Melanconium fuligineum* in culture. Mycologia **43**:625–634.

Tinline, R. D., and B. H. MacNeill. 1969. Parasexuality in plant pathogenic fungi. Ann. Rev. Phytopathol. **7**:147–170.

Trione, E. J., and C. M. Leach. 1969. Light-induced sporulation and sporogenic substances in fungi. Phytopathology **59**:1077–1083.

Turian, G. 1966. Morphogenesis in Ascomycetes. *In:* Ainsworth, G. C., and A. S. Sussman, Eds., The fungi—an advanced treatise. Academic Press, New York, **2**: 339–385.

———, 1974. Sporogenesis in fungi. Ann. Rev. Phytopathol. **12**:129–137

———, 1975. Differentation in *Allomyces* and *Neurospora.* Trans. Brit. Mycol. Soc. **64**:367–380.

———, and D. B. Bianchi. 1972. Conidiation in *Neurospora.* Botan. Rev. **38**:119–154.

———, and E. C. Cantino. 1959. The stimulatory effect of light on nucleic acid synthesis in the mould *Blastocladiella emersonii.* J. Gen. Microbiol. **21**:721–735.

Tuveson, R. W., and E. D. Garber. 1961. Genetics of phytopathogenic fungi. IV. Experimentally induced alterations in nuclear ratios of heterokaryons of *Fusarium oxysporum* f. *pisi.* Genetics **46**:485–492.

Werkman, T. A., and H. van den Ende. 1973. Trisporic acid synthesis in *Blakeslea trispora*—interaction between plus and minus mating types. Arch. Mikrobiol. **90**:365–374.

Wheeler, H. E. 1954. Genetics and evolution of heterothallism in *Glomerella.* Phytopathology **44**:342–345.

———, and J. W. McGahen. 1952. Genetics of *Glomerella.* X. Genes affecting sexual reproduction. Am. J. Botany **39**:110–119.

Whitehouse, H. L. K. 1949. Heterothallism and sex in the fungi. Cambridge Phil. Biol. Rev. **24**:411–447.

Wilkins, W. H., and G. C. M. Harris. 1946. The ecology of the larger fungi V. An investigation into the influence of rainfall and temperature on the seasonal production of fungi in a beechwood and a pinewood. Ann. Appl. Biol. **33**:179–188.

———, and S. H. M. Patrick. 1940. The ecology of the larger fungi IV. The seasonal frequency of grassland fungi with special reference to the influence of environmental factors Ann. Appl. Biol. **27**:17–34.

Wurtz, T., and J. Jockusch. 1975. Sexual differentiation in *Mucor:* trisporic acid response mutants and mutants blocked in zygospore development. Develop. Biol. **43**:213–220.

TEN

Spores

Fungal spores occupy a unique position in the life cycle as they terminate both the reproductive and developmental cycles and have the inherent potentialities to develop into a new generation. The spore is much less susceptible to adverse environmental conditions than the vegetative thallus is, and it often enables the fungus to endure unfavorable conditions. More important, however, the fungus spore is a reproductive unit capable of being dispersed for great distances.

Events directly involving spores are their formation by the parent sporocarp or mycelium, discharge or release from the parent, dispersal to new locales, possible period of dormancy, and finally germination into a new thallus.

There are many hazards met by the spores along the way—discharge or dispersal mechanisms may not be effectively operating, a suitable substratum for germination and growth may not be encountered, and, finally, environmental conditions may not be suitable for either germination or further development. The laws of probability work against the success of any single spore, and the majority of the spores are wasted. In order to realize this, we need only consider that the giant puffball, *Calvatia gigantea,* produces 7×10^{12} basidiospores in a basidiocarp of 232 grams; if each spore were to germinate and form an equally large basidiocarp, their combined mass would be 800 times the size of our planet (Buller, 1924).

SPORE RELEASE

After spores are produced by the parent, they must be released in order to reach a new locale. Spores may be freed through the intervention of physical aspects of the environment during which the fungus plays a passive role, or they may be forcibly discharged by the fungus.

374

Passive Mechanisms of Spore Release

A large number of fungi, especially conidial forms, produce dry spores that are freely exposed to the surrounding air. Spores of this type are especially adapted to removal from the parent by air currents which blow the spores off. Removal of spores by air currents is much more effective in a cupulate sporocarp than on a flat surface because an eddy system is produced within the cup which carries the spores upward where they can join a second eddy stream which carries the spores away from the cup (Brodie and Gregory, 1953). An equivalent type of eddy system is not produced over a flat surface.

Falling raindrops fragment into smaller drops upon impact and, if they land on a fungus, will dislodge the spores and carry them away in their splatter. This type of spore release may occur in dry-spored types in which wind dislodgement occurs, but it is especially significant for fungi which produce their spores in a slime, making wind dislodgement impossible.

Fungi bearing spores within a slime include conidial fungi and some ascomycetes that lack an explosive ascus. Ascomycetes of this nature include some bitunicate members in which ascospore liberation occurs through gelatinization of the ascocarp and in some members of the Ophiostomatales (such as *Chaetomium*) in which the asci deliquesce. In the latter type, the disintegrating asci form a slime which contains the freed ascospores. The slime oozes out of the perithecial neck, much as paste would ooze out of a squeezed tube. Liberation of the spores is accomplished by splashing rain. In some instances, the entire sporocarp may be dislodged by the falling raindrop and splashed away. Perithecia of *Chaetomium* were found to be dispersed to a distance of 33 centimeters in this manner (Dixon, 1961).

Raindrops may also dislodge spores through bombardment of the sporocarp. Raindrops landing on a Gasteromycete peridium may depress it like a bellows, forcing air and spores out through the ostiole in a puff which rises to a height of 2 to 3 centimeters (Gregory, 1949).

Structures especially adapted to utilize the force of landing raindrops are the "splash cups" found in the Nidulariales, the bird's nest fungi (see Chapter 5) (Brodie, 1951, 1956). We consider *Cyathus striatus* as an example of this mechanism. Large raindrops are about 4 millimeters in diameter, only slightly smaller than the 6 to 8 millimeter diameter of the mouth of the funnel-shaped basidiocarp of *C. striatus*. The large drops have a terminal velocity of 6 meters/second, and the displacement of water in the funnel creates a strong thrust along the sides, inclined 60° to 70° horizontally (through the use of models, this was demonstrated to be the most effective inclination). The strong upward thrust of the water dislodges the peridioles and ejects them to a distance of up to 1 meter. While in flight, the cablelike funiculus trails along behind the peridiole until its adhesive ending, the hapteron, attaches itself to an object. The peridiole is suddenly checked in its flight and is jerked backward, and the funiculus becomes wound around the retaining object (Fig. 10-1). This object is frequently some form of vegetation that may be eaten by an animal, enabling the fungus to complete its life cycle.

Active Mechanisms of Spore Release

All active mechanisms of spore release share the characteristic of flinging the spore into the air with considerable force. The distance to which the spore will travel is determined by its initial velocity, shape, size, and density. A simplified formula for the distance (d) that a spore will travel may be expressed as $d = Kr^2$, where K is a constant

Fig. 10-1. Splashing of a peridiole from a basiodiocarp of *Cyathus striatus:* (a) and (b) Raindrop lands in a cup; (c) The periodiole is splashed out with the hapteron extended; (d) Hapteron sticks to a plant as the priodiole is carried forward by its own momentum and the funiculus is extended by a pull; (e) Peridiole is jerked backwards when the funiculus is extended to its full length; (f) Jerk causes the periodiole to swing around its point of attachment and the cord is wrapped around the plant stem as another raindrop lands in the splash cup. [Reproduced by permission of the National Research Council of Canada from H. J. Brodie, *Can. J. Botany,* 1951, **29:** 224–234.]

and r is the radius of the spore. As the radius of the spore increases, so does the distance to which it will travel as a result of the forceful discharge. Therefore we find that the smaller spores can be thrown only short distances, while the larger ones can be thrown for comparatively large distances (Ingold, 1960, 1966, 1971).

EXPLOSIVE MECHANISMS

A common means of forcible spore discharge is that in which a structure containing a large vacuole becomes turgid through an increase in osmotic concentration and suddenly bursts, carrying the spores away in a jet of water. This mechanism occurs in *Pilobolus* (Mucorales), in *Basidiobolus ranarum* and *Entomophthora muscae* (Entomophthorales), and in the majority of the ascomycetes.

Pilobolus. Species of *Pilobolus* forcibly eject an entire sporangium, thus earning their name which literally means "hat-thrower" (Greek). These fungi occur in abundance on horse dung. If fresh horse dung is placed in a moist chamber which is exposed to the light, numerous sporangia of *Pilobolus* will usually appear within a few days. Successive crops of sporangia appear each morning and complete their discharge by early afternoon. *Pilobolus* bears a single sporangium on each sporangiophore. The sporangium has a dense, black wall, is lenticular in shape, and contains from 15,000 to 30,000 spores. The sporangiophore is a single cell consisting of three parts: (1) a basal

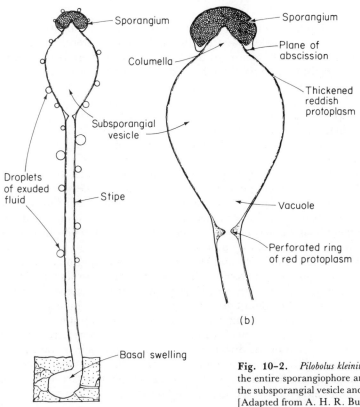

(a)

(b)

Fig. 10-2. *Pilobolus kleinii:* (a) A longitudinal section through the entire sporangiophore and sporangium. × 3.3; (b) Detail of the subsporangial vesicle and sporangium. Approximately × 10. [Adapted from A. H. R. Buller, 1909, *Researches on fungi*, Vol. 1, Longmans, Green & Co. Ltd., London.]

swelling which attaches it firmly to the substratum, (2) an elongate stipe, and (3) a subsporangial vesicle (Fig. 10-2). The subsporangial vesicle is pyriform and considerably larger than the sporangium, which it supports on a nipple-shaped columella. The subsporangial vesicle contains a single vacuole which completely fills it except for a thin layer of protoplasm adjacent to the wall. This protoplasm is specialized in two sites where it is thickened and reddish, owing to the presence of carotene. One such site occurs near the top of the subsporangial vesicle, and the second occurs at its base where the reddish protoplasm forms a perforated ring delimiting the vesicle from the stipe.

The sporangiophore is phototropic, and the stipe will bend toward a source of light until the sporangium is aimed directly at the light. This phototropic response is controlled by the subsporangial vesicle. The black sporangium blocks any light rays which strike it, but the protruding subsporangial vesicle is fully exposed to the light that strikes the thickened, reddish protoplasm near the top of the vesicle. This mass of protoplasm functions in the same manner as a biconcave lens does and refracts the light rays so that they converge in a spot on the opposite side of the vesicle. If the sporangium is already pointing directly toward the source of light (that is, light rays are parallel to the long axis of the subsporangial swelling), the refracted light will strike the light-sensitive, reddish protoplasmic ring at the base of the subsporangial vesicle in a symmetrical manner. If the light strikes in a symmetrical fashion, no change will occur. If the sporangium is not pointing toward the source of light, however, the refracted light will strike the subsporangial wall (and not the ring) and will be asymmetric for the vesicle as a whole (Fig. 10-3). A stimulus is then exerted on the protoplasm which is conveyed to the motor region of the stipe (beneath the light-sensitive ring), causing the stipe wall on the side away from the light to grow faster than the wall toward the light. Growth of the wall continues until the sporangiophore is parallel to the light rays, which is registered when the refracted light strikes the light-sensitive ring.

Meanwhile, certain changes are occurring as a prelude to sporangium discharge. The lower wall of the sporangium is colorless and breaks away from the subsporangial vesicle. A ring of mucilage surrounding the base of the sporangium is thus exposed. The subsporangial vesicle is extremely turgid (about 5.5 atmospheres of pressure) and finally a weak abscission zone (surrounding the columella) ruptures. The pressure within the subsporangial vesicle is suddenly relieved, and the vacuolar fluid surges upward as a forceful squirt, carrying the sporangium with it. The speed of the discharge is about 14 meters/second, and the distance of the horizontal throw is as great as 2 meters (Buller, 1934). The ring of mucilage is wetted by the fluid and causes the sporangium to adhere to any object on which it lands (Fig. 10-4). This object may be a blade of grass which will be eaten by a horse.

The phototropic response and forcible discharge of the sporangium are adaptations to the environment in which species of *Pilobolus* occur. They sporulate on horse dung, which is deposited in irregular and overlapping piles. It is necessary that the sporangia reach open spaces and blades of grass to again be consumed by a horse: the sporangia would be wasted if they were discharged onto overhanging dung. *Pilobolus* species are well adapted to these constraints because their sporangiophores are directed towards the open spaces (detected by the direction of the light), and their sporangia are forcibly ejected in that direction. The *Pilobolus* gun is deadly accurate, as can be determined by placing dung bearing sporangia within a dark box with a target of white paper with a small illuminated hole at the top of the box.

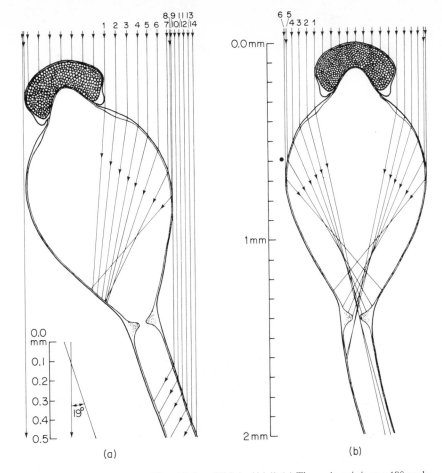

Fig. 10-3. *Pilobolus kleinii:* (a) The main axis is at a 19° angle to the source of strongest illumination. This is a physiologically unstable position; (b) The main axis of the subsporangial vesicle and sporangium is now redirected toward the source of strongest illumination, a physiologically stable position. [From A. H. R. Buller, 1934, *Researches on fungi,* Vol. 6, Longmans, Green & Co. Ltd., London.]

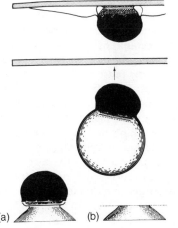

Fig. 10-4. *Pilobolus kleinii:* (a) A sporangium shortly after dehiscence of the lower sporangial wall. The base of the sporangium is surrounded by a gelatinous band through which the sporangiospores can be distinguished; (b) Flight and landing of the sporangium after discharge. Part of the subsporangial wall remains attached to the sporangium, and it is accompanied by a large drop of fluid. This fluid enables the sporangium to adhere to a surface upon landing (top). [Adapted from A. H. R. Buller, 1909, *Researches on fungi,* Vol. 1, Longmans, Green & Co. Ltd., London.]

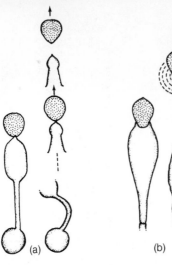

Fig. 10-5. Sporangium discharge by *Basidiobolus ranarum* (a) and *Entomophothora muscae* (b). [From C. T. Ingold, 1934, *New Phytologist* **33**:274–277.]

Basidiobolus Ranarum and Entomophthora Muscae. *Basidiobolus ranarum,* which occurs on frog excrement, and *Entomophthora muscae,* a parasite of flies, rely upon the sudden rupture of the sporangiophore to eject the single sporangium (Fig. 10-5).

Basidiobolus ranarum forms minute sporangiophores from either the mycelium or a discharged sporangium. The sporangiophore is enlarged at the apex into a subsporangial bulb that supports the single sporangium on a columella. There is a weak circumsessile zone near the base of the subsporangial bulb, which suddenly ruptures when the turgor increases. The wall of the subsporangial bulb above the line of dehiscence is elastic, while that below the line is rigid. When dehiscence occurs, the elastic upper portion contracts and squirts the vacuolar fluid backward. The upper portion of the subsporangial bulb is shot off together with the sporangium, which is shot to a distance of 1 to 2 centimeters (Ingold, 1934).

The sporangium of *Entomophthora muscae* is also borne on an inflated sporangiophore, but, the sporangiophore, unlike that of *Basidiobolus ranarum,* ruptures immediately beneath the sporangium. The resulting jet of sap then propels the sporangium to a distance of 1 to 1.5 centimeters (Ingold, 1953).

Ascomycotina. An almost universal characteristic of the ascus is its explosive discharge of the ascospores. The ascus is lined with a thin layer of cytoplasm which surrounds a large vacuole. The ascospores are suspended in the vacuole, hanging together by their own adhesiveness and apically supported by a strand of mucilage. A great deal of pressure is then built up in the ascus. This buildup of pressure is usually attributed to a large quantity of glycogen (detectable by chestnut-brown staining in iodine) in the vacuole and its subsequent conversion to soluble sugars. The soluble sugars, in combination with any salts which may be present, increase the osmotic concentration of the ascus contents. It has been estimated that the concentration of the fluid within the ascus of *Sordaria fimicola* is about 10 to 30 atmospheres (Ingold, 1971). In some species, pressure in the asci may be generated primarily by the direct absorption of rain water (Zoberi, 1973), or by the swelling of a mucilage (Ingold, 1971). Regardless of the mechanism for pressure buildup, the resulting pressure causes the ascus tip to open at a predetermined pore or slit and forces the ascospores out violently. The asci within an ascocarp burst singly, in succession, or simultaneously. The distance to which the ascospores may be discharged may be less than a millimeter or up to 40 centimeters but is usually 0.5 to 2.0 centimeters (Ingold, 1971).

The morphology of the ascocarp bearing explosive asci may place special requirements on ascus behavior before ascospore discharge, and we may distinguish three principal types of situations: those encountered in a cleistothecium with a non-disintegrating wall, in an ostiolate ascocarp, and in an apothecium. These types are considered in greater detail.

Cleistothecia with Nondisintegrating Wall. Cleistothecia with a non-disintegrating wall are found in the Erysiphales. In these fungi, the ascus enlarges and protrudes through the confining ascocarp wall. The wall which is forced open in this way is under some strain and suddenly snaps shut, ejecting the entire ascus to a distance of several centimeters. The ascus then later forcibly ejects its ascospores (Woodward, 1927).

Ostiolate Ascocarps. Enclosed ascocarps opening by only a small ostiole occur in the locular and perithecial fungi. They can release their ascospores only if the asci can protrude through the ostiole. Some species have a neck that is positively phototropic, bending towards the light (Ingold, 1971). The ostiole is so narrow that only one ascus can discharge its ascospores at a time.

In most ostiolate ascomycetes, the asci remain attached at their bases and reach the ostiole by elongation. If the ascus is unitunicate, the entire ascus elongates, whereas if the ascus is bitunicate, only the inner ascus wall elongates. The asci apparently elongate by their own growth and by pressure of the surrounding asci. Each ascus elongates, reaches the ostiole, and discharges its ascospores. The empty ascus shrinks back into the ascocarp and disintegrates (Ingold, 1971).

Alternatively, some asci become detached at their bases, float upward to the ostiole in a mucilage, individually lodge between the lips of the ostiole, and then forcibly discharge their ascospores. Those Pyrenomycetes having detached asci are found in some members of the Sphaeriales with very long perithecial necks. In these fungi, an elongating ascus could not reach the exterior (Ingold, 1933).

Physiological aspects of ascospore discharge in ostiolate ascocarps with elongating asci have been studied, and some interesting periodic responses to light have been demonstrated. Ascospore discharge by *Sordaria fimicola* occurs primarily during the light if the fungus is cultured in alternating cycles of light and dark (Ingold and Dring, 1957). The perithecial necks of *S. fimicola* are positively phototropic, and both the ascospore discharge and phototropism respond to wavelengths of light in the blue range (less than 520 nanometers). In contrast, ascospore discharge by *Hypoxylon fuscum* and *Daldinia concentrica* is inhibited by light and therefore is greater in darkness (Ingold and Marshall, 1963; Ingold, 1956, respectively). A pronounced endogenous rhythm exists in *Daldinia concentrica,* which will continue to release its ascospores during the night even if perithecia are maintained in total darkness (Ingold and Cox, 1955).

Returning to *Sordaria fimicola,* it has been demonstrated that light and temperature may interact in controlling the rate of ascospore discharge. In some experiments, cultures with perithecia were subjected to three alternating 24-hour cycles of low and high temperatures (8°C and 20°C, 20°C and 25°C, or 15°C and 25°C) in the dark. In all three experiments, the high temperatures stimulated ascospore discharge during the initial 8 to 10 hours of that temperature period, but the rate of ascospore discharge subsequently declined, becoming equivalent, or nearly so, to that occurring at the lower temperature. When the temperature again rose, ascospore discharge again peaked (Fig. 10-6). This temperature-dependent cycle could be repeated several times in the dark. However, if alternating periods of light and dark were provided, the fungus would respond to the light cycles and not to the temperature in those cycles utilizing 15°C and 25°C or 20°C and 25°C. Light was in-

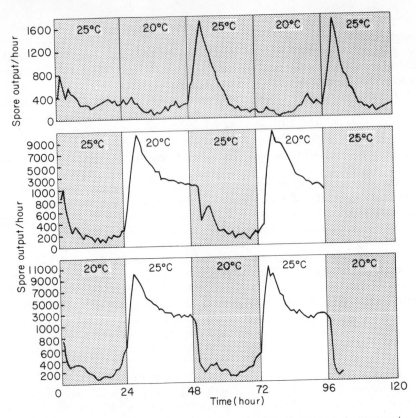

Fig. 10-6. *Sordaria fimicola.* Upper curve: rate of spore discharge from culture grown in the dark at 20°C and then subjected to alternate days at 25°C and 20°C in dark. Lower two curves: two parallel experiments on cultures reared in light at 20°C and then treated as shown with alternating days at 20°C and 25°C and alternating days of darkness and light (1,000 lux, "daylight" fluorescent): in one relatively high temperature and darkness coincide, in the other relatively high temperature and light. In the lower two curves note change of scale above 1,000 spores per hour. [Figure and legend from C. T. Ingold, 1971, *Fungal spores—their liberation and dispersal.* Clarendon Press, Oxford, 302 pp.]

capable of stimulating discharge in cultures maintained at 8°C. Ingold (1971) suggests that light is the limiting factor at high temperatures, while temperature itself is limiting under cold conditions. In nature, both higher temperatures and maximum light occur during the day, presumably contributing to maximum ascospore discharge by *S. fimicola.*

    Apothecia. Ascospore discharge from apothecia is more easily accomplished than from perithecia because all asci are exposed to the exterior. If the apothecium has a flat or convex hymenium, discharge of the ascospores is completely unhindered. There are problems encountered, however, in a deeply cupulate apothecium. In such an apothecium, the asci line the inside of the cup and asci on the sides would aim their ascospores toward the opposite hymenium if the line of discharge were at right angles to the hymenium, as we would normally expect. An adaptation avoiding this potential problem exists in the form of positive phototropism, bending of the individual asci toward the strongest source of light. Phototropic asci in a deeply cupulate apothecium curve toward the light, resulting in a different degree of curvature in the different parts of the apothecium (Fig. 10-7). This orients the asci in such a way that each is pointing toward open space, with the result that the ascospores will be discharged into open air.

As the entire hymenium is exposed in the Discomycetes, there is no restriction on the number of asci that can simultaneously discharge their ascospores (unlike the former types). Simultaneous discharge of ascospores is a common occurrence in the Discomycetes and is especially likely to occur when there is a change in humidity or temperature or if the apothecium is touched. In *Ascobolus,* this discharge may be induced by transferring the apothecium from darkness to high-intensity light in the blue range (400 to 460 nanometers); however, if red light (greater than 600 nanometers) is provided along with the blue light, no discharge occurs (Ingold and Oso, 1968). The simultaneous discharge may be detected as a sudden puffing forth of a cloud of spores (Fig. 10-8) and produces an audible hiss if the puffing apothecium is placed closed to one's ear. The sudden puffing of the spores creates a minute blast of air above the apothecium that increases the normal turbulence and aids in the dispersal of the ascospores. In *Ascobolus viridulus,* the increased distance to which spores are discharged during puffing is five times that achieved without puffing (Ingold and Oso, 1968).

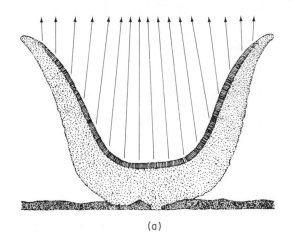

(a)

Fig. 10-7. Phototropism of asci in the Discomycete *Peziza vesiculosa:* (a) Direction of spore discharge from the apothecium. × 1.7; (b) Phototropic curvature of asci from three sections of the apothecium. ×71. [From A. H. R. Buller, 1934, *Researches on fungi,* Vol. 6, Longmans, Green & Co. Ltd., London.]

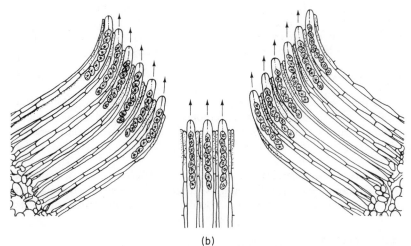

(b)

Fig. 10-8. Ascospore discharge by apothecia of *Ciboria aestivalis.* Ejection of the ascospore cloud was stimulated by a sudden change in atmospheric conditions. [Courtesy Plant Pathology Department, Cornell University.]

Ascospore discharge in the Discomycete *Cookeina sulcipes* is favored by rain. The apothecia are oriented towards the light owing to positive phototropism of their asci. The cuplike apothecium catches the falling raindrops and retains the water until the cup is full. The asci absorb the water, becoming turgid and developing hydrostatic pressure. As the water in the cup evaporates, the ascus tips above the water's surface become desiccated, producing a negative vapor pressure which causes the ascus tips to be ejected. The ascus tips dry in succession as the water level goes down, and finally there is a big puff when all the water has evaporated (Zoberi, 1973).

ROUNDING-OFF MECHANISM

The normal shape of a turgid cell would be spherical if it were completely unconfined. An analogy may be found in soap bubbles, which are spherical when independently floating through air but assume polyhedral shapes when compressed by neighboring bubbles. Sudden rounding out of a previously compressed cell to discharge spores occurs in some fungi.

A fungus of this type is *Entomophthora coronata* (= *Conidiobolus villosus*), a member of the Entomophthorales and a parasite of aphids and termites. A single sporangium is borne on each sporangiophore. Both have a double wall, with the outer wall of the sporangium continuous with that of the sporangiophore. In addition, the sporangium is separated from the sporangiophore by a double wall: one layer belongs to the sporangium and the second layer belongs to the sporangiophore. This cross wall forms a columella and bulges into the sporangium. Prior to spore discharge, the outer wall joining the sporangiophore with the sporangium ruptures, and there is a sudden eversion or rounding out of the depressed portion of the sporangium (Fig. 10-9). The sporangium is suddenly shot away when the tension on it is released and travels a distance of 0.5 to 4.0 centimeters. Spore discharge is favored by high relative humidity, and transfer of the fungus from high to low relative humidity will greatly reduce spore discharge (Ingold, 1971).

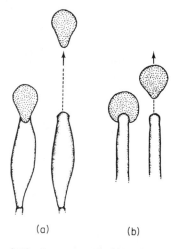

Fig. 10-9. Sporangium discharge by the rounding out mechanism in *Entomophthora grylii* (a) and *Entomophthora coronata* (b). [From C. T. Ingold, 1934, *New Phytologist* **33**:274–277.]

(a)   (b)

A similar type of discharge mechanism occurs in the downy mildews when both the sporangiophore tips and the sporangium suddenly round out and is also observed in aeciospore release in rusts. Rust aeciospores are tightly packed in the aecium and tend to assume a polyhedral form from the pressure of adjacent spores. When damp conditions prevail, the cells become more turgid and round out, and the terminal cells are shot off.

THE BUBBLE MECHANISM OF THE BASIDIOMYCOTINA

Violent discharge of basidiospores occurs in the majority of the basidiomycetes, probably occurring in all the Hymenomycetes, but being absent in the Gasteromycetes. A marked periodicity in spore discharge in response to day and night cycles has been noted under natural conditions in many Hymenomycetes (Haard and Kramer, 1970).

Basidia that forcibly eject their spores bear sterigmata on which the basidiospores are poised at approximately a 45° angle and have a minute kneelike projection, the hilar appendix (Fig. 10-10). A bubble begins to accumulate at the hilar appendix about 5 to 10 seconds before spore discharge occurs. This bubble consists of gas that accumulates between the basidiospore wall layers, causing a blister to form (Ingold and Dann, 1968; Van Niel et al., 1972). The gas accumulates after the basidiospore is separated from the sterigma by a septum, and the bubble forms because the outer wall weakens prior to its separation from the sterigma. The surrounding wall layer suddenly bursts, and the basidiospore is shot away violently, propelled by the escaping jet of gas that results from the sudden release of pressure. This propulsion has been calculated to require a pressure of about 5 atmospheres in a bubble of 2 microns (Van Niel et al., 1972). The basidiospore is accompanied by a water drop. Although not proven, it has been suggested that some concentrated aqueous solutions occur in the hilar region and, immediately upon rupture and exposure to the air, these solutions absorb water to form the water droplet on the surface of the spore which accompanies the spore in its flight (Van Niel et al., 1972).

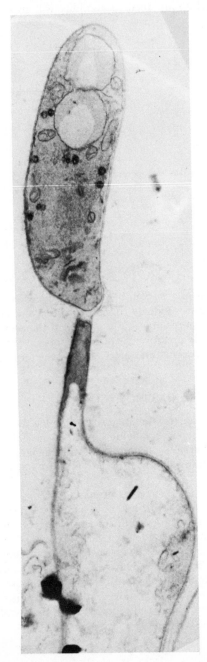

(a)

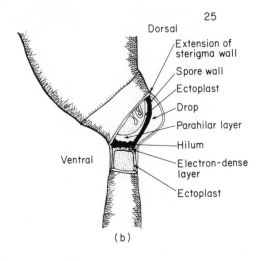

(b)

Fig. 10-10. The basidiospore and hilar appendix of *Schizophyllum commune*. (a) An electronmicrograph of a basidiospore attached to the sterigma on a basidium. Approximately × 12,000; (b) A diagrammatic interpretation of the hilar appendix. [Courtesy K. Wells, 1965, *Mycologia* **57**:236–261.]

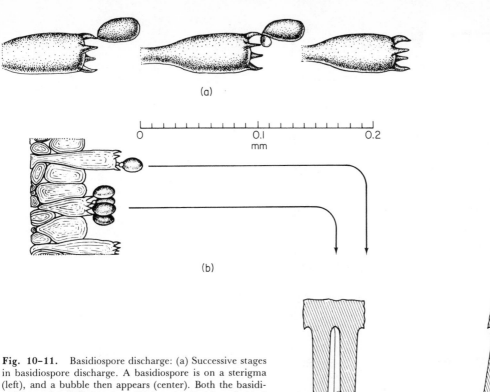

(a)

(b)

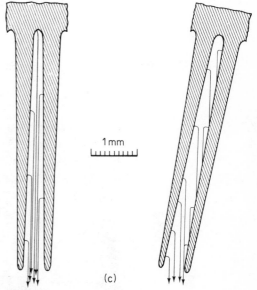

Fig. 10-11. Basidiospore discharge: (a) Successive stages in basidiospore discharge. A basidiospore is on a sterigma (left), and a bubble then appears (center). Both the basidiospore and bubble are lost, leaving the basidium behind (right); (b) A section of a hymenium showing the relative distance of the horizontal distance and path (sporobola) followed by a discharged basidiospore; (c) Gills in their normal, vertical position which allows basidiospores to escape (left), and gills tilted in a position which would cause some of the basidiospores to be trapped. [Adapted from A. H. R. Buller, (a) and (b) 1924, (c) 1909, *Researches on fungi,* (a) and (b) vol. 3, (c) vol. 1, Longmans, Green & Co. Ltd., London.]

     The basidiospore is propelled to a distance of less than 1 millimeter (Ingold, 1953), a very short distance when compared with the distance to which an ascus can eject its ascospores. This distance is sufficient to eject the spore to approximately the midway point of the free passage between the hymenia lining either the teeth, lamellae, or tubes; then the spore changes its course and begins to fall in response to gravity (Fig. 10-11). This course of events causes the spore to travel in a precise curve (the *sporobola*) (Buller, 1909).

     For the mechanism to be effective in dispersal of spores from a pileus, the hymenophore must be vertically oriented. Then the basidiospore, once shot horizontally clear of the hymenium, is free to fall through the narrow free space between the

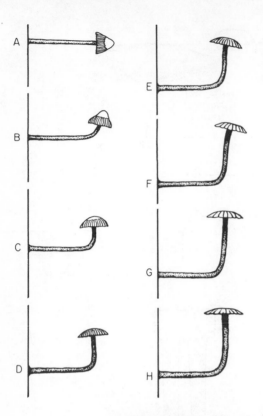

Fig. 10–12. Geotropic swinging and adjustment of the pileus in space. A basidiocarp was placed in the position shown in (a), and after $2\frac{1}{2}$ hours it assumed the position shown in (b). The remaining drawings were made at intervals of 1 hour, until finally the basidiocarp assumed the final position in (h) in which the gills are perpendicular to the ground. [Adapted from A. H. R. Buller, 1909, *Researches on fungi,* vol. 1, Longmans, Green & Co. Ltd., London.]

Fig. 10–13. A distorted basidiocarp of the bracket fungus *Ganoderma applanatum.* The log on which this fungus was growing was moved several times, and the newly forming tissue always reoriented itself in such a manner that the tubes would be perpendicular to the ground. About $\frac{7}{18}$ natural size. [Courtesy Plant Pathology Department, Cornell University.]

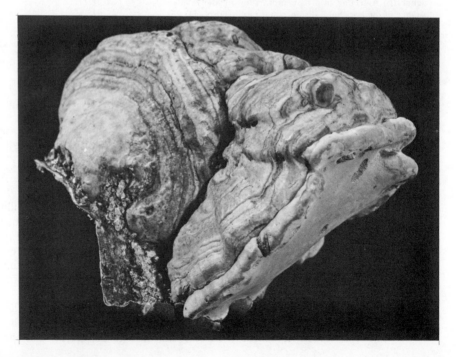

hymenia until it reaches open air, where it will be caught up in the turbulent air currents and disseminated. If the hymenophore were tilted, the falling spore would be captured on a neighboring hymenium that obstructs its path. The vertical orientation of the hymenium is maintained by positive *geotropism* of the hymenophore and negative geotropism of the stipe. The stipe may grow in various directions although the lamella, tubes, or teeth will remain on the required vertical axis. For example, an agaric growing on an inclined surface produces a curved stipe that holds the gills vertically (Fig. 10-12), and a bracket fungus that is growing on a log that is accidentally overturned reorients its growth (no matter how grotesque the end result) (Fig. 10-13) and forms its tubes vertically. In addition to exhibiting positive geotropism, the hymenia must be separated from the ground by open air into which the spores can fall. This is accomplished either by forming a stipe which supports the pileus above the ground or by growing on a tree or similar object which serves the same function. Further, the pileus protects the hymenium from wetting, which would make it impossible for the spores to be discharged (Ingold, 1971).

EVERSION MECHANISM

Sphaerobolus is a Gasteromycete that may occur in large groups on rotting wood or on dung of herbivorous animals: it is usually 1 to 2 millimeters in diameter. During the first half of its development, exposure to blue light is required, while during maturation exposure to red light is required. The fungus is not responsive to light during its final day of development. Meanwhile the basidiocarp is becoming oriented towards the light; this positive phototropism is controlled by blue light (Alasodura, 1963; Ingold and Nawaz, 1967; Nawaz, 1967).

At maturity, the gleba is exposed when the apex of the peridium splits in a stellate fashion and folds backward. The glebal mass is suddenly ejected with a catapultlike mechanism (Fig. 10-14) and is the largest fungal projectile known. Opening of the peridium and ejection of the gleba occur in the early daylight hours.

To more fully explore this mechanism, we discuss *Sphaerobolus stellatus*. The gleba of *S. stellatus* is surrounded by six distinct peridial layers, the inner three of which are of particular interest in this discussion (Fig. 10-14). The innermost peridial layer (6) which surrounds the gleba is a thin layer of pseudoparenchyma; the adjacent layer (5) is a palisade layer made up of prismatic cells; no. 5 is surrounded by a layer (4) of fibrous cells. Just before the peridium opens, the thin pseudoparenchymatous layer (6) adjacent to the gleba deliquesces, separating the gleba from the outer peridial layers and forming a watery pool which serves as a lubricant. The outermost portion of the fibrous layer (4) also deliquesces, producing a space between the remainder of layer 4 and the outer layers of the peridium (3, 2, and 1) but remaining solidly affixed to these outer peridial layers at the teeth of the stellate fissures. The prismatic cells of the fifth layer then absorb water made available from the deliquescence that has occurred and become increasingly turgid. At this time, there is a decrease in glycogen content of the prismatic cells, and an increase in soluble sugars, greatly increasing the osmotic concentration of these cells. As the cells of layer 5 increase in turgidity, the rim of the opened peridium bends outward as a result of the continual expansion of this turgid layer. Only the upper portion of the layer 5 is free to expand, as the fibrous layer (4) is attached to the bottom and is comparatively inelastic. A tension is exerted on these layers, and a point is reached when the layers suddenly yield to the strain and turn inside out, remaining attached at the teeth to the rigid outer three peridial layers, and sling away the gleba with great force. The actual discharge of the gleba is too rapid to perceive, but the initial velocity may be calculated to be as great as 9.3 meters/second.

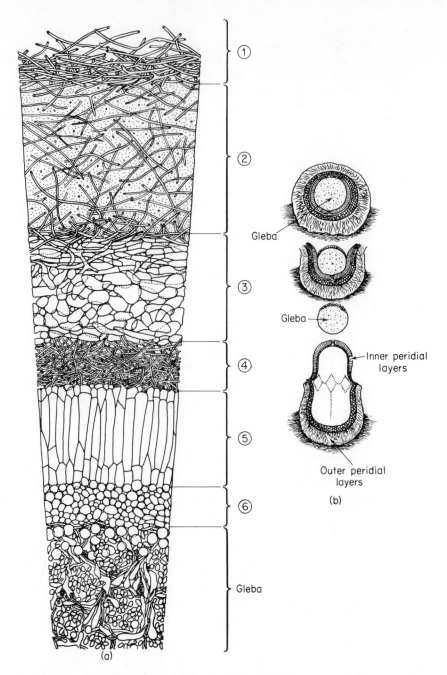

Fig. 10–14. *Sphaerobolus stellatus:* (a) Detail of the peridial layers. × 63; (b) Successive stages in discharge of the gleba top to bottom. See text for additional explanation. [(a) From A. H. R. Buller, 1933, *Researches on Fungi,* Vol. 1, Longmans, Green & Co., Ltd., London. (b) From A. Engler and K. Prantl, 1900, *Die natürlichen Pflanzenfamilien.* Engelmann, Leipzig.]

The horizontal distance to which a gleba may be ejected has been observed to be as great as 6.1 meters, the greatest distance known for any fungus projectile (Buller, 1933).

TENSILE WATER MECHANISM

A mechanism of apparently common occurrence in the Hyphomycetes operates when the atmospheric humidity is decreasing. Evaporation of cellular moisture creates a negative pressure within the cell (conidium, conidiophore, or both). The cohesion of water molecules to each other and the adhesion of the water molecules to the cell wall pull the cell walls inward, placing them under considerable tension. Suddenly a rupture of the tensile water occurs, possibly because either the adhesion or cohesion can no longer be maintained. This rupture is marked by the sudden appearance of a gas bubble and by the sudden relief of the tension, allowing the walls to spring back to their original position. The conidium is catapulted away (Fig. 10–15).

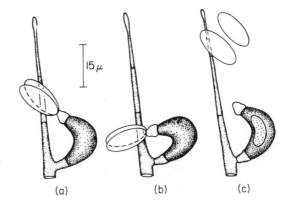

Fig. 10–15. Conidium discharge by *Zygosporium oscheoides* by rupture of tensile water: (a) Conidia borne on swollen, curved vesicle, and in its normal form before its first movement; (b) Curvature of the vesicle in reponse to internal negative pressure; (c) The sudden return of vesicle to its normal position after catapulting the conidia. A gas bubble has appeared. [From D. S. Meredith, 1962, *Ann. Botany London* II, **26**:233–241.]

(a) (b) (c)

SPORE DISPERSAL

Spores are passively disseminated by air, water, and animals or by transport of their plant hosts. The fungi depend upon these mechanisms for the eventual deposit of their spores in a satisfactory environment and for their continued existence. Dispersal is important because it enables a fungus species to occupy diverse habitats.

Dispersal by Air

The majority of fungus spores are released (or simply dislodged by wind) into and disseminated by the air. A sample of air may contain as many as 200,000 fungus spores per cubic meter, although 10,000 spores per cubic meter would be a more usual figure (Gregory, 1952). Hyphal fragments are also found in the air. The number of spores in the air fluctuates with prevailing conditions and is particularly great after a rainfall.

In some species the spores that may be collected from the air may differ markedly at various times of the day (e.g., morning as opposed to the afternoon). This periodicity may be in response to rhythms in the environment: for example winds are often stronger or gustier near noon than at other times of the day, which would in turn dislodge more mature spores. Some fungi respond to light and dark cycles, and their spores may follow a periodicity in ripening. The urediniospores of many rusts are diurnal, maturing in the day. Therefore at midday when the winds are the highest, there will be an exceptionally large number of urediniospores in the air. In contrast, discharge of aeciospores tends to be nocturnal because the high relative humidity favors the buildup of turgor required for their discharge. Numbers of spores in the air at a particular time may be affected by the rhythmic response of a particular fungus to light and dark cycles (Ingold, 1971) or to relative humidity. As an example of the latter, *Drechslera turcica* discharges its spores during the morning when the high relative humidity generated by dew deposition is decreasing. Leach (1976) measured large voltage changes in the diseased portion of the maize leaf that were a result of changes in relative humidity and were correlated with spore discharge. He suggests that the conidiophores and conidia become charged with similar electrical charges during periods of high relative humidity, thereby creating an electrostatic repulsion between the conidia and conidiophores and resulting in the discharge of the spore as the relative humidity decreases.

The number and kind of spores in the air may be determined by a device such as the Hirst automatic volumetric suction trap, through which a known volume of air is drawn and directed onto a slide made sticky with glycerol or petroleum jelly. The spores then become impacted on the slide and can be counted microscopically. Simpler methods involve (1) the exposure of a similar sticky slide to the air (perhaps for 24 hours), after which the slide can be examined for spores and (2) the exposure of a petri dish with nutrient medium to the air, which allows some of the impacted spores to germinate on the medium. Disadvantages of these methods are that they measure spore deposition and do not accurately account for the total spore flora of the air. A disadvantage of the second method using nutrient medium is that not all spores will germinate on any single medium, and only those spores actually germinating are accounted for.

In perfectly still air, a spore would fall in response to gravity. The rate of sedimentation of the spore fall may be calculated using Stokes' law, which describes the rate of fall of a small sphere in a viscous fluid.* (For the purposes of this calculation, the density of spores is assumed to be approximately equal to that of water.) The rate of sedimentation of fungus spores ranges from approximately 0.003 to 2.78 centimeters/second with the usual order of magnitude about 1.0 centimeter/second. Asymmetry and roughness of the spore would presumably affect the rate of fall (Gregory, 1973).

Although the rate of fall of a spore in still air may be determined, still air is virtually never met in nature. A microscopically thin layer of perfectly still air does exist at the surface of the ground, but immediately above this still layer is the *lamellar layer* of the air. The lamellar layer is the lowest layer of moving air, and the air moves parallel to the nearest surface. The thickness of the lamellar layer is usually about 1 millimeter

---

* $V + \frac{2}{9} \cdot \dfrac{\sigma - P}{\mu} \cdot gr^2$

V = terminal velocity (velocity of sedimentation), centimeters/second
σ = density of sphere in grams/cubic centimeter (water = 1.00) approximately equal to spore
P = density of medium (air = 1.27×10^{-3} grams/cubic centimeter)
g = acceleration of gravity (981 centimeters/second^2)
μ = viscosity of medium (air at 18°C = 1.8×10^{-4} gram/centimeter^2)
r = radius of sphere in centimeters

but varies with the wind speed and with the roughness of the adjacent surface. It is thin in high wind speeds and much thicker under calm conditions. Above the lamellar layer is the *turbulent layer,* where airflow is complex, the result both of wind and local eddies. One can detect turbulence by watching smoke from a bonfire or chimney. The turbulent layer increases in thickness with increasing wind.

The actual rate of sedimentation of a spore, and the path it will travel, is determined by its velocity in response to gravitation (Stokes' law) and the direction and rate of airflow. A spore in the lamellar layer falls in response to gravity, but its path is determined by both its velocity and the rate of lamellar air movement. A spore in the turbulent layer may be carried both upward and laterally by the air currents and may remain in the turbulent layer indefinitely. The flight of a spore in the turbulent layer may last hours, days, or even years, and the distance traveled may be tens or hundreds of miles.

It is evident that if a spore is to be disseminated for great distances, it must be discharged directly into the turbulent layer and not into the lamellar layer. Those fungi which have become adapted to air transport of their spores have mechanisms which release their spores directly into the turbulent layer. This may be seen in Discomycetes which expel their spores with such force that they are shot through the lamellar layer and into the turbulent layer and in basidiomycetes which form the hymenium well above the ground so that spores drop directly into the turbulent layer (Fig. 10-16).

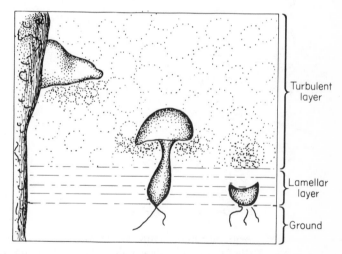

Fig. 10-16. Discharge of spores of two Hymenomycetes and a Discomycete directly into the turbulent air layer.

Turbulent layer

Lamellar layer

Ground

Air dissemination of the spore ends with deposition, or the landing of the spore, on a substratum that may or may not be favorable for its germination. Spore deposition may occur (1) by the impaction of a spore on a protruding surface (such as a blade of grass) which interrupts the flow of air, (2) by sedimentation from the air in response to gravity, (3) by washing out of the air by rain or snow, (4) by electrostatic deposition, which may occur if a spore bearing a minute charge (either positive or negative) encounters a substratum bearing an opposite charge, or (5) by boundary-layer exchange in which spores pass from the turbulent layer into the lamellar layer, whereupon they sediment out. Spore deposition is a comparatively inefficient process as it is random. Many spores must be produced, but only a few accidentally land on a suitable substratum. There is a great wastage of spores.

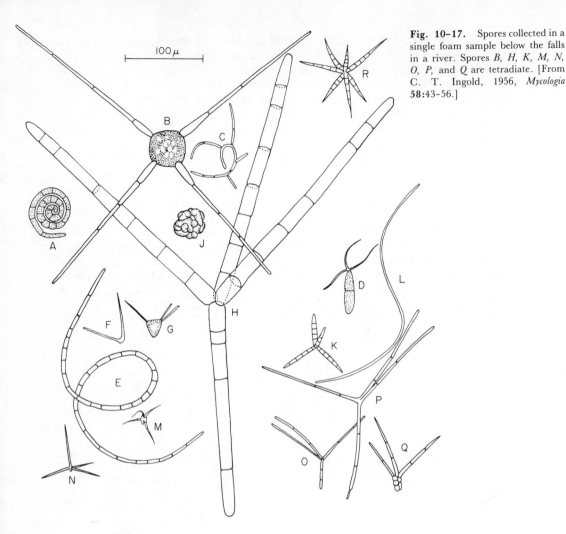

Fig. 10-17. Spores collected in a single foam sample below the falls in a river. Spores *B, H, K, M, N, O, P,* and *Q* are tetradiate. [From C. T. Ingold, 1956, *Mycologia* **58**:43–56.]

100 μ

Dispersal by Water

Aquatic fungi include those which form zoospores, such as chytrids, their allies, and the Oomycetes. Fewer than 2% of the fungi are known to be aquatic (Ingold, 1975). The zoospores of these fungi swim feebly and randomly about and are also carried by strong water currents or by any motion of the water. Passive transport by water currents is undoubtedly of greater significance in the dispersal of these fungi than their own rather limited locomotion.

A second type of aquatic fungi are those which do not produce a motile zoospore stage. Abundantly occurring fungi of this latter type include numerous species of hyphomycetous deuteromycetes growing on leaves and debris at the bottom of a swiftly moving, well-aerated stream. The majority of these fungi produce four-armed (tetraradiate) spores (Fig. 10-17), which may be collected in abundance in the foam and scum that collects behind barriers in the stream, especially in the late autumn or winter. The spores are transported in swiftly moving currents. For spores in swiftly moving water, coming to rest on a substratum can pose a problem, and it is probable that the irregular form of the tetraradiate spore enables it to anchor or entangle itself on the substratum more easily than a rounded spore. It has been demonstrated experimentally that a tetraradiate spore is caught more efficiently on a flat surface than a rounded spore is, thus tending to confirm the hypothesis. Such a spore makes contact at three points, like a tripod, which gives a stable form of attachment (much more so than one point). Further, a branched spore can be trapped in air bubbles better than nonbranched spores, thus favoring the capture of the branched spores in the foam which removes them from the swiftly moving current (Iqbal and Webster, 1973). Although most of the tetraradiate spores are produced by Hyphomycetes, a few are produced by basidiomycetes. Ingold (1975) suggests that this is an adaptive morphological form which was achieved by convergent evolution.

Spores of many terrestrial fungus species are distributed by water, even though the fungi may not be primarily adapted for water dispersal of their spores. Such a situation may be found when raindrops strike a spore-bearing structure and splatter, carrying the spores in the splash droplets (as we have seen in the slime-spored fungi). This provides an effective means of dispersal as a water drop of 5 millimeters (the size of a large raindrop) falling from a height of 7.4 centimeters may produce over 5,000 splash droplets which travel an average distance of 20 centimeters (Gregory, et al., 1959). Running water that passes from one locale to another may carry spores with it. This may occur in the trickling of rain down an object bearing fungi, such as a tree; the seepage of water through soil; and the overflowing of creeks or rivers over their normally dry banks.

Dispersal by Animals

Animals of all types, including humans, are often important agents of spore dispersal. The animal may accidentally come into contact with fungus spores by brushing against the plants or other substrata bearing fungi and thereby pick up spores on the integument, hair, claws, or clothing. Dispersal is completed when the spores are again deposited. A survey of 36 birds shot while resting in chestnut trees revealed that 19 of them were carrying spores of the chestnut blight fungus, *Endothia parasitica* (Heald and Studhalter, 1914). Also the animal may consume vegetation bearing spores of the fungus or consume a fleshy sporocarp, and the spores then pass through the alimentary canal of the animal, sometimes unharmed and capable of germination when defecated. Talbot (1952) made a survey of 72 small animals (such as wood lice, mites, grubs, centipedes, slugs, springtails, and worms) and found that a large portion of the contents of the gut and feces were of fungal origin and that only one animal did not contain fungus spores. Many of the spores were viable. In a study in which the shoes of airplane passengers were swabbed and nutrient agar streaked, at least 65 species of fungi were recovered (Baker, 1966).

Virtually any fungus spore may be transported by an animal if the animal picks up the spore on its body or ingests it. The relative importance of this type of dispersal depends on the distance traveled by the animal, the viability of the spores under the conditions imposed, and the extent to which the fungus is dependent on animal dispersal in comparison with wind and water dispersal.

Numerous fungi depend almost entirely on dispersal by animals and utilize a number of devices to ensure transport by animals. This increases efficiency of dispersal and decreases spore wastage. Examples of these include the following:

1. *Occupation of a mutual habitat.* Phloem-feeding bark beetles of the Scolytidae bore out brood galleries under the bark, where eggs are laid and the larvae develop (Fig. 10-18). Phloem-feeding fungi such as species of *Ceratocystis* (which cause a blue staining of the wood) and yeasts occupy these brood galleries. The bark beetles feed on the same plant material on which the fungi feed and also feed on the fungi; so it is inevitable that the slimy spores of the fungi adhere to the surface of the beetle. The beetle migrates to new substrata, carrying the spores with it, and prepares a new tunnel in which the spores are rubbed off or deposited in the feces where they remain viable. This process not only provides for the dispersal of the fungus but

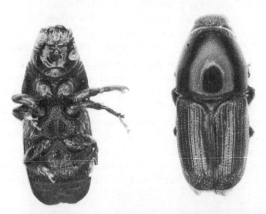

Fig. 10-18. Brood chambers beneath the bark of an elm tree formed by the beetle *Scolytus multistriatus* (top). [Courtesy Plant Pathology Department, Cornell University.]

also causes it to be inoculated directly into a suitable substratum. This greatly eliminates the hazards of a random spore flight. Many similar situations exist whereby a plant pathogenic fungus is regularly transported by insects. An example is the transportation of smut spores by insects which are involved in pollination (Chapter 5).

2. *Odor attracting animals.* A rather conspicuous example is found in the stinkhorns (Phallales) which produce a foul-smelling, slimy gleba containing the basidiospores. Flies (especially blowflies and green bottle flies) and slugs visit the basidiocarp, feed on the foul-smelling liquid, and leave, carrying basidiospores on the outside of their bodies.

A similar phenomenon may be found in the rust fungi which produce their spermatia, pycniospores, in pycnia. The pycnia exude droplets of a sweet fluid, which is attractive to flies. Flies feed on the fluid, pick up pycniospores, and, in subsequent visits to other pycnia, leave behind pycniospores which they have accumulated. This is the means by which cross-fertilization is accomplished in rusts, and the sexual nature of pycnia was discovered only after Craigie (1927) observed the regular visitation by flies and was prompted to carry out cross-inoculation studies.

Hypogeous fungi such as *Endogone* and members of the Tuberales and some Gasteromycetes produce an odor which attracts animals (especially rodents) that dig up the sporocarps, consume them, and later deposit dung containing spores. Spores of *Endogone* can be recovered regularly from the alimentary canals of rodents that have fed on the sporocarps (Dowding, 1955).

3. *Ejection of spores onto edible vegetation.* Coprophilous fungi must have dung as a substratum on which to sporulate and, therefore, both their dispersal and further livelihood depend on the deposition of their spores onto vegetation which will be tempting food for an herbivorous animal. Assuming that the spore is consumed along with the supporting vegetation, it will pass through the alimentary canal where it is activated and then be deposited in dung where it can again sporulate.

Adaptations which help ensure the ejection of spores onto vegetation include phototropism (discussed in connection with *Pilobolus* but also occurring in the ascomycetes *Dasyobolus immersus, Saccobolus* spp., and *Sordaria* spp.) and forceful discharge mechanisms that fling the projectile to great distances (as we have seen in the coprophilous fungi *Basidiobolus ranarum, Sphaerobolus* spp., bird's nest fungi, and *Pilobolus* spp.). In addition, coprophilous fungi often have a very large projectile (a large sporangium, gleba, peridiole, or all eight ascospores stuck together with mucilage) which makes the distance traveled upon ejection rather large (recall that $d = Kr^2$). Those ascomycetes which fling their spores the greatest distance are coprophilous species in which the ascospores cling together (25 centimeters in *Sordaria curvula* and 60 centimeters in *Dasyobolus immersus*) (Ingold, 1953).

SPORE DORMANCY

Some spores may be mature and able to germinate immediately upon their introduction into a suitable environment. Many spores, however, undergo a *dormancy* of varying lengths during which they remain inactive, and germination occurs only when dormancy is broken. Dormancy is of varying significance to different fungi and is controlled by different factors.

Dormancy may be *constitutive* and under the innate control of the spore. This dormancy may be an enforced rest period during which the spore completes its maturation, or it may occur in spores which are already mature.

If spores are deposited in an environment unsuitable for germination and growth of the thallus, some spores may die while others are forced into dormancy that may tide the spore over until the environmental conditions improve to the extent that the vegetative stage could survive. This type of dormancy is of an *exogenous* nature and is under control of external factors (Sussman, 1966-a, 1966-b; Sussman and Halvorson, 1966).

Constitutive Dormancy

Constitutive dormancy may be under control of self-inhibitors or of the innate metabolism of the spore.

SELF-INHIBITORS

The role of self-inhibitors in imposing dormancy is frequently apparent when high concentrations of spores fail to germinate, while the frequency of germination increases if the spore concentration is decreased. One of the earliest observations of this phenomenon was that of Edgerton (1910) who noted that when more than 12 to 15 conidia of *Colletotrichum lindemuthianum* were included in 1 cubic millimeter of medium, the percentage of germination was reduced. In some cases, this inhibition is due to the effects of specific chemical inhibitors, but in other cases, this inhibition may simply be due to an inadequate supply of oxygen or nutrients.

Although the chemical nature of the inhibitor(s) is not known for many fungal species, we are learning that there is a great deal of variability among the different inhibitors in terms of their chemical nature and mode of action. Some of the inhibitors are volatile compounds that accumulate in high concentrations as a result of the crowded conditions. Volatile inhibitors include ethanol and acetaldehyde, which may be the major inhibitory compounds (Robinson et al., 1968). Other inhibitors may be removed from the spores simply by washing them in water, which suggests that these particular inhibitors are small molecules occurring on or near the surface of the spores. Inhibitors of urediniospore germination in some rust fungi have been extensively studied (Macko et al., 1976). In the rust fungi, the inhibitor present in many species is methyl *cis*-3,4-dimethoxycinnamate, while that formed by *Puccinia graminis* is methyl *cis*-ferulate (Fig. 10–19). These inhibitors are probably located either on or next to the surface of the urediniospores and can be extracted by floating the urediniospores in water. The inhibitory action is exerted against the initiation and growth of the germ tube prior to its emergence from the spore, but no inhibition is exerted after the emergence from the spore. Methyl *cis*-ferulate prevents digestion of the wall material in the pore area where the germ tube normally emerges upon germination. The effects of this inhibitor are reversible, and upon its removal, digestion of the wall to form the pore will occur.

If a self-inhibitor is responsible for dormancy, some self-activating mechanism must be present to counteract the inhibitor and allow germination to proceed. Although the nature of the self-activating mechanism is unknown, it presumably could operate by a direct competition with the inhibitor or by interfering with either the production or release of the inhibitor (Sussman and Halvorson, 1966).

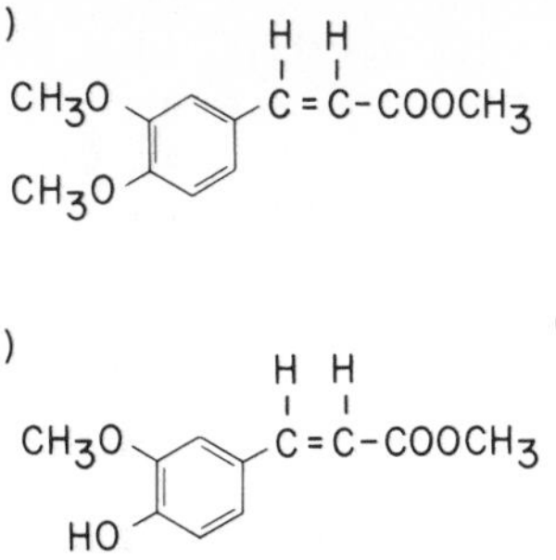

Fig. 10–19. (a) Methyl *cis*-3,4-dimethoxycinnamate; (b) Methyl *cis*-ferulate.

Self-inhibitors may confer an ecological advantage to those spores having this mechanism, as they prevent the germination of spores in the sporocarp. This is important because an ungerminated spore can be efficiently dispersed, unlike a germinated spore, and is more resistant to unfavorable conditions (Cochrane, 1958). In addition, the germination of spores is prevented when the density of spores is so high that large numbers of mycelia would compete for the limited food available and undoubtedly face starvation.

METABOLIC CONTROL

Dormancy of some spores may be terminated and germination initiated (*activated*) by certain treatments or the addition of nutrients which are *not* required to support vegetative growth once initiated. These activators serve as a "trigger," which modifies the spore's metabolism in such a way as to cause germination to take place. Most of our information about metabolic control of dormancy is inferred from those triggering devices that can activate spores (Sussman and Halvorson, 1966).

Activators may include any aspect of either the physical or chemical environment such as temperature, light, moisture, composition of the substratum or any combination of these. Examples of activating treatments include those discussed below.

1. *Exposure of spores to high heats.* A requirement for exposure of spores to high heats is especially common among saprophytic ascomycetes. For example, ascospores of species of *Neurospora* require exposure to temperatures in the range of 50° to 60°C for about 20 minutes to germinate. The requirement for exposure to heat for spore activation is shared by many coprophilous fungi. Heat activation for coprophilous fungi may be accomplished during passage through the alimentary canal and by the high heat generated by microbial activity in the deposited dung. The effects of heat on the spore may be exerted on some proteins, which are denatured, and also on the lipids in the cellular membranes, which undergo changes in some properties such as permeability. The manner in which such changes as these are translated into actual activation of spore germination is not known. One hypothesis is that denaturation of an inhibiting enzyme may remove its inhibition, allowing germination to occur (Sussman, 1976).

2. *Exposure to cold temperatures.* The majority of the fungi whose spores are stimulated to germinate by exposure to cold temperatures are the basidiomycetes and especially plant parasites such as the smuts belonging to the genus *Tilletia*. *Tilletia* spores may require exposure to temperatures as low as 5° to 10°C for an indefinite period of time to germinate. The requirement for cold temperatures provides a timing

mechanism whereby the spores of the parasites remain dormant over the winter and germinate only when their hosts renew their growth.

3. *Light.* Some spores are activated by light, and the majority of these are plant parasites. A large number of the parasites are rust and smut species; however, not all rust and smut species require exposure to light to germinate, and only spore stages may have a light requirement (for example, the urediniospores of *Cronartium ribicola* require light while the teliospores do not). As in the case of cold activation, a requirement for exposure to light for spore germination may correlate the activities of the parasite and host.

4. *Chemicals.* Under laboratory conditions, spore germination may be triggered by the addition of a diverse range of chemical agents, including inorganic ions, solvents (alcohols, ethers, chloroform, acetone, furans, pyrroles, and thiophenes), aliphatic esters, amino acids, carbohydrates, lipids, and vitamins (Fig. 10-20). Spores are often not specific for any one chemical, and various chemicals or chemical combinations may be effective.

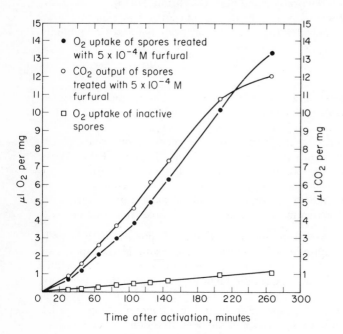

- O_2 uptake of spores treated with 5×10^{-4} M furfural
- CO_2 output of spores treated with 5×10^{-4} M furfural
- O_2 uptake of inactive spores

Time after activation, minutes

Fig. 10-20. Respiratory changes induced after activation of ascospores of *Neurospora crassa* by furfural. [From A. S. Sussman, 1953, *Am. J. Botany* **40**:401-404.]

Not unexpectedly, spores may sometimes be triggered to germinate by the addition of extracts of natural materials, often from the same environmental niche in which the organism would occur in nature. For example, spores of coprophilous fungi may be stimulated to germinate in media containing dung or a decoction of dung; spores of saprophytic fungi living on wood or leaves in nature may be stimulated to germinate in the presence of an extract of wood or leaves; spores of many plant parasites are stimulated to germinate in the presence of extracts of their plant hosts; and the spores of saprophytic soil fungi may be stimulated by the addition of an extract of soil to the medium. Passage through the digestive tract of an animal may be required to break the dormancy of the spores of some fungi; for example, the conidia of *Basidiobolus ranarum* will germinate only after consumption by a frog or lizard. The spores of some fungi may require the presence of another organism in order to germinate; for example, the basidiospores of many puffballs will germinate only in the presence of a yeast such as *Rhodotorula* (Fries, 1966).

Triggering by extracts of natural materials may be assumed to be due to the presence of required nutrients or metabolites contained in these materials. Germination in response to the presence of needed nutrients tends to assure that the fungus spores will germinate in an environment that will further support vegetative growth.

Mechanisms of Metabolic Control. Activated spores often show an abrupt increase in the rate of respiration and other physiological changes before the morphological changes of germination take place. These changes, together with activating mechanisms, indicate that a complex metabolism within the spore is responsible for "timing" spore germination. Failure of a spore to germinate may be a function of a metabolic block or of a specific deficiency within the spore that can be overcome by a particular activator. In general, neither the mechanism of the metabolic control of dormancy nor the ability of the activators to trigger germination is understood.

The mechanism for the metabolic control of dormancy has been studied in *Fusarium solani phaseoli*. Spores of this organism require exogenous sources of carbon and nitrogen and yeast extract in order to germinate. Either ethyl alcohol or acetone will replace the yeast extract in the activation process (J. C. Cochrane, et al., 1963). Ethyl alcohol depresses endogenous respiration and accelerates the rate of glucose oxidation. A suggestion for the role of ethyl alcohol is that spore germination is limited by the availability of amino acids (which will partially replace the requirement for yeast extract) and the synthesis of amino acids may depend on a supply of active acetaldehyde which is synthesized from the ethyl alcohol. Acetate will not replace ethyl alcohol as an activator, although metabolism of ethyl alcohol and acetate in yeast and fungi is usually believed to proceed by the same terminal pathways (V. W. Cochrane, et al., 1963).

Exogenous Dormancy

Spores, either with or without a constitutive dormancy, may have a dormancy imposed upon them by environmental conditions unfavorable for vegetative growth. Exogenous dormancy may be imposed by inhibitors in the soil or marine environment or by inhibitors produced by plants. Inhibition in soil is discussed in Chapter 11. Temperatures lower than those permitting germination or vegetative growth will often impose a dormancy on the spore, while the spore will usually die at temperatures higher than that permitting vegetative growth (Sussman and Halvorson, 1966). The majority of the spores require a high relative humidity (usually 80% to 100%) (Cochrane, 1958) and will become dormant if the relative humidity is too low.

Generally, dormancy will be broken and the spores activated when the restraining environmental condition is alleviated. Unlike constitutive dormancy, these factors allowing the germination of the spores are required for continued vegetative growth and are not triggering mechanisms.

Like constitutive dormancy, exogenous dormancy may play an ecologically important role in the life of the fungus. It may serve as a timing mechanism in the case of overwintering spores and, in general, allows the spores to germinate only when conditions are such that their vegetative growth will be supported.

SPORE GERMINATION

The irreversible chain of events transforming the self-contained spore (sometimes dormant) to its vegetative form constitutes germination. These changes include both morphological and physiological events. Spore germination is usually marked by a swelling of the spore and a number of cytological changes. These typically include swelling and then division of the nucleus, an increase in the quantity of endoplasmic reticulum and number of ribosomes and mitochondria, disappearance of lipid bodies, and sometimes vacuole formation. If the spore is to give rise to a mycelium (and not to zoospores), one or more germ tubes are formed. A germ tube may appear from any point on the surface of the spore or from a predetermined pore, marked initially by a thin wall. The formation of the germ tube is often first visible when the outer wall of the spore cracks and the germ tube protrudes through it as a papilla. The wall of the germ tube usually appears to be continuous with the inner wall of the spore (if an inner wall is present), which is often extensible and increases in size through the synthesis of additional wall materials. Migration of a nucleus and other cellular organelles into the germ tube takes place, and subsequently the germ tube grows by apical extension to give rise to a mycelium.

Many physiological changes take place upon germination, all more or less involved with a general increase in metabolic rate which converts the dormant spore with a low metabolic rate to an actively growing vegetative state. These changes may include the synthesis of macromolecules (such as DNA, RNA, protein, carbohydrates, and lipids); an excretion of metabolites, enzymes, or vitamins; an increase in dry weight and volume; absorption of water; and activation of enzyme systems. The most notable change, however, is that which occurs in respiration. The dormant spore characteristically respires at a low rate, apparently utilizing the Embden-Meyerhof (EM) pathway in most cases. When germination begins, there is a great increase in the respiratory rate marked by an increased oxygen intake and carbon dioxide output and a shift from anaerobic respiration to the aerobic respiration characteristic of fungal mycelia (Figs. 10-20 and 10-21). For *Neurospora tetrasperma,* the respiratory rate

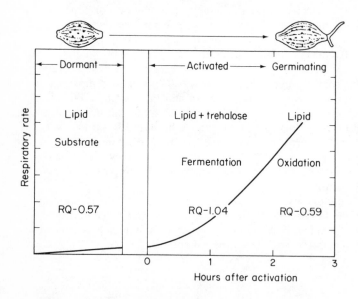

Fig. 10-21. Metabolic events occurring during germination of *Neurospora* ascospores. Lipids are utilized as an endogenous energy source supporting fermentative respiration during dormancy. Activation is accompanied by the ability of trehalase to hydrolyze the sugar, trehalose, which together with lipids serves as a substrate for respiration. Germination is accompanied by the oxidative respiration of lipids. [From A. S. Sussman, 1961, *Spores II,* Burgess Publishing Co., Minneapolis.]

may increase 20 to 30 times upon germination (Sussman, 1953). Some spores require an exogenous source of carbon or nitrogen during germination, while others do not require exogenous sources of nutrients but utilize their reserve materials (such as lipids, carbohydrates, or protein) as a substrate for respiration.

References

Alasodura, S. O. 1963. Fruiting in *Sphaerobolus* with special reference to light. Ann. Botany **27**:125–145.

Baker, G. E. 1966. Inadvertent distribution of fungi. Can. Microbiol. **12**:109–112.

Broadbent, L. 1960. Dispersal of inoculum by insects and other animals, including man. *In:* Horsfall, J. G., and A. E. Dimond, Eds., Plant pathology—an advanced treatise. Academic Press, New York. **3**:97–135.

Brodie, H. J. 1951. The splash-cup dispersal mechanism in plants. Can. J. Botany **29**:224–234.

——. 1956. The structure and function of the funiculus of the Nidulariaceae. Svensk. Botan. Tidsskr. **50**:142–162.

——. and P. H. Gregory. 1953. The action of wind in the dispersal of spores from cup-shaped plant structures. Can. J. Botany **31**:402–410.

Buller, A. H. R. 1909. Researches on fungi. Vol. 1. Longman, Inc., New York, London. 287 pp.

——. 1922. Upon the ocellus function of the subsporangial swelling of *Pilobolus*. Trans. Brit. Mycol. Soc. **7**:61–64.

——. 1924. Researches on fungi. Vol. 3. Longman, Inc., New York, London. 611 pp.

——. 1933. Researches on fungi. Vol. 5. Longman, Inc., New York, London. 416 pp.

——. 1934. Researches on fungi. Vol. 6. Longman, Inc., New York, London. 513 pp.

Cochrane, J. C., V. W. Cochrane, F. G. Simon, and J. Spaeth. 1963. Spore germination and carbon metabolism in *Fusarium solani*. I. Requirements for spore germination. Phytopathology **53**:1155–1160.

Cochrane, V. W. 1958. Physiology of fungi. John Wiley & Sons, Inc., New York. 524 pp.

——. 1966. Respiration and spore germination. *In:* Madelin, M. F., Ed., The fungus spore. Butterworth & Co., Ltd., London. pp. 201–215.

——. J. C. Cochrane, J. M. Vogel, and R. S. Coles, Jr. 1963. Spore germination and carbon metabolism in *Fusarium solani*. IV. Metabolism of ethanol and acetate. J. Bacteriol. **86**:312–319.

Craigie, J. H. 1927. Discovery of the function of the pycnia of the rust fungi. Nature **120**:765–767.

Dixon, P. A. 1961. Spore dispersal in *Chaetomium globosum* (Kunze). Nature **191**:1418–1419.

Dodge, B. O. 1924. Aecidiospore discharge as related to the character of the spore wall. J. Agr. Res. **27**:749–756.

Dowding, E. S. 1955. *Endogone* in Canadian rodents. Mycologia **47**:51–57.

Edgerton, C. W. 1910. The bean anthrocnose. Louisiana State University Agr. Expt. Sta. No. 119. 1–55.

Forsyth, F. R. 1955. The nature of the inhibiting substance emitted by germinating urediospores of *Puccinia graminis* var. *tritici*. Can. J. Botany **33**:363–373.

Fries, N. 1966. Chemical factors in the germination of spores of basidiomycetes. *In:* Madelin, M. F., Ed., The fungus spore. Butterworth & Co., Ltd., London. pp. 189–200.

Gottlieb, D. 1966. Biosynthetic processes in germinating spores. In: Madelin, M. F., Ed., The fungus spore. Butterworth & Co., Ltd., London. pp. 217–234.

——. 1976. Carbohydrate metabolism and spore germination. *In:* D. J. Weber, and W. M. Hess, Eds., The fungal spore—form and function. John Wiley & Sons, Inc., New York. pp. 141–163.

Gregory, P. H. 1949. The operation of the puff-ball mechanism of *Lycoperdon perlatum* by raindrops shown by ultra-high-speed schlieren cinematography. Trans. Brit. Mycol. Soc. **32**:11–15.

——. 1952. Spore content of the atmosphere near the ground. Nature **170**:475–477.

——. 1966. Dispersal. *In:* Ainsworth, G. C., and A. S. Sussman, Eds., The fungi—an advanced treatise. Academic Press, New York. **2**:709–732.

——. 1973. The microbiology of the atmosphere. Second ed. Leonard Hill, Aylesbury. 377 pp.

——, E. J. Guthrie, and M. E. Bruce. 1959. Experiments on splash dispersal of fungus spores. J. Gen. Microbiol. **29**:328–354.

——, and O. J. Stedman. 1958. Spore dispersal in *Ophiobolus graminis* and other fungi of cereal foot rots. Trans. Brit. Mycol. Soc. **41**:449–456.

Haard, R. T., and C. L. Kramer. 1970. Periodicity of spore discharge in the Hymenomycetes. Mycologia **62**:1145–1169.

Hawker, L. E. 1966. Germination: morphological and anatomical changes. *In:* Madelin, M. F., Ed., The fungus spore. Butterworth & Co., Ltd., London. pp. 151–163.

Heald, F. D., and R. A. Studhalter. 1914. Birds as carriers of the chestnut-blight fungus. J. Agr. Res. **2**:405–422.

Hirst, J. M. 1952. An automatic volumetric spore trap. Ann. Appl. Biol. **39**:257–265.

——, and G. W. Hurst. 1967. Long-distance spore transport. Symp. Soc. Gen. Microbiol. **17**:307–344.

Ingold, C. T. 1932. The sporangiophore of *Pilobolus*. New Phytologist **31**:58–63.

——. 1933. Spore discharge in the ascomycetes. I. Pyrenomycetes. New Phytologist **32**:175–196.

——. 1934. The spore discharge mechanism in *Basidiobolus ranarum*. New Phytologist **33**:274–277.

——. 1953. Dispersal in fungi. Clarendon Press, Oxford. 208 pp.

——. 1956. The spore deposit of *Daldinia*. Trans. Brit. Mycol. Soc. **39**:378–380.

——. 1958. On light-stimulated spore discharge in *Sordaria*. Ann. Botany London II, **22**:129–135.

——. 1959. Submerged aquatic Hyphomycetes. Quekett Microscop. Club, IV, **5**:115–130.

——. 1960. Dispersal by air—the take-off. *In:* Horsfall, J. G., and A. E. Dimond, Eds., Plant pathology—an advanced treatise. Academic Press, New York. **3**:137–168.

——. 1966. Spore release. *In:* Ainsworth, G. C., and A. S. Sussman, Eds., The fungi—an advanced treatise. Academic Press, New York. **2**:679–707.

——. 1969. Effect of blue and yellow light during the later developmental stages of *Sphaerobolus*. Am. J. Botany **56**:759–766.

——. 1971. Fungal spores—their liberation and dispersal. Clarendon Press, Oxford. 302 pp.

——. 1975. Convergent evolution in aquatic fungi: the tetraradiate spore. Biol. J. Linnaean Soc. **7**:1–25.

——, and V. J. Cox. 1955. Periodicity of spore discharge in *Daldinia*. Ann. Botany London II. **19**:201–209.

——, and V. Dann. 1968. Spore discharge in fungi under very high surrounding air-pressure and the bubble-theory of ballistospore release. Mycologia **60**:285–289.

——, and V. J. Dring. 1957. An analysis of spore discharge in *Sordaria*. Ann. Botany London II, **21**:465–477.

——, and B. Marshall. 1963. Further observations on light and spore discharge in certain Pyrenomycetes. Ann. Botany London II. **27**:481–491.

——, and M. Nawaz. 1967. Sporophore development in *Sphaerobolus:* effect of blue and red light. Ann. Botany **31**:791–802.

——, and B. A. Oso. 1968. Increased distance of discharge due to puffing in *Ascobolus*. Trans. Brit. Mycol. Soc. **51**:592–594.

——, and ——. 1969. Light and spore discharge in *Ascobolus*. Ann. Botany **33**:463–471.

Iqbal, S. H., and J. Webster. 1973. The trapping of aquatic hyphomycete spores by air bubbles. Trans. Brit. Mycol. Soc. **60**:37–48.

Leach, C. M. 1976. An electrostatic theory to explain violent liberation by *Drechslera turcica* and other fungi. Mycologia **68**:63–86.

Macko, V., R. C. Staples, Z. Yaniv, and R. R. Granados. 1976. Self-inhibitors of fungal spore germination. *In:* D. J. Weber and W. M. Hess, Eds., The fungal spore—form and function. John Wiley & Sons, Inc., New York. pp. 73–98.

Meredith, D. S. 1961. Spore discharge in *Deightoniella torulosa* (Syd.) Ellis. Ann. Botany London. II. **25**:271–278.

——. 1962. Spore discharge in *Cordana musae* (Zimm.) Höhnel and *Zygosporium oscheoides* Mont. Ann. Botany London II. **26**:233–241.

——. 1965. Violent spore release in *Helminthosporium turicum*. Phytopathology **55**:1099–1102.

Nawaz, M. 1967. Phototropism in *Sphaerobolus*. Biologia **13**:5–14.

Olive, L. S. 1964. Spore discharge mechanism in basidiomycetes. Science **146**:542–543.

Robinson, P. M., D. Park, and M. K. Garrett. 1968. Sporostatic products of fungi. Trans. Brit. Mycol. Soc. **51**:113–124.

Rockett, T. R., and C. L. Kramer. 1974. Periodicity and total spore production by lignicolous basidiomycetes. Mycologia **66:**817-829.

Smith, J. E., K. Gull, J. G. Anderson, and S. G. Deans. 1976. Organelle changes during fungal spore germination. *In:* D. J. Weber, and W. M. Hess, Eds., The fungal spore—form and function. John Wiley & Sons, Inc., New York. pp. 301-352.

Sussman, A. S. 1953. The effect of furfural upon the germination and respiration of ascospores of *Neurospora tetrasperma.* Am. J. Botany **40:**401-404.

——. 1961. The role of endogenous substrates in the dormancy of *Neurospora. In:* Halvorson, H. O., Ed., Spores II. Burgess Publishing Co., Minneapolis. pp. 198-217.

——. 1966-a. Dormancy and spore germination. *In:* Ainsworth, G. C., and A. S. Sussman, Eds., The fungi—an advanced treatise. Academic Press, New York. **2:**733-764.

——. 1966-b. Types of dormancy as represented by conidia and ascospores of *Neurospora.* In: Madelin, M. F., Ed., The fungus spore. Butterworth & Co., Ltd., London. pp. 235-257.

——. 1976. Activators of fungal spore germination. *In:* D. J. Weber and W. M. Hess, Eds., The fungal spore—form and function. John Wiley & Sons, Inc., New York. pp. 101-137.

——. and H. O. Halvorson. 1966. Spores—their dormancy and germination. Harper & Row, Publishers, Inc., New York. 354 pp.

Talbot, P. H. B. 1952. Dispersal of fungus spores by small animals inhabiting wood and bark. Trans. Brit. Mycol. Soc. **35:**123-128.

Van Niel, C. B., G. E. Garner, and A. L. Cohen. 1972. On the mechanism of ballistospore discharge. Arch. Mikrobiol. **84:**129-140.

Van Sumere, C. F., C. van Sumere-de Preter, L. C. Vining, and G. A. Ledingham. 1957. Coumarins and phenolic acids in the uredospores of wheat stem rust. Can. J. Microbiol. **3:**847-862.

Walker, L. B., and E. N. Andersen. 1925. Relation of glycogen to spore-ejection. Mycologia **17:**154-159.

Walkey, D. G. A., and R. Harvey. 1966. Spore discharge rhythms in Pyrenomycetes. I. A survey of the periodicity of spore discharge in Pyrenomycetes. Trans. Brit. Mycol. Soc. **49:**583-592.

——, and ——. 1968. Spore discharge rhythm in Pyrenomycetes. IV. The influence of climatic factors. Trans. Brit. Mycol. Soc. **51:**779-786.

Webster, J. 1959. Experiments with spores of aquatic Hyphomycetes. I. Sedimentation and impaction on smooth surfaces. Ann. Botany London II. **23:**595-611.

Woodward, R. C. 1927. Studies on *Podosphaera leucotricha* (Ell. & Ev.) Salm. I. The mode of perennation. Trans. Brit. Mycol. Soc. **12:**173-204.

Zoberi, M. H. 1961. Take-off of mould spores in relation to wind speed and humidity. Ann. Botany London II. **25:**53-64.

——. 1973. Influence of water on spore release in *Cookeina sulcipes.* Mycologia **65:**155-160.

Part Three

ECOLOGY
AND UTILIZATION
BY HUMANS

ELEVEN

Fungi as Saprophytes

Like other saprophytes, saprophytic fungi are decay-causing organisms that feed on nonliving organic materials. The *substratum,* or food source, may be the bodies of dead animals or plants; any separate parts of the plant or animal such as leaves, fallen branches, or hair; waste products from plant or animal sources; some synthetic compounds; and any soluble products diffusing from the above. The individual organic compounds that comprise the substratum are the actual materials used as a food source, and it is by the individual degradation or decomposition of these compounds that the larger mass is eliminated, or *decayed.* These organic compounds are the polysaccharides, organic acids, lignins, aromatic and aliphatic hydrocarbons, sugars, alcohols, amino acids, purines, lipids, proteins, and nucleic acids that are characteristic of life. Any species of microorganism is capable of decomposing only certain components in the plant or animal residue, and a number of species of microorganisms are required to bring about the complete decomposition of the substratum. Fungi, bacteria, actinomycetes, and protozoa often are involved in decomposition of a single substratum, the relative importance of each varying with the prevailing conditions.

To the microorganism, the process of degradation is simply a means of obtaining nutrients by digestion. Bacteria, actinomycetes, and fungi secrete digestive enzymes into the environment which degrade a specific molecule into its simpler components, making it go into solution. The food substances then enter the metabolic pathways, such as those described in Chapter 8. Although some compounds are broken down to respiratory carbon dioxide or released as metabolites, much of the substratum is incorporated into the protoplasm of the decay-causing saprophyte and will be liberated only when the microorganism itself dies and is decayed by other microorganisms.

Decomposition is complete when the organic compounds are returned to the environment in their inorganic or mineral form. For carbon, that form is carbon dioxide; for nitrogen, it is ammonia; and for phosphorus, it is phosphate.

DECOMPOSITION

Methods for Study of Decomposition

In studying the activities of saprophytic fungi, one often needs to know the rates at which a substratum is being decomposed, how these rates are affected by environmental conditions or nutrients, and the course of decomposition of the various chemical components of the substratum. The disappearance of the substratum may be followed by calculating the loss in total weight, but this method cannot account for any synthesis of new protoplasm by the microorganisms or the rate or course of decomposition of specific materials. Alternatively, an analysis of metabolic or waste materials formed during decomposition may be more helpful. For example, a measurement of carbon dioxide evolution may be used as an index of carbohydrate decomposition (Fig. 11–1), while ammonia evolution may simultaneously serve as an index of decomposition of nitrogenous substances. Measurement of metabolic products, like a measurement of total weight loss, cannot account for the decomposition of single components. This may be accomplished by determining the chemical composition of a substratum before decomposition begins and then by comparing the original composition with the varying chemical composition during the course of decomposition.

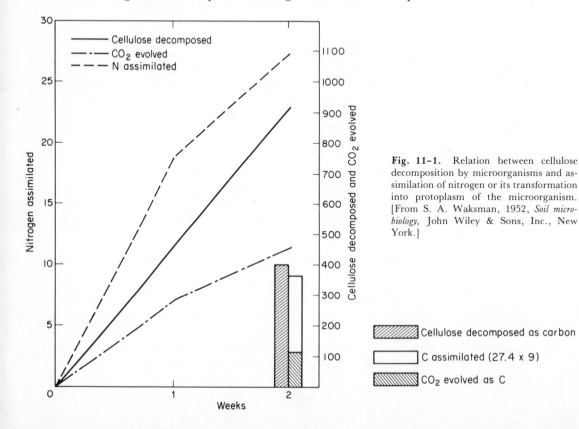

Fig. 11–1. Relation between cellulose decomposition by microorganisms and assimilation of nitrogen or its transformation into protoplasm of the microorganism. [From S. A. Waksman, 1952, *Soil microbiology,* John Wiley & Sons, Inc., New York.]

The Decomposition Process

Constituents of the substratum are attacked at different rates. Readily soluble sugars and amino acids are apparently absorbed directly and metabolized within the microorganism and are among the first components to disappear. Simple carbohydrates (sugars, starches, hemicelluloses, etc.) and some of the proteins are decomposed rapidly by a great variety of microorganisms. Cellulose, fats, and oils are decomposed more slowly and usually by specific microorganisms; lignins, waxes, and tannins are the most resistant to decomposition and are attacked by comparatively few fungus species. It may take years for some leaves and stems to decay; a pine needle may undergo decay for 8 to 9 years before it can no longer be recognized (Kendrick, 1959).

Maximum decomposition will occur only as long as there is an abundant supply of nitrogenous and carbon compounds and other essential nutrients (especially phosphorus) present in the substratum or soil. The ratio of carbon to nitrogen is especially important, as it primarily determines the amount of microbial growth that can occur and, therefore, indirectly the amount of decomposition that can take place

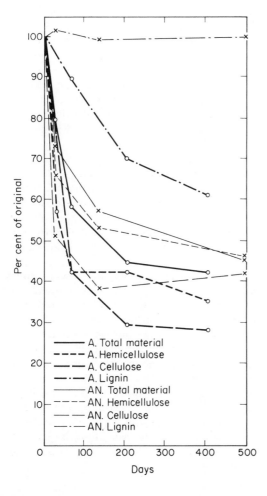

Fig. 11-2. Course of decomposition of an alfalfa plant. *A* indicates aerobic, *AN* indicates anaerobic. [From S. A. Waksman, 1952, *Soil microbiology,* John Wiley & Sons, Inc., New York.]

(Fig. 11-1). Nitrogen is needed by microorganisms to synthesize protoplasm and, in the absence of adequate nitrogen, fungi will be considerably impaired in their ability to decompose available carbohydrates. If the nitrogen supply is in excess of the amount required for protein synthesis (generally, more than 1.5% to 1.7% of the substratum), the excess is liberated as ammonia. If the carbon supply (and therefore the energy source) is inadequate, as much as 50% to 80% of the total nitrogen may be released as ammonia. Under favorable conditions, fungi will assimilate an average of 30% to 40% of the carbon of the substratum that has been decomposed (Waksman, 1952).

Young, green plants or animal residues high in nitrogen and phosphorus will undergo rapid and almost complete decay without added nutrients. In contrast, materials low in nitrogen (such as straw, stubble, and forest litter) decompose slowly and incompletely, perhaps leaving 50% to 60% of their original weight after 3 to 10 months of decomposition (Fig. 11-2) (Waksman, 1952). Agriculturalists often add fertilizers containing nitrogen and phosphorus to the soil to increase the decomposition of fibrous or woody plant remains consisting primarily of cellulose or lignin, such as straw or stubble.

COURSE OF DECOMPOSITION

It is often difficult to determine the chemical steps involved in decomposition in nature, where substrata are present as complex mixtures of compounds and environmental conditions are quite variable. It is difficult to duplicate these conditions in the laboratory, but an approach may be made by utilizing media which contain purified components and inoculating these with pure or mixed cultures of microorganisms. In addition, studies of enzyme systems may be utilized. Information obtained in this manner is generally equivalent to that presented in Chapter 7, but it is not necessarily representative of activities of fungi in soil. A fungus may be able to degrade a substance (cellulose, for example) in culture, but this does not indicate that it can establish itself and feed on this complex compound in nature under intense competitive conditions. The chemical pathway of decomposition may also be modified according to the particular organisms that are involved and by the prevailing environmental conditions.

Although at least some fungi are capable of degrading most substrata in nature, fungi are especially active in the decomposition of celluloses, hemicelluloses, and lignins.

Cellulose Decomposition. Cellulose, a glucose polymer, occurs widely in nature in plant cell walls. Actinomycetes, bacteria, fungi, protozoa, and some insects are cellulose decomposers but, of these, the fungi are particularly active as cellulose decomposers. The cellulose-decomposing fungi include members of the Ascomycotina, Basidiomycotina, and Deuteromycotina as well as some chytrids that may be trapped by using filter paper as a bait. Some members of the Holobasidiomycetidae are the most active in the decomposition of cellulose, and the majority of the members of this group are probably able to decompose cellulose.

The first enzymes involved in cellulose decomposition are the cellulases. The cellulases may occur as a complex, and the particular combination of enzymes in "cellulase" may be different from organism to organism. Cellulose is first cleaved into long, linear chains and then into disaccharide units (cellobiose) by enzymes in the cellulase complex. Cellobiose is then hydrolyzed to glucose by cellobiase. The overall scheme of the decomposition is as follows:

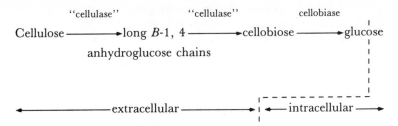

Decomposition of glucose depends on the nature of the organisms and the environmental conditions, as they may favor the growth of one group of organisms over another and may also determine the speed of decomposition. Among the environmental conditions, oxygen supply is especially important. If aerobic conditions prevail, glucose may be converted to organic acids or carbon dioxide and water by filamentous fungi. If, on the other hand, anaerobic conditions prevail, yeasts or bacteria may produce fermentative end products of respiration.

Hemicellulose Decomposition. The hemicelluloses occur abundantly in plant walls interspersed through cellulose and lignin. Next to cellulose, hemicelluloses are the most abundant compound in the plant cell wall.

The hemicelluloses are a heterogeneous group of polysaccharides and, upon hydrolysis, yield hexoses, pentoses, and frequently uronic acids. Decomposition of the hemicelluloses involves hydrolysis of the polymer to smaller units by specific enzymes. The actual enzymes involved and the biochemical steps depend on the particular hemicellulose which is being degraded.

Hemicelluloses tend to disappear rather rapidly during the early stages of decomposition and are decomposed by aerobic and anaerobic bacteria, actinomycetes and fungi. Fungi are more active in the early stages of decomposition, while the actinomycetes are more active in the later stages. The hemicellulose-decomposing fungi belong to all major groups, and the total number of species which can decompose hemicellulose is far greater than those capable of decomposing cellulose. Generally, the microorganisms that degrade hemicellulose do not exhibit a high degree of substratum specificity and often may use organic acids and simple sugars.

Lignin Decomposition. Lignin is decomposed almost entirely by fungi and especially by basidiomycetes, such as members of the Aphyllophorales and Agaricales. Lignin occurs abundantly in plant material, where it is bound with polysaccharides in the secondary cell wall and also occurs in the intercellular lamella. The lignin content of the plant increases both as the plant ages and with an increasing abundance of woody tissues. There are many types of lignins, with different types occurring in different species or even coexisting in the same plant. The lignin molecule is a highly complex polymer with the nature and arrangement of the monomers subject to variation. Although the structure of the lignin molecule is incompletely known, it is believed that the basic unit of the polymer is a phenylpropane structure consisting of a 3-carbon side chain on a benzene ring:

$$\langle\!\!\!\bigcirc\!\!\!\rangle\!-\!C\!-\!C\!-\!C$$

Although the biochemical details of lignin decomposition are not known, it has been suggested that the lignin-decomposing basidiomycetes secrete extracellular enzymes which release the aromatic units from the polymer. Further breakdown of the aromatic units probably can be carried out by a wide range of fungi other than the primary lignin decomposers.

Lignin and Humus Formation. Organic matter in the soil often is found in two principal fractions. The surface litter layer, prevalent in forests, consists primarily of recently fallen or deposited organic materials of which the origin may still be recognized. Beneath the litter is a dark humus layer in which the organic material is in later stages of decomposition, and the origin of the organic matter is no longer recognizable. In addition to the partially decomposed organic matter, the humus layer contains *humic* acid, a compound similar to lignin. The structure of humic acid is incompletely known, although numerous phenolic compounds may be chemically isolated from it which are generally similar to those in lignin. Humic acid is produced mainly under aerobic conditions and involves polymerization reactions which are poorly understood (Flaig et al., 1975). Decomposition of the original organic material and humic acid synthesis take place concurrently.

Humus is important in maintaining soil fertility (the ability of the soil to support plants). Humus has larger quantities of nitrogen and carbon than is found in the bodies of organisms. Through the comparatively slow but sustained decomposition of humus, these elements and others are constantly available to plants as nutrients. Humus also provides a physically favorable medium for plant growth as it is possible for the humus to modify the pH of the soil, to absorb substances which are injurious to plants, and to retain water and air. A high humus content is invariably associated with a high degree of soil fertility.

Fungi play a role in maintaining the humus level and, therefore, the soil fertility. Fungi decompose organic matter which is used in the formation of the humic acid molecule. The lignin molecule and the humic acid molecule bear a large number of structural similarities, suggesting that the decomposition of lignin may provide the monomers required for humic acid synthesis.

The extent to which fungi play a role in the polymerization of these monomers in nature is unknown although such polymerization may occur in cultures (Flaig et al., 1975). Fungi can also degrade humus, thereby releasing nutrients trapped within it.

ROLE OF SAPROPHYTES IN NATURE

The majority of the substrata that decay-causing organisms attack represent the ''garbage'' of nature that litters the habitats of living creatures. The sheer abundance of this debris could reach astronomical proportions. The weight of leaves formed in a single season by an elm tree may be calculated to be 400 pounds (Gray, 1959), or 2 tons over a 10-year period. In a hardwood forest, an estimated 1 to 2 tons of leaves and branches are dropped per acre annually; in a tropical rain forest the figure rises to 60 tons of debris (Burges, 1965). Farmers usually harvest only a small amount of a crop, such as the grain or fruit alone, leaving the remainder of the crop plant to decay in the field. The amount of wastepaper decomposed annually averages about 150 pounds/square mile over the land area of the continental United States (Gray, 1959). Without decay of this debris, the earth would become covered with plant and animal bodies, probably excluding most living organisms from their natural habitats.

The real significance of decay, however, lies in the return of those chemical compounds locked in the organisms to nature's cycles. There is a finite amount of materials available for the construction of living organisms, and life would eventually cease if chemicals were indefinitely bound in dead organisms and unavailable to living creatures. Carbon compounds are of special significance in this connection.

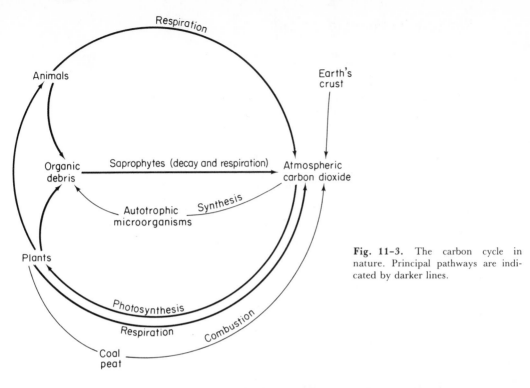

Fig. 11-3. The carbon cycle in nature. Principal pathways are indicated by darker lines.

In nature, carbon regularly cycles through its inorganic form (carbon dioxide) and organic forms that are required by all living organisms (Fig. 11-3). Autotrophic organisms (chemosynthetic bacteria, photosynthetic bacteria, and green plants) are the only organisms that are capable of converting inorganic carbon (carbon dioxide) into organic compounds. All other life is ultimately dependent upon these autotrophic organisms as a source of organic compounds. Especially significant are the photosynthetic green plants, which serve as the base of the food chain by directly supporting herbivorous animals and, in turn, supporting carnivorous animals.

The principal source of carbon dioxide required by plants during photosynthesis is the atmosphere, which contains carbon dioxide in the comparatively low concentration of 0.03% or about 2.1 billion tons. Land plants fix annually about 80,000 million (8×10^{10}) tons of carbon dioxide. The amount of carbon dioxide in the atmosphere would only be sufficient to last a few decades unless replenished (Kononova, 1966).

Carbon dioxide may be restored to the atmosphere as an end product of cellular respiration of plants and animals; from the decomposition of limestone and other rocks; as a result of volcanic eruptions; or from the combustion of coal, petroleum, and gas or other materials. The major source, however, is from respiration of microorganisms, which provides approximately 80% of the carbon dioxide required by land plants. Soil microorganisms have been estimated to yield 63.9×10^9 tons of carbon dioxide annually, and fungi contribute approximately 13% of this amount (Kononova, 1966).

Other materials, including nitrogen, sulfur, phosphorus, potassium, calcium, and iron, are required by photosynthetic plants and are absorbed in organic forms and converted into organic compounds. Like carbon, these chemicals are cycled through inorganic and organic forms.

Fungi as Saprophytes   **415**

Saprophytic Fungi in the Soil

From the previous section, it is evident that nature's cycles are largely initiated through the autotrophic green plants. The plants assimilate inorganic chemicals and convert them to organic compounds used by other organisms; eventually, the organic compounds are returned to the inorganic form by decay-causing microorganisms. These cycles take place in lakes, rivers, oceans, and on land. Fungi are involved in causing decay in all of these environments, but the greatest amount of decay by fungi occurs in the soil.

Soil is a complex medium, consisting of inorganic minerals and organic debris which serve as boundaries to channels and pores containing air and water. Substrata for decay-causing fungi are abundant in soil. Plant roots slough off their outer tissues, excrete organic materials (principally carbohydrates, amino acids, vitamins, and organic acids), and drop their leaves and dead branches upon the soil. Green manure crops such as clover and unharvested remains of crops are plowed under. Animals migrate through and over the soil and leave their excrement, outer sloughed tissues, shed hair and feathers, and their bodies when death overcomes them. This abundance of organic material provides nourishment either directly or indirectly for an extensive and varied microflora and microfauna. A conservative estimate of the numbers of microorganisms in 1 gram of fertile soil is (Nicholas, 1965)

True bacteria	10^6 to 10^9
Actinomycetes	10^5 to 10^6
Protozoa	10^4 to 10^5
Algae	10^1 to 10^3
Fungi	10^4 to 10^5

This estimate does not include larger soil animals such as nematodes, earthworms, mites, ants, or other worms and arthropods.

METHODS FOR STUDY OF SOIL FUNGI

Investigators of soil fungi may like to determine the extent of mycelial activity in the soil, the species present, and the relative distribution of fungi in the soil. Such determinations are necessary for the study of effects of soil treatments (for example, the addition of lime or fertilizer), seasonal changes, close association with roots, or different soil types. Direct visual observation of fungi in the soil is hindered by the difficulty of making microscopical observations in an opaque medium and the irregularity with which spores or reproductive structures may be found. Many highly specialized methods have been developed for study of soil fungi, but generally each method is highly selective for certain types of fungi and several methods must be used in combination to obtain an approximate picture of soil fungus activities at any given time (Fig. 11–4). Some frequently used techniques follow.

Rossi-Cholodny Contact Slide Method. A technique useful for studying soil microflora in its natural distribution in soil, or in situ, is the Rossi-Cholodny contact slide method. A glass microscope slide is buried in the soil and allowed to remain in place for several days or weeks. Microorganisms coming into contact with the slide adhere to it and, after the appropriate amount of time, the slide may be removed from the soil, air-dried, fixed with gentle heat, rinsed to remove the soil particles, and stained. The fungi observed represent those which came into contact with the slide through the entire period it was buried. Changes in the fungal population may be

	Method of isolation						
			Isolation from				Total number of species
	Dilution plates	Soil plates	Hyphae	Rhizo-morphs	Sclerotia	Fruc-tifications	
Lower fungi	18	17	11	0	0	11	26
Ascomycetes (b) Discomycetes	0	0	0	0	3	4	5
(b) all others	8	10	3	0	1	5	18
Basidiomycetes	0	0	8	8	7	12	30
Fungi imperfecti	81	60	16	0	2	24	95
Sterile mycelia	29	23	59	1	2	0	85
Total number of species	136	110	97	9	15	56	—

(a)

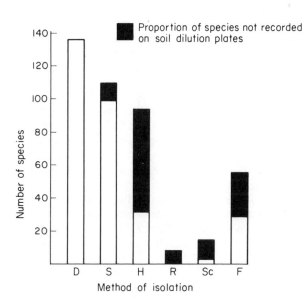

Proportion of species not recorded on soil dilution plates

Number of species

Method of isolation

(b)

D = Soil dilution plates   R = Isolation from rhizomorphs
S = Warcup soil plates   Sc = Isolation from sclerotia
H = Hyphal isolation   F = Isolation from fructifications

Fig. 11-4. A comparison of fungi obtained from wheat-field soil by different isolation techniques: (a) Relative numbers of fungi in each group; (b) Numbers of fungi obtained and the proportion of species different from those isolated on soil-dilution plates. [From J. H. Warcup, 1965, *Ecology of soil-borne plant pathogens—prelude to biological control.* Reprinted by permission of the Regents of the University of California.]

determined by burying a number of slides, which are removed at intervals during the test period. The Rossi-Cholodny technique is especially useful for determining the abundance of fungal mycelia in different soils at different times. Owing to the general lack of sporulating structures formed, it is not useful for the identification of soil fungi.

Dilution Plates. Culture techniques have been widely used to survey soil microflora, especially for identification purposes. The usual technique is to prepare a dilute soil solution by shaking a known amount of soil in sterile water or medium and progressively diluting it to yield a final concentration such as 1:100, 1:1,000, or 1:10,000. A 1-milliliter aliquot is placed in a petri dish, melted agar (cooled to just above solidification) is added, and the dish swirled to distribute the sample evenly. The spores and mycelial fragments are distributed throughout the medium and may give rise to sporulating colonies which can be identified. Usually the medium is acidified, or an agent such as streptomycin or rose bengal is added to the medium to retard bacterial growth. The composition of the medium may vary from a simple water agar to a medium enriched with a particular substratum such as cellulose that favors the growth of cellulose-decomposing fungi, often to the exclusion of others. The fungi isolated are often determined by the composition of the medium, and no single medium will support the growth of all soil fungi. This technique is useful, but cannot be relied upon to give an accurate survey of mycelial activity in the soil as a distinction cannot be made between a colony arising from an active mycelium or from a dormant spore. In addition, some fungi such as *Penicillium* and *Aspergillus* sporulate abundantly in the soil and, upon culturing a soil sample with such a fungus present, that species would be greatly overrepresented although its abundance in the soil as mycelium may have been slight. It is evident that the number of times a given fungus is recovered is often a function of its ability to survive under a given set of cultural conditions and not necessarily a function of its prevalence in nature. Members of the Deuteromycotina are recovered most frequently in soil dilution plates, whereas the basidiomycetes and nonsporulating fungi (Mycelia Sterilia) are rarely recovered although they are of widespread occurrence in the soil.

Warcup Soil Plate. The previous method often fails to recover nonsporulating fungi in the Mycelia Sterilia (such as *Rhizoctonia solani*) because mycelial fragments cling to the soil particles which are discarded. The standard dilution methods may be modified by placing a small amount of soil in the empty petri dish, crushing it with a spatula, and then adding cooled medium. Generally the method is similar to the previous one and also favors the heavily sporulating species. It has the advantage of being somewhat simpler to use.

Direct Isolation. The previous techniques generally fail to recover basidiomycetes, although a casual glance at the many basidiocarps would indicate that basidiomycete mycelia do exist in the soil. Warcup (1951-a, 1959) found it possible to obtain basidiomycete mycelium in culture through germination of their sclerotia or from fragments of rhizomorphs.

THE SOIL FUNGI

Fungi are most abundant in the upper 4 inches of soil and are chiefly limited to the upper 12 inches. Their numbers and types decrease with increasing depth. The top 6 inches of arable land that contains 60,000 pounds per acre of dry organic matter may be estimated to contain 300 pounds dry weight of fungus mycelium (Nicholas, 1965).

Most types of fungi from chytrids to agarics, from saprophytes to parasites of plants and animals (including man), and from predators of soil animals to symbiotic mycorrhiza-forming fungi may be detected at one time or another in the soil. Representatives of all fungal groups occur in the soil. Those most frequently isolated are members of the Deuteromycotina often *Aspergillus, Penicillium,* and *Trichoderma* (Fig. 6-1).

Only some of the many fungi occurring in the soil lead active saprophytic lives there and, for these fungi, the soil is their usual habitat. Other fungi are transient organisms that invade the soil on the plant or animal hosts that they parasitize. The transient fungi may not be able to survive in the absence of their hosts or may be able to lead a saprophytic existence of a limited duration. Often the transient fungi are capable of temporarily colonizing substrata in the soil but are unable to compete successfully with the native fungi. Numerous fungi are represented in soil as dormant structures; these include sporangiospores, conidia, oospores, and ascospores and probably basidiospores, chlamydospores, and sclerotia. For many of these fungi, occurrence in soil is incidental; they do not play a significant role in saprophytic activities in the soil. Active soil fungi exist as similar dormant structures, however, when food is lacking.

The dormant condition is more or less typical for fungi in the soil, and there is comparatively little fungus activity in the absence of an available food source. A suitable substratum is often as small as a single cellulose fiber that supports its own rather typical microflora. Many such substrata, each constituting a microhabitat, occur in the soil in a three-dimensional array or mosaic of separate ecological niches. Between the individual ecological niches are spaces where fungi are comparatively inactive or dormant (Garrett, 1955). Fungus structures in the less active areas include spores, chlamydospores, sclerotia, rhizomorphs, and mycelia (often dead or inactive). When a suitable substratum becomes available, dormant fungus structures in the immediate vicinity of the new substratum burst into activity; colonize, exploit, and exhaust the substratum; and then become dormant again (Garrett, 1951; Warcup, 1967). The mycelium may lyse when nutrients are no longer available, making survival dependent on the formation of dormant spores (Lockwood, 1964).

Fungistasis. In the absence of a suitable substratum, fungal mycelia and spores are generally maintained in a dormant condition. This dormancy is maintained by a widespread fungistatic effect of the soil *(soil fungistasis)*. Fungistasis is found in practically all soils (except in some deep subsoils, highly acid soils, or soils exposed to cold temperatures). Fungistasis may be reversed by the addition of readily decomposable organic materials such as glucose or by the growth of roots into the region and subsequent secretion of organic compounds by the roots. Roots may secrete a wide variety of organic compounds, including amino acids, soluble sugars, and vitamins. With such a reversal of fungistasis, fungus spores germinate and the fungal mycelium resumes active growth.

Fungistasis is invariably associated with the presence and activity of soil microorganisms. Soil sterilization reverses fungistasis, and subsequently fungistasis may be restored by the addition of a small amount of unsterilized soil. Those soil types in which fungistasis is lacking are those without a substantial microbial population. Fungistasis is generally assumed to be a result of the general microbial activity in the soil and not necessarily a product of any specific group of microorganisms.

There is little knowledge or agreement concerning the mechanism involved in fungistasis. One possible mechanism is that the living microorganisms compete for nutrients required by fungus spores and the germination of the fungus spores is inhibited until nutrients become available from a newly introduced substratum or from roots (Lockwood, 1964). Alternative suggestions are that fungistasis results from the accumulation of staling substances produced by soil microorganisms (Jackson, 1965; Park, 1960) or that specific inhibitors are produced by microorganisms (Park, 1967; Watson and Ford, 1972). Volatile compounds such as acetaldehyde, ethanol, and ethylene are produced in culture by soil fungi and will inhibit spore germination, which leads some investigators to suggest that the inhibitors are volatile compounds which may be produced by fungi as well as by bacteria (e.g., Gray, 1976). Watson and Ford (1972) propose that fungistasis is maintained by volatile inhibitors that are primarily of microbial origin but may also depend on nonbiotic factors (perhaps the pH or the iron or calcium carbonate content of the soil), and that this inhibition may be released by stimulators, which are primarily biotic in origin and are mostly nutrients but may also include volatiles emitted by plant roots.

Conditions Affecting Growth in Soil. Factors affecting the growth of mycelia in the soil include temperature, moisture, carbon dioxide, oxygen, and acidity.

Soil moisture is a fairly complex environmental parameter and involves free water as well as relative humidity. Levels of soil moisture indirectly affect soil aeration, movement of solutes, and diffusion of nutrients or inhibitors (Griffin, 1969). Fungi are generally able to tolerate a wide range of moisture conditions in the soil, but if the moisture level becomes extremely low, both the available free water and the relative humidity are reduced, which results in little or no fungal activity. Fungal activities tend to be inhibited generally by relative humidity levels less than 75% and favored by high relative humidities near 100% (Griffin, 1963). If the free-water level becomes very high (to the extent that aeration is hindered), the number of active mycelia is also reduced. This effect of reduction in aeration is greatest for strongly aerobic fungi but may be negligible for many species, some of which may be abundant in waterlogged soils. A similar reduction in aeration may be caused by extreme compaction of the soil.

Soil fungi are relatively insensitive to reductions in oxygen, and there is typically little change in the growth rate until the oxygen concentration is lowered to about 4% in the gas phase. After this low level is reached, fungal species will respond differently to a further lowering (Griffin, 1972).

The carbon dioxide concentration in the soil atmosphere usually is between 0.002 and 0.02 atmospheres, but values of 0.1 atmosphere or more may occur at greater depths or in soil with a high water content (Griffin, 1972). High carbon dioxide concentrations are associated with lowered oxygen partial pressure. A study of fungi originating from different levels within the soil usually reveals that those from the surface levels of the soil are intolerant of carbon dioxide while those isolated from depths where carbon dioxide can accumulate are more tolerant of high carbon dioxide levels.

Temperatures may fluctuate greatly in the soil. Soil temperature is affected by moisture content, slope, color, and the nature of the vegetation which it supports. Typically, there is daily fluctuation of temperature, the lowest temperatures occurring at dawn and the highest temperatures an hour or two after noon. Daily fluctuations of as much as 35°C may be noted in the temperate zone. Seasonal variations may be even more extreme. Soil fungi are generally tolerant of wide fluctuations in temperature. This tolerance is probably chiefly a position effect; that is, those fungi killed by extreme temperatures at one level (as by freezing at the surface) are replaced by other fungi surviving at a different level where temperatures are less extreme (Raney, 1965). Fungi are apparently active in the humus layers even during the winter months, although their activity may decrease to approximately 43% of the summer level (Waid, 1960). Fungi may respond to the effects of temperature and relative humidity in combination: for example, they may survive longer in extremely hot (56%C) soils if the relative humidity is low (85% to 90%) rather than high (100%) (Chen and Griffin, 1966).

Fungi as a group can tolerate a wide pH range, but some are confined to either alkaline or acid soils, while other species can tolerate a wide pH range. Soil pH affects the availability of nutrients, but it is not known whether this effect is significant for soil fungi. Altering the soil pH by addition of lime will in turn alter fungal activity.

Antagonism between Soil Microorganisms. A soil fungus is in close proximity with other soil microorganisms and its ability to live in the soil is partially determined by its ecological interrelationships with other forms of soil life. These interrelationships are often *antagonistic* in nature. Antagonisms result when one or more of the organisms involved in the interrelationship are harmed or have their activites curtailed. In the case of soil fungi, antagonism may take the form of exploitation, antibiosis, and competition.

Exploitation. A soil fungus may be directly attacked by parasitic or predatory organisms. Bacteria may cause lysis of the mycelium and are often responsible for the early disappearance of fungus mycelia in the soil. Some soil fungi are capable of parasitizing other fungi in culture (see Chapter 12), but the extent to which parasitism occurs in nature is unknown. In addition, various members of the soil fauna feed on fungi in the soil.

Antibiosis. Fungi modify the environment by excreting metabolic wastes or by-products (staling products). These metabolic by-products are often of a potentially dangerous composition and, in high concentrations, may be toxic to the fungus that produced them or to any other soil microorganisms that might be sensitive to them (Fig. 11-5). This phenomenon is known generally as *antibiosis*. The metabolic products that can cause antibiosis are varied and include (1) respiratory carbon dioxide, alcohol, acids, and other simple compounds that have a nonspecific effect; and (2) the antibiotics, higher-molecular-weight byproducts (often a weak acid), some of which have specific toxic effects on susceptible microorganisms. Antibiosis may make an environment generally toxic and, therefore, antagonistic to some fungi present. The

Fig. 11-5. Antibiosis between *Trichoderma lignorum* (left) and *Mucor heterogamous* (right). *T. lignorum* produces antibiotic(s) which inhibit the growth of *M. heterogamous*, resulting in a clear zone with little growth where the fungi meet. [Courtesy Plant Pathology Department, Cornell University.]

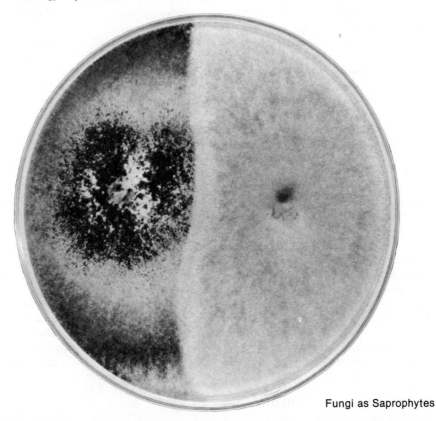

degree of antagonism experienced is largely a matter of tolerance of the individual organisms and the ability to withstand toxic conditions favors the survival of a microorganism. Antibiosis is not necessarily a product of specific antibiotic production, although in some instances this is the case. Antibiosis differs from fungistasis in that the effects of antibiosis are localized and usually specific, whereas those of fungistasis are widespread and general.

ANTIBIOTICS. Antibiotics are produced by bacteria, actinomycetes, and fungi and may be toxic to members of any group. Production of antibiotics is widespread among the fungi, occurring in comparatively few of the lower fungi but in many members of the Ascomycotina, Basidiomycotina, and Deuteromycotina. The majority of the fungi producing antibiotics are soil fungi, which often gives these fungi an ecological advantage. Antibiotics may be isolated in small quantities from soil samples but may be inactivated by alterations in pH, absorption on clay or humus, or degradation by microorganisms (Gray, 1976). The role of antibiotics in competition will be discussed more completely in the next section.

Antibiotic-producing fungi which are effective against bacteria may be isolated from the soil. An agar plate is inoculated with bacteria which are allowed to develop for 12 to 24 hours; then small particles of soil are added to the plate. A fungus producing antibiotics (effective against the test bacteria) may be detected when the bacteria surrounding the growing hyphae become lysed, producing a clear zone around the hyphae. Portions of the hyphae may be cut out and cultured individually.

Most antibiotics are toxic to a broad spectrum of microorganisms, and the antibiotic producer may not be immune to its effects but may have a high tolerance. Many antibiotics are toxic to only certain organisms or types of organisms. Types of toxicity include (1) a specific toxicity to Gram-positive bacteria, (2) semispecific toxicity to acid-fast bacteria, affecting Gram-positive bacteria at higher doses, and (3) a specific toxicity to fungi.

Resistance to antibiotics by microorganisms may take several paths: (1) An organism may produce an enzyme which will destroy the inhibitor; (2) production of metabolites may antagonize or neutralize the effects of the antibiotic; (3) possession of biosynthetic pathways that are not inhibited by the antibiotic; and (4) possession of barriers may prevent the movement of antibiotics into the cell. The last is believed to be especially significant in fungi.

Commercial production of antibiotics is discussed in Chapter 14.

Competition. When some commodity of the environment is in short supply and cannot meet the demands of all organisms present, the organisms will attempt to obtain enough of that commodity to support themselves. This rivalry between organisms for a commodity is termed *competition.* Soil fungi most commonly must compete for a food source, and competition is especially keen in substratum colonization. This interplay may become antagonistic if a species is eliminated or curtailed, but the rivalry in itself is not necessarily antagonistic.

The physiological ability of a fungus to compete successfully for a substratum may be termed *competitive saprophytic ability* (Garrett, 1950, 1963). Competition for a substratum is indirect, usually taking the form of differential growth or ability to withstand prevailing conditions in the soil. Characteristics that may give a fungus a favorable competitive ability are (1) a high growth rate and rapid germination of spores in the presence of a substratum, (2) a high degree of metabolic efficiency in production of enzymes and utilization of substrata, (3) ability to produce substances (including antibiotics) that are toxic to other soil microorganisms, and (4) a high degree of tolerance to antibiosis. A successful competitor may have any combination of these traits, but probably it is endowed with at least one. However, no single factor has been conclusively demonstrated to be the sole cause for differential survival in the soil. Rapid germination of spores, a rapid growth rate, and an ability to quickly consume the available nutrients are comparatively obvious when one considers that the fungus that can get to the food source first and consume the most will get more of the food. Although the production of toxic byproducts is an incidental excretion of waste products by the producer, it can give that organism an advantage by keeping others away. Likewise, a resistance to toxic substances simply means that an organism is not likely to be deprived of the food by producers of the toxic substances.

Although antibiotic production may greatly benefit a given fungus, this is not always the case. In nature there is interaction between a large number of organisms and often a delicate system of checks and balances that determine the extent of activity of any given organism. The potential benefit of antibiotic production by a given fungus, *A*, may be modified in nature by (1) the relative numbers of organisms sensitive to *A*'s antibiotic; (2) the relative numbers of organisms able to inhibit *A* and whether or not they are sensitive to *A*'s antibiotic; and (3) the proportion of fungi eliminated by *A* that would otherwise have eliminated *A*'s antagonists (Park, 1960). (See Fig. 11-6.)

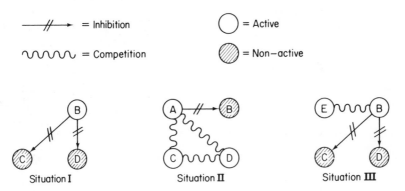

Fig. 11-6. A hypothetical situation where antibiotic production is a disadvantage to the producer, *A;* Left, Antibiotic producer *A* is absent and antibiotic producer *B* eliminates sensitive competitors *C* and *D;* Center, Antibiotic producer *A* arrives, eliminates sensitive *B*, and now must compete with both nonsensitive *C* and *D;* Right, Nonantibiotic-producing *E*, in the same situation as *A* in the center, must compete only with *B*.

Substratum Colonization and Succession. Studies of substratum colonization and succession were pioneered by Garrett (1951, 1963). The concepts and terms presented below are drawn from his work.

Upon the renewal of activity, the fungus is able to penetrate and decompose any of the nutrients within the substratum that it is capable of using. While the nutrients that it can utilize are available, the fungus grows actively; but it will become dormant when those particular nutrients are exhausted. The substratum may still have many nutrients which other fungi can utilize, and these fungi may then colonize the substratum and exploit it of those nutrients that they can use. This *succession* of fungi will continue until the substratum is completely decomposed or exhausted (Fig. 11-7). In principle, succession of fungi on a substratum is similar to that of higher plants on land because in each case environmental conditions are changed by the organisms.

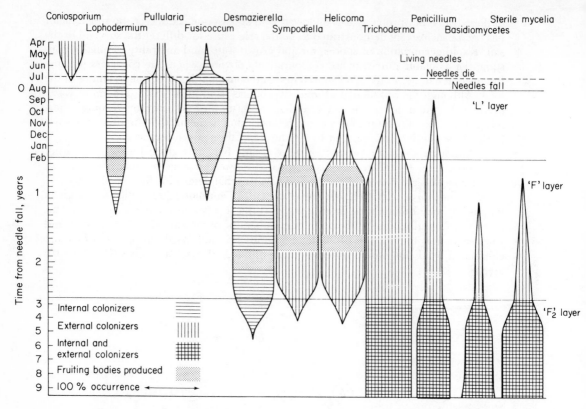

Coniosporium   Pullularia   Desmazierella   Helicoma   Penicillium   Sterile mycelia

Lophodermium   Fusicoccum   Sympodiella   Trichoderma   Basidiomycetes

Fig. 11-7. Succession of fungi on leaves of *Pinus sylvestris* in litter, showing the relative proportion of each genus and the temporal and spatial relationships. The sugar fungi are lacking in this succession. [From W. B. Kendrick and A. Burges, 1962, *Nova Hedwigia* 4:313-344.]

Unlike higher plant succession, succession by fungi depletes the environment and makes it less satisfactory for growth by other organisms, whereas higher plant succession constantly improves the environment for growth by other plants.

Succession of fungi occurs commonly in nature and may be observed on plant parts which remain temporarily above ground (standing dead trees or herbaceous stems), are on the surface of the ground (logs, nut hulls, leaves), or are initially present within the soil (roots) (Hudson, 1968).

Fungi are especially important in the breakdown of plant remains and, therefore, succession of fungi on plant substrata has special significance. A simultaneous succession of bacteria and soil fauna occurs on the substratum alongside of the fungi. Fungal succession of plants usually begins even before death occurs. Unlike animals, plants do not ordinarily die a sudden death except in the case of accidents such as floods or fires, but individual roots or branches may die and leaves are shed. Rather than a sudden death, the plant typically undergoes a period of gradual decline, perhaps due to attack by a highly specialized plant parasite or to unfavorable growing conditions. In these instances, the plant is moribund or undergoing a period of gradual death. The tissues are highly susceptible to invasion by fungi that are comparatively unspecialized as parasites. Virtually no senescent plant tissue is free of parasites, as they occur in dying roots and even in autumn leaves before their fall, even though only isolated parts of the plant may be moribund. With death of the entire plant or plant part, the parasite may continue to live saprophytically. In almost all cases, the parasites are the initial or pioneer invader of the plant tissue substrate.

Either overlapping with the parasites or following them next in the succession are the *sugar fungi,* which suddenly invade the substratum after death of the plant is complete, and which occur in or on soil.

The sugar fungi are predominantly members of the Mucorales and are capable of metabolizing the available sugars and simple carbon compounds. These invading fungi compete not only for the carbon source but also for the available oxygen and other nutrients such as nitrogen, phosphorus, and potassium. These fungi are able to dominate the environment because their spores germinate rapidly in the presence of a substratum and their hyphae have a rapid growth rate, both resulting in a sudden burst of activity. Eventually this group eliminates itself from the substratum because it is sensitive to the accumulation of its metabolic by-products, especially carbon dioxide. Formation of dormant structures and quiescence follows. Evidence from *Pythium mamillatum* indicates that these fungi may be restricted to the early stages of colonization before toxic metabolites accumulate. *P. mamillatum* actively colonizes virgin substrata, but colonization is reduced if the substratum has been precolonized (Barton, 1960).

The next stage of the succession consists of colonization by *cellulose-decomposing fungi* which either overlap with the sugar fungi or follow them. Cellulose-decomposing fungi are predominantly members of the Basidiomycotina, but they include chytrids and members of the Deuteromycotina and Ascomycotina. These fungi have the disadvantage of having to compete for a substratum that is somewhat depleted of readily available simple sugars and of nutrients such as nitrogen. The cellulose decomposers hydrolyze the cellulose molecule to cellobiose and then to glucose, which they can assimilate. The hydrolysis of cellulose occurs externally, and the simpler forms are free in the vicinity of the mycelium before their intake and assimilation. Many *secondary sugar fungi,* unable to degrade cellulose, exist in the vicinity of cellulose-decomposing fungi and feed upon the simpler carbohydrates alongside of the cellulose decomposers. The secondary sugar fungi and cellulose decomposers physiologically could have been the first to colonize the fresh substratum but could not compete with the rapid colonization characteristic of the primary sugar fungi. The fungi in this stage of succession decline with the exhaustion of available food.

The predominant material left after the decline of the cellulose-decomposing fungi is lignin. Lignin is the principal constituent of humus (comprising 30% to 60% of it, in contrast to the 10% to 30% in the original plant). Lignin tends to accumulate as there are comparatively few fungi which are capable of strongly decomposing it; all of them are higher basidiomycetes (Fig. 11–8). There is comparatively little competition at this stage in succession as the substratum is practically devoid of readily obtainable food that other fungi can use, thus giving the lignin fungi an advantage. The lignin fungi may combat the shortage of food by forming an extensive mycelial system through which food may be transported from one portion to another or one food base to another. Unlike the ephemeral sugar fungi that become dormant until food is available, the lignin decomposers (and some cellulose decomposers) form either mycelial strands or rhizomorphs that travel actively from one substratum to another. Evidence that a lignin-decomposing fungus can establish itself earlier only with dif-

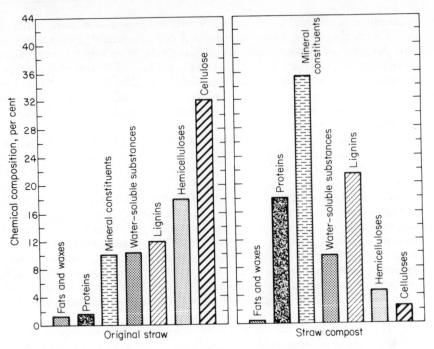

Fig. 11-8. Comparative chemical composition of oat straw and of a compost prepared from it. [From S. A. Waksman, 1952, *Soil microbiology,* John Wiley & Sons, Inc., New York.]

ficulty comes from *Agaricus brunnescens.* Commercially, *A. brunnescens* is grown only on compost that is in a late stage of decomposition; the proportion of lignin is increased and the readily decomposable materials (sugars, hemicelluloses, etc.) are decreased, thus making the compost unfavorable for would-be competitors (Fig. 11-8).

DETERIORATION OF MATERIALS BY FUNGI

Many of the materials used by people to make products are of plant or animal origin. These include wood and wood products, paper, cotton and woolen fabrics, leathers, and food products. Any of the plant or animal parts used to make these goods are commonly encountered by fungi in nature, and any of these products may support the growth of saprophytic fungi if environmental conditions are favorable. In addition, other manufactured products may not be directly produced from plant or animal parts but may nevertheless support growth of saprophytic fungi. Products of this type include certain plastics and paints.

The growth of fungi on these products is a substratum-saprophyte relationship, with fungi attacking goods that can provide nourishment as long as favorable environmental conditions prevail. Generally, fungus attack is favored by high humidity or abundant moisture and warm temperatures. For many goods, the usually low humidity and moisture alone is adequate to prevent fungal growth.

We have established that growth of saprophytic fungi leads to decomposition in nature and that decomposition of natural materials is necessary to remove waste products and to recycle the components of these materials. In the same manner, decay of human goods which often find their way to the garbage dump is essential for the removal of the products of civilization. Fungi and other microorganisms play a role in the decomposition of paper, plastics, and other materials used by humans. However, if saprophytic fungi become established on goods that have not been discarded, *deterioration* results. Deterioration may be considered as a reduction in the useful properties of these goods due to their partial decomposition.

The role of fungi in deteriorating goods has long been familiar. It is not unusual to find mildew or molds (usually a member of the Deuteromycotina) on shoes or books that have been left in a damp place or to observe basidiocarps on a decaying log. During World War II, deterioration by fungi in the tropics became such a severe problem that much interest was further stimulated in this topic. Many troops were in the tropics where high humidity, frequent rainfall, and high temperatures prevailed. These conditions favored the often luxurious growth of fungi on all types of goods. Fungi damaged or destroyed canvas tents, leather goods such as shoes and cases, clothing, packing boxes, electrical equipment (Fig. 11–9), rubber and plastic products, and optical equipment. The "kerosene fungus," *Cladosporium,* has been found to be capable of growing in storage and aircraft tanks containing kerosene-type fuels and of causing corrosion of the aluminum (Hendey, 1964). In many instances, superficial fungus growth did not seriously impair the product, although it was often responsible for lowered morale of the troops. In other instances, the goods were unusable.

Fig. 11–9.   Fungal growth (arrows) on electrical equipment. [Courtesy U.S. Army Frankford Arsenal, Philadelphia, Pennsylvania.]

It is evident from the previous paragraph that fungi may cause deterioration of a wide variety of materials. Damage by fungi is commonly a problem on wood and wood products, paper, fabric, and foods. These are considered in greater detail below.

Wood and Wood Products

Forests constitute one of our most important natural resources, contributing wood for innumerable uses by humans. Of the types of materials affected by fungi, the greatest amount of damage occurs to wood in the form of standing timber, rough lumber, or finished wood products, including paper. Deterioration is caused by abrasion, termites, marine borers in salt water, but fungi are the most serious.

Fungi often attack the heartwood of standing, living trees. These fungi are saprophytes in that they attack and decay the nonliving heartwood of the tree, which may not interfere with the functions of the living sapwood. Such infected trees are structurally weakened and may meet an early death because of the heartwood destruction. In any event, heartwood is usually rendered unfit for lumber. Fungi attacking the heartwood are basidiomycetes, usually Aphyllophorales; they may be divided into (1) the "brown rot" fungi that remove the structural carbohydrates (cellulose and hemicellulose) and leave the lignin relatively unchanged, and (2) the "white rot" fungi that primarily attack the lignin but also remove a small amount of the structural carbohydrates, leading to a lightening of the wood.

Felled timber, in the form of logs or sawed lumber, is seasoned by air-drying or kiln-drying to rid it of excess moisture. Logs and lumber are subject to attack by a large number of fungi, especially during the early stages of seasoning by air-drying while the wood has a high moisture content. Air-drying is usually accomplished by stacking the wood outdoors. Care must be taken to stack it in such a way that air may freely circulate throughout and to avoid as much contact with the ground as possible. If moisture accumulates, either within the stack or next to the ground, favorable conditions are created for the growth of fungi. Fungi growing under these conditions are often deuteromycetes and ascomycetes that live chiefly on cell contents and often cause a discoloration (staining) of the wood. The stain may be yellow, brown, or blue in color and the color may be a result either of the mycelium of the fungus within the wood or of pigmented by-products. Similar fungi may produce superficial (mold-like) colonies on the wood without staining and may simply be wiped away. Wood invaded by either of these superficial types of fungi may be used for most purposes, excluding those where the unsightly stains would be undesirable (furniture, etc.) or where the remaining mycelium and spores could contaminate food. In addition, the wood may be invaded by decay-causing basidiomycetes that decompose the wood, making it unsound for structural purposes. Generally the fungi that attack the wood are similar in their physiology to those causing heartrot, but they are different organisms. The sapwood is particularly vulnerable to decay in these stages (Fig. 11–10).

Vulnerability to fungus attack decreases when the moisture content of the wood becomes lower than the point at which free water within the cells is gone, usually between 25% to 30% moisture content. With complete drying to 20% or less moisture, the wood is ordinarily immune to fungus attack. Exceptions to the above are (1) invasion by dry rot fungi, such as *Merulius lacrymans,* which translocate water from a moist substratum to its mycelium, causing decay of dry timbers in buildings; and (2) wood that is immersed in water or mud where anaerobic conditions prevail, as in piling for wharves, which prevents fungal growth.

Fig. 11–10. Rot of an oak log caused by *Polyporus versicolor.* Note the basidiocarps on the surface of the log and the rotting sapwood in the region penetrated by the mycelium. [Courtesy Plant Pathology Department, Cornell University.]

For many wood products that would normally be maintained under dry conditions (such as packing crates, furniture, and wood for indoor construction) no further treatment beyond adequate seasoning is necessary to prevent destruction by fungi. Development of any existent mycelium is arrested as long as the moisture content remains under 20%, although it may become active if sufficient moisture becomes available. Wood intended for many uses where the product will be exposed to high humidity and moisture (especially next to the ground) requires additional treatment with a chemical preservative to prevent decay. In 1950, an estimated 289 million cubic feet of wood was treated in the United States (Hunt and Garratt, 1953). Treatment may consist of applying coal-tar creosote under pressure to wood intended for use in contact with the ground or water (as telephone poles, fence posts, railroad ties, foundation piles, etc.). Creosote impregnation is generally considered to be the most effective treatment for prevention of decay, but it cannot be used where it might interfere with the application of paint or where the odor or appearance would be objectionable. For wood to be used for such things as station wagons, airplane parts, boats, and timber for construction, an effective treatment consists of applying a mixture of 5% pentachlorophenol and 95% oil carrier to the wood, usually by immersing the wood

Fig. 11-11.   Fungal damage to paint on wood. Left, typical appearance of damage to paint. Right, germinating spores of *Pullularia pullulans* on a latex paint film. *P. pullulans* is the fungus most often associated with paint damage. × 55. [Courtesy the Nuodex Division of Tenneco Chemicals, Inc.]

Fig. 11-12.   Spotting of finished paper resulting from slime carried along during processing. [Courtesy of the Institute of Paper Chemistry, Appleton, Wisconsin.]

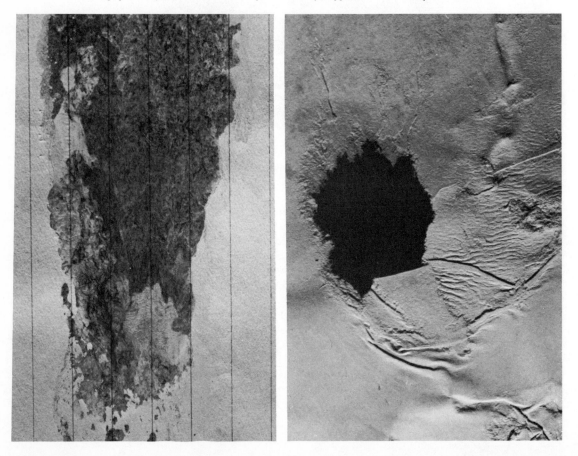

into the chemical for a short period of time or occasionally under pressure. Ordinary paints, varnishes, oils, and stains used as a wood finish have little, if any, value in protecting the wood from decay as moisture can penetrate through the surface (Fig. 11–11).

Paper

Raw materials and paper are vulnerable to fungus attack at every stage of paper manufacture. Pulpwood (90% of the raw material for paper) must be handled carefully during outdoor storage at the paper plant to avoid fungus attack. Decay of the pulpwood lowers the yield of pulp and destroys fibers which may lead to a 75% reduction in strength of the finished product if it is used in the paper (Gascoigne and Gascoigne, 1960). During manufacture, pulpwood is converted into pulp by mechanical grinding or through chemical treatments with sulfite, soda, or sulfate that separate the cellulose fibers. Pulp is vulnerable to fungus attack during storage by the usual mold fungi and wood-decaying fungi encountered in lumber. The most severe problem is the formation of slime in a watery suspension of pulp by bacteria and, to a lesser extent, by fungi. The slime, probably a product of hemicellulose digestion, damages paper made from the pulp by discoloring it or by producing shiny, translucent spots on it (Fig. 11–12). Fungi frequently isolated from slimes include *Penicillium*, *Cladosporium*, *Trichoderma* (Fig. 6–1), and *Mucor*. Additional damage during paper manufacture occurs from mildewing and discoloration of paper stocks.

Once the paper is finished, it provides a suitable substratum for cellulolytic fungi that may cause yellow, brown, or black discoloration and spotting through mildewing or destruction of the paper. Paper usually escapes attack by these organisms as it is not ordinarily used or maintained under very humid conditions, unlike some lumber. Numerous species of fungi have been isolated from paper, the majority of them members of the Deuteromycotina. *Aspergillus* and *Penicillium* are frequently encountered. Paper to be used under adverse conditions may be protected from fungi by incorporating a small amount of fungicide into the paper during manufacture.

Textiles

Textiles and cord made of linen, cotton, manilla, or jute consist almost entirely of cellulose and are subject to the attack by cellulolytic fungi, especially outdoors or under conditions of high humidity. Cellulolytic fungi attacking textiles include *Aspergillus* spp., *Chaetomium globosum*, *Cladosporium herbarum*, and *Humicola*. The marine fungus, *Zalerion*, a member of the Deuteromycotina, is a serious assailant on twines and ropes used in seawater. Principal damage to textiles results from musty odors, spotting and discoloration, loss in water repellency, loss in strength, and decrease in flexibility. Staining, both from pigment production and presence of fungus structures, is often the most objectionable effect. Fabric which is expected to be exposed to adverse environmental conditions (tents, cord, sandbags, netting, etc.) where it would be subject to fungus attack may be treated during manufacture with a fungicide, such as an aqueous solution of copper napthenate.

Woolens may be attacked by fungi, but comparatively few fungi attack wools and the damage is usually minor in comparison to fungus damage on cottons. *Microsporum* and *Trichophyton* and other wool decomposers are close allies of those fungi that cause skin diseases (see Chapter 12).

Food Spoilage and Mycotoxins

Every type of food is subject to infestation by fungi of one type or another. Food may be simply made unpalatable by a discoloration (as in rice), by production of off-flavors (as mustiness in coffee), or by the simple distaste of eating food with an obvious fungus growth. Food value is usually lessened, especially that of fats, carbohydrates, and proteins. Most often, food is made unfit for consumption by outright spoilage or through production of toxins. Losses during storage and marketing of food crops are due to mechanical damage, aging, and decay. Fungi are the agents most often responsible for decay.

Fruits and vegetables suffer from fungus infestation during storage, transport, and marketing. The fungi attacking fruits and vegetables may be field parasites, in addition to those fungi which occur predominantly in storage. Common examples of storage fungi are *Penicillium digitatum* on citrus fruits and *Rhizopus stolonifer* on sweet potatoes and other vegetables.

Microorganisms found in dairy products are usually bacteria, although fungi also occur and on occasion may cause deterioration of the product. Fungi in milk, butter, and cheese are primarily yeasts and deuteromycetes (*Geotrichum candidum, Penicillium, Alternaria,* and *Cladosporium*) (Figs. 6-1, 6-2). The predominant problem caused by these fungi is the development of a rancid or off-flavor.

Dried fruits are especially likely to support growth of yeasts on their surface where a sugary substance often accumulates. Infestation by yeasts leads commonly to souring of dried prunes both in storage and in the package and in the souring of dates after packaging.

Perhaps the most familiar example of fungi on food is that of *Rhizopus stolonifer,* the "bread mold," on bread. Numerous other fungi occur on bread, quickly making it unfit for consumption if the prevailing conditions are favorable for growth of fungi. The baker attempts to keep the bakery as free as possible of contaminating spores and to produce the bread under conditions as nearly aseptic as possible. In addition, preservatives are added to retard fungal growth.

MYCOTOXINS IN FOOD

Grains and forage plants may become infected by fungi as they stand and dry in the field or after they are harvested and stored. Infestation is most commonly by species of *Aspergillus* and *Penicillium,* which do not ordinarily infect the plants while living but are soil saprophytes. Fungal growth is encouraged if the grains are allowed to remain in the field during rainy weather or to overwinter in the field or, in the case of stored grains, if the grain is inadequately dried before storage, if it is damaged, or if the moisture or temperature levels are too high during storage. Nearly all the fungi on grains primarily attack the embryos, often completely destroying 14% to 15% of the embryos while not attacking the endosperm. These fungi have undesirable effects on the grains; including a decrease in the percentage of germination; discoloration of the embryos which appear as black specks in milled flour and may result in off-flavors; or the production of toxins.

The toxins may cause severe illness or death in animals or humans that eat the moldy grains, hay, or straw. For example, species of *Penicillium* and *Aspergillus* may invade standing or fallen corn in the late summer or early autumn, especially if the weather is rainy. This moldy corn is responsible for moldy corn toxicosis in pigs and cattle that forage on it in the field or consume animal feed prepared with moldy corn.

Poisoning due to moldy corn toxicosis results in hemorrhaging, liver damage, and frequently death. A second example is the poisoning of horses that are confined to a stall and fed damp straw on which the saprophyte *Dendrodochium toxicum* has grown. The poison acts principally on the central nervous and cardiovascular systems and results swiftly in death (Brook and White, 1966).

The best-known example of poisoning of humans by moldy grains results in a condition known as alimentary toxic aleukia, which results in abnormalities of the blood system. This condition has been observed sporadically in rural populations in Siberia and was especially widespread from 1941 to 1947 when epidemics resulted in poisoning of up to 10% of the population in some areas. Poisoning occurs when millet is allowed to overwinter in the field before it is harvested. The mild winters accompanied by numerous cycles of freezing and thawing favor growth of fungi that are resistant to cold and are even able to grow at $-10°C$. Toxin production is favored by temperatures between $1°C$ and $4°C$. Although various fungi may be involved, the principal causal agent is *Fusarium sporotrichoides,* which produces several toxins. Milling of the grain removes the outer infected portions of the grain and most of the toxins, and therefore poisoning is virtually unknown in the urban parts of Russia, where refined grain products are used. About 1.5 kilograms of contaminated grain eaten over a period of about 6 weeks induces pathological changes in the blood (Feuell, 1969). Symptoms of the poisoning are somewhat variable, but typically include fever, headache, diarrhea, nausea, and vomiting during the first few days. As the condition progresses, there is a destruction of the blood-forming elements of the bone marrow. Erythrocytes, platelets, hemoglobin, and leucocytes decrease and other abnormalities in the blood occur. There is a pronounced tendency for hemorrhaging to occur, and bleeding may occur from the nose, throat, gums or internally in organs such as the lungs, liver, or brain. Hemorrhagic rashes occur on the skin as a result of friction, and bleeding ulcers appear on the mucous membranes. Swelling of the lymph glands occurs, and may be so pronounced that death occurs by strangulation. This condition is frequently fatal, but recovery can result if the symptoms are recognized early and the patient does not consume additional contaminated millet. Control also involves educating the public about the necessity for handling grain properly and sampling grain to determine if it is contaminated. (Forgacs and Carll, 1962).

Aflatoxins. The toxins produced by fungi on grains include a subgroup known as the aflatoxins, all lactones with a similar molecular structure (Fig. 11–13) and capable of causing extensive pathological changes in the liver. The aflatoxins are produced by *Aspergillus flavus* and by some other closely related species of *Aspergillus* on stored wheat, corn, rice, barley, bran, flour, soybeans, and peanuts. These fungi occur commonly in the soil as saprophytes and do not generally seem to invade seeds prior to harvest.

The aflatoxins were discovered in the early 1960s when veterinarians, pathologists, microbiologists, nutritionists, and chemists teamed together to study an apparently new disease, turkey X disease. In 1960, more than 100,000 turkeys died within a few months within a radius of less than 100 miles of London but widespread death of turkeys did not occur elsewhere. The disease was characterized by a loss of appetite, lethargy, and a weakness of the wings. The turkey died within a week and at death assumed a characteristic position with the head drawn back and the legs extended backwards. Postmortem studies showed that the liver had hemorrhaged and had necrotic lesions, and frequently the kidneys were swollen. The scientists were unable to demonstrate that a pathogenic microorganism was responsible for the disease and concluded that the turkeys were being poisoned. It was finally established

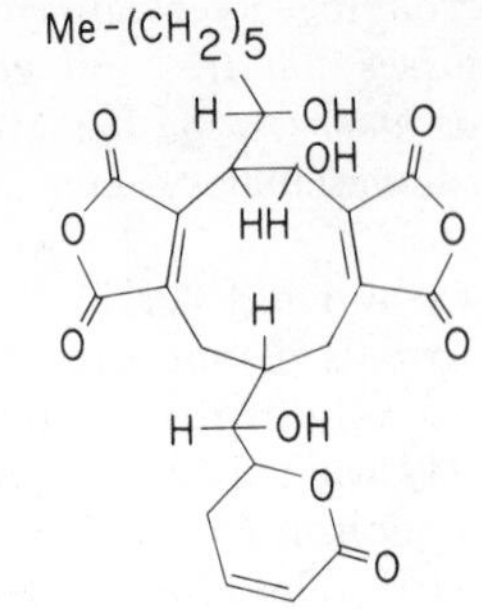

(a) Rubratoxin B

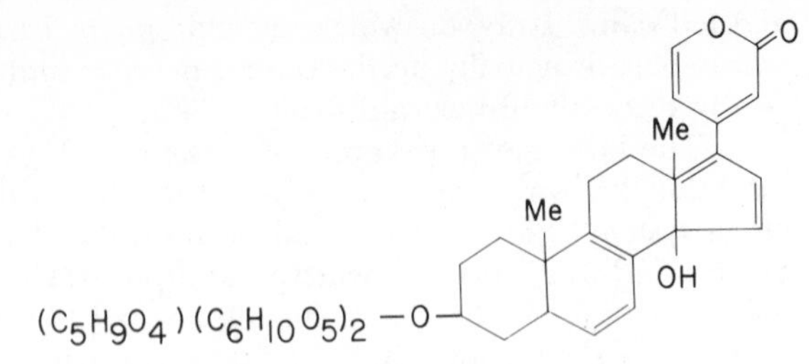

$(C_5H_9O_4)(C_6H_{10}O_5)_2 - O$

(b) Sporofusariogenin glycoside

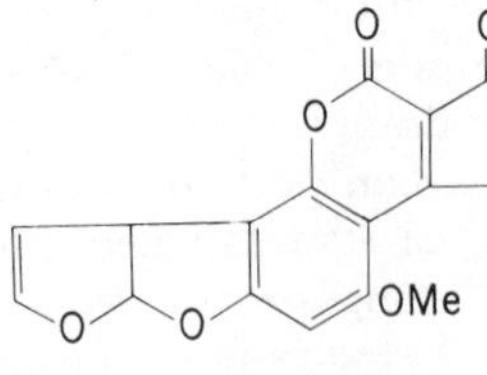

(c) Aflatoxin B_1

Fig. 11-13. (a) The principal toxin causing moldy corn toxicosis produced by *Penicillium rubrum;* (b) one of the toxins causing alimentary toxic aleukia; (c) the most commonly produced aflatoxin.

that all birds affected had been fed with feed produced at one particular mill in London. Peanut meal from that mill caused similar symptoms in poultry which were experimentally fed with it. A toxin was extracted from the meal and was found to be active in the purified form. The toxin was named *aflatoxin* as it was determined that it had been produced by *Aspergillus flavus*. Meanwhile in the United States a similar situation was developing. Trout in hatcheries had been characteristically fed with fresh animal by-products from slaughter houses, but their diet was changed to a dry food including grain derivatives. After two years, these fish developed tumors of the liver which has subsequently been attributed to aflatoxins in the feed (Goldblatt, 1969; Halver, 1969).

Similar sporadic outbreaks of poisoning due to aflatoxins (aflatoxicosis) have been observed in farm animals and have been attributed to aflatoxins in their feed. Studies with aflatoxins show that a wide variety of farm animals are susceptible to them but differ in their reactions. Young animals are more susceptible than older ones. For example, calves are more susceptible than mature cattle. Day-old ducklings are among the most susceptible of the animals, while mature sheep are perhaps the most resistant. The first clinical signs of aflatoxicosis are a loss of appetite and weight. A few days before death, the animals lose muscular coordination, appear to be dull, and become inactive. Pathological changes occur in the liver and possibly in other organs and vary somewhat from species to species.

Not all animals die when they consume food containing aflatoxins, and it is important to know whether human food produced from the farm animals contains aflatoxins. The aflatoxins are excreted in the urine and are not retained in the tissues that are normally used as meat or incorporated in the eggs. One aflatoxin, B_1, is metabolized and converted into a closely related toxin or toxins, the *M* toxins, found in milk. The *M* toxins are stable when the milk is dried, and because they are bound to the casein fraction, they are incorporated into cheese if cheese is made from the milk

(Allcroft, 1969). Rats fed primarily on milk containing the M toxin subsequently developed necrosis of the liver (Kraybill and Shapiro, 1969).

A high correlation can be shown between the consumption of food contaminated with aflatoxin and a high incidence of liver cancer in certain populations (Fig. 11-14). However it is not known to what extent the aflatoxin is responsible for causing the liver cancer or operates in conjunction with other predisposing factors such as malnutrition or parasitism. There are certain populations in parts of Africa south of the Sahara, Southeast Asia, Japan, and southern India where there is a high incidence of primary liver cancer and also a high utilization of grain products that may become contaminated by *Aspergillus flavus*. For example, a survey was conducted in Ethiopia, where it is the custom to store grain from one harvest to another, a period including the rainy season which encourages fungal growth. Gruels made from grain are eaten by nursing mothers, and these gruels and other grain products are an important part of the diet generally. Analysis of several samples of grains and grain products from a local market showed that *Aspergillus flavus* and aflatoxin B_1 were normally present (Kraybill and Shapiro, 1969). The highest rate of liver cancer occurs among the Bantu population in Mozambique, where primary liver cancer occurs at a rate approximately 500 times that in the United States. These people consume large quantities of corn meal. By drawing samples from food prepared for normal meals, it was found that 9.3% of the food sampled contained aflatoxin. The average daily intake of aflatoxin for each adult was calculated as 15 milligrams, the highest aflatoxin dosage known to occur in the human diet (Van Rensburg et al., 1974).

A practical problem is the production of animal or human foods that are free of aflatoxins. Contamination of peanuts is an especially important problem, as they are widely used for human and animal feed and aflatoxin production in them is commonplace. Therefore we will concentrate on peanuts to illustrate how aflatoxin levels in food may be controlled. The fungi are virtually absent in unblemished, immature peanuts just removed from the soil but invade the peanuts in the postharvest period while they are still moist and continue to do so until the moisture level drops to 10% or below. Growth of *Aspergillus flavus* on stored peanuts is encouraged by damage to the peanuts, maturity of the peanuts, high relative humidity (over 85%), and high temperatures (optimum 36°C to 38°C) during storage (Diener and Davis, 1969). Therefore control partially depends on harvesting immature peanuts in such a manner

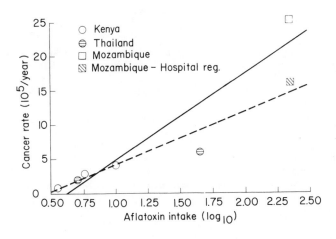

Fig. 11-14. Relationship between the level of aflatoxin intake of population and the primary liver cancer rate, obtained by pooling international data. Solid line: The cancer rate for Mozambique was calculated by equating the incidence rate for goldminers to that for the general population. Broken line: Hospital registrations were used to calculate the cancer rate for Mozambique. [Figure and legend from S. J. Van Rensburg, J. J. Van Der Watt, I. F. H. Purchase, L. P. Coutinho, and R. Markham, 1974, *South African Med. J.* **48**:2508a–2508d.]

that they are not damaged, culling out damaged pods, drying the peanuts quickly after harvest, and maintaining them at low temperatures and low relative humidity whenever they are stored. *A. flavus* is more likely to invade defective kernels that are broken, shriveled, or discolored than those in good condition. After harvest, batches of peanuts can be assayed for their possible contamination level by sorting out the peanuts which are undersized or not in good condition. Weighing is used to determine what percentage these defective peanuts represent of the total batch, and samples are examined with a microscope to determine whether or not *A. flavus* is visible. If either *A. flavus* is visible or the percentage of damaged peanuts is high, the peanuts are used to make oil, as the aflatoxins are not retained in the oil but are left behind in the pressed cake, which is then used for fertilizers. Assuming that the batch of peanuts has passed this initial inspection, it can be further processed for use as food in the form of salted nuts, confections, peanut butter, or animal food. The peanuts are subsequently shelled, and samples removed at the time of shelling and analyzed chemically for the presence of aflatoxins. If aflatoxins are absent in this test, the peanuts are processed further. At each stage, defective or broken kernels are culled out. By following these procedures, it is possible to produce a final product such as peanut butter in which aflatoxins cannot be detected even though present in low levels in the unprocessed batch of peanuts (Kensler and Natoli, 1969).

References

Alexander, M. 1977. Introduction to soil microbiology, Second ed. John Wiley & Sons, Inc., New York. 467 pp.

Allcroft, R. 1969. Aflatoxicosis in farm animals. *In:* Goldblatt, L. A., Ed., Aflatoxin—scientific background, control and implications. Academic Press, New York. pp. 237–264.

Barton, R. 1960. Antagonism amongst some sugar fungi. *In:* Parkinson, D., and J. S. Waid, Eds., The ecology of soil fungi: an international symposium. Liverpool University Press, Liverpool. pp. 160–167.

Birch, L. C., and D. P. Clark. 1953. Forest soil as an ecological community with special reference to fauna. Quart. Rev. Biol. **28**:13–36.

Brian, P. W. 1951. Antibiotics produced by fungi. Botan. Rev. **17**:357–430.

——, 1960. Antagonistic and competitive mechanisms limiting survival and activity of fungi in soil. *In:* Parkinson, D., and J. S. Waid, Eds., The ecology of soil fungi: an international symposium. Liverpool University Press, Liverpool. pp. 115–129.

Broadbent, D. 1966. Antibiotics produced by fungi. Botan. Rev. **32**:219–242.

Brook, P. J., and E. P. White. 1966. Fungus toxins affecting mammals. Ann. Rev. Phytopathology **4**:171–194.

Burges, A. 1965. The soil microflora—its nature and biology. *In:* Baker, K. F., and W. C. Snyder, Eds., Ecology of soil-borne plant pathogens—prelude to biological control. University of California Press, Berkeley, Los Angeles, Calif. pp. 21–32.

——. 1967. The decomposition of organic matter in the soil. *In:* Burges, A., and F. Raw, Eds., Soil biology. Academic Press, New York. pp. 479–492.

——, and E. Fenton. 1953. The effect of carbon dioxide on the growth of certain soil fungi. Trans. Brit. Mycol. Soc. **36**:104–108.

——, and P. Latter. 1960. Decomposition of humic acid by fungi. Nature **186**:404–405.

Chen, A. W., and D. M. Griffin. 1966. Soil physical factors and the ecology of fungi. VI. Interaction between temperature and soil moisture. Trans. Brit. Mycol. Soc. **49**:551–561.

Chesters, C. G. C., and R. H. Thornton. 1956. A comparison of techniques for isolating soil fungi. Trans. Brit. Mycol. Soc. **39**:301–313.

Christensen, C. M., and H. H. Kaufmann. 1965. Deterioration of stored grains by fungi. Ann. Rev. Phytopathol. **3**:69–84.

Clark, F. E. 1965. The concept of competition in microbial ecology. *In:* Baker, K. F., and W. C. Snyder. Eds., Ecology of soil-borne plant pathogens—prelude to biological control. University of California Press, Berkeley, Los Angeles, Calif. pp. 339–347.

Diener, U. L., and N. D. Davis. 1969. Aflatoxin formation by *Aspergillus flavus. In:* Goldblatt, L. A., Ed., Aflatoxin—scientific background, control and implications. Academic Press, New York. pp. 13–54.

Dobbs, C. G., W. H. Hinson, and J. Bywater. 1960. Inhibition of fungal growth in soils. *In:* Parkinson, D., and J. S. Waid, Eds., The ecology of soil fungi: an international symposium. Liverpool University Press, Liverpool. pp. 130–147.

Feuell, A. J. 1969. Types of mycotoxins in foods and feeds. *In:* Goldblatt, L. A., Ed., Aflatoxin—scientific background, control and implications. Academic Press, New York. pp. 187–222.

Flaig, W., H. Beutelspacher, and E. Rietz. 1975. Chemical composition and physical properties of humic substances. *In:* Gieseking, J. E., Ed., Soil components—organic components. Springer-Verlag, New York. **1**:1–211.

Forgacs, J., and W. T. Carll. 1962. Mycotoxicoses. Advan. Veterinary Sci. **7**:274–382.

Garrett, S. D. 1950. Ecology of the root-inhabiting fungi. Biol. Rev. **25**:220–254.

———. 1951. Ecological groups of soil fungi: a survey of substrate relationships. New Phytologist **50**:159–166.

———. 1955. Microbial ecology of the soil. Trans. Brit. Mycol. Soc. **38**:1–9.

———. 1963. Soil fungi and soil fertility. The Macmillan Co., New York. 165 pp.

Gascoigne, J. A., and M. M. Gascoigne. 1960. Biological degradation of cellulose. Butterworth & Co., Ltd., London. 264 pp.

Goldblatt, L. A. 1969. Introduction. *In:* Goldblatt, L. A., Ed., Aflatoxin—scientific background, control and implications. Academic Press, New York. pp. 1–11.

Golumbic, C. 1965. Fungal spoilage in stored food crops. *In:* Wogan, G. N., Ed., Mycotoxins in foodstuffs. Massachusetts Institute of Technology Press, Cambridge, Mass. pp. 49–67.

Gray, W. D. 1959. The relation of fungi to human affairs. Holt, Rinehart & Winston, New York. 510 pp.

Gray, T. R. G. 1976. Survival of vegetative microbes in soil. Symp. Soc. Gen. Microbiol. **26**:327–364.

Greathouse, G. A., B. Fleer, and C. J. Wessel. 1954. Chemical and physical agents of deterioration. *In:* Greathouse, G. A., and C. J. Wessel, Eds., Deterioration of materials—causes and preventive techniques. Reinhold Publishing Co., Stamford, Conn. pp. 71–174.

Griffin, D. M. 1963. Soil physical factors and the ecology of fungi III. Activity of fungi in relatively dry soil. Trans. Brit. Mycol. Soc. **46**:373–377.

———. 1969. Soil water in the ecology of fungi. Ann. Rev. Phytopathol. **7**:289–310.

———. 1972. Ecology of soil fungi. Syracuse University Press, Syracuse. 193 pp.

Halver, J. E. 1969. Aflatoxicosis and trout hepatoma. *In:* Goldblatt, L. A., Ed., Aflatoxin—scientific background, control, and implications. Academic Press, New York. pp. 265–306.

Harley, J. L. 1960. The physiology of soil fungi. *In:* Parkinson, D., and J. S. Waid, Eds., The ecology of soil fungi: an international symposium. Liverpool University Press, Liverpool. pp. 265–276.

Henderson, M. E. K. 1960. Studies on the physiology of lignin decomposition by soil fungi. *In:* Parkinson, D., and J. S. Waid, Eds., The ecology of soil fungi: an international symposium. Liverpool University Press, Liverpool. pp. 286–296.

Hendey, N. I. 1964. Some observations on *Cladosporium resinae* as a fuel contaminant and its possible role in the corrosion of aluminum alloy fuel tanks. Trans. Brit. Mycol. Soc. **47**:467–475.

Hiscocks, E. S. 1965. The importance of molds in the deterioration of tropical foods and feedstuffs. *In:* Wogan, G. N., Ed., Mycotoxins in foodstuffs. Massachusetts Institute of Technology Press, Cambridge, Mass. pp. 15–26.

Hudson, H. J. 1968. The ecology of fungi on plant remains above the soil. New Phytologist Co., **67**: 837–874.

Hunt, G. M. 1954. Wood and wood products. *In:* Greathouse, G. A. and C. J. Wessel, Eds., Deterioration of materials—causes and preventive techniques. Reinhold Publishing Co., Stamford, Conn. pp. 308–354.

———, and G. A. Garratt. 1953. Wood preservation. McGraw-Hill Book Company, New York. 417 pp.

Hurst, H. M., and N. A. Burges. 1967. Lignin and humic acids. *In:* McLaren, A. D., and G. H. Peterson, Eds., Soil biochemistry. Marcel Dekker, Inc., New York. pp. 260–286.

——, and ——, and P. Latter. 1962. Some aspects of the biochemistry of humic acid decomposition by fungi. Phytochemistry **1:**227–231.

Jackson, R. M. 1965. Antibiosis and fungistasis of soil microorganisms. *In:* Baker, K. F., and W. C. Snyder, Eds., Ecology of soil-borne plant pathogens—prelude to biological control. University of California Press, Berkeley, Los Angeles, Calif. pp. 363–373.

Joffe, A. Z. 1971. Alimentary toxic aleukia. *In:* S. Kadis, A. Ciegler, and S. J. Ajl, Eds., Microbial toxins. Academic Press, New York. **7:**139–189.

Kendrick, W. B. 1959. The time factor in the decomposition of coniferous leaf litter. Can. J. Botany **37:**907–912.

Kensler, C. J., and D. J. Natoli. 1969. Processing to ensure wholesome products. *In:* Goldblatt, L. A., Ed., Aflatoxin—scientific background, control and implications. Academic Press, New York. pp. 334–359.

Kevan, S. K. McE. 1965. The soil fauna—its nature and biology. *In:* Baker, K. F., and W. C. Snyder, Eds., Ecology of soil-borne plant pathogens—prelude to biological control. University of California Press, Berkeley, Los Angeles, Calif. pp. 33–51.

Kirk, T. K. 1971. Effects of microorganisms on lignin. Ann. Rev. Phytopathol. **9:**185–210.

Kononova, M. M. 1966. Soil organic matter—its nature, its role in soil formation and in soil fertility, Second Engl. ed., Translated by Nowakowski, T. Z., and A. C. D. Newman. Pergamon Press, New York. 544 pp.

Kraybill, H. F., and R. E. Shapiro. 1969. Implications of fungal toxicity to human health. *In:* Goldblatt, L. A., Ed., Aflatoxin—scientific background, control and implications. Academic Press, New York. pp. 401–441.

Lockwood, J. L. 1964. Soil fungistasis. Ann. Rev. Phytopathol. **2:**341–362.

Nicholas, D. J. D. 1965. Influence of the rhizosphere on the mineral nutrition of the plant. *In:* Baker, D. F., and W. C. Snyder, Eds., Ecology of soil-borne plant pathogens—prelude to biological control. University of California Press, Berkeley, Los Angeles, Calif. pp. 210–217.

Park, D. 1960. Antagonism—the background to soil fungi. *In:* Parkinson, D., and J. S. Waid, Eds., The ecology of soil fungi: an international symposium. Liverpool University Press, Liverpool. pp. 148–159.

——, 1967. The importance of antibiotics and inhibiting substances. *In:* Burges, A., and F. Raw, Eds., Soil biology. Academic Press, New York. pp. 435–447.

Raney, W. A. 1965. Physical factors of the soil as they affect soil microorganisms. *In:* Baker, K. F., and W. C. Snyder, Eds., Ecology of soil-borne plant pathogens—prelude to biological control. University of California Press, Berkeley, Los Angeles, Calif. pp. 115–119.

Robinson, P. M., D. Park, and M. K. Garrett. 1968. Sporostatic products of fungi. Trans. Brit. Mycol. Soc. **51:**113–124.

Rovira, A. D. 1965. Plant root exudates and their influence upon soil microorganisms. *In:* Baker, K. F., and W. C. Snyder, Eds., Ecology of soil-borne plant pathogens—prelude to biological control. University of California Press, Berkeley, Los Angeles, Calif. pp. 170–186.

St. George, A. A., T. E. Snyder, W. W. Dykstra, and L. S. Henderson. 1954. Biological agents of deterioration. *In:* Greathouse, G. A., and C. J. Wessel, Eds., Deterioration of materials—causes and preventive techniques. Reinhold Publishing Co., Stamford, Conn. pp. 175–233.

Sarkanen, K. V. 1963. Wood lignins. *In:* Browning, B. L., Ed., The chemistry of wood. John Wiley & Sons, Inc., New York. pp. 249–311.

Schuerch, C. 1963. The hemicelluloses. *In:* Browning, B. L., Ed., The chemistry of wood. John Wiley & Sons, Inc., New York. pp. 191–243.

Siu, R. G. H., and E. T. Reese. 1953. Decomposition of cellulose by microorganisms. Botan. Rev. **19:**377–416.

Van Rensburg, S. J., J. J. Van Der Watt, I. F. H. Purchase, L. P. Coutinho, and R. Markham. 1974. Primary liver cancer rate and aflatoxin intake in a high cancer area. South African Medical Journal **48:**2508a–2508d.

Waid, J. S. 1960. The growth of fungi in soil. *In:* Parkinson, D., and J. S. Waid, Eds., The ecology of soil fungi: an international symposium. Liverpool University Press, Liverpool. pp. 55–75.

Waksman, S. A. 1938. Humus: origin, chemical composition and importance in nature. The Williams & Wilkins Company, Baltimore. 526 pp.

——. 1944. Antibiotic substances, production by microorganisms—nature and mode of action. Am. J. Public Health **34:**358-364.

——. 1952. Soil microbiology. John Wiley & Sons, Inc., New York. 356 pp.

——, and E. S. Horning. 1943. Distribution of antagonistic fungi in nature and their antibiotic action. Mycologia **35:**47-65.

——, and W. Nissen. 1932. On the nutrition of the cultivated mushroom, *Agaricus campestris,* and the chemical changes brought about by this organism in the manure compost. Am. J. Botany **19:**514-537.

Warcup, J. H. 1951-a. Studies on the growth of basidiomycetes in soil. Ann. Botany London, II. **15:**305-317.

——, 1951-b. The ecology of soil fungi. Trans. Brit. Mycol. Soc. **34:**376-399.

——, 1959, Studies on basidiomycetes in soil. Trans. Brit. Mycol. Soc. **42:**45-52.

——. 1967. Fungi in soil. *In:* Burges, A., and F. Raw, Eds. Soil biology. Academic Press, New York. 532 pp.

Watson, A. G., and E. J. Ford. 1972. Soil fungistasis—a reappraisal. Ann. Rev. Phytopathol. **10:**327-348.

Wessel, C. J. 1954-a. Paper. *In:* Greathouse, G. A., and C. J. Wessel, Eds., Deterioration of materials—causes and preventive techniques. Reinhold Publishing Co., Stamford, Conn. pp. 355-407.

——, 1954-b. Textiles and cordage. *In:* Greathouse, G. A. and C. J. Wessel, Eds., Deterioration of materials—causes and preventive techniques. Reinhold Publishing Co., Stamford, Conn. pp. 408-506.

White, W. L., R. T. Darby, G. M. Stechert, and K. Sanderson. 1948. Assay of cellulolytic activity of molds isolated from fabrics and related items exposed in the tropics. Mycologia **40:**34-84.

——, G. R. Mandels, and R. G. H. Siu. 1950. Fungi in relation to the degradation of woolen fabrics. Mycologia **42:**199-223.

Wood, E. J. F. 1965. Marine microbial ecology. Reinhold Publishing Co., Stamford, Conn. 243 pp.

TWELVE

Fungi as Predators and Parasites

At least some fungi are capable of exploiting virtually every living organism as a source of food. Exploiting fungi are either predators or parasites. *Predators* actively trap other organisms, which they then kill and consume for food. *Parasites* invade a living plant or animal, feeding and multiplying within it at the host's expense but not contributing to the welfare of the host.

FUNGI AS PREDATORS

Several fungi obtain their livelihood by capturing animals (amoebae, rotifers, other protozoa, and nematodes) in a specialized trap and then feeding on them after their death. These are the *predacious* fungi; they are common in soil, dung, rotting wood, aquatic habitats, and especially mosses. Predacious fungi constitute the order Zoopagales entirely and are also numerous in the family Moniliaceae of the Hyphomycetes. These two groups represent the majority of the predacious fungi, although predacious fungi occur in diverse taxonomic groups.

Predacious fungi may be obtained in culture by pouring cornmeal agar* into petri dishes, allowing it to solidify, and then adding a small amount of substratum such as leaves, rotting wood, or moss plants with a small amount of soil clinging to their roots. Nematodes crawl from the debris, and the predacious fungi may often be

* Mix 20 grams of cornmeal with 1 liter of tap water; heat mixture at 70°C for 1 hour. Decant supernatant liquid through a glass wool filter, solidify with 20 grams agar, and autoclave.

440

found after an incubation period of 1 week to 2 months at room temperature (Duddington, 1955).

All the members of the Zoopagales are predacious fungi. As discussed in Chapter 3, these fungi are obligate predators on amoebae and other protozoa and may occur either exogenously or endogenously.

Predacious Hyphomycetes

Predacious members of the Moniliaceae are predominantly captors of nematodes. Unlike the Zoopagales, these fungi have a comparatively broad or unrestricted host range and are not obligate predators, meaning that they may live saprophytically in the absence of prey. The predacious Hyphomycetes are predominantly exogenous, capturing their prey through a large variety of mechanisms.

ADHESIVE ORGANS

A comparatively simple mechanism for the capture of nematodes is the formation of short lateral hyphal branches that curl and anastomose with similar branches or with the parent mycelium, forming single loops or a network of loops. Nematodes are captured in the sticky secretion of these loops or by entanglement in the network. An example of this type of mechanism may be found in *Arthrobotrys oligospora* (Fig 12-1). *A. oligospora* may be isolated by placing horse manure in a glass chamber and then culturing some of the milky fluid which rises on the sides of the container. This fluid is rich in both predacious fungi and nematodes (Duddington, 1956). The presence of nematodes stimulates the formation of traps in *A. oligospora,* which at first are single small loops formed by lateral branches that recurve and anastomose with the parent hypha. Additional loops are formed until a network results. All loops are at right angles to each other, so that some loops may be flat upon the substratum while others are perpendicular to the substratum, making it extremely difficult for a nematode to escape the network. Copious amounts of an adhesive viscous material are secreted from the loops when contact is made with a nematode, enabling the fungus to trap large nematodes that are 500 to 600 microns in length. After a nematode is trapped by

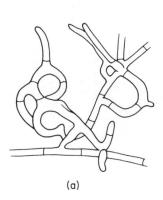

(a)

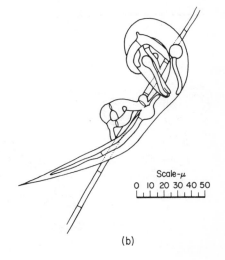

Scale-μ
0 10 20 30 40 50

(b)

Fig. 12-1. *Arthrobotrys oligospora:* (a) Network of loops; (b) Captured nematode with endogenous hyphae. [From C. Drechsler, 1937, *Mycologia* **29**:447–552.]

the loops, the dying nematode's integument is perforated by the fungus and a large globular organ practically equal to the width of the nematode's cavity is formed within the nematode. A number of hyphae arise from the globular organ and fill the cavity of the now deceased nematode, depleting the contents of the carcass. Death occurs within about 2 hours after capture (Drechsler, 1937; Duddington, 1955).

A subspherical knob borne on a short lateral branch is the organ of capture in certain members of the genera *Dactylaria* and *Dactylella* (Fig. 12-2) (Drechsler, 1937, 1935-c; Duddington, 1955). In *Dactylella ellipsospora*, the stalks bearing globose knobs are formed at regular intervals along the hyphae and are usually vertical so that the knob is held slightly above the surface of the substratum. These knobs may be formed in the absence of nematodes but are more prevalent in their presence. When a nematode comes into contact with a sticky knob, it is caught and struggles to get away, coming into contact with more knobs and being held more firmly. When the nematode has become quiescent, an outgrowth from the knob penetrates its body and forms a globose bulb from which trophic hyphae that fill its body cavity are formed.

Fig. 12-2.   (a) Knobbed hypha and nematode parasitized by *Dactylella ellip-sospora*. (b)-(e) *Dactylella lysipaga;* (b) Living nematode encircled by eight rings, none of them having yet penetrated the animal; (c) Hyphae with rings and knobs; (d) Infected nematode; (e) Conidiophore with conidia. [From C. Drechsler, 1935, *Mycologia* **27**:447-552.]

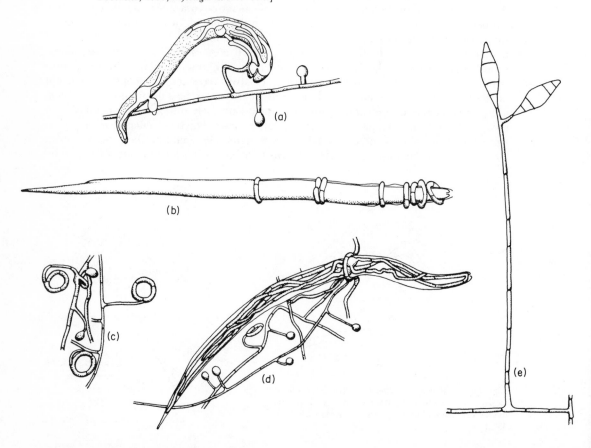

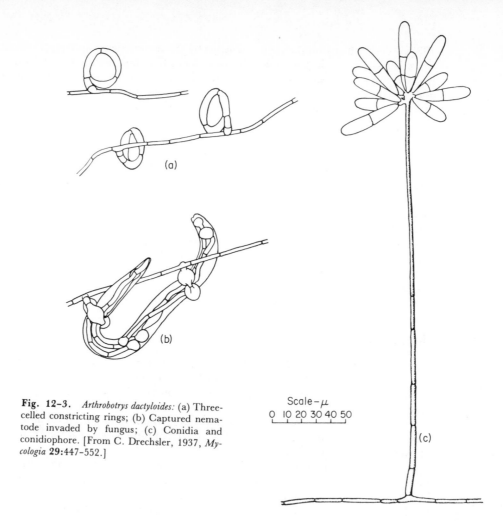

Fig. 12–3. *Arthrobotrys dactyloides:* (a) Three-celled constricting rings; (b) Captured nematode invaded by fungus; (c) Conidia and conidiophore. [From C. Drechsler, 1937, *Mycologia* **29**:447–552.]

Scale–μ
0 10 20 30 40 50

(c)

RINGS

Some predacious fungi form curved cells which join to form a closed ring. A nematode is trapped when it tries to pass through a ring in its wanderings and wedges itself in the ring when trying to squeeze through. Often the ring is broken off from the parent mycelium, and an active nematode may accumulate several rings before its activity is impaired. In *Dactylella lysipaga* and *D. leptospora,* fungi of this type, penetration of the nematode is accomplished by an outgrowth from the ring (even though isolated from the parent mycelium). This outgrowth is a globose knob which gives rise to trophic hyphae within the nematode (Drechsler, 1937)

Numerous predacious fungi, including *Arthrobotrys dactyloides* (Fig. 12-3), form rings similar in structure to the above but differing in that they are not passive. When a nematode pushes its head through a ring, the cells of the ring suddenly enlarge in diameter and shorten in length and thus firmly ensnare the nematode in the constriction. In *A. dactyloides,* constriction is so severe that the nematode's internal organs are frequently severed at the point of strangulation, resulting in the paralysis and death of the nematode. Hyphae penetrate the nematode, ending in curious knobs (Drechsler,

1937). Mechanical stimulation of the rings by the nematodes is apparently responsible for ring closure, as it was found that contact with a micromanipulator would cause ring closure in *Dactylaria brochophaga,* in which the ring increased to three times its size and closed in less than 0.1 second. Ring closure in *D. brochophaga* is a result of swelling of the vacuoles (Comandon and de Fonbrune, 1938-b).

TRAP INDUCTION

In the absence of nematodes, the fungal trapping organs are sometimes entirely absent or present in only small numbers. The addition to a vegetative culture of a small amount of sterile water in which nematodes have lived is sufficient to stimulate trap formation, indicating that one or more diffusible substances produced by the nematode are responsible for initiating this response (Comandon and de Fonbrune, 1938-a). A variety of biological materials will also induce a similar response: they include human blood serum, extracts from earthworms, yeast extract, and the amino acids valine, leucine, and isoleucine (Barron, 1977). Traps may be induced to form in *Arthrobotrys dactyloides* by water or nutrient deficiencies in culture, which leads to the suggestion that in nature both the presence of the prey and the generally unfavorable conditions lead to trap formation (Balan and Lechevalier, 1972).

Biological Control

Some studies have been made to determine whether it is feasible to utilize predacious fungi as biological agents to control plant-parasitic nematodes causing disease of crop plants. Addition of fungal spores alone to the soil does not increase the population of predacious fungi that occurs naturally. It is believed that agricultural soils ordinarily have a high population of indigenous nematode trapping fungi, and that as these are poor competitors, they trap nematodes under conditions adverse for their survival. However, if large quantities of green manure (fresh plant material that can be readily decomposed) or other organic matter is added to the soil along with the fungal spores, there may be a decrease in the nematode population. The green manure enhances the buildup of bacteria and free-living nematodes, which in turn encourages growth of the predacious fungi and trap formation, resulting in increased trapping of plant-parasitic nematodes (Barron, 1977).

FUNGI AS PARASITES

Parasitic fungi are those that invade a living plant or animal (the host), feeding and multiplying within it at the host's expense while not contributing to the welfare of the host. Some parasites may cause virtually no damage to the organisms on which they are feeding. The majority of parasitic fungi, however, are *pathogens* and cause an abnormal physiology in the host or a destruction of the host that develops concurrently with the parasitic relationship. This abnormal physiology culminates in *disease,* marked by detectable changes in function or morphology, *symptoms.*

Some parasitic fungi lead predominantly saprophytic lives (often in the soil) and become parasites only when they accidentally gain entrance into a potential victim, which is often weakened by injury or disease. These are the *facultative parasites.* The facultative parasites as a group are not specialized as parasites and often are so virulent as parasites that they cause the rapid death of their hosts. The early death of the host deprives the parasite of a source of food and may lead to starvation of the parasites.

In contrast to the facultative parasites, some fungi are *obligate parasites* and grow exclusively in living tissues. These fungi are highly specialized as parasites and establish a delicately balanced physiological relationship with their host. The host may show only mild symptoms and live indefinitely, although it is parasitized. In this manner, the life of the host is preserved and the parasite is ensured of a food supply for a long period of time.

Fungi as Parasites of Fungi

Fungi frequently grow upon other fungi in nature and perhaps occur exclusively in this partnership. Examples include the Pyrenomycete *Cordyceps,* which occurs on ascocarps of *Elaphomyces,* a member of the subterranean Tuberales; the Pyrenomycete *Hypomyces,* which may occur on agarics; and *Boletus parasiticus,* which arises from the base of a puffball (Figs. 12–4, 12–5). The relationship of these fungi to each other and many similar associations remain uncertain, as positive demonstrations of a parasitic relationship are largely lacking. Association does not imply a parasitic relationship; for example, Fitzpatrick (1915) studied the association of the agaric *Claudopus subdepluens* on *Polyporus perennis* (Fig. 12–6) and found that the mycelium of *C. subdepluens* could be traced through the tissues of the polypore and through the stipe, and it probably extended into a common substratum from which both fungi were deriving their nourishment. In addition, no physical connections (feeding structures such as haustoria) or damage to the host could be detected.

A true parasite-host relationship exists between many fungi in which the parasite relies directly on the host for its maintenance at the expense of the host. There are two fundamentally different groups of these parasites. A treatment of these two groups follows.

Fig. 12–4. *Cordyceps capitata,* a parasite on *Elaphomyces granulatus,* a subterranean fungus. [Courtesy Plant Pathology Department, Cornell University.]

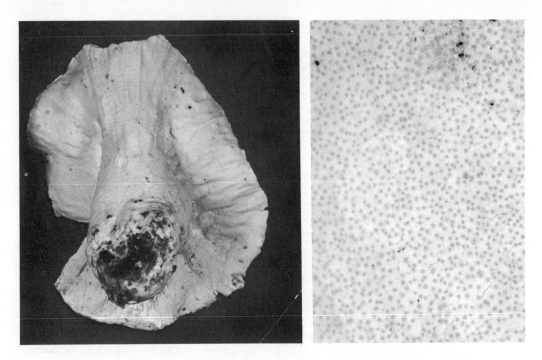

Fig. 12-5. Left, a deformed agaric parasitized by *Hypomyces*. Note the absence of gills. Right, an enlarged view of the diseased agaric showing perithecia of *Hypomyces* which appear as black dots. [Courtesy Plant Pathology Department, Cornell University.]

Fig. 12-6. *Claudopus subdepluens* growing on, but apparently not parasitizing, *Polyporus sp.* [Courtesy Plant Pathology Department, Cornell University.]

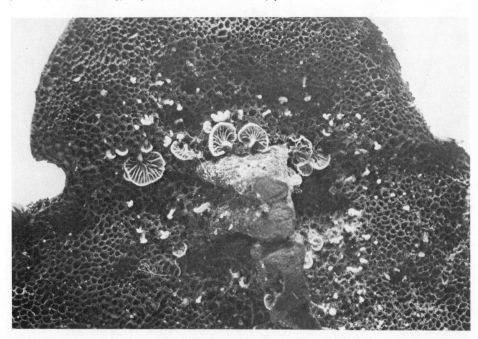

BIOTROPHIC PARASITES

The majority of the fungal parasites, the *biotrophic* parasites, develop a delicate physiological balance with their host and feed from the living host cell. These parasites vary greatly in the manner in which they parasitize their host, ranging from those that contact the host without penetrating it to those that develop within the host. Effects on the host may be rather insignificant or may eventually kill the host. A high degree of host specificity exists within this group, and a parasite may be restricted to a single host species or to a few closely related species. Many of these fungi are obligate parasites. These parasites include members of the chytrids, Mucorales, and Deuteromycotina.

More than 30 chytrid species are known to parasitize other chytrids and filamentous Oomycetes. The parasitic chytrids penetrate and develop within or upon their hosts and finally destroy the host protoplasm as maturity is reached. Many of the chytrid parasites belong to the genus *Rozella*, which causes a marked local enlargement of the host, a septation of the host cell, or both. A three-membered host-parasite relationship exists in which *Chytridium parasiticum* is parasitic on *Chytridium suburceolatum*, which parasitizes another chytrid, *Rhizidium richmondense* (Willoughby, 1956).

Many members of the Mucorales are parasites, the majority occurring on other Mucorales. These fungi penetrate the host and typically produce a haustorium with a number of slender, short branches (Fig. 12-7). *Piptocephalis* is an example of this type of parasite. Under laboratory conditions, parasitism by *Piptocephalis virginiana* causes no significant weight loss in the host (Berry, 1959). No species of *Piptocephalis* is known to grow indefinitely in the absence of a host and will develop normally only on a host. The degree of parasitism by *Piptocephalis* is correlated with the quantity of soluble nitrogen in the host mycelium, and is favored by a high nitrogen-to-carbon ratio and

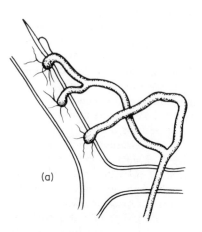

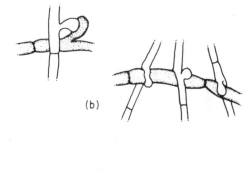

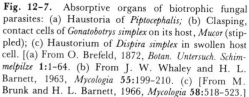

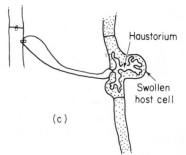

Fig. 12-7. Absorptive organs of biotrophic fungal parasites: (a) Haustoria of *Piptocephalis;* (b) Clasping, contact cells of *Gonatobotrys simplex* on its host, *Mucor* (stippled); (c) Haustorium of *Dispira simplex* in swollen host cell. [(a) From O. Brefeld, 1872, *Botan. Untersuch. Schimmelpilze* 1:1-64. (b) From J. W. Whaley and H. L. Barnett, 1963, *Mycologia* 55:199-210. (c) [From M. Brunk and H. L. Barnett, 1966, *Mycologia* 58:518-523.]

Haustorium

Swollen host cell

certain amino acids (Barnett and Binder, 1973). A few members of the Mucorales cause an enlarged gall to form at the point of contact with the host cell. This gall consists of both an enlarged or branched ending of the parasite hypha and an outgrowth of the host hypha.

Some members of the Deuteromycotina are parasites on other members of the Deuteromycotina, Oomycetes, Mucorales, or Ascomycotina. These parasites do not penetrate the host, but instead they form somewhat specialized hyphae which contact the host. The contacting hyphae may have flattened tips that adhere firmly to the host, globose or fingerlike projections that partially clasp the host hyphae, or other modifications (Fig. 12-7). The cytoplasm of the host and parasite may be in contact either through a large pore that forms in the wall at the interface of the two fungi or through plasmodesmata transversing that interface (Hoch, 1978). The biotrophic parasites may alter the morphology of the host or cause a slight reduction in the growth rate. *Calcarisporium parasiticum* and *Gonatobotrys simplex* are representatives of this group, and are discussed below.

Physiological Aspects of the Contact Parasites. If a parasite is to establish itself upon a host, its spores must be able to germinate in the presence of the host and then make contact with the host. Although the spores of some parasites are able to germinate in water, those of other parasites require exogenous substances to germinate. These exogenous substances may be provided by the host or by natural products in the substrate (e.g., nutrients or excretions of the host fungus). Contact with the host may be a result of trophic response. Germ tubes of *Calcarisporium parasiticum* and *Gonatobotrys simplex* secrete stimulants which cause host hyphae within a distance of about 40 microns to branch or to change directions and grow directly toward the parasite (Barnett and Lilly, 1958; Whaley and Barnett, 1963).

Next, a successful parasite must be able to establish a nutritional relationship with the host. This is usually preceded by penetration of the host or by formation of cytoplasmic contact. Host fungi differ in susceptibility to a parasite. Resistance of a potential host to a parasite may be due to the parasite's inability to establish a nutritional relationship with the host or inability to penetrate the host walls or to make cytoplasmic contact. Resistance to infection by *Calcarisporium parasiticum* may serve as an example. This organism parasitizes species of *Physalospora:* a large pore is formed in the parasite and host walls at the point of contact and the cytoplasm becomes confluent (Hoch, 1977-b). Resistance to infection may be due either to the inability of *C. parasiticum* to establish this cytoplasmic connection or to a deficiency within the host of a specific growth factor required by the parasite. To elucidate this situation further, *C. parasiticum* cannot be grown in the absence of its host unless an extract of the host mycelium is added to the culture. Ordinarily, *C. parasiticum* can parasitize *Physalospora obtusa* vigorously while it is unable to parasitize *P. rhodina*. Water extracts of both *P. obtusa* and *P. rhodina* are able to support growth of the parasite, indicating (1) that the required growth factor is retained within the mycelium (not secreted into the environment) and (2) that *C. parasiticum* cannot parasitize *P. rhodina* owing to an inability to establish the initial contact (Barnett and Lilly, 1958). The growth factor is believed to be a specific, vitaminlike compound. This growth factor was found in extracts from numerous fungi from the Ascomycotina, Basidiomycotina, and Deuteromycotina (in which immunity would be due to an inability by the parasite to obtain the growth factor), but it was not found in some fungi examined (in which case immunity would be due to a lack of a growth factor required by the parasite). Parasites other than *C. parasiticum* require the same growth factor.

After the parasite is established, nutrition of the host may influence the degree of parasitism, although the host is not noticeably affected. Parasites differ in their response to different nutritional conditions of the host. For one group of parasites, a high ratio of carbon to nitrogen in the host's medium may decrease growth of the parasite. In these instances, the internal composition of the host is varied and becomes less favorable for growth of the parasite. Perhaps the high carbon concentration leads to a depletion of the nitrogen reserve in the host, leading to nitrogen starvation of the parasite (Shigo, et al., 1961). In contrast, another group of parasites may not be favored by these same conditions.

NECROTROPHIC PARASITES

Unlike the biotrophic parasites, the *necrotrophic* parasites destroy their fungal hosts (apparently by means of enzymes or other toxic substances) and then absorb the nutrients from the dead host cells. Although these fungi are nutritionally dependent on their host fungi, they do not have a requirement for any specific nutrient. The host range of these parasites is rather broad, which indicates that their toxic substances are probably nonspecific. The majority of these fungi are capable of growing saprophytically on common laboratory media as well as on dead fungus structures. These fungi include several members of the Deuteromycotina and several wood-rotting basidiomycetes.

Rhizoctonia solani serves as an example of this type of parasitism (Butler, 1957). This fungus is a common plant parasite that may become parasitic on members of the Mucorales, Peronosporales, and Deuteromycotina. There are two basic types of infection. The first is that in which the parasite coils about the host but never penetrates the host. The second is that in which the parasite penetrates the host and establishes internal hyphae, sometimes forming external coils in addition (Fig. 12–8). *R. solani* may be induced to form coils about fibers of glass wool or to penetrate fibers of cotton in the absence of a host. The type of infection (with or without penetration) depends on the host species.

Coiling of parasite hyphae without penetration occurs when *Pythium* is the host. As *R. solani* overgrows the *Pythium* colony, tight coils are formed around the host hyphae. The protoplasm of the encircled hyphae either coagulates into oil-like globules or becomes progressively less dense and disappears. Finally the walls of the host collapse.

Penetration of the host occurs when *Gilbertella persicaria* is parasitized (Fig. 12–8). Short, lateral infection hyphae produced by *R. solani* penetrate the host. After penetration, the parasite grows rapidly within the host where it branches and enlarges to fill the host mycelium. Some of the internal hyphae of *R. solani* may penetrate the host's wall and grow outward again. Oil-like globules may appear in the host's cytoplasm, and the host's cytoplasm is diminished.

A given host may resist the invasion by *R. solani*. This resistance may be by (1) a disintegration or lysis of the internal hyphae of the parasite by the host or (2) a formation of walls separating parasitized portions of the host mycelium from those portions which are yet unparasitized, thus preventing the spread of the parasite within the host (Fig. 12–9).

Maximum infection of hosts by *R. solani* occurs under those environmental conditions which favor growth of both the host and parasite, but apparently no factor plays a disproportionate role in determining susceptibility.

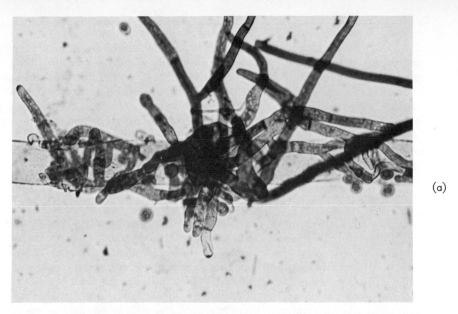

(a)

Fig. 12-8. Parasitism of fungi by *Rhizoctonia solani:* (a) *R. solani* coiling about a host, *Rhizopus stolonifer.* Internal infection hyphae will develop from this coiled hyphal mass; (b) *R. solani* within a sporangiophore of *Gilbertella persicaria;* (c) A later stage of infection of *R. solani* within *G. persicaria.* Note the hyphal network of the parasite, the diminished host cytoplasm, and oil-like droplets. Courtesy E. E. Butler.]

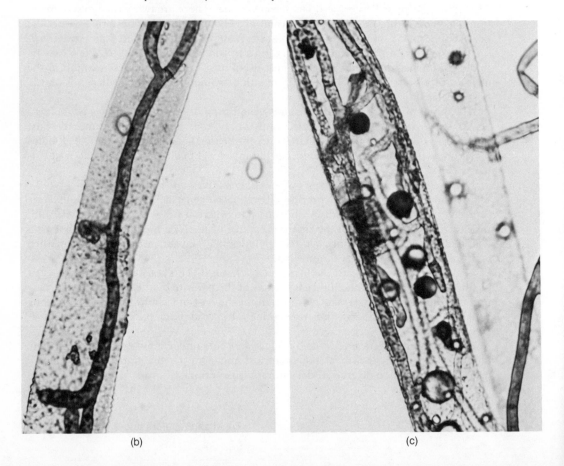

(b)   (c)

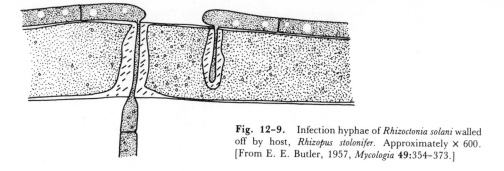

Fig. 12–9. Infection hyphae of *Rhizoctonia solani* walled off by host, *Rhizopus stolonifer*. Approximately × 600. [From E. E. Butler, 1957, *Mycologia* **49**:354–373.]

Fungi as Parasites of Animals

Virtually no animal is free from the possibility of incurring an invasion by a parasitic fungus, thus serving as its host and source of nutrients. Fungus diseases of animals vary from those that cause only a slight, superficial irritation that gives the host little, if any, discomfort (such as the Laboulbeniales on insects) to those that are quickly lethal.

ARTHROPOD PARASITES

A large number of fungi are parasitic on insects, other small arthropods such as mites and spiders, and larger arthropods such as crayfish and lobsters. Insect parasitic fungi include some chytrids, almost all members of the Entomophthorales, a few yeasts, numerous members of the Ascomycotina and Deuteromycotina, and the basidiomycete *Uredinella*. Crayfish and lobsters are attacked by members of the Oomycetes and *Fusarium*. Specific examples of insect diseases will be discussed.

Entomophthora Infections. Most insects are susceptible to infection caused by one of the more than 40 species of *Entomophthora*. An *Entomophthora* disease which is familiar to many laypersons is that of houseflies caused by *E. muscae* (Fig. 12–10). Flies killed by this fungus may commonly be found attached to window panes, surrounded by a halo of conidia forcibly discharged from conidiophores which emerge between the abdominal segments of the flies.

Infection of the host by the parasite and subsequent development of the parasite are more or less similar for all species of *Entomophthora* (MacLeod, 1963). A sporangium comes into contact with the host; a germ tube formed from the sporangium then penetrates the integument of the insect, especially in the thinner intersegmental areas or joints of the appendages, and enters the body cavity. Inside the host, the fungus usually grows by budding and fission until the body cavity is almost completely filled with short, thick hyphal bodies containing a fatty protoplasm (Fig. 12–11). In most instances, these parasites invade the entire head, thorax, and abdomen and may even be found within the tarsi. Finally all internal structures of the host are destroyed, leaving only the fungus-filled exoskeleton. During the early stages of the infection, there may be no noticeable effects of the disease, but later the insect becomes restless. The females may become sterile, which leads to a reduction in the insect population. Just before death, usually 5 to 8 days after infection, the host may crawl upward on blades of grass, settle on the underside of branches or leaves, or seek other elevated positions. Death characteristically occurs in the afternoon between 3 and 7 P.M. Upon death, the host is attached to its final substratum by rhizoids produced by the fungus which

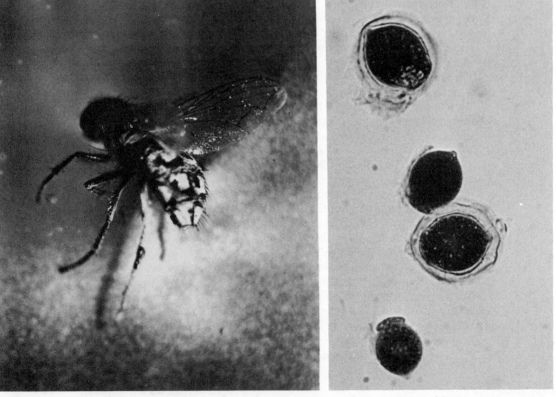

(a)

Fig. 12-10. *Entomophthora muscae* on insect hosts: (a) Note the cushionlike ring of sporangio-phores which protrude between the segments of the insect host. × 4; (b) A white mass of sporangia is distributed around the dead fly. × 5; (c) Discharged sporangia. [Courtesy J. Weiser, 1969, *An atlas of insect disease, Academia, Prague.*]

(b)

(c)

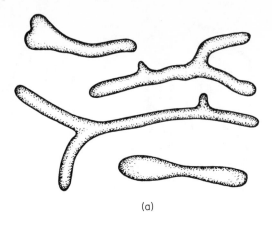

(a)

Fig. 12-11. *Entomophthora sphaerosperma:* (a) Hyphal bodies from a live, adult insect; (b) Stages in formation of zygospores (resting spores). Both × 400. [From A. G. Dustan, 1927, *J. Econ. Entomol.* **20**:68–75.]

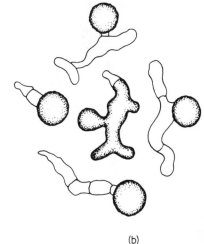

(b)

simultaneously forms sporangiophores and sporangia. The sporangiophores usually emerge through the intersegmental areas, forming cushion-like rings surrounding portions of the insect. The sporangiophores form overnight and discharge their sporangia the next morning while dew is still present. Sporangium formation may be interrupted, with sexual reproduction following, which leads to the formation of zygospores within the body cavity (Fig. 12-11).

Massospora Infections. The genus *Massospora* is also in the Entomophthorales but, unlike *Entomophthora,* consists of relatively few species. The best known member of this group is *Massospora cicadina* (Fig. 12-12), a parasite on the 17-year locust (MacLeod, 1963). It is possible that infection of the locust occurs underground at some time during its long dormancy. Infection is confined to the softer tissues in the posterior portion of the locust's body and, in some instances, is confined to male insects. Unlike *Entomophthora* species, this parasite forms oval sporangia in clusters within cavities in the abdomen that are not forcibly discharged. During the later stages of infection, posterior portions of the abdomen drop away successively. The exposed segment becomes filled with sporangia that burst through the confining chitinous ring as a creamy mass. As the mass of sporangia dries, it drops away, eventually followed by the entire abdominal segment. The insects will continue to fly around when left with only a head and thorax. Dark brown resting spores (zygospores) may be formed within the abdomen after the decline of the sporangial stage.

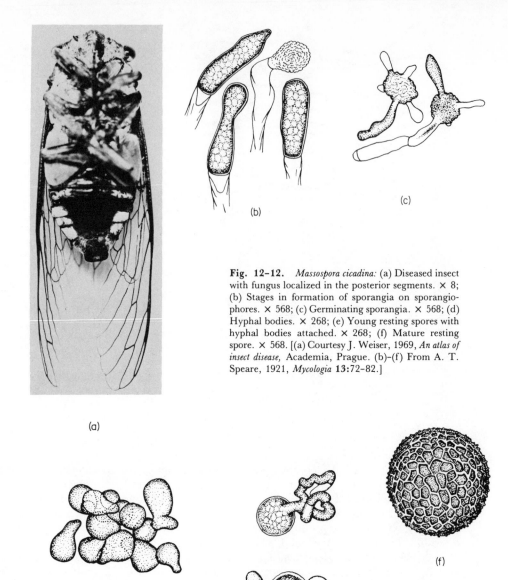

Fig. 12–12. *Massospora cicadina:* (a) Diseased insect with fungus localized in the posterior segments. × 8; (b) Stages in formation of sporangia on sporangiophores. × 568; (c) Germinating sporangia. × 568; (d) Hyphal bodies. × 268; (e) Young resting spores with hyphal bodies attached. × 268; (f) Mature resting spore. × 568. [(a) Courtesy J. Weiser, 1969, *An atlas of insect disease,* Academia, Prague. (b)-(f) From A. T. Speare, 1921, *Mycologia* **13:**72–82.]

Cordyceps Infections. The largest single genus of insect parasites in the ascomycetes is the genus *Cordyceps* (Hypocreales), which has about 200 species. Insects infected by *Cordyceps* are chiefly in the orders Hemiptera, Diptera, Lepidoptera, Hymenoptera, and Coleoptera. The larvae are most commonly infected. A few species attack spiders.

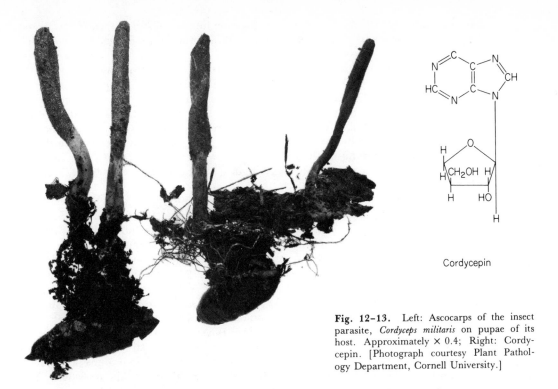

Cordycepin

Fig. 12–13. Left: Ascocarps of the insect parasite, *Cordyceps militaris* on pupae of its host. Approximately × 0.4; Right: Cordycepin. [Photograph courtesy Plant Pathology Department, Cornell University.]

Species of *Cordyceps* were the first insect-attacking fungi to be recognized, owing to their conspicuous, colored clublike stroma that arises from the diseased insect (Fig. 12–13). These infected insects were originally described as "vegetable wasps," "plant worms," or "trees" growing from insects (McEwen, 1963). In China, caterpillars infected with *Cordyceps* are considered to be both a tonic and a delicacy and are prepared in a broth. As they are expensive, only the middle classes or well-to-do can afford this delicacy (Hoffman, 1947).

Cordyceps militaris is perhaps the best known member of this group and will serve as an example of the life history of these fungi. (McEwen, 1963). The discharged ascospore fragments transversely into a number of smaller fragments, which may then land on the moist skin of a larva. The spore fragment becomes rounded, and a germ tube is put forth which directly penetrates the integument presumably by hydrolyzing the chitin. Within the larva, the rather thick hyphae ramify between the muscles and fatty tissue throughout the body cavity. The hyphae break down into a number of cylindrical segments, and they pass into the blood, where they elongate up to several times their original size. The segments divide repeatedly by transverse wall formation and by budding to form additional single-celled bodies. These cells are dispersed throughout the larvae until they fill the body cavity, meanwhile replacing or digesting the blood and softer tissues. At this point, the larva becomes soft and relaxed and dies. After death of the insect, the entire body cavity (except the alimentary canal) becomes filled with the budded cells, and the corpse expands to its former size and turgidity. In a few days, most of the insect tissues within the exoskeleton are replaced by fungus tissues, and the entire mass becomes hardened. The insect tissues are extremely resistant to decay due to the production of an antibiotic, *cordycepin* (Fig. 12–13), by the parasite (Madelin, 1966). The perfect stage of the fungus, a club-shaped yellow or red fleshy stroma, arises from the stromatized larva. Perithecia are formed in the swollen apical portion of the stroma.

Muscardine Diseases. "Muscardin" in French means a bonbon, and the term "muscardine" as adapted from it by French biologists is a descriptive term for a number of fungus diseases which transform the insects into a white mummy, somewhat reminiscent of a bonbon or sugared almond. The fungus responsible for causing a large number of the so-called muscardine diseases is *Beauveria bassiana,* a member of the Deuteromycotina. *B. bassiana* may cause disease in the silkworm, European corn borer, coddling moth larvae, and other insects (Fig.12–14). This fungus is also weakly pathogenic on rodents and humans (MacLeod, 1954; Petch, 1925; Steinhaus, 1949).

Muscardine diseases caused by *B. bassiana* are similar and the well-known infection of silkworm larvae serves as an example (Beauverie, 1914; Steinhaus, 1949). In silkworms, conidia of *B. bassiana* are transmitted from one individual to another through direct contact. The conidia germinate by means of a germ tube that directly penetrates the intact integument. Although direct penetration through the integument is the preferred route in the insects studied, Gabriel (1959) showed that conidia placed in the alimentary canal by means of a micropipette will in some instances germinate and penetrate through the lining of the alimentary canal. Presumably penetration of the silkworm occurs through a combination of mechanical pressure and enzyme secretion, as in the case of the corn borer (Lefebvre, 1934). The mycelial filament advances through the silkworm larvae and digests chitin in its path, destroying the chitin and causing a molt. As the filament penetrates through the cuticle and into the body cavity of the insect, the number of blood cells increases in the vicinity of the mycelium. The hypodermal tissue is destroyed, and the fungus multiplies in the blood in the form of short filaments or hyphal bodies that are disseminated to all parts of the insect's body, eventually destroying all internal organs.

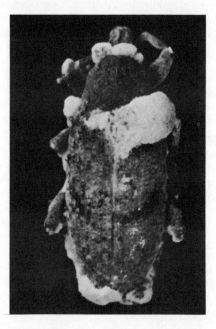

Fig. 12–14. *Beauveria bassiana* on dead beetle. × 2. [Courtesy J. Weiser, 1969, *An atlas of insect disease,* Academia, Prague.]

As the disease progresses, the volume of blood and number of blood cells decrease, while the parasite shows a corresponding increase. Certain changes also take place in the composition of the blood: normal acidity changes to near neutrality, certain amino acids disappear (especially after death), and oxalic acid crystals are formed. A toxin is also formed in the blood and, if injected into healthy larvae, will cause a swelling and stiffening of the body. In later stages of infection, blood circulation slows and then stops, and the consistency of the blood becomes pasty. The actual killing of the insect may be partly attributed to the increased viscosity of the blood (Madelin, 1963).

After death, the insect's body becomes progressively more hardened and reddish. Hyphae of *B. bassiana* emerge within 24 to 48 hours from the corpse to the exterior where they produce conidia under humid conditions. Meanwhile there is also the formation of a crystalline inflorescence, perhaps magnesium and ammonium oxalates.

Muscardine of silkworms may present a serious economic threat to growers of silkworms. This disease has been particularly severe in France and Italy, where infested silkworm nurseries are quarantined and workers from such a nursery are not allowed to visit healthy ones. Contaminations of healthy individuals with conidia of *B. bassiana* comes primarily from contact with diseased individuals. A silkworm is easily contaminated if conidia are present in the rearing cages or rooms, and the longevity of the conidia (up to 5 years in dry air) contributes greatly to this problem. If the disease becomes apparent during the rearing season, the infected larvae may be separated by hand or, more conveniently, all larvae are separated from food by a barrier of wire mesh. The healthy silkworms will be able to crawl through the barrier to obtain food, while the diseased ones will not be able to do so and can be culled out. The diseased larvae and litter should then be burned. Between rearing seasons, the surroundings may be disinfected with a fungicide.

Fungus Parasites as Biological Control Agents. Naturally occurring infections may be responsible for limiting insect populations. For example, the citrus mealybug is not a serious pest in Florida because it is held in check by *Entomophthora fumosa,* which is favored by the rainy, warm growing season. By contrast, the citrus mealybug is a serious pest in California, where the rainy season coincides with the cooler months, therefore not favoring growth of *E. fumosa* and allowing the insect pest to multiply in great numbers (Steinhaus, 1949). If a population of an insect pest is to be substantially reduced by a fungus in the field, there must generally be a high population density of the host, high levels of the fungal spores to serve as inoculum, and favorable climatic conditions. Intermittent rains that maintain a high relative humidity together with a water film are generally favorable. The humidity maintains the fungal inoculum in a viable condition, while the water film favors germination. Temperature requirements vary, but the optimum generally falls between 20°C and 30°C (Ferron, 1978; Hagen and Van den Bosch, 1968; Mackauer and Way, 1976).

Considerable effort has been expended to learn how to utilize infection of insects by pathogenic microorganisms to control insect pests. An advantage of biological control is that many microorganisms are host-specific and could be utilized to control specific insect pests while not injuring beneficial species. In contrast, insecticides kill both useful and harmful insects and also leave chemical residues on plant parts that will be used for food. Many research programs are being actively pursued in countries around the world (Ferron, 1978). In general, there has been some measure of success, especially with viral and bacterial parasites. Success with fungal parasites has been somewhat sporadic.

The use of a parasitic fungus to control insects may involve (1) manipulation of the environment by agricultural practices to enhance the rate of naturally occurring infection, for example, spraying crops with water to increase the humidity, or (2) the introduction of a suitable insect parasite into a population of insects. Most emphasis has been placed on the second method. Pioneer research of this type was carried out by Dustan (1927), who reared diseased insects in cages and then released them in orchards to infect healthy insect pests. Partial control of the green apple bug and the European apple sucker was obtained. In more recent studies, strains of pathogenic fungi are selected in the laboratory, grown in culture to obtain large numbers of spores, and then artificially disseminated in the field. Appropriate strains must be selected that are virulent for the insect pests and are host-specific so that they not only will be restricted to the insect pests but also will not become pathogenic on vertebrates. At present, there is little evidence that any substantial danger exists from a lack of specificity for insect hosts. Once a suitable strain is isolated and cultured, it must be released in the field under appropriate conditions. This is frequently difficult, as success often depends upon density of the insect host population, humidity, or rainfall (Ferron, 1978). For example, although species of *Entomophthora* infect aphids and reduce their populations naturally, it is difficult to obtain a higher kill rate by using introduced spores because the aphids do not generally occur in dense enough populations to enhance the activity of the fungus (Hagen and Van den Bosch, 1968).

Field studies indicate that certain insects may be controlled by pathogenic fungi under experimental conditions. For example, up to 60% of a population of mosquito larvae hatching from disease-free eggs can become infected by *Coelomomyces* in nature, which suggests that the artificial spread of *Coelomomyces* by humans may be useful in mosquito control (Chapman, 1974). In the Soviet Union, suspensions of conidia of *Beauveria bassiana* have been used to control the Colorado beetle and coddling moth. Utilization of *Metarhizium anisopliae* has been effective in reducing population of the coconut palm rhinoceros beetle in Tonga and of leafhoppers and froghoppers in pastures and on sugar cane in Brazil. A preparation containing conidia of *M. anisopliae* is now available to farmers in Brazil, who are beginning to use biological control in their everyday practice. In general, more progress is required before insect diseases are used as a matter of course in agriculture to control insect pests (Ferron, 1978).

FUNGI AS PARASITES OF BIRDS AND MAMMALS

Birds and mammals, including humans and their domesticated animals, are susceptible to attack by numerous fungi that cause disease. A disease of animals is frequently termed a *mycosis* and the name for the disease is generally derived by adding -*osis* or -*mycosis* to the name of the fungal agent. For example, the disease sporotrichosis, especially common among florists, gardeners, and berry pickers, is caused by *Sporotrichum schenckii,* a fungus which occurs on plants and enters the human body at the site of an injury, perhaps caused by the penetration of a splinter. Coccidioidomycosis, a lung disease of rodents, other animals, and humans, is caused by *Coccidioides immitis.* Some diseases, such as athlete's foot or ringworm, are simply known by their common descriptive name.

Mycotic diseases of higher animals and humans are ubiquitous, and it is probably rare that an individual escapes mild infection by a fungus throughout life. Common diseases such as ringworm of the scalp often occur in epidemic proportions among schoolchildren or among farm animals. Many individuals have had mild diseases from which they have recovered spontaneously and may learn that they had such a disease only when they are screened by appropriate tests (see further discussion on p. 460). Some of the diseases are especially likely to occur among people in certain occupations. For example, histoplasmosis occurs frequently among poultry farmers and explorers of caves because the causal fungus occurs in soil and in the eliminated feces of chickens, other birds, and bats and the spores can be inhaled from the air. In spite of the common occurrence of fungus diseases, there are comparatively few deaths among humans that are attributed directly to these diseases. In 1973, only 530 deaths in the United States were attributed to fungal diseases (Emmons et. al., 1977).

The Fungi. The fungi causing diseases of higher animals belong to different fungal groups, ranging from a few possible representatives in the Chytridiales to some members of the Agaricales. Diseases of cattle are caused by members of the Mucorales (*Mucor, Absidia,* and *Rhizopus*), which may invade the central nervous system or cause abortion. The majority of the diseases are caused by conidial stages of members of the Ascomycotina or by members of the Deuteromycotina. In the Ascomycotina, the pathogenic fungi include yeasts, Plectomycetes, and Pyrenomycetes. Members of the Basidiomycotina known to be parasitic on animals include *Ustilago maydis* (causal agent of corn smut), which was isolated from an epithelial infection of a farm laborer (Fischer, 1965), *Ustilago hypodytes,* which causes skin lesions among sugarcane workers and basket makers (Bereston, 1974), and the agarics *Schizophyllum commune* and *Coprinus cinereus* (Rippon, 1978).

Host ranges are comparatively broad in this group, and a parasitic fungus typically can infect a number of animal species. Most of the diseases that attack humans also occur on domesticated and wild animals. *Entomophthora coronata,* a member of the Entomophthorales that causes insect disease, also causes a disease in horses and humans characterized by nasal polyps, subcutaneous granulomas of the face, and infection of the sinuses. A degree of host specificity may be found in certain instances. For example, coccidioidomycosis occurs most frequently among dark-skinned males (Negro, Mexican, Filipino) between the ages of 30 and 50 years, while light-skinned females enjoy the greatest immunity to this disease.

Most fungi that may cause diseases of the higher animals are facultative parasites and these fungi may be isolated with great regularity from the soil. They are opportunists and if entrance into an animal through a wound or the respiratory tract is gained, they may become parasitic. *Coccidioides immitis, Aspergillus fumigatus,* and *Rhizopus* are common saprophytes that are widely distributed in the environment but may cause a fatal disease. Although spores from many hundreds of species of saprophytic fungi must be breathed in by humans and animals, come into contact superficially with the skin and nails, or be introduced in wounds, a very small number of species are apparently capable of causing infection. Only about 60 species of fungi may be regularly isolated from human subjects, and at least double that number of species have been isolated in rare instances (Emmons, 1960). With the exception of the dermatophytes, the soil serves as the source of inoculum for pathogenic fungi. Animal-to-animal or person-to-person contact resulting in the spread of a pathogen is rare. Obligate parasitism is rare or absent among the fungi that cause diseases of higher animals, and no fungus is known to be an obligate parasite in humans.

Fungi causing disease of animals fall into three major groups: (1) those fungi which cause systemic infections deep within the tissues (involving the vital organs or nervous system) and which are often fatal; (2) those fungi (the dermatophytes) which cause superficial and usually mild infections of the skin, nails, or subcutaneous tissues; and (3) those fungi causing infections intermediate in severity between the two extremes above.

Systemic Diseases. The fungi which cause systemic disease exist in nature as sporulating saprophytes and often gain entrance to the body through inhalation of spores (direct transmission from individual to individual rarely occurs). If spores are inhaled, the lungs are the organs which are first infected. Symptoms are similar to those of tuberculosis and include nodule or cavity formation within the lungs or inflammation of the lungs. Alternatively, the spores may be introduced into the body in a superficial wound caused by a splinter or other object. In this latter case, open pitlike ulcers are formed in superficial tissues. The parasite is disseminated to other parts of the body in the blood and lymph, and additional infections occur in other internal organs or in the nervous system. Although some infections are comparatively mild and may not be noticed, an alarming percentage are chronic or fatal.

Once the fungus has made contact with the body, it may or may not be able to invade the tissues as a parasite. Some of the fungi that cause systemic infections are especially successful as parasites. They have the ability to adapt to the higher temperatures of the host's body and to the lower oxidation-reduction potential of the tissues. The systemic fungi are also able to overcome the host's defense mechanisms by a combination of a rapid growth rate and relative insensitivity to these defense mechanisms. Unlike many pathogenic fungi, the most successful systemic fungi do not depend upon a debilitated host or injuries to become established. One of the unique features that the systemic fungi share with a few other pathogenic fungi is that they are *dimorphic* (Rippon, 1978, 1980).

Dimorphic fungi may occur in two distinct morphological forms. One of these is typical of the saprophytic phase and the second is typical of the parasitic phase. While parasitic within tissues, these fungi form unicellular bodies which resemble yeasts or are thick-walled spores. If these fungi are cultured on artificial media at moderate temperatures (about 25°C), colonies consisting of mycelium and conidia are formed. At higher temperatures (about 37°C), the cell form typical of the parasitic phase is produced (Fig. 12–15). If a parasitic fungus that ordinarily would become dimorphic and be disseminated in a warm-blooded animal is injected into a cold-blooded animal such as a frog or lizard, which is then maintained at temperatures of approximately 25°C to 27°C, the fungus remains in the mycelial stage and is incapable of producing a systemic infection (Rippon, 1980). Unlike the dimorphic fungi that cause systemic infections, those fungi that attack superficial tissues are not dimorphic but occur within the lesion in their mycelial form. If a fungus which ordinarily causes a superficial infection can be induced in culture to form a unicellular form, the unicells may be able to cause a systemic infection if injected into an animal. Such evidence indicates that the ability of a fungus to cause systemic infection is correlated with its ability to become dimorphic and form unicells.

An example of a systemic disease is coccidioidomycosis, which is also known as San Joaquin Valley fever and is caused by *Coccidioides immitis*. This dimorphic fungus produces mycelium and arthrospores in the soil. The arthrospores are inhaled and become spherical within a few hours or days. The enlarged arthrospore functions as a sporangium when mature and cleaves internally into sporangiospores. The sporangiospores are released when the outer wall ruptures, and after they have been disseminated within the body and have become mature, each sporangiospore may function as a sporangium and produce additional sporangiospores by cleaving internally. Although sexual stages are unknown, this fungus resembles the Zygomycetes because its mode of sporangiospore cleavage is similar to that occurring in members of the Zygomycetes (Emmons et al., 1977).

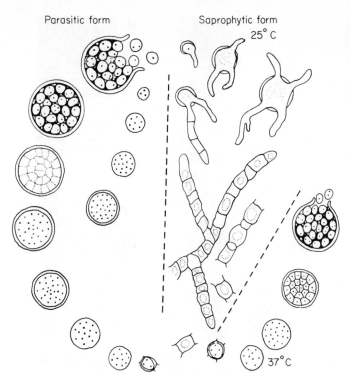

Parasitic form　　　　Saprophytic form
　　　　　　　　　　　　　　25° C

Fig. 12-15. Life cycle and dimorphism of *Coccidioides immitis,* showing both parasitic forms and saprophytic forms at 25°C and 37°C. [From J. W. Wilson and O. A. Plunkett, 1965, *The fungous diseases of man.* Reprinted by permission of the Regents of the University of California.]

37°C

Coccidioidomycosis occurs in rodents, dogs, cattle, horses, and sheep as well as in humans. It occurs principally in the southwestern parts of the United States, where dry dusty conditions prevail and the fungal spores are readily blown about. In that region, virtually all human residents have had the disease at one time or another; the infection may have been so mild that its existence was determined only by a hypersensitivity to an antigen injected intradermally (Emmons, 1960). Mild infections may go unnoticed or produce symptoms similar to those of pneumonia or tuberculosis. In severe cases, the fungus may be disseminated throughout the body in the blood stream, which leads to the formation of lesions of the skin, subcutaneous tissues, bones, joints, internal organs, and brain. Symptoms include fever, chills, prostration, progressive emaciation, and usually death in those cases where dissemination has occurred. In domestic animals, the infection by *Coccidioides immitis* is generally not severe and causes no apparent illness. Past or present infection is often not detected unless the animal is tested with an antigen or lesions in the lung or lymph nodes are detected after slaughter (Blood and Henderson, 1960). In a sample of sheep tested in California, 52% yielded a positive skin test for coccidioidomycosis, while up to 20% of the slaughtered cattle examined in Arizona had the characteristic lesions (Maddy, 1960).

Usually antibodies and a subsequent immunity result from an infection by a systemic fungus. This immunity helps the individual combat the original infection and prevents further infections by the same species. An immunity developed during the course of the disease may be adequate to cure the disease spontaneously without the benefit of additional treatment.

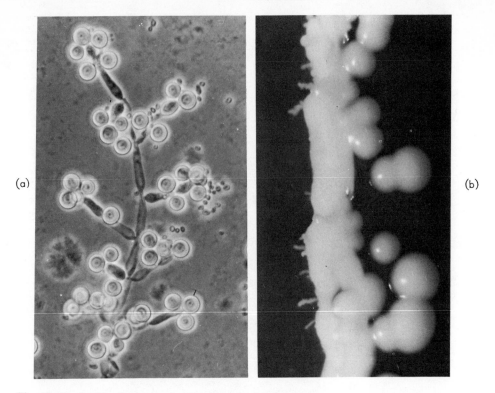

(a)

(b)

Fig. 12–16. *Candida albicans* and the disease it causes, candidiosis: (a) Conidia and conidiophores of *C. albicans*. × 600; (b) Yeastlike colonies of *C. albicans*. × 5; (c) Candidiosis of the tongue. [Courtesy P. M. D. Martin.]

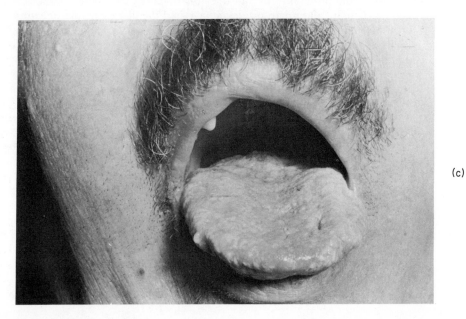

(c)

Treatment of systemic diseases includes bed rest and supplemental nutrients. These enable natural immunity to develop and cure the disease. Certain antifungal drugs may be used with varying degrees of success, and the effectiveness of a particular drug varies with the particular disease. The antibiotic amphotericin B is frequently helpful. In some instances, an infected body part may be amputated or a lesion surgically removed if the infection is localized.

Intermediate Diseases. Some parasitic fungi occur on the skin, subcutaneous tissue, lungs, or mucous membranes and may extend to a considerable depth within the tissue. Unlike the systemic diseases, the infection is not regularly distributed to other parts of the body. Infection may be acquired by inhalation of spores, by introduction of spores into a wound, or possibly by superficial contact with the fungus or by ingestion of spores. Unlike the systemic diseases, a specific immunity against these diseases is difficult to demonstrate. Examples of the intermediate diseases include candidiosis and aspergillosis.

Candidiosis (also known as "thrush") is caused by various species of *Candida*, most commonly by *Candida albicans* (Fig. 12–16). This fungus is a member of the Deuteromycotina and predominantly forms yeastlike cells, but there is some formation of mycelium. *C. albicans* is a normal saprophytic inhabitant of the intestinal tract or vagina but quickly becomes parasitic under debilitating circumstances such as malnutrition, unsanitary conditions, the presence of other diseases, or prolonged antibiotic therapy. Candidiosis may affect the skin around the fingernails or between the fingers of people who keep their hands wet for prolonged periods of time (e.g., dishwashers) or may affect the lungs. However, candidiosis is primarily a disease of the mucous membranes, on which soft, gray-white lesions will form (Fig. 12–16). In humans, candidiosis may occur as a vaginal infection, often appearing during pregnancy, when the glycogen content of the mucous membranes in the vagina increases, favoring the growth of the fungus. During birth, the infant is infected by fungal cells from the vagina and often develops candidiosis in the mouth, where lesions occur on gums, tonsils, tongue, or mucous membranes. In some cases, gasteroenteritis will develop in humans and may occur when infants swallow some of the *C. albicans* cells from their mouth. The fungus may be disseminated in the body in a small percentage of the cases, and may be lethal, especially in patients who are suffering from other diseases. Candidiosis also occurs in other mammals and birds (Small, 1971). It is common in caged pet birds such as canaries and parakeets when cages are not maintained in a sanitary condition and debilitating factors such as vitamin B deficiencies are present. The disease is most commonly localized in the crop of the bird. Baby pigs may develop candidiosis if they are reared on an artificial diet (Blood and Henderson, 1960). The piglets vomit frequently, become emaciated by 2 weeks of age, and have a more or less continuous white lining which covers the back of the tongue and esophagus. Candidiosis has also been observed in cattle in a feedlot, where they had diarrhea and symptoms similar to pneumonia, and resulted in the death of some of the cattle.

Treatment of candidiosis varies according to the severity and location of the infection. Systemic infection may be treated by injections of the antibiotic amphotericin B. For less severe localized infections, a second antibiotic, nystatin, may be administered orally or incorporated into topical creams or ointments for application to the infected area.

Aspergillosis, caused by *Aspergillus fumigatus,* is a disease that is common among newly hatched fowl, often occurring in epidemic proportions. The lungs of infected chicks become covered by a felt of mycelium and spores to the extent that breathing is impossible. This fungus grows abundantly on decomposing vegetation, and compost piles often serve as an important source of inoculum. Although primarily a parasite of fowl, *A. fumigatus* may cause abortion in cattle and sheep, corneal ulcer of the eye if introduced with an injury, and a disease of the honeybee (Emmons, 1960). Normal humans are immune to infection by *A. fumigatus;* but if infection does occur, symptoms are often necrosis and cavity formation in the lung.

Superficial Diseases. Superficial diseases are caused by fungi that attack the skin or its appendages (nails, feathers, and hair). These fungi are known as *dermatophytes* and are members of the Deuteromycotina or conidial stages of the Ascomycotina. They are almost entirely confined to superficial tissues, surviving in deeper tissues only in rare instances. Diseases of this type include various ringworms, athlete's foot, and epidermophytosis (Fig. 12–17).

Dermatophyte diseases are found in all parts of the world. The relative importance of any given disease varies from region to region, influenced by the local climatic conditions, nutrition, hygiene, or local clothing habits that tend to make perspiration accumulate in a particular part of the body. These fungi are widely distributed as soil saprophytes that may cause a disease upon contact of a potential host with the soil; as infective agents on dogs, cats, horses, cattle, sheep, and monkeys that will cause infection of humans if given the opportunity; or as infective agents that occur preferentially on humans. They are passed from person to person upon contact. The last type may

Fig. 12–17. A dermatophytic disease, epidermophytosis: (a) Symptoms on the hand; (b) The fungus, *Epidermophyton,* within the tissue of the hand prepared by biopsy. × 480. [Courtesy P. M. D. Martin.]

(a)

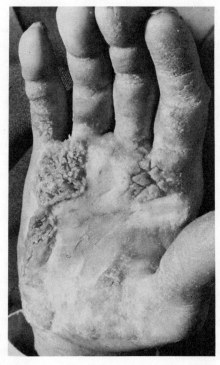

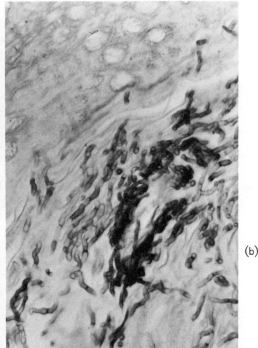

(b)

be responsible for epidemics among humans. Spores and mycelia are abundant in the environment in skin, hairs, and feathers shed from diseased individuals. Repeated contact with dermatophytes is inevitable, and infections may be severe under certain conditions. For example, during the Vietnam War, some American soldiers developed severe dermatophytic infections on the body and in the groin area. These infections were favored by the hot, damp conditions as these particular soldiers frequently walked in paddies and swamps and were immersed in water to the waist or higher. Most people have at least one dermatophyte infection during their lifetime and sometimes repeated infections. For example, athlete's foot is of extremely common occurrence, sometimes clinically diagnosed in more than 50% of the individuals of a given sample (Emmons, 1940). In spite of the ubiquity of the parasites and the occasional occurrence of an epidemic, the actual number of dermatophytic infections remains comparatively low, considering the abundance of inoculum of many types.

The low incidence of infection by contacted dermatophytes may be attributed in part to the fact that a high proportion of humans are apparently immune to attack by dermatophytes. In other instances, infection never takes place because the inoculum is washed away with soap and water and favorable circumstances for infection are not present. Circumstances favoring infection by dermatophytes include (1) frequent wetting of a body part, such as the hands of a dishwasher; (2) natural occurrence of a moist, oily skin, which is much less resistant than a dry skin, which continually sheds its outer cells; (3) close-clipping of hair, as commonly done at the nape of the neck in men so that the highly susceptible hair follicles which otherwise would be protected by a covering of resistant, inert hair are exposed; (4) wearing of tightly fitting, poorly ventilated shoes which do not allow perspiration to evaporate; and (5) obesity, which leads to folds and crevices in the flesh where moisture accumulates (Wilson and Plunkett, 1965).

Any dermatophyte may infect the skin, hair, or nails although usually a preference is shown for one of these. Symptomatology is primarily a function of the body part infected and, to a lesser extent, of the fungus causing the disease. For this reason, dermatophytoses are usually named for the body part infected and not for the fungal agent. Typical symptoms may include inflammation or redness of the infected part, scaling, brittleness and fissures of the nails, and loss of hair in the part infected.

As an example of a dermatophytic disease, we briefly examine ringworm. Ringworm is a dermatophytic disease that is important in humans as well as in dogs, cats, horses, and cattle. Among the animals, this disease may impair the growth of calves or damage the coat of fur-bearing animals or the leather produced from infected animals. The disease may be transmitted from one animal species to another or to humans by direct contact or by contact with contaminated bedding, harnesses, grooming aids, or feeding utensils. In rural areas, it has been estimated that 80% of the ringworm on humans results from infection by animals (Blood and Henderson, 1960). Ringworm is caused by species of *Trichophyton* and *Microsporum* that invade the keratinized layers of the skin, the hair fibers, or both. Autolysis of the hair occurs and it breaks off. There is also the production of a characteristic, circular dry, scabby lesion. The fungus invades the tissue and causes exudation to take place, and the exudate combines with the shedding epithelial material to produce the scab. Once a scab is established, the fungus cannot survive underneath it near the center because anaerobic conditions are established, and therefore the fungus dies at the center but survives at the periphery. The fungus then continues to grow centrifugally, extending the circular lesion that gives this disease its name. The infection is favored by a slightly alkaline pH. In humans, children are more readily infected because the pH of their skin is about 6.5, but the pH drops to about 4.0 at puberty when fatty acids are secreted. Both the lower pH and the fungistatic action of the fatty acids contribute to a lower incidence of ringworm in adults. Spontaneous cure may occur in both animals and humans. Ringworm may be treated by removing the scabby lesions by scraping or brushing and then applying an antifungal agent. Systemic treatment with griseofulvin is useful in both veterinary and human medicine.

The majority of the dermatophytic diseases may be effectively treated by the oral administration of griseofulvin (Fig. 12–18), an antifungal antibiotic prepared from *Penicillium griseofulvum*. Griseofulvin is incorporated into the epidermal cells in high enough concentrations to inhibit the growth of the dermatophyte, although it does not kill the fungus. Topical application of fungicidal materials is often ineffective because the medication cannot penetrate through the infected tissue. However, localized infections which are not extensive may be treated by the application of ointments or solutions containing antifungal agents such as selenium sulfide or sodium thiosulfate.

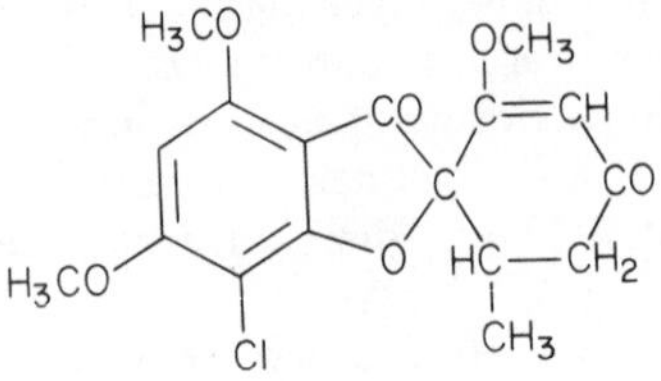

Fig. 12–18. Griseofulvin.

Fungi as Parasites of Plants

Diseases of plants may be caused by a number of organisms, including fungi, bacteria, algae, nematodes, viruses, and mites. The number of plant parasitic fungi is greater than the number of all other parasites combined.

All plant species are susceptible to attack by fungi, and examples of plant diseases may be found among the algae, which frequently suffer from chytrid infections in epidemic proportions; among mosses infected by the basidiomycete *Eocronartium muscicola;* among ferns, which are parasitized by a number of rusts; and among

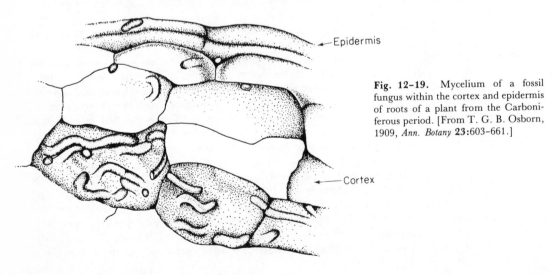

Epidermis

Cortex

Fig. 12–19. Mycelium of a fossil fungus within the cortex and epidermis of roots of a plant from the Carboniferous period. [From T. G. B. Osborn, 1909, *Ann. Botany* **23**:603–661.]

the higher vascular plants, which may have diseases such as rusts, smuts, powdery mildews, and others. The most ancient examples of fungi are plant-inhabiting fungi fossilized together with their vascular plant hosts (Fig. 12–19). It is undetermined whether these fungi were parasites or symbionts.

The most familiar fungi are those that have caused disease of food crops and have caused hunger or monetary loss since the dawn of agriculture. There are numerous passages in the Bible referring to "blasts," "mildews," and "blights" which often led to famine. As an example (I Kings 8:36, 37, 39),

> Then hear Thou in heaven, and forgive the sin of Thy servants . . . . If there be in the land famine, if there be pestilence, blasting, mildew, locust . . . or if there be caterpillar; if their enemy besiege them in the land of their cities; whatsoever plague, whatsoever sickness there be . . . then hear Thou in heaven Thy dwelling place, and forgive . . . .

The ancient Hebrews apparently believed that plant diseases were punishments by an angry God for their sins and thought that the situation could be alleviated through prayer. In a similar manner, around 700 B.C. and into the Christian era the Romans believed that the rust god, Robigus, was displeased when a 12-year-old boy tied straw to the tail of a fox (guilty of raiding his father's hen roost) and then lighted the straw before setting the fox free. Robigus punished the Romans on behalf of the fox by inflicting their wheat and barley with rust. Annual rituals were held to placate Robigus. These rituals often included white-robed processions, the sacrifice of a yellow dog or other yellow animal, and occasionally the tying of a lighted torch to the tail of a fox which was then chased through the fields (Stakman and Harrar, 1957).

Starvation due to plant diseases may be found in even more recent times. In the years 1845 to 1860, a million people died in Ireland as a result of illness and starvation after their potato crops were ravaged by late blight of potatoes (caused by *Phytophthora infestans).* An additional $1\frac{1}{2}$ million people (about one-fourth of the total population) emigrated to other countries. The misery caused by this disease is reflected in the following passage (Large, 1940):

> "On July 27th," wrote Father Matthew, "I passed from Cork to Dublin, and this doomed plant bloomed in all the luxuriance of an abundant harvest. Returning on August 3rd I beheld with sorrow one wide waste of putrefying vegetation. In many places the wretched people were seated on the fences of their decaying gardens, wringing their hands and wailing bitterly at the destruction which had left them foodless."

The massive emigration from Ireland affected both the future of Ireland and the countries into which the people immigrated.

At the present time, plant disease epidemics may still wipe out an entire crop if conditions favorable for disease develop. The yield of cereal crops (which supply 80% of human food) may be reduced by 80% during a single epidemic of the stem rust disease. In an advanced country, where agriculture is sufficiently diversified and there is a sound economy, such an epidemic may not result in starvation, as it would have in previous centuries, but even lesser epidemics would spell starvation and disaster for an underdeveloped country.

Although sudden loss of an entire crop from epidemics may pose a serious threat, the majority of losses result from smaller outbreaks of disease which often occur with greater regularity. These losses are due to the reduction in crop yield and also to the expense of utilizing control measures as a regular agricultural practice. In the United States the average annual loss has been estimated to be in the billions of dollars for all types of plant disease, while the average loss in yield from each crop as a result of disease is 10%.

Since 1974, a combination of factors has reduced the world's grain reserves to extremely low levels. Factors contributing to this shortage are the increasing populations in developing countries; the rising consumption of beef, which results in an increased demand for feed grains; shortages and high costs of pesticides, fertilizers, and fossil fuels; adverse weather conditions; and finally, the cumulative effects of a number of plant disease and pest problems (Apple, 1977). The unrelenting pressures on the world's food supplies mean that productivity from crop plants must be maximized and plant diseases controlled.

FUNGI THAT CAUSE DISEASE

Fungi that cause plant disease may be found in all major groups of fungi, ranging from the morphologically simple *Plasmodiophora brassicae,* which causes clubroot of cabbage, to the complex members of the Agaricaceae such as *Armillaria mellea* (the "honey fungus"), which causes root rots of standing trees. The groups with the largest number of plant pathogens are the Deuteromycotina and Ascomycotina. Such a diverse group of fungi has little in common other than the ability to parasitize living plants.

Plant parasitic fungi range from those that are facultative parasites and lead predominantly saprophytic lives in the soil to those that are obligate parasites of plants.

Pathogenicity. The plant parasites vary greatly in their *pathogenicity,* the ability to cause disease in different host plants. Some fungi, especially the facultative parasites, have comparatively wide host ranges. An example is *Verticillium albo-atrum* which has been isolated from more than 70 host genera. Many parasites will infect several host species that are closely related; we may cite *Venturia inaequalis* (causal agent of apple scab) which attacks species in the host genus *Malus.* Some parasites such as the Erysiphales are so specialized that they are capable of attacking only one host species. Host specificity is carried to its ultimate by some races of plant parasites which can attack only certain genetic races or lines of a single host species, as the rusts do. A high degree of host specificity is typical among the obligate parasites.

It is known for many host-parasite combinations that both resistance of the host and the alternative state, susceptibility, and the pathogenicity of the fungus and its alternative state, nonpathogenicity, are genetically controlled. One of the better understood relationships is the *gene-for-gene* mechanism which occurs between the flax rust *Melampsora lini* and its host. For each gene controlling pathogenicity in *M. lini,* there is a corresponding gene in the flax controlling resistance. Resistance of a specific host race to a specific pathogen race occurs only where the complementary genes of both the host and parasite are dominant (Fig. 12-20). A similar gene-for-gene mechanism exists between several plant hosts and their pathogens, including several smut, rust, and powdery mildew fungi (Flor, 1971), as well as the apple scab fungus *Venturia inaequalis* (Keitt et al., 1959).

New physiologic varieties, races, or biotypes of pathogenic fungi may arise through mutation or hybridization of sexually reproducing fungi (as, for example, in the rusts, Chapter 5), and new genetic lines of nonsexually reproducing pathogens may arise through mutation, heterokaryosis, and the parasexual cycle, previously discussed in Chapter 9. Race formation provides for a flexibility of genotypes, and the resultant phenotype may be endowed with greater pathogenicity, allowing the establishment of a fungus on a new host in nature.

Fungus race no.	Fungus race genotype			Reaction[a] of plants possessing genes							
				llmmnn	L	M^a	N^1	LM^a	LN^1	M^aN^1	LM^aN^1
108	$a_L a_L$	A_M^a	A_N^1	S	S	R	R	R	R	R	R
123	A_L	$a_M^a a_M^a$	A_N^1	S	R	S	R	R	R	R	R
52	A_L	A_M^a	$a_N^1 a_N^1$	S	R	R	S	R	R	R	R
156	$a_L a_L$	A_M^a	$a_N^1 a_N^1$	S	S	R	S	R	S	R	R
192	A_L	$a_M^a a_M^a$	$a_N^1 a_N^1$	S	R	S	S	R	R	S	R
154	$a_L a_L$	$a_M^a a_M^a$	A_N^1	S	S	S	R	S	R	R	R

[a] S = susceptible; R = resistant

Fig. 12-20. Host reaction to different combinations of three sets of host-parasite complementary genes in the gene-for-gene mechanism of *Melampsora lini* on flax. [From H. H. Flor, 1955, *Phytopathology* **45**:680–685.]

THE PLANT DISEASES

Penetration. Parasitic fungi may enter the plant through (1) wounds, (2) natural openings, or (3) direct penetration. A fungus may be capable of utilizing only one means of entrance or any combination. Wounds are created when the plant is damaged, often during cultivation, by insects or animals, or by growth when roots penetrate the cortex. Natural openings such as lenticels or stomates provide a means of entrance for some fungi (Fig. 12–21). A third major means of entrance is by direct penetration of the intact plant epidermis. Those fungi that directly penetrate the epidermis typically form an enlarged, flattened *appressorium* that adheres firmly to the cell wall. Penetration is effected by a slender outgrowth from this, the stylet-like infection peg, which forces its way into the cell through application of pressure, probably aided by partial enzymatic digestion of the cuticle and the wall.

Fig. 12-21. Mycelium of *Thielaviopsis basicola* entering a leaf through a stomate. [Courtesy Plant Pathology Department, Cornell University.]

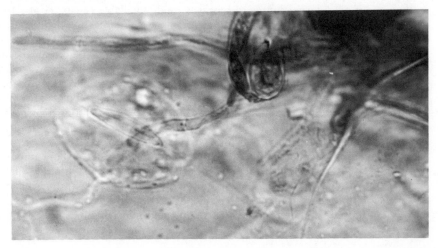

Generally younger plants are more susceptible to invasion through natural openings and direct penetration than older plants, which may have developed additional cutin, cork, or lignin, all barriers to the parasite, or a toughened epidermal cell wall. A plant may be resistant to a parasite if the fungus cannot penetrate the plant because of a structural barrier in the plant; for example, *Cercospora beticola* enters sugar beet leaves only through open stomata. Young leaves in which stomatal movement is not active escape infection; so do old leaves which are not invaded because stomatal movement is slight. Mature leaves with actively functioning stomata are seriously infected (Pool and McKay, 1916).

Establishment Within the Host. The parasite may or may not be able to live within the host after having successfully penetrated it. A prime requisite is that the fungus be able to utilize the immediately available food supply, which requires the presence of appropriate enzymes. These may include a wide variety of hydrolytic enzymes capable of cleaving pectin, cellulose, lipid, and protein into their monomers (a root-rotting fungus that enters the leaves of a susceptible plant by accident may not survive because it cannot utilize the nutrients immediately available in the leaves). A second requisite is that the fungus be able to survive and penetrate normally through the tissues of the plant in order to maintain a source of food (for example, the host must not be resistant). The fungus may not be able to do this if (1) the host protoplasm provides an unfavorable environment, (2) a hypersensitive reaction by the host occurs, or (3) mechanical barriers to the growth of the fungus are formed by the host.

The nature of protoplasmic factors that make a cell incompatible for the growth of a fungus are poorly understood, but they constitute the primary line of defense for the plant. Possible mechanisms include the presence of a substance that is inhibitory to the pathogen. An example of this may be found in colored onions resistant to *Colletotrichum circinans* (the causal agent of onion smudge) owing to the presence of protocatechuic acid, which causes a bursting of the spores or young hyphae of the pathogen. This compound is lacking in nonpigmented susceptible varieties (Walker, et al., 1929). In other instances, an inhibitory metabolite may be induced to form after infection, the pH or osmotic concentration of the protoplasm may be unfavorable, or the protoplasm may not be able to provide essential nutrients for the fungus: an example of the last situation is found in *Venturia inaequalis,* which has specific nutritional requirements that must be met by the host. A resistant host may lack biotin, for example, and a nonpathogenic fungus strain which has a biotin requirement may become pathogenic if an external source of biotin is made available (Keitt, et al., 1959).

Hypersensitive plants are those that are so sensitive to infection by the parasite that the invaded cells die quickly, thus isolating the parasite in a few killed cells where it can no longer live. Although a few cells are killed, the plant remains healthy. A hypersensitive host may produce large quantities of fungal-inhibiting metabolites in response to the fungal infection, thus killing the invading fungus. A rapid browning and collapse of the host cells accompanies this reaction. The hypersensitive reaction typically occurs for incompatible pathogen and host combinations controlled by the gene-for-gene mechanism and is likely to occur when the first haustorium penetrates the plant cell or when several haustoria are formed (Bushnell, 1972).

Mechanical barriers simply are cellular components that make it impossible for the pathogen to penetrate further into the host tissue. Many diseased plants have localized hemispherical thickenings that often appear to be part of the walls but differ in their composition and may be callose deposits. Such thickenings are formed by the host cell at the site of penetration by the fungus and may actually encase the fungus in a sheath. Although it is difficult to prove, it has been suggested that these sheaths may function as a mechanical barrier (Aist, 1976). Other mechanical barriers include (1) the formation of wound cork or a gum deposit in response to infection; (2) the natural occurrence of cell walls that are difficult to penetrate (often owing to secondary thickenings); and (3) the natural occurrence of middle lamellae that are often unusually resistant to enzymatic degradation.

Infection. Assuming that the parasite is able to establish itself within the host, we may consider the means by which it develops within the tissues and the means by which it causes damage to the host.

One group of parasites (those causing soft rots, a watery disintegration of tissues) kills tissues in advance of their spread by secreting pectic enzymes that destroy the middle lamellae and the pectins of the cell walls, causing the cells to separate and hence the death of the tissue. Death is associated with rapid permeability changes resulting from plasmalemma damage. Although the exact mechanism of plasmalemma damage is not known, it may be related to the swelling due to osmotic pressure exerted in the absence of the restraining walls and permitting electrolyte leakage to occur (Mount, 1978). It is possible that nonenzymatic toxins are also involved in killing the cells. The fungi then live saprophytically on the dead cell material. Rotting of sweet potatoes by *Rhizopus stolonifer* occurs by this mechanism.

A second major group of pathogens consists of those that feed directly from the living plant cells, either with or without eventually killing them. They must either penetrate the cells or live in close conjunction with them. These fungi may be (1) intercellular with haustoria, (2) intercellular with no means of penetrating the cells, (3) intracellular, or (4) any intergradation of these. Pathological damage to the host may be a disintegration of the tissue or interference with normal growth, reproduction, or physiological processes (as water transport, or respiration).

Tissue Disintegration. Tissue and cell disintegration may occur by the destruction of the middle lamellae by pectic enzymes or of the cell walls by cellulolytic enzymes or both. This is equivalent to that type of tissue disintegration occurring in advance of the fungus, as found in the soft rots (above). The extent of wall breakdown or softening may vary, and the effect, unlike that found in the soft rots, may simply result in a weakening of the tissue rather than its maceration.

A second type of cell damage is that in which the protoplast is attacked directly. Although comparatively little may be known about the manner in which the protoplast is attacked for a particular disease, it is possible that the primary damage may be (1) a destruction or alteration of the semipermeable membrane, with a subsequent loss of water and metabolites from the cell and an unrestricted movement of molecules into the cell; or (2) a direct action of a proteolytic enzyme or nonenzymatic toxin on the protoplast, disturbing or inhibiting normal metabolic processes. Toxins are produced by a wide variety of pathogenic fungi, and the damaging effects may be experimentally exerted by toxins in the absence of the fungi which produced them. Many toxins are known that affect the plasmalemma; for example, the toxin *victorin* produced by *Helminthosporium victoriae* causes the host cells to become leaky and lose large quantities of electrolytes and other materials, while uptake and accumulation of mineral salts and other materials are inhibited. Eventually the oat cells infected by *H. victoriae* undergo false plasmolysis and die. Although many toxins from other fungi have similar effects, unique effects such as the increase in the uptake of nitrate may be induced by other toxins (Wheeler, 1968). The fungus benefits from these morbid activities because it is able to use the damaged cell contents for growth.

Fungi as Predators and Parasites   **471**

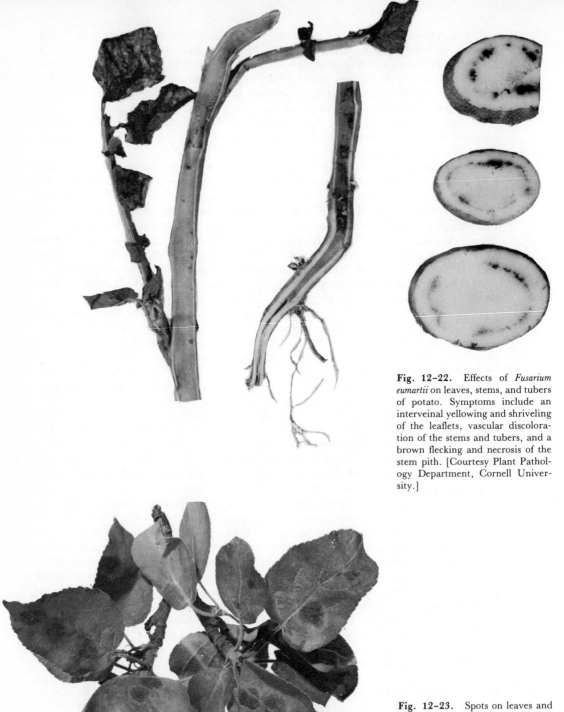

Fig. 12-22. Effects of *Fusarium eumartii* on leaves, stems, and tubers of potato. Symptoms include an interveinal yellowing and shriveling of the leaflets, vascular discoloration of the stems and tubers, and a brown flecking and necrosis of the stem pith. [Courtesy Plant Pathology Department, Cornell University.]

Fig. 12-23. Spots on leaves and fruit caused by the apple scab pathogen, *Venturia inaequalis.* [Courtesy Plant Pathology Department, Cornell University.]

Tissue disintegration impairs the normal physiological processes of the plant part. For example, a leaf may no longer be able to produce food or a root may no longer be able to absorb water. Also it results in the death of a plant part, such as areas of the leaves or stems, otherwise termed a *necrosis*. Necrotic symptoms include *dry rots,* shrinkage and collapse of the tissue with no release of fluids (found in dry rot of potato caused by *Fusarium coeruleum*); *canker,* a wound or localized lesion that is often sunk beneath the surface of the stem and is surrounded by healthy tissue (found in chestnut blight caused by *Endothia parasitica*); *damping-off,* in which there is a watery disintegration of the cortical tissues of young seedlings at the point of emergence from soil (found in most species of seed plants and caused by various fungi, including several species of *Pythium*); *root rots,* the rotting of cortical and phloem tissue of roots and basal stem portions (caused by various species of *Fusarium* and *Rhizoctonia* (Fig. 12–22); *leaf spots,* localized and limited lesions of collapsed cells on the leaf (found in apple scab caused by *Venturia inaequalis*) (Fig. 12–23); and *blight,* a generalized and rapid browning of the tissue that results in death (found in raspberry cane blight caused by *Leptosphaeria coniothyrium*) (Fig. 12–24).

Fig. 12–24. Raspberry cane blight caused by *Leptosphaeria coniothyrium.* Note the widespread death of the plant parts. [Courtesy U.S. Department of Agriculture.]

Interference with Normal Growth. The parasite may not cause an outright destruction of the cells and tissues of the host but rather qualitative or quantitative changes from the normal development pattern. Development in plants is controlled by auxins and by inhibitory substances or systems that control the synthesis and response of the cells to the growth-promoting auxins. Some metabolites formed by pathogenic fungi are similar in their action to those regulatory substances occurring naturally in plants. Other metabolites are capable of eliciting unusual morphogenic responses from the plant, unlike those given by the auxins and regulatory mechanisms ordinarily present. A third possible mechanism is that the pathogenic fungus may not release a morphogenic metabolite but may interfere with the regulatory system of the host that would ordinarily limit or control development.

A symptom developing from an interference with normal growth is typically a *hypertrophy,* an overdevelopment of a plant part. Examples include the curling of the leaf due to excess tissue formation and excess anthocyanin development in peaches

Fig. 12-25. Galls on ear of corn resulting from the abnormal enlargement of the kernels in the corn smut disease caused by *Ustilago maydis.* The black teliospores are visible through the ruptured host tissue. [Courtesy U.S. Department of Agriculture.]

infected by the peach leaf curl pathogen, *Taphrina deformans;* adventitious root formation in tomato plants with the *Fusarium* disease; and galls (enlarged tumor-like masses
of tissue) such as those formed in cabbage by the clubroot organism, *Plasmodiophora
brassicae* or by the corn smut organism, *Ustilago maydis* (Fig. 12-25). Generally, the
abnormal development in itself is not harmful to the plant, but it invites malfunction
of other physiological processes, such as water movement or reproduction, and, especially, diverts and uses food which would otherwise be available to support normal
growth.

One of the best studied examples of growth interference by a pathogen is that of
the "bakanae" or "foolish seedling" disease of rice caused by the ascomycete *Gibberella fujikuroi (= Fusarium moniliforme)*. In rice, the leaves yellow and curl inward and
often growth of the affected seedlings is arrested. The yield is often reduced. In some
instances, the affected plants grow more rapidly than healthy ones and are conspicuous because of their greater size, etiolated appearance, and early flowering. The
biologically active compounds produced by the pathogen were found to be gibberellic
acids (Fig. 12-26). When introduced into test plants, these gibberellic acids were
similar in their biological activity to growth-regulating compounds that occur
naturally in higher plants (Braun, 1959).

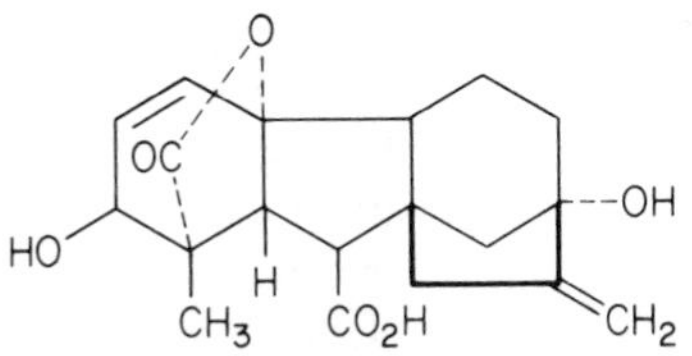

Fig. 12-26.   Gibberellic acid.

Interference with Physiological Processes. Many plant parasitic fungi interfere
with the normal processes of the plant, such as nutrition, uptake and movement of
water and minerals through the plant, and respiration. Tissue may or may not be
disintegrated in these cases. Disturbance in one physiological process may alter the
course of another physiological process, so that ultimately these disturbances may have
far-reaching effects in the plant.

A parasite may interfere with the nutritional processes of the plant (1) by impeding photosynthesis, (2) by preventing or impeding translocation of photosynthates
to nongreen portions of the plant (perhaps by damage to the phloem), or (3) by causing a redistribution of nutrients or by assimilating them, thus preventing their normal
utilization by the plant. In reference to the last point, those nutrients most often utilized by the parasite are carbohydrates and are probably intermediate compounds in
photosynthesis or glycolysis. These include glyceraldehyde, phosphoglyceric acid,
dihydroxyacetone, oxaloacetic acid, or ribulose diphosphate. Symptoms resulting
from an interference with nutrition often include an underdevelopment or dwarfing of
the plant parts involved, resulting ultimately in a reduced yield as well as premature
ripening or senescence. Others lead to a degeneration of the plant; for example, both
rust and powdery mildew diseases impair photosynthesis and the normal distribution
of photosynthates. In these diseases, nutrient "sinks," areas where nutrients
accumulate in unusually high concentrations at the expense of surrounding tissue, first
develop in the vicinity of the haustoria of the parasite. The cells in the vicinity of the
haustoria may sometimes enlarge and their life span is temporarily prolonged. Eventually, there is a reduction of chlorophyll in all parts of the leaf and all parts become
yellowed (Bushnell, 1972; Huber, 1978; Kosuge, 1978).

Water and mineral absorption may be impaired by a number of conditions, including root damage or disease, a generalized disintegration of tissue through which water must pass, and gum formation in the xylem region. Some toxins produced by pathogenic fungi cause the stomates to remain permanently open, resulting in loss of large quantities of water through excessive transpiration. Pathogens such as *Venturia inaequalis* and many of the rust fungi develop in a subcuticular position, and the localized rupturing of the impermeable cuticle results in water loss by transpiration at the site of the rupture. Losses in water are generally first expressed as a turgor loss of the cells most directly affected, followed by a wilting or drooping of leaves and stem tips, and finally by a reduction in growth. Many plant diseases indirectly cause a disruption of the water flow, but the wilt diseases, such as *Fusarium* wilt of tomato (caused by *Fusarium lycopersici*) and Dutch elm disease (caused by *Ceratocystis ulmi*), are primarily diseases of the vascular system. The wilt disease fungi cause a local necrosis in the xylem. Pectic enzymes produced by the parasite act on the middle lamellae that are exposed in the pits, releasing pectic acid and other products of hydrolysis, which are then converted into calcium gels and upon further hydrolysis into gums. The fungi may also produce mucilaginous polysaccharides that form a viscous mass. Although the exact fate of these mucilaginous polysaccharides is not known, it has been suggested that they become dislodged from the hyphae and occlude the vessels or form a suspension or solution in the water in the xylem, increasing its viscosity and resistance to flow. The water flow may also be reduced to some extent by the hyphae if large quantities are present in a heavily infected cell, but the hyphae themselves are usually not considered to be a factor in the decreased water flow. *Tyloses* may form as a result of the infection: these are protoplasmic, balloonlike protrusions extending through the pits from adjacent parenchyma cells into the vessels. Tyloses may become so large and numerous that they become packed to form a tissuelike mass completely blocking the vessels (Talboys, 1978). In addition to the actual blockages in water flow, flow of minerals within the plant is severely disrupted. Movement of radioactive phosphorus can be reduced by 96% to 98% in the xylem of tomato plants afflicted by *Fusarium* wilt (Huber, 1978). Toxic metabolites such as ethylene and phenol are also released. The phenols become intermixed with the gums, and upon oxidation the phenols become red-brown, causing the characteristic discoloration of diseased tissues. The ethylene is carried upward to the leaves, where it causes them to droop. As the disease progresses, the leaves become yellow, die, and then drop off. The actual severity of the wilt diseases may vary greatly. If only a few cells or a small section of the vascular tissue is infected, the effects may be minimal and involve only those cells to which water transport is decreased. In contrast, an extensive infection resulting in massive blockage will result in the death of the plant (Talboys, 1978).

The respiratory rate of diseased tissues generally increases, possibly depleting the plant of available energy reserves (Fig. 12–27). A rise in temperature usually accompanies the increase in respiration. The increase in respiration is apparently due to loss of control at one or more points in the metabolic pathway (perhaps the uncoupling of oxidative phosphorylation) (Kosuge, 1978). Many ATP-utilizing reactions (primarily synthetic, defense reactions) of the host tissue are increased in response to infections. Such reactions include cork formation, increased protoplasmic streaming, hypertrophic cell growth, increased synthesis of compounds, and accumulation and mobilization of compounds to the site of infection (Allen, 1959; Uritani and Akazawa, 1959). It has been suggested that the respiratory increase is a prerequisite for the initiation and continuation of the resistant reactions of the plants (Uritani and Akazawa, 1959).

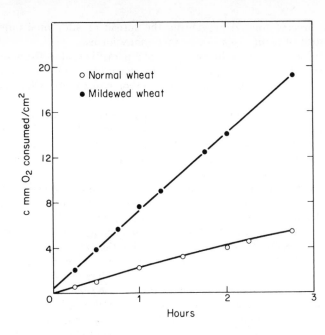

Fig. 12-27. Increased respiration in tissues of wheat infected by the mildew pathogen, *Erysiphe graminis,* over that in normal tissues. Oxygen uptake is used as an index for the rate of respiration. [From P. J. Allen and D. R. Goddard, 1938, *Am. J. Botany* **25**:613–621.]

ENVIRONMENT

Plants lack the complex regulatory systems found in higher animals, and the growth of the plant is more closely related to the prevailing environmental conditions than is the growth of an animal. The extent of plant disease, unlike animal disease, is therefore more directly controlled by external environmental conditions. As a plant disease is the product of interaction between parasite and host, any environmental factor is potentially capable of affecting the parasite, the host, and the manner in which they interact. Both the parasite and host have individual reactions to each environmental factor; for example, a parasitic fungus has a minimum, maximum, and optimum temperature for growth (see Chapter 7) and this range may be quite different from that for the host plant. A fungus has an ideal temperature for infection, while the plant has a temperature range in which it is best able to withstand infection. Therefore a plant disease develops at its maximum in that narrow range where the environmental conditions are favorable to the fungus but unfavorable to the host plant.

Environmental conditions exerting a profound influence on plant diseases include pH of the soil, available nutrients in the soil, soil moisture, aeration of soil, rainfall, atmospheric humidity, temperature range and fluctuation, and light intensity. Each parasitic fungus-host combination reacts uniquely to environmental conditions, and those conditions favoring one disease would be inhibitory to another disease. The relative importance of the various environmental factors varies from disease to disease, but temperature and moisture variations are those which are most often the critical factors.

CONTROL

In the control of plant disease it is of comparatively little value to attempt to save an individual plant (unless it is a valuable ornamental plant such as an elm tree).Instead, attention is directed toward preventing the spread or additional outbreaks of the disease. Control of plant disease may take many forms.

One type of control measure involves the exclusion of a parasitic fungus from an area where the fungus does not occur. Exclusion takes the form of quarantine, and plants that might bear the parasite are not transferred into a region or country which is free of that pest.

A second means of control is the eradication of a susceptible host species from an area, thus indirectly eliminating the parasite. This type of control is useful for heteroecious rust fungi which require two hosts to complete their life cycle. The host which is not of economic importance is eliminated; for example, eradication of the barberry may be used to control stem rust of wheat.

Frequently disease control is achieved by cultural manipulation or protection of the plant. These may include controlling the environment, sanitation, crop rotation, and application of fungicides to growing plants. As indicated above, environmental factors may influence the course of disease. Agriculturalists may control specific plant diseases by modifying certain aspects of the environment so that they are favorable for plant growth but are unfavorable for the development of diseases. For example, the moisture level may be varied by irrigating, either by use of sprinklers or by flooding, while temperature may be varied by shading the soil or altering the time at which the crops are planted. Sanitation procedures include the destruction of diseased plant parts and sterilization of equipment that might carry the parasite and facilitate its spread from field to field. Crop rotation is the cyclical planting of unrelated plants in a field (for example, soybeans, potatoes, and rye may be alternated). Crop rotation prevents the buildup of parasitic fungi by depriving them of a susceptible crop and causing their starvation. Chemical fungicides (such as copper, sulfur, and dithiocarbamates) may be applied externally to a plant as a spray or dust. The fungicides kill fungi by inhibiting chitin synthesis or by causing disturbances in the plasmalemma, respiration, sterol metabolism, nuclear division, or synthesis of DNA, RNA, and protein. A single fungicide may impair any one or several of these functions.

Disease forecasting is becoming increasingly important in the control of disease. Records are kept of fluctuating biological conditions (e.g., spore release) and of the climate and are correlated with known factors favoring a particular disease. Computers may be used to store and correlate the data and also to simulate models of plant growth and disease. Local and regional forecasting centers then advise the agriculturalists if an unusual outbreak of a disease is predicted. In turn, appropriate preventive measures such as additional sanitation measures, application of a fungicide, or a change of the time of planting may be made in the field.

The control measures outlined above are expensive to apply and are often ineffective, especially in the midst of an epidemic. The ideal mode of control is to plant crops that are immune to the parasites and which will not become diseased. Development of immune varieties involves extensive plant breeding programs. Such breeding programs are complicated by the fact that often large numbers of physiological varieties, races, and biotypes of a fungus species exist which differ in their pathogenicity. This makes it comparatively difficult to develop a host which is resistant to all forms of the fungus. Once a resistant variety is developed, there remains the probability that new genetic lines of the parasitic fungus will arise and that new pathogenic gene combinations will be established through natural selection. Formation of new pathogenic fungus types makes it mandatory to frequently introduce new genetic lines of crop plants to achieve control of the disease.

References

Ainsworth, G. C. 1968. Fungal parasites of vertebrates. *In:* Ainsworth, G. C., and A. S. Sussman, Eds., The fungi—an advanced treatise. Academic Press, New York. **3**:211–226.

Aist, J. R. 1976. Papillae and related wound plugs of plant cells. 1976. Ann. Rev. Phytopathol. **14**:145–163.

Allen, P. J. 1954. Physiological aspects of fungus diseases of plants. Ann. Rev. Plant Physiol. **5**:225–248.

——. 1959. Metabolic considerations of obligate parasitism. *In:* Holton, C. S., Ed., Plant pathology—problems and progress 1908-1959. University of Wisconsin Press, Madison, Wisc. pp. 119–129. 119–129.

Apple, J. L. 1977. The theory of disease management. *In:* Horsfall, J. G., and E. B. Cowling, Eds. Plant disease—an advanced treatise. Academic Press, New York. **1**:79–101.

Baker, E. D., and L. P. Cadman. 1963. Candidiasis in pigs in northwestern Wisconsin. J. Am. Vet. Med. Assoc. **142**:763–767.

Balan, J., and H. A. Lechevalier. 1972. The predacious fungus *Arthrobotrys dactyloides:* induction of trap formation. Mycologia **64**:919–922.

Barnett, H. L. 1963. The nature of mycoparasitism by fungi. Ann. Rev. Microbiol. **17**:1–14.

——. 1964. Mycoparasitism. Mycologia **56**:1–19.

——, and F. L. Binder. 1973. The fungal-host parasite relationship. Ann. Rev. Phytopathol. **11**:273–292.

——, and V. G. Lilly, 1958. Parasitism of *Calcarisporium parasiticum* on species of *Physalospora* and related fungi. West Virginia University Agr. Expt. Sta. Bull. 420T:1–37.

——, and ——. 1962. A destructive mycoparasite, *Gliocladium roseum.* Mycologia **54**:72–77.

Barron, G. L. 1977. The nematode-destroying fungi. Canadian Publications Ltd., Guelph. 140 pp.

Bateman, D. F., and R. L. Millar. 1966. Pectic enzymes in tissue degradation. Ann. Rev. Phytopathol. **4**:119–146.

Beauverie, J. 1914. Les Muscardines—le genre *Beauveria* Vuillemin. Rev. Gen. Botan. **26**:81–105, 157–173.

Beckman, C. H. 1964. Host responses to vascular infection. Ann. Rev. Phytopathol. **2**:231–252.

Bereston, E. S. 1974. Occupational fungus infections: *In:* Robinson, H. M., Ed., The diagnosis and treatment of fungal infections. Charles C Thomas, Springfield, Ill. p. 99–111.

Berger, R. D. 1977. Application of epidemiological principles to achieve plant disease control. Ann. Rev. Phytopathol. **15**:165–183.

Berry, C. R. 1959. Factors affecting parasitism of *Piptocephalis virginiana* on other Mucorales. Mycologia **51**:824–832.

——, and H. L. Barnett. 1957. Mode of parasitism and host range of *Piptocephalis virginiana.* Mycologia **49**:374–386.

Blood, D. C., and J. A. Henderson. 1960. Veterinary medicine. Williams and Wilkins Co., Baltimore. 1008 pp.

Boosalis, M. G. 1964. Hyperparasitism. Ann. Rev. Phytopathol. **2**:363–376.

Braun, A. C. 1959. Growth is affected. *In.* Horsfall, J. C., and A. E. Dimond, Eds., Plant pathology—an advanced treatise. Academic Press, New York. **1**:189–248.

——. 1965. Toxins and cell-wall dissolving enzymes in relation to plant disease. Ann. Rev. Phytopathol. **3**:1–18.

Bushnell, W. R. 1972. Physiology of fungal haustoria. Ann. Rev. Phytopathol. **10**:151–176.

Butler, E. E. 1957. *Rhizoctonia solani* as a parasite of fungi. Mycologia **49**:354–373.

Buxton, E. W. 1959. Mechanisms of variation in *Fusarium oxysporum* in relation to host-parasite interactions. *In:* Holton, C. S., Ed., Plant pathology—problems and progress 1908-1958. University of Wisconsin Press, Madison, Wis. pp. 183–191.

——. 1960. Heterokaryosis, saltation, and adaptation. *In* Horsfall, J. G., and A. E. Dimond, eds., Plant pathology—an advanced treatise. Academic Press, New York. **2**:359–405.

Cameron, J. W. MacB. 1963. Factors affecting the use of microbial pathogens in insect control. Ann. Rev. Microbiol. **8**:265–286.

Chapman, H. C. 1974. Biological control of mosquito larvae. Ann. Rev. Entomol. **19**:33–59.

Comandon, J., and P. de Fonbrune. 1938-a. Recherches expérimentales sur les champignons prédateurs de nématodes du sol. Conditions de formation des organes de capture. Compt. rend. Soc. biol. Paris. **129**:619–620.

––, and ––. 1938-b. Recherches expérimentales sur les champignons prédateurs de nématodes du sol. Les pièges garrotteurs. Compt. rend. Soc. biol. Paris. **129**:620–622.

––, and––. 1938-c. Recherches expérimentales sur les champignons prédateurs de nématodes du sol. Les glaux ou pièges collants. Compt. rend. Soc. biol. Paris. **129**: 623–625.

Couch, J. N., and C. J. Umphlett. 1963. *Coelomomyces* infections. *In:* Steinhaus, E., Ed., Insect pathology—an advanced treasise. Academic Press, New York. **2**:149–188.

Cridland, A. A. 1962. The fungi in Cordaitean rootlets. Mycologia **54**:230–234.

Cruickshank, I. A. M. 1963. Phytoalexins. Ann. Rev. Phytopathol. **1**:351–374.

Dimond, A. E. 1970. Biophysics and biochemistry of the vascular wilt syndrome. 1970. Ann. Rev. Phytopathol. **8**:310–322.

Drake, T. E., and H. I. Maibach. 1974. Cutaneous candidiasis. *In:* Robinson, H. M., Ed., The diagnosis and treatment of fungal infections. Charles C Thomas, Springfield, Ill. pp. 5–28.

Drechsler, C. 1937. Some Hyphomycetes that prey on free-living terricolous nematodes. Mycologia **29**:447–552.

Duddington, C. L. 1955. Fungi that attack microscopic animals. Botan. Rev. **21**:377–439.

––. 1956. The friendly fungi—a new approach to the eelworm problem. Faber & Faber, Ltd., London. 188 pp.

––, and C. H. E. Wyborn. 1972. Recent research on the nematophagous Hyphomycetes. Botan. Rev. **38**:545–565.

Dustan, A. G. 1927. The artificial culture and dissemination of *Entomophthora sphaerosperma* Fres. a fungous parasite for the control of the European apple sucker (*Psyllia mali* Schmid B). J. Econ. Entomol. **20**:68–75.

Eide, C. J. 1955. Fungus infection of plants. Ann. Rev. Microbiol. **9**:297–318.

Emmons, C. W. 1940. Medical mycology. Botan. Rev. **6**:474–514.

––. 1942. Coccidioidomycosis. Mycologia **34**:452–463.

––. 1943. Coccidioidomycosis in wild rodents—a method of determining the extent of endemic areas. Public Health Rept. 58:1–5.

––. 1947. Biology of *Coccidioides. In:* Nickerson, W. J., Ed., Biology of pathogenic fungi. Ann. Cryptogam Phytopath. **6**:71–82.

––. 1951. The isolation from soil of fungi which cause disease in man. Trans. N.Y. Acad. Sci. Ser. II, **14**: 51–54.

––. 1955. Mycoses of animals. Advan. Vet. Sci. **2**:47–63.

––. 1960. The Jekyll-Hydes of mycology. Mycologia **52**:669–680.

––, C. H. Binford, J. P. Utz, and K. J. Kwon-Chung. 1977. Medical mycology, Third ed. Lea and Febiger, Philadelphia. 592. pp.

Federici, B. A., and D. W. Roberts. 1975. Experimental laboratory infection of mosquito larvae with fungi of the genus *Coelomomyces* I. Experiments with *Coelomomyces psorophorae* var. in *Aedes taeniorhynchus* and *Coelomomyces psorophorae* var. in *Culiseta inornata*. J. Invertebrate Pathol. **26**:21–27.

Ferron, P. 1978. Biological control of insect pests by entomogenous fungi. Ann. Rev. Entomol. **23**: 409–442.

Fischer, G. W. 1965. The romance of the smut fungi. Mycologia **57**:331–342.

Fisher, W. S., E. H. Nilson, and R. A. Shleser. 1975. Effect of the fungus *Haliphthoros milfordensis* on the juvenile stages of the American lobster *Homarus americanus,* J. Invertebrate Pathol. **26**:41–45.

Fitzpatrick, H. M. 1915. A parasitic species of *Claudopus*. Mycologia **7**:34–37.

Flor, H. H. 1971. Current status of the gene-for-gene concept. Ann. Rev. Phytopathol. **9**:275–296.

Foister, C. E. 1946. The relation of weather to fungus diseases of plants. II. Botan. Rev. **12**:548–591.

Gabriel, B. P. 1959. Fungus infection of insects via the alimentary canal. J. Insect Pathol. **1**:319–330.

Griffith, N. T., and H. L. Barnett. 1967. Mycoparasitism by basidiomycetes in culture. Mycologia **59**:149–154.

Hagen, K. S. and van den Bosch, R. 1968. Impact of pathogens, parasites and predators on aphids. Ann. Rev. Entomol. **13**: 325–384.

Hoch, H. C. 1977-a. Mycoparasitic relationships: *Gonatobotrys simplex* parasitic on *Alternaria tenuis.* Phytopathology **67**:309–314.

———. 1977-b. Mycoparasitic relationships. III. Parasitism of *Physalospora obtusa* by *Calcarisporium parasiticum.* Can. J. Botany **55**: 198–207.

———. 1978. Mycoparasitic relationships. IV. *Stephanoma phaeospora* parasitic on a species of *Fusarium.* Mycologia **70**:370–379.

Hoffman, W. E. 1947. Insects as human food. Proc. Entomol. Soc. Wash. **49**:233–237.

Huber, D. M. 1978. Disturbed mineral nutrition. *In:* Horsfall, J. G., and E. B. Cowling, Eds., Plant disease—an advanced treatise. Academic Press, New York. **3**:163–181.

Ingham, J. L. 1972. Phytoalexins and other natural products as factors in plant disease resistance. Botan. Rev. **38**:343–424.

Karling, J. S. 1942. Parasitism among the chytrids. Am. J. Botany **29**:24–35.

———. 1948. Chytridiosis of scale insects. Am. J. Botany **35**:246–254.

———. 1960. Parasitism among the chytrids. II. *Chytriomyces verrucosus* sp. nov. and *Phylctochytrium synchytrii.* Bull. Torrey Botan. Club **87**:326–336.

Keitt, G. W., D. M. Boone, and J. R. Shay. 1959. Genetic and nutritional controls of host-parasite interactions in apple scab. *In:* Holton, C. S., Ed., Plant pathology—problems and progress 1908–1958. University of Wisconsin Press, Madison, Wis. pp. 157–167.

Kirkham, D. S. 1959. Host factors in the physiology of disease. *In:* Holton, C. S., Ed., Plant pathology—problems and progress 1908–1958. University of Wisconsin Press, Madison, Wis. pp. 110–118.

Kligman, A. M., and E. D. DeLamater. 1950. The immunology of the human mycoses. Ann. Rev. Microbiol. **4**:283–312.

Kosuge, T. 1978. The capture and use of energy by diseased plants. *In:* J. G. Horsfall and E. B. Cowling, Eds., Plant disease—an advanced treatise. Academic Press, New York. **3**:85–116.

Large, E. C. 1940. The advance of the fungi. Dover Publications, Inc., New York. 488 pp.

Lefebvre, C. L. 1934. Penetration and development of the fungus, *Beauveria bassiana,* in the tissues of the corn borer, Ann. Botany, London **48**:441–452.

Lightner, D. V., and C. T. Fontaine. 1973. A new fungus disease of the white shrimp *Penaeus setiferus.* J. Invertebrate Pathol. **22**:94–99.

Lyr, H. 1977. Mechanism of action of fungicides. *In:* Horsfall, J. G. and E. B. Cowling, Eds., Plant disease—an advanced treatise. Academic Press, New York. **1**:239–261.

Mackauer, M., and M. J. Way. 1976. *Myzus persicae* Sulz. an aphid of world importance. *In:* Delucchi, V. L., Ed., Studies in biological control. Cambridge University Press, Cambridge. Intern. Biol. Progr. **9**:51–119.

MacLeod, D. M. 1954. Investigations on the genera *Beauveria* Vuill. and *Tritirachium* Limber, Can. J. Botany **32**:818–890.

———. 1963. Entomophthorales infections. *In:* Steinhaus, E. A., Ed., Insect pathology—an advanced treatise. Academic Press, New York. **2**:189–231.

Maddy, K. T. 1960. Coccidiomycosis. Advan. Vet. Sci. **6**:251–286.

Madelin, M. F. 1963. Diseases caused by hyphomycetous fungi. *In:* Steinhaus, E. A., Ed., Insect pathology—an advanced treatise. Academic Press, New York. **2**:233–271.

———. 1966. Fungal parasites of insects. Ann. Rev. Entomol. **11**:423–448.

———. 1968-a. Fungal parasites of invertebrates 1. Entomogenous fungi. *In:* Ainsworth, G. C., and A. S. Sussman, Eds., The fungi—an advanced treatise. Academic Press, New York. **3**:227–238.

———. 1968-b. Fungi parasite on other fungi and lichens. *In:* Ainsworth, G. C., and A. S. Sussman, Eds. The fungi—an advanced treatise. Academic Press, New York. **3**:253–269.

Mains, E. B. 1948. Entomogenous fungi. Mycologia **40**:402–416.

———. 1957. Species of *Cordyceps* parasitic on *Elaphomyces.* Bull. Torrey Botan. Club **84**:243–251.

———. 1958. North American entomogenous species of *Cordyceps.* Mycologia **50**:169–222.

Malkinson, F. A., and R. W. Pearson, Eds. 1974. The year book of dermatology, 1974. Year Book Medical Publishers, Inc. Chicago. 446 pp.

Malthre, D. E. 1978. Disrupted reproduction. *In:* Horsfall, J. G., and E. B. Cowling, Eds. Plant disease—an advanced treatise. Academic Press, New York. **3**:257–278.

Manocha, M. S., and R. Golesorkhi. 1979. Host-parasite relations in a mycoparasite. V. Electron microscopy of *Piptocephalis virginiana* infection in compatible and incompatible hosts. Mycologia **71:** 565–576.

McEwen, F. L. 1963. *Cordyceps* infections. *In:* Steinhaus, E. A., Ed., Insect pathology—an advanced treatise. Academic Press, New York. **2:**273–290.

Mount, M. S. 1978. Tissue is disintegrated. *In:* Horsfall, J. G., and E. B. Cowling, Eds. Plant disease—an advanced treatise. Academic Press, New York. **3:**279–297.

Müller, K. O. 1959. Hypersensitivity. *In:* Horsfall, J. G., and A. E. Dimond, Eds., Plant pathology—an advanced treatise. Academic Press, New York. **1:**469–519.

Mullins, J. T., and A. W. Barksdale. 1965. Parasitism of the chytrid *Dictyomorpha dioica.* Mycologia **57:** 352–359.

Nickerson, W. J. 1953. Medical mycology. Ann. Rev. Microbiol. **7:**245–272.

Nilson, E. H., W. S. Fisher, and R. A. Schleser. 1976. A new mycosis of larvae lobster (*Homarus americanus*). J. Invertebrate Pathol. **27:**177–183.

Orlob, G. B. 1971. History of plant pathology in the Middle Ages. Ann. Rev. Phytopathol. **9:** 1–20.

Paddock, W. C. 1967. Phytopathology in a hungry world. Ann. Rev. Phytopathol. **5:**375–389.

Pool, V. W., and M. B. McKay. 1916. Relation of stomatal movement to infection by *Cercospora beticola.* J. Agr. Res. **5:**1011–1037.

Petch, T. 1925. Studies in entomogenous fungi. VIII. Notes on *Beauveria.* Trans. Brit. Mycol. Soc. **10:**244–271.

Rippon, J. W. 1978. Mycosis (pathogenesis and epidemiology). *In:* Vinken, P. J., and G. W. Bruyn, Eds. Handbook of clinical neurology. **35:**371–381.

——. 1980. Dimorphism in pathogenic fungi. CRC Critical Reviews in Microbiology. pp. 49–97.

——, and G. H. Scherr. 1959. Induced dimorphism in dermatophytes. Mycologia **51:** 902–914.

Shigo, A. L. 1960. Parasitism of *Gonatobotryum fuscum* on species of *Ceratocystis.* Mycologia **52:** 584–598.

——, C. D. Anderson, and H. L. Barnett. 1961. Effects of concentration of host nutrients on parasitism of *Piptocephalis xenophila* and *P. virginiana.* Phytopathology **51:**616–620.

Shrum, R. D. 1978. Forecasting of epidemics. *In:* Horsfall, J. G., and E. B. Cowling, Eds. Plant disease—an advanced treatise. Academic Press, New York. **2:**223–238.

Slifkin, M. E. 1963. Parasitism of *Olpidiopsis incrassata* on members of the Saprolegniaceae. II. Effect of pH and host nutrition. Mycologia **55:**172–182.

Small, E. 1971. Candidiasis (moniliasis). *In:* Kirk, R. W., Ed. Current veterinary therapy IV. Small animal practice. W. B. Saunders, Philadelphia. pp. 288–289.

Stakman, E. C., and J. J. Christensen. 1953. Problems of variability in fungi. In: Plant diseases—the yearbook of agriculture. U.S. Department of Agriculture, Washington D.C. pp. 35–62.

——, and ——. 1960. The problem of breeding resistant varieties. *In:* Horsfall, J. G., and A. E. Dimond, Eds. Plant Pathology—an advanced treatise. Academic Press, New York **3:** 567–624.

——, and J. G. Harrar. 1957. Principles of plant pathology. Ronald Press, New York. 581 pp.

Steinhaus, E. A. 1949. Principles of insect pathology. First ed. McGraw-Hill Book Company, New York. 757 pp.

——.1957. Microbial diseases of insects. Ann. Rev. Microbiol. **11:**165–182.

Talboys, P. W. 1978. Dysfunction of the water system. *In:* Horsfall, J. G., and E. B. Cowling, Eds., Plant disease—an advanced treatise. Academic Press, New York. **3:**141–162.

Tanada, Y. 1959. Microbial control of insect pests. Ann. Rev. Entomol. **4:**277–295.

Tomiyama, K. 1963. Physiology and biochemistry of disease resistance of plants. Ann. Rev. Phytopathol. **1:**295–324.

Uritani, I., and T. Akazawa. 1959. Alteration of the respiratory pattern in infected plants. *In:* Horsfall, J. G., and A. E. Dimond, Eds., Plant pathology—an advanced treatise. Academic Press, New York. **1:**349–390.

Vey, A., and J. Fargues. 1977. Histological and ultrastructural studies of *Beauveria bassiana* infection in *Leptinotarsa decemlineta* larvae during ecdysis. J. Invertebrate Pathol. **30:** 207–215.

Walker, J. C., K. P. Link, and H. R. Angell. 1929. Chemical aspects of disease resistance in the onion. Proc. Natl. Acad. Sci. **15:**845–850.

——, and M. A. Stahmann. 1955. Chemical nature of disease resistance in plants. Ann. Rev. Plant Physiol. **6:**351–366.

Webster, R. K. 1974. Recent advances in the genetics of plant pathogenic fungi. Ann. Rev. Phytopathol. **12:**129-137.

Whaley, J. W., and H. L. Barnett. 1963. Parasitism and nutrition of *Gonatobotrys simplex.* Mycologia **55:** 199-210.

Wheeler, B. E. J. 1968. Fungal parasites of plants. *In:* Ainsworth, G. C., and A. S. Sussman, Eds., The fungi—an advanced treatise. Academic Press, New York. **3:**179-210.

Williams, P. H. 1979. How fungi induce disease. *In:* Horsfall, J. G., and E. B. Cowling, Eds. Plant disease—an advanced treatise. Academic Press, New York **4:**163-179.

Willoughby, L. G. 1956. Studies on soil chytrids. I. *Rhizidium richmondense* sp. nov. and its parasites. Trans. Brit. Mycol. Soc. **39:**125-141.

Wilson, J. W., and O. A. Plunkett. 1965. The fungous diseases of man. University of California Press, Berkeley, Calif. 428 pp.

Wood, R. K. S. 1960. Pectic and cellulolytic enzymes in plant disease. Ann. Rev. Plant Physiol. **11:** 299-322.

Wright, E. T. 1974. Coccidioidomycosis: mycology—pathology—immunology. *In:* Robinson, H. M., Ed. The diagnosis and treatment of fungal infections. Charles C Thomas, Springfield, Ill. pp. 341-353.

Zentmyer, G. A., and J. G. Bald. 1977. Management of the environment. *In:* Horsfall, J. G., and E. B. Cowling, Eds. Plant disease—an advanced treatise. Academic Press, New York **1:**121-144.

Zimmerman, M. H., and J. McDonough. 1978. Dysfunction in the flow of food. *In:* Horsfall, J. G., and E. B. Cowling, Eds. Plant disease—an advanced treatise. Academic Press, New York **3:**117-140.

THIRTEEN

Fungi as Symbionts

Some fungi regularly form partnerships with other organisms, sometimes to the exclusion of an independent existence. Such a regular association is termed *symbiosis,* and the partners are termed *symbionts.* Symbiosis includes all relationships in which there is a regular association, and this may include parasitism. As we learned from the preceding chapter, a parasitic relationship harms one partner. Alternatively, both partners may gain from the symbiotic relationship, or the partners may share food but may be neither harmed nor benefited from the relationship. In this chapter, we explore the relationships of some fungal symbionts that cause little or no harm to their hosts. In all cases, the fungus benefits nutritionally from the relationship, but the benefits (if they do indeed occur) for the nonfungus member are varied.

FUNGI AS PARTNERS OF PLANTS

Fungi are found in dual associations with plants ranging from the algae (in which case the dual plant is a *lichen*) to the bryophytes or fern prothalli and finally to an association with roots of vascular plants (termed a *mycorrhiza*). It is generally unknown to what extent these analogous relationships are similar in their biology. Attention has been focused primarily on the economically important higher plants. We discuss the two better known types, lichens and mycorrhizae.

Lichens

An association between a fungus and an alga that develops into a unique morphological form that is distinct from either partner is termed a *lichen*. Excluded from this concept are associations in which an alga and fungus enter into limited interactions that do not form a new morphological entity, such as an alga overgrowing a mushroom or limited parasitization of algal cells by fungi.

Lichen associations may be preserved by nonsexual reproduction, and their morphology is so constant that they were thought to be genetically autonomous plants until the Swiss botanist Simon Schwendener described their dual nature in 1868. Even at the present, the more than 16,000 lichen "species" may be classified in their own form class, the Lichenes, by using characteristics peculiar to the lichen association. As the lichens are a biological group and not a genetically related group, some taxonomists are correctly giving scientific names to both the fungus and algal partners and then integrating these into their proper respective positions in the mycological and phycological classifications.

Lichens are able to survive under severe conditions that cannot be tolerated by the majority of plants. They may be found in virtually every environment from the desert to cold sandy beaches, from bare rock to fertile soil or living leaves, and from the tropics to the arctic regions. Lichens often are the dominant vegetation of some sites, such as mountains above the tree line or in northern locations such as Greenland, Iceland, or Alaska. Lichens are conspicuously absent from the flora in and surrounding large cities because of their inability to withstand air pollution. They are especially sensitive to sulfur dioxide and fluorine, which are common pollutants.

Lichens have been important to humans in various ways. In the arctic region and to a lesser extent in the antarctic, lichens are harvested as fodder for reindeer. Various people have used lichens as food. For example, people in Iceland have prepared soups, desserts, and breads from lichens. It is thought that the manna referred to in the Bible may have been *Lecanora esculenta,* a lichen that grows in the mountains of Israel and may be blown loose into the lowlands. Desert tribes still eat this lichen, which is dried and mixed with dry meal to form a flour. An important way of using lichens has been to prepare blue, red, brown, or yellow dyes for cloth. Lichens have also been used as a source of medicines, poisons, cosmetics, perfumes, and essential oils by various peoples (Perez-Llano, 1948). The indicator pigment used to prepare litmus paper is derived from a lichen.

BIOLOGY OF THE SYMBIONT

Ordinarily the dominant member of the lichen symbiont is the fungus, which controls the morphological form that the lichen will assume (that is, lichens formed with a single fungus species but with different algal species may have the same form). In some lichens the situation is reversed and the alga determines the final form. The thallus may be leaflike and flat on the substratum (*foliose*), crustaceous and flat on the substratum (*crustose*), or upright and branched or pendulous (*fruticose*) (Figs. 13–1, 13–3). The lichen consists of tissues that are generally similar to those which have been described in Chapters 1, 4, and 5. The most characteristic tissue is one consisting of loosely interwoven, branched hyphae that form a netlike structure. The algal cells may be evenly distributed among the hyphae or, most commonly, may occur in a thin layer. Anatomically, the most complex lichens are some of the foliose lichens, which are typically organized into the following layers (from top to bottom): (1) an epider-

(a)

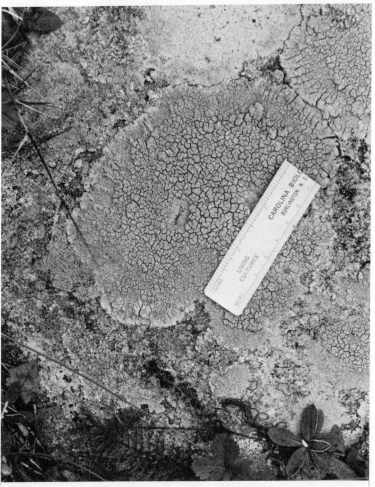

(b)

Fig. 13-1. (a): An example of a foliose lichen, *Parmelia caperata.* Approx. × 1; (b) An example of a crustose lichen, *Diploschistes scruposus.* Approximately ½ ×. [Courtesy M. E. Hale.]

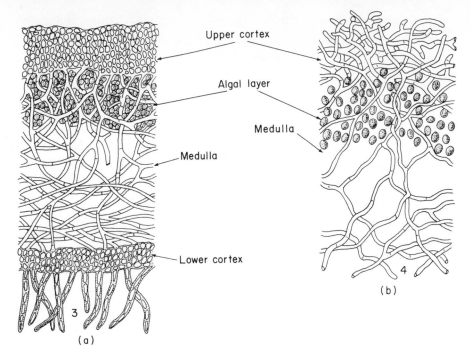

Fig. 13–2.   Longitudinal section through the vegetative thallus of a foliose lichen (a) and a crustose lichen (b). Compare with the fruticose lichen in Fig. 13-3. [From A. Schneider, 1897, *A text-book of general lichenology*. W. N. Clute & Co., Binghampton, N.Y.]

Fig. 13–3.   A fruticose lichen. (a) Habit of the thallus. (b) Section through a thallus showing a portion of an apothecium at the left and and a vegetative portion at the right. [Adapted from A. Schneider, 1897, *A text-book of general lichenology*, W. N. Clute & Co., Binghampton, N.Y.]

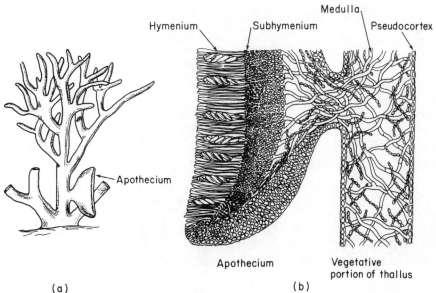

487

mislike upper cortex of pseudoparenchymatous tissue, (2) a usually thin algal layer in which the algae cells are penetrated by fungal haustoria, (3) a medullary layer of loosely interwoven hyphae, and (4) a lower cortex. Crustose thalli are not divided into as many layers, as they lack a lower cortex, and some exceptionally simple ones may consist only of a single homogeneous layer containing the fungal hyphae and algal cells (Fig. 13-2). Fruticose forms consist of a medulla surrounded by a pseudocortex. The algal cells are distributed either evenly or in patches in the medulla (Fig. 13-3). Pores that allow the exchange of air may be present on the undersurface. Other modifications of the thallus may include loosely attached scales, columnar outgrowths (*isidia*), pockets where the alga occurs alone, and *soredia* (discrete structures consisting of a few algal cells surrounded by hyphae).

Reproduction may occur by nonsexual or sexual means. Nonsexual reproduction occurs by fragmentation of the thallus and is especially likely to occur by separation of the columnar isidia, which appear to be an adaptation for nonsexual reproduction. Also the soredia may become detached and disseminated in a manner similar to spores and then develop into a new thallus. The fungus may also be propagated nonsexually by conidia formed in pycnidia. As the majority of the lichen fungi are in the Ascomycotina, the sexual reproductive structures are usually ascocarps (Figs. 13-3, 13-4). Ascospores are formed in asci and are discharged. Although direct evidence is lacking, it is assumed that mycelium from germinating conidia or ascospores may come into contact with a suitable alga and reestablish the lichen relationship.

Growth of lichens is generally favored by periods of high humidity, cool temperatures, and low light intensities. Of these, moisture is the most critical factor because lichens apparently have no special mechanisms for uptake or conservation of water. In respect to water, they behave very much like an agar gel, rapidly absorbing water when it is available (sometimes up to 100% to 300% of its dry weight) and also rapidly losing water during a drought (to as low as 2% to 15% of its dry weight) (Smith, 1962). Both respiration and photosynthesis are favored by optimum water

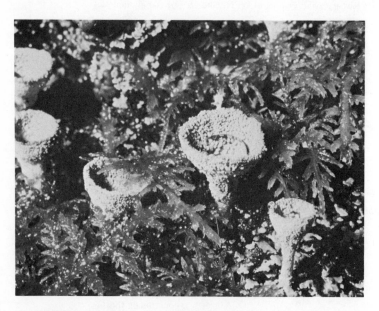

Fig. 13-4. Apothecia of the pixie-cup lichen, *Cladonia* sp. [From E. J. Moore and V. N. Rockcastle, 1963, *Fungi*, Cornell Science leaflet.]

content of the thallus; they decrease with drying and sometimes with excess saturation. Moisture in the form of fog is especially favorable for lichen growth because it provides the necessary moisture while allowing photosynthesis to take place in the optimum low light intensity. Moisture availability is the critical factor in determining lichen distribution in nature.

Lichens grow extremely slowly, partially because they are exposed to optimal growing conditions for only a few hours a morning when there is still sufficient moisture from fog, dew, or recently melted snow to allow photosynthesis to occur at its optimum rate. Average annual increments for many lichens are less than 1 millimeter, and the greatest known increment is 4 centimeters. Lichens may reach maturity and produce ascocarps between 4 and 8 years of age. Longevity of lichens is often measurable in tens or hundreds of years. The age of some lichens in the alpine-arctic areas has been estimated to be between 1,000 to 4,500 years. This longevity can partially be attributed to the ability of lichens to survive long periods (up to 2 to 3 months) of drought unharmed.

Physiological studies of the composite lichen in culture are valuable but are difficult to carry out for extended periods of time because the lichen association tends to break down under the luxuriant conditions of moisture and nutrition inadvertently provided in culture. It is advantageous to study the physiology of the fungus and alga partners individually to better understand the lichen association. In order to carry out such studies, it is necessary to separate the components from each other and culture them independently.

The algal component may be isolated by first washing the lichen thallus in running tap water for 15 minutes, crushing the thallus, and removing the algal cells with attached hyphae with a sterile micropipette under a microscope. (This technique eliminates the possibility of culturing a transient contaminant.) These algal cells are then passed through four or five drops of sterile water and introduced onto slants of a solid inorganic mineral medium, which are incubated in the light. The fungus will not grow under such conditions. A simpler but less reliable method consists of peeling off the cortex of the lichen, fragmenting the remainder of the thallus, and then incubating a fragment on the illuminated medium. If this second method is employed, it is likely that foreign algal cells will be cultured; the algal partner should be distinguished by comparison with a microscopic mount of the lichen and possibly purified by subculturing (Ahmadjian, 1967-b).

The fungal partner may be isolated by using standard mycological techniques. Since the majority of lichen fungi are in the Ascomycotina, one may take advantage of spore discharge by preparing a petri dish with a 2% agar medium, soaking the lichen ascocarps in cold water for 15 minutes, and then attaching the inverted ascocarp to the underside of the petri dish lid with a dab of petroleum jelly. As the wet lichen dries, it will discharge large numbers of spores onto the agar. After germination, a block of the agar bearing mycelium is transferred to an enriched medium (such as malt-yeast agar) for further culturing (Ahmadjian; 1967-a).

THE COMPONENTS OF THE SYMBIONT

The Algal Partner. At least 26 alga genera are members of the lichen association. These include 8 blue-green algae and 1 yellow-green alga; the remainder are green algae. *Nostoc* is the blue-green alga most frequently found in lichens, while the green alga *Trebouxia* is found in 75% or more of the lichens of the temperate zones (Fig. 13-5) (Ahmadjian, 1966-a, 1966-b). All the algal partners are probably able to exist independently in nature.

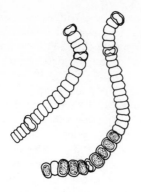

Fig. 13-5. Algae that are commonly partners in lichens: (a) *Nostoc;* (b) *Trebouxia.*

Numerous nutritional and physiological studies have been made of the algae (most often of *Trebouxia*) in pure culture. The majority of the isolates make improved growth on a medium with added carbohydrates or organic nitrogen, and some can grow heterotrophically in the dark with these added nutrients. The algae apparently do not require an exogenous source of vitamins and are extremely slow growers. Their slow rate of growth partially accounts for the slow growth rate of the lichen as a whole.

Lichens with *Nostoc* as the algal partner are able to fix atmospheric nitrogen. The isolated *Nostoc* not only excretes considerable amounts of nitrogenous compounds into the culture medium, but it also excretes polysaccharides and vitamins (biotin, thiamine, riboflavin, and nicotinic acid).

The Fungal Partner. It is not known whether the fungal partners have an extensive independent existence in nature, although it is not unlikely that both a saprophytic mycelial stage and a nonlichenized perfect state might occur. Most of the lichen fungi are members of the Ascomycotina, while a few others are in the Basidiomycotina or Deuteromycotina. Taxonomically, the lichenized fungi are widely distributed in the Ascomycotina, and many of these fungi can form lichens. Lichenized and nonlichenized fungi may occur in the same order. For example, most of the members of the Order Dothideales (Loculoascomycetes, p. 174) are nonlichenized but some may form lichens. In contrast, the Orders Caliciales and Lecanorales in the Discomycetes are comprised almost entirely of lichenized fungi. When isolated, the fungus retains the scientific name of the lichen.

Like the alga and the composite lichen, the fungi grow extremely slowly in culture and may achieve a size of only 1 to 2 millimeters within a year. All are either wholly or partially dependent on an exogenous source of thiamine and/or biotin and, to a lesser extent, the other vitamins. The nonlichenized perfect state of *Cladonia cristatella* has been obtained in culture (Ahmadjian, 1966-c).

RESYNTHESIS OF THE LICHEN FROM THE PARTNERS

Attempts to resynthesize the lichen from the isolated partners have met with limited success. These studies are of value primarily in indicating those conditions most favorable for resynthesis of the lichen, thereby giving additional clues about their biology. The lichen association can only be established if growing conditions are unfavorable for the independent growth of both partners. For example, resynthesis of the lichen thallus by *Endocarpon pusillum* was achieved only on soil that had been subjected to alternating periods of drying and wetting and also light and darkness. Low nutrient levels were also required. The most critical condition favoring resynthesis by *E.*

pusillum was slow drying, a feature generally shared with other lichens that have been resynthesized (Ahmadjian, 1973). Fluctuating environmental conditions, particularly light and moisture, also favor the growth of the lichen once it is established (Harris and Kershaw, 1971; Pearson, 1970). A formed lichen will dissociate into its components if favorable conditions of moisture, light, and adequate nutrition are made available (this may occur in nature), or even if the environmental conditions are nonfluctuating. These observations indicate that the lichen association is a forced one that depends on adverse conditions for its maintenance.

NATURE OF THE INTERACTION

The physiological relationship of the partners in the lichen association is a matter of speculation. The most widely held view is that the photosynthetic alga provides organic material for both partners and that the fungus protects the alga from desiccation, high light intensities, and injury and provides it with water and minerals. This interpretation may be regarded as either (1) controlled parasitism, if the alga is exploited and is not dependent on the "benefits" that the association yields, or (2) a relationship in which the alga gains something that it would not ordinarily have if living in the free state. This broad interpretation has been only partially supported experimentally and, further, it cannot be assumed that all lichen relationships are similar in view of the many different algae and fungi that may be involved. We explore below some specific points that may aid in interpreting the nature of the interaction.

Possible Advantages for the Fungal Partner. It is well established that the fungus gains nutritionally from the alga in the lichen relationship (Fig. 13-6). In culture, the alga produces excess vitamins corresponding to the vitamin deficiencies in the cultured fungus, which suggests that the alga provides the fungus with necessary growth factors in the lichen. In addition, the excretion of nitrogenous compounds by some nitrogen-fixing algal partners may benefit the fungus in a similar fashion.

Carbon compounds manufactured by the algal partner during photosynthesis move into the medullary tissues of the lichen where they are absorbed by the fungus. This can be confirmed experimentally by incubating lichen discs in a medium containing isotopic carbon, C^{14}. Clearly the carbohydrates produced by the algae from photosynthesis are in excess of the amount that they require, but they cannot divert this food surplus to reproduction or additional growth as they would if living in the free state.

The flow of nutrients from the alga to the fungus is substantial as the algal cells have been estimated to occupy only 3% to 10% of the total mass of the lichen by weight, but yet must supply carbohydrates to the entire lichen. As much as 70% to 80% of the carbon fixed by *Nostoc* or *Trebouxia* is probably passed to the fungus (Farrar, 1976). The flow of nutrients into the fungus is at its maximum when the thallus is saturated with moisture and is in the light. Conversely, fungal activity may be inhibited by low moisture levels in either the light or dark, and apparently little of the carbohydrate flows into the fungus under these conditions (Harris and Kershaw, 1971).

Possible Advantages for the Algal Partner. It is not so easy to assess the possible advantages of the lichen relationship for the algae. The algae may be at a disadvantage as they cannot grow and reproduce at their maximum rates because they are unable to utilize all the carbohydrates that they produce. The algae apparently retain all or most of the carbohydrates that they produce only when water levels are low (thereby inhibiting fungal activity) and light is present so that photosynthates can be

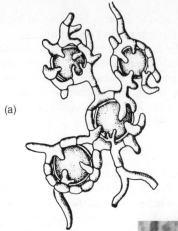

(a)

Fig. 13-6. Haustoria of the fungus partner which penetrate the algal cells: (a) As seen with the light microscope; (b) As seen with the electron microscope. Key: *Al W* indicates algal cell wall; *FW* indicates fungal cell wall; *N* indicates nucleus of fungal cell; *H* indicates haustorium; *Ch* indicates chloroplast of algal cell. The lichen in (b) is *Cladonia cristatella*. × 17,000. [(a) From A. Schneider, 1897, *A text-book of general lichenology*. W. N. Clute & Co., Binghampton, N.Y.; (b) Courtesy of R. T. Moore and J. H. McAlear, 1960, *Mycologia* **52**:805–807.]

(b)

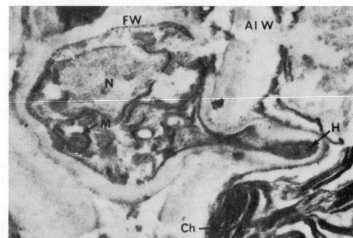

produced. Such conditions may provide the principal periods during which the alga can store photosynthates and perhaps grow (Harris and Kershaw, 1971).

An interesting view has been given by Ahmadjian (1966-b). He noted that *Trebouxia,* the most common algal partner, is rarely found in the free-living state in nature. However, the most widely distributed free-living alga, *Pleurococcus vulgaris,* is never found as a member of a lichen association. This may be because *P. vulgaris* (and many other algae) cannot survive the initial parasitic contacts with a lichen fungus, while *Trebouxia* can withstand these encounters. *Trebouxia* has characteristics that place it at a competitive disadvantage in the free state (a low optimum light intensity, slow growth rate, and a preference for organic nitrogen) but allow it to thrive in the lichen association. The lichen association is favorable for this alga because it is removed from competition with free-living algae and is placed in its own ecological niche.

We may conclude that the lichen association is usually of doubtful value to the algal component. Although the alga may be able to inhabit new environments because of this association, it sacrifices its ability to grow and reproduce at its maximum rate and is not dependent on this relationship for its welfare. An exception is perhaps *Trebouxia* which is unable to compete effectively when free-living, and for this alga the lichen association is beneficial.

Mycorrhizae

A dual association of a root of a higher plant and a fungus that is not disease-producing is termed a *mycorrhiza*. The fungus lives as an invader of the root and derives nutrients from it. It has been a matter of controversy whether the plant is a beneficiary of this relationship. Theories concerning this interaction have varied from those in which the fungus is thought to be a parasite while the plant derives little or no advantage from the relationship to those in which the partners both benefit from the association. Recent evidence indicates that mycorrhizal fungi are often beneficial to their plant partners, particularly under some commonly prevailing conditions, and only rarely become parasites when the normal balance is overthrown. Mycorrhizae constitute such a diverse group that few generalizations can be made concerning the physiology of the group as a whole, especially as the relationship may vary with the prevailing conditions or with the fungus-plant combination.

Mycorrhizae may be divided into two major types: (1) *ectomycorrhizae* in which the fungus occurs only outside of the host cells, and (2) *endomycorrhizae* in which the fungus occurs entirely within the host cells. A third intermediate type also exists, but few mycorrhizal associations are of this type and comparatively little is known of their biology. Ectomycorrhizae and endomycorrhizae are discussed further below.

ECTOMYCORRHIZAE

Ectomycorrhizae are those in which the fungus forms an external pseudoparenchymatous sheath up to 40 microns thick and constituting up to 40% of the dry weight of the entire organ. The fungus often penetrates between the cells of the epidermis and the first few cells of the cortical region, forming an intercellular network of hyphae (Fig. 13–7).

Fig. 13–7. Cross section of an ectomycorrhiza, showing fungal sheath and intercellular hyphae. [From J. L. Harley, 1965, *Ecology of soil-borne plant pathogens—prelude to biological control,* University of California Press, Los Angeles. Reprinted by permission of The Regents of the University of California.]

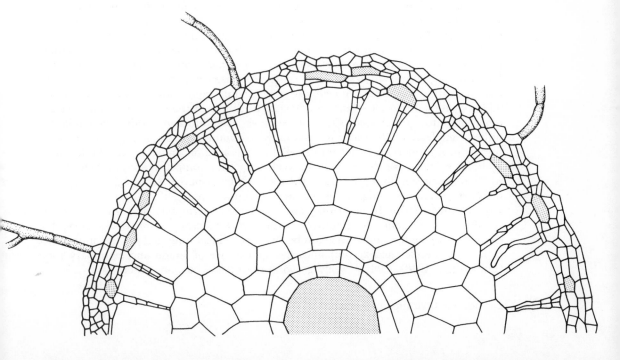

The Ectomycorrhizal Plants. Ectomycorrhizae are typical of forest trees of the temperate region, where they occur more commonly on roots in the humus layer than at greater depths. This association between the fungus and root is established on actively growing secondary roots. Some forest trees are obligate ectomycorrhizal formers and will fail to grow in the absence of mycorrhizal fungi: these include species of pine, oak, beech, and spruce. Other trees are facultative ectomycorrhizal formers: these may thrive in the absence of mycorrhizal fungi, although they may form ectomycorrhizae if the fungi are present. Facultative ectomycorrhizal trees include species of maple, juniper, willow, and elm, which are also often among the first invaders of wasteland, and are pioneers in forest succession. When moving into a wasteland, the facultative ectomycorrhizal plant species can survive in the absence of the established mycorrhizal fungus flora. As these plants grow, the mycorrhizal fungi eventually become established, and later the obligate ectomycorrhizal plant species may eventually become established (Meyer, 1973). It is sometimes desirable to establish trees in unfavorable locations such as barren wastelands or sites where there has been an extensive disruption resulting from construction of roadways. Either facultative ectomycorrhizal plant species or nursery-grown trees with established ectomycorrhizae may often be successfully planted in such sites.

The Ectomycorrhizal Fungi. Fungi that form ectomycorrhizae are usually members of the Basidiomycotina, especially members of the Agaricales and the Gasteromycetes. Some members of the Ascomycotina in the orders Eurotiales and Tuberales also form ectomycorrhizae. Sporocarps of certain fungi are frequently found under trees of a single species, and the assumption is often made that these are symbionts. Examples include the regular association of the agaric *Lactarius deliciosus* with *Pinus pinea* and also the bolete *Suillus granulatus* and the agaric *Russula emetica* with *Pinus pinaster*. The hypogeous genus *Rhizopogon* is also regularly associated with conifers. Some ectomycorrhizal fungi have a much broader host range and will apparently form the symbiotic relationship with several genera of host trees or sometimes live saprophytically. The actual number of ectomycorrhizal fungi is unknown, and many species apparently living entirely as saprophytes may actually be ectomycorrhizal.

In many instances, the degree to which the fungus is dependent on the ectomycorrhizal relationship remains obscure. Attempts to culture the fungus partner of ectomycorrhizae are sometimes futile, which implies the existence of intricate nutritional relationships with the host. Some fungi are apparently obligate ectomycorrhizal formers and are not known to sporulate in the absence of the mycorrhizal association; it is not known whether they exist saprophytically apart from their hosts. Melin (1953) found in his cultural studies of some ectomycorrhizal fungi that they cannot utilize a complex carbohydrate such as cellulose or lignin but require soluble carbohydrates. In addition, these fungi all require an exogenous source of thiamine, and some require other vitamins, growth factors in root exudates, or amino acids. Although inability to utilize complex carbohydrates is apparently of common occurrence among the ectomycorrhizal formers, it is not universal. In general, a dependency on soluble carbohydrates would make the ectomycorrhizal fungi poor competitive saprophytes in the soil and dependent on the sugars in the root as a food source. As explained further in the next section, the fungi do obtain sugars from the root.

Nature of the Interaction. The plant seeds germinate and give rise to an uninfected plant, and the ectomycorrhizae are established some weeks later. For example, in pine the initial infection by the mycorrhizal fungi follows the appearance of the first needles and accompanies the appearance of lateral and secondary roots (Huberman, 1940). Only actively growing secondary roots are generally invaded.

Compounds are secreted from the root that stimulate the germination and growth of fungi in the vicinity of the root. These include a highly active but unidentified *M* factor and B vitamins, especially thiamine (Melin, 1963). The fungi invade the root by penetrating between the cells of the cortex. Further invasion of the fungus into the root seems to be inhibited by endogenous volatile compounds (Melin and Krupa, 1971).

As the fungi become established in the roots, the fungi secrete a variety of growth regulators, including auxins, cytokinins, and gibberellins. These compounds are homologous to those formed normally by the host plant and which regulate cell division, growth, and other physiological processes such as the mobilization and control of nutrient translocation. It has been suggested that secretion of growth regulators by the fungus may be beneficial to the plant and may partially account for the enhanced growth and development generally noted in plants with ectomycorrhizae (Slankis, 1973). These growth regulators are present in above normal concentrations and influence root development. Typically the root is maintained in a juvenile condition and fails to develop suberin, root hairs fail to develop, the root is extremely short and sometimes widened, and additional branching occurs (Fig. 13–8). Branching is usually dichotomous in the pines while it is pinnate to irregular in other hosts. Anatomical studies of beech mycorrhizae show that the root continues to grow although the meristematic region is considerably reduced in length, owing both to a decrease in the division of the meristematic cells and to cell elongation. The surface area of the mycorrhizal roots is increased and their life prolonged.

After the fungus becomes established within the root, nutritional dependence on the plant can be demonstrated. Melin and Nilsson (1957) determined that car-

Fig. 13–8. Roots of pine seedlings with (left) and without (right) the ectomycorrhizal fungus *Pisolithus tinctorius*. These are roots from the seedlings shown in Fig. 13–9. [Courtesy D. H. Marx, U.S. Department of Agriculture, Forest Service.]

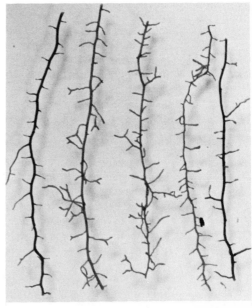

bohydrates move from the root into the fungus. They grew pine seedlings in an atmosphere containing C^{14}-labeled carbon dioxide and showed that the C^{14} fixed in photosynthesis could be detected in organic compounds that were translocated from the root into the fungal sheath. In ectomycorrhizae generally, the soluble carbohydrates that pass from the root into the fungal sheath are converted into insoluble storage polysaccharides (mannitol, trehalose, and glycogen) within the sheath. These polysaccharides are available to the fungus but not to the plant. These events help maintain a flow of carbohydrates to the roots by maintaining a gradient in the concentration of soluble carbohydrates in the root and by having a "sink" into which the carbohydrate moves.

Ectomycorrhizae aid the plant in mineral uptake. The ectomycorrhizae have the ability to accumulate ions from a very dilute solution. These ions accumulate primarily in the fungus sheath, while only a small amount passes into the root. For phosphates, as much as 90% of the phosphate may remain in the fungus sheath, while only 10% passes into the root. The quantity of phosphate that passes into the root increases when the external supply of phosphate becomes deficient (Harley and McCready, 1950; Harley et al., 1953). Under woodland conditions, quick uptake of ions by ectomycorrhizae is significant, as soluble ions are quickly flushed from the seasonal litter fall, passing immediately into the deeper mineral layers of the soil. These ectomycorrhizae can accumulate ions during the brief period when they are in luxurious supply and regulate their passage into the plant during the remainder of the year when ions are in short supply (Harley, 1969). In addition, mineral uptake is enhanced because of the increase in surface area of the ectomycorrhizal roots and extension of mycelium into the soil.

The greater absorptive capacity of ectomycorrhizal roots was demonstrated in a field experiment by Hatch (1937). Some pine seedlings were potted in prairie soil (which lacks mycorrhizal fungi), and mycorrhizal fungi were introduced into one test series. Those plants having ectomycorrhizae were large and dark green and contained normal quantities of mineral salts. The check plants without ectomycorrhizae were small and yellow and contained only minute quantities of potassium, phosphorus, and nitrogen. This experiment demonstrates that ectomycorrhizae aid in absorption of mineral salts and that pines are incapable of absorbing sufficient quantities of nutrients to maintain normal growth in some soils where the ectomycorrhizal fungi are lacking (Fig. 13–9). The increased ability to accumulate ions and to selectively pass them to the host plant may confer an advantage to the host plant, especially when it is growing in nutrient-deficient soils. Probably all soils but the most fertile agricultural soils are somewhat deficient in nutrients, which makes ectomycorrhizal formation obligate for most plants in woodlands (Harley, 1969).

Environmental factors affect the establishment of the ectomycorrhizal relationship. Generally, ectomycorrhizal formation is favored by a deficiency of inorganic ions, especially nitrogen and phosphorus. Ectomycorrhizae are usually more abundant in soils that contain low quantities of nitrogen and phosphorus, and an increase in these ions can reduce or prevent ectomycorrhizal formation. Further, an increase in nitrogen or phosphorus concentration can cause an ectomycorrhizal root to revert to the nonmycorrhizal form. Light intensities play an important role in ectomycorrhizal formation, which is generally favored by high light intensities and does not occur at low light intensities. Most of the above information was derived from laboratory or greenhouse studies, but woodland studies of beech trees also indicate that the greatest number of ectomycorrhizae occur on beech trees growing in open areas where high light intensity prevails and on nutrient-deficient soil. The manner in which nutrient

Fig. 13–9. Enhanced growth occurs in the pine seedling that has established an ectomycorrhizal relationship with *Pisolithus tinctorius.*[Courtesy D. H. Marx, U.S. Department of Agriculture, Forest Service.]

deficiencies and high light intensities favor ectomycorrhizal formation is poorly understood, as a number of complex and interrelating factors may be altered by these conditions. These include effects on root development that in turn influence the number of ectomycorrhizae formed, the type and quantity of exudates produced by the root, and the levels of some soluble compounds such as sugars or auxins within the plant cells (Harley, 1969). The effects of light on auxin concentration are believed to be especially important (Slankis, 1971).

ENDOMYCORRHIZAE

Endomycorrhizae have little in common except the usual absence of an external sheath and presence of intracellular hyphae within the root. Virtually any nonpathogenic root invader complies with this description. As Harley (1969) points out, it would be more unusual to find a root free of fungi in the soil than to find roots with internal fungi. In the absence of clearly distinguishing features of endomycorrhizae and in the absence of obvious damage to the host, endomycorrhizae may have negligible effects or they may have some selective advantage to the plant under certain ecological conditions.

Endomycorrhizae may be divided into two principal types: (1) those with septate fungi and (2) those with nonseptate fungi (Harley, 1969).

Septate Fungi. Many of the septate fungi are clamp-bearing basidiomycetes or members of the genus *Rhizoctonia,* nonsporulating members of the Deuteromycotina with basidiomycetous affinities. Plants forming associations with these fungi include a

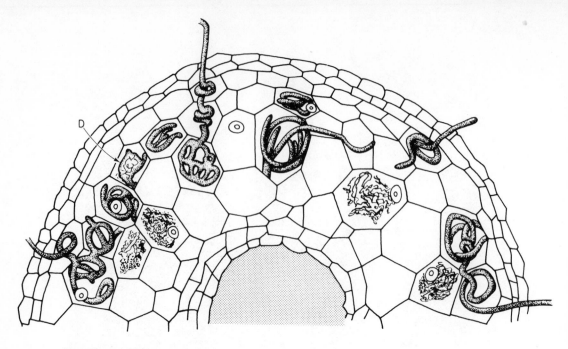

Fig. 13–10. Cross section of an orchid endomycorrhiza showing funeral penetration and intracellular digestion (*D*). From J. L. Harley. 1965, *Ecology of soil-borne plant pathogens— prelude to biological control*. University of California Press, Los Angeles. Reprinted by permission of The Regents of the University of California.]

number of families in the Ericales (especially the Ericaceae, including *Rhododendron, Vaccinum*, and the Indian pipe *Monotropa*) and virtually all members of the orchid family Orchidaceae. The orchid endomycorrhizae serve as an example of this type of endomycorrhiza (Fig. 13–10) although it should not be considered typical of this imperfectly understood group.

The orchids constitute a diverse group of temperate and tropical members, including both terrestrial and epiphytic forms. Orchids produce a very large number of minute seeds in each capsule, sometimes up to several million. The seeds contain a very small embryo and an infinitesimal food reserve that is depleted after the first few divisions of the embryo. After the food reserves in the seed are used and before the plant becomes completely autotrophic, virtually all orchids exist temporarily as symbionts. The length of the symbiotic period varies from a few months or years to their entire lifetime in the case of some non-chlorophyll-containing orchids. Prior to the symbiotic period, orchid seeds under natural conditions are invaded by hyphae of a basidiomycete or *Rhizoctonia* species that forms endomycorrhizae with the developing roots or other absorbing organs. The fungi are intercellular, forming coils of hyphae within the cells.

Endomycorrhizal formation is obligate for the orchids under naturally occurring conditions. In the absence of fungal invasions, orchids fail to grow unless they are supplied with an external source of organic carbon compounds and sometimes vitamins. The fungi provide carbohydrates and possibly other accessory metabolites such as vitamins to the orchids, so in reality the orchids are parasitic on the fungi. Unlike those fungi that form ectotrophic mycorrhizae, the orchid fungi are able to utilize complex carbohydrates, often including lignin, pectins, and cellulose. The fungus hyphae digest organic materials in the surrounding environment (either the soil or supporting tree, in the case of the epiphytes), and assimilate these nutrients to support their own growth and to produce glucose, ribose, and other simple carbohydrates. The carbohydrates and other nutrients are translocated within the hyphae and finally released into the orchid host. This release is partially accomplished by the host's digestion of some of the intracellular coils, during which the coils swell and then disintegrate. In addition, it is likely that metabolites pass through the hyphal membranes and into the host cells without digestion of the fungus. Although the major direction of nutrient flow is from the fungus to the orchid, it is important to remember that the fungus is also deriving nutrients from the orchid. The fungi are apparently sometimes dependent upon the orchid as a source of amino acids or exogenous vitamins, such as thiamine or perhaps only one of the moieties of thiamine. However, experiments utilizing radioactive carbon have indicated that carbon is not transferred from the orchids to the fungus (Arditti, 1979).

The prime importance of the digestive process is apparently to control the degree of invasion by the fungus and to prevent it from parasitizing the seedling. Frequently invasion by a particularly virulent fungus completely parasitizes and kills the host plant. Conversely, the orchid seedling may so actively digest the fungus that an endomycorrhizal relationship is not established, which again leads to the death of the host. An additional resistance mechanism is the formation of at least three antifungal substances that are produced by some orchids in response to fungal invasion. These antifungal substances include *orchinol,* a high-molecular-weight phenolic compound, which inhibits fungal growth.

The fungi forming orchid endomycorrhizae are different from the ectomycorrhizal fungi in several important respects. The orchid fungi are widely distributed and have an extensive saprophytic existence apart from the orchids. Specificity for mycorrhizal formation with the orchids is low, and some of the fungi are virulent plant parasites. For example, some pathogenic strains of *Rhizoctonia solani* isolated from wheat, cauliflower, and tomato were capable of forming an endomycorrhiza with an orchid (Downie, 1957).

Nonseptate Fungi. The nonseptate fungi in endomycorrhizae are abundantly represented in fossils from the Devonian and later periods. They occur in several genera of the gymnosperms, the majority of the angiosperms, and abundantly in herbaceous plants grown in cultivated soils and grasses. Many of the plants in which these endomycorrhizae occur are commercially important: these include the legumes, grasses, tomatoes, apples, strawberries, peaches, and coffee. Overall, these particular endomycorrhizae are more common than the other mycorrhizal types and occur almost universally in plants lacking other types, including some forest trees. Unlike the ectomycorrhizae, the morphology of the roots usually remains essentially unchanged although sometimes they may become darker, slightly thickened, or more brittle.

These fungi are members of the genus *Endogone* or closely related members of the Endogonaceae (p. 95). Attempts to isolate the fungi and to grow them in pure culture have been unsuccessful, and therefore studies of the nutritional requirements or other aspects of the fungi in the absence of the roots have not been made. Endomycorrhizal formation is apparently obligate for these fungi.

The endomycorrhizal fungi grow as individual strands or loose wefts over the roots and extend into the soil. Unlike the ectomycorrhizae, they never form a distinct fungal sheath (Fig. 13–11). The fungi penetrate the host cells by forming cellulolytic enzymes that digest small portions of the cell walls. The fungus may be limited to the

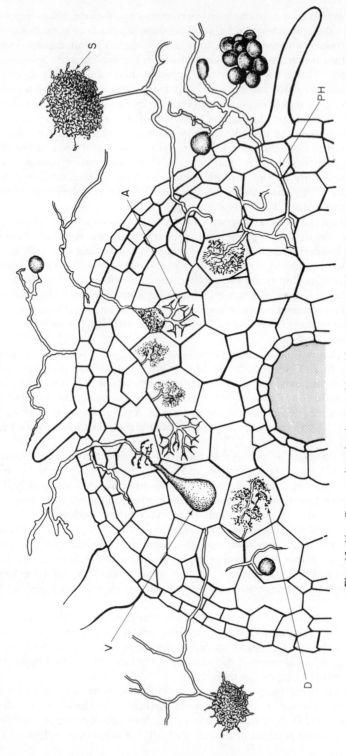

Fig. 13-11. Cross section of endomycorrhiza with a nonseptate fungus. Note the penetrating hypha (*PH*), branched arbuscules (*A*), globular vesicles (*v*), digestion of haustoria by plant (*D*), and external sporocarps (*S*) with spores. [Adapted from J. L. Harley. 1965. *Ecology of soil-born plant pathogens—a prelude to biological control.* University of California Press, Los Angeles. Reprinted by permission of the Regents of the University of California.]

root hairs or other epidermal cells but also may penetrate cells of the cortex. Within the cells, the fungus produces hyphal coils, enlarged ovoid or globular vesicles, and clusters of dichotomously branched hyphal endings (the *arbuscules*). Because of the regular occurrence of vesicles and arbuscules, these endomycorrhizae are often termed the *vesicular-arbuscular* type. Although the function of the vesicles is unknown, it has been suggested that they function as storage organs and may also function as reproductive structures under adverse conditions (Mosse, 1973). The vesicles may be formed among the hyphae on the outside of the root as well as within the cells. The branched habit of the arbuscules increases the surface area in direct contact with the host cytoplasm. It is generally accepted that transfer of minerals from the fungus to the host and of carbohydrates from the host to the fungus probably occurs largely across the plasmalemma of the arbuscules. The arbuscules collapse when they become senescent, which also releases nutrients into the host cell (Tinker, 1975). As with the ectomycorrhizae, formation of the endomycorrhizae is generally favored by high light intensities and soils with low fertility.

Interrelationships of these fungi with their host plants may be important in determining the welfare of the host plants (Mosse, 1973; Tinker, 1975). The host plants vary greatly in their response to the endomycorrhizal fungi; this variation in part is determined by the morphology of the root system and various conditions of the soil. In poor soils or in aged roots, invasion is typically more intense than in favorable soils or young roots. Small quantities of nutrients (especially nitrogen) are required for endomycorrhizal formation but larger quantities will not favor their formation or cause the association to break down. Growth of many plant species may be enhanced in poor soils by adding spores of the endomycorrhizal fungus to the soil and encouraging development of the endomycorrhizae by adding small quantities of nutrients. The enhanced growth is most often attributed to the increased uptake of phosphates. If radioactive phosphate is added to the soil, the uptake may be followed through the fungus and into the root and ultimately into the shoot. The phosphate levels are much higher in plants with than in those without endomycorrhizae. Generally, the endomycorrhizal fungi often enable the host plants to live on phosphate-deficient soil because the fungus can accumulate phosphate when it is present in such low concentrations that it is practically unavailable to nonmycorrhizal roots. If phosphate is adequate or abundant in supply, the inoculation by endomycorrhizal fungi will not further enhance the already vigorous growth of the host plant and may even cause a decrease, perhaps due to competition for carbohydrates. Under certain conditions, endomycorrhizal formation may also increase uptake of other minerals, including potassium, iron, copper, calcium, and zinc. Endomycorrhizal roots have a greater capacity to take up water and may withstand the stress of low levels of available moisture better than nonmycorrhizal roots. For example, they recover from wilting better than nonmycorrhizal plants and are better able to withstand the shock of being transplanted.

FUNGI AS INSECT SYMBIONTS

A number of fungi have symbiotic relationships with insects. Insect symbionts may be either *endosymbionts,* which are harbored within the insect, or *ectosymbionts,* which occur externally to the insect.

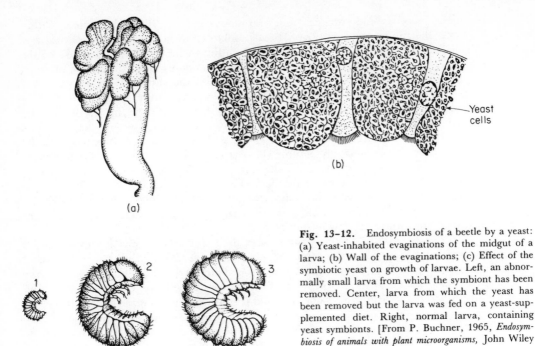

(a)

(b)

Yeast
cells

1

2

3

(c)

Fig. 13–12. Endosymbiosis of a beetle by a yeast: (a) Yeast-inhabited evaginations of the midgut of a larva; (b) Wall of the evaginations; (c) Effect of the symbiotic yeast on growth of larvae. Left, an abnormally small larva from which the symbiont has been removed. Center, larva from which the yeast has been removed but the larva was fed on a yeast-supplemented diet. Right, normal larva, containing yeast symbionts. [From P. Buchner, 1965, *Endosymbiosis of animals with plant microorganisms,* John Wiley and Sons, Inc., New York.]

Endosymbionts

Insect endosymbionts are analogous to the intestinal flora of man. These insect endosymbionts include numerous species of bacteria and some yeasts (predominantly species of *Candida* but also *Torulopsis* and *Taphrina*) that occur routinely in the gut, fat bodies, Malpighian tubules, or other internal organs of some insect species. Endosymbionts are especially common in insects that live on a restricted diet such as wood, grains, or humus. These are rich in cellulose but poor in nitrogen and deficient in vitamins.

The endosymbionts may be harbored in specialized pouches of the gut or in modified cells (Fig. 13–12). The endosymbionts may be transmitted to offspring in a number of ways, including the following: (1) The offspring may suck or lick endosymbiont-containing droplets from the mother's body; (2) the eggs are smeared with material containing the bacteria or yeast in their passage through the ovipositor, and the young consume fragments of the shell upon hatching; or (3) the endosymbiont may be included within the egg and thereby included within the developing embryo.

Endosymbionts are provided with a relatively constant, favorable environment and are richly supplied with nutrients. From studies of nutritional requirements of symbiont-free insects, it has been determined that those endosymbionts studied produce vitamins (especially B vitamins) and some amino acids needed by the host to grow and develop normally and, in some instances, to remain alive. In some experiments, the symbiont-free insects throve if provided with amino acids alone but not with vitamins alone, which suggests that the dependence on amino acids as a source of nitrogen may be greater than the dependency on vitamins.

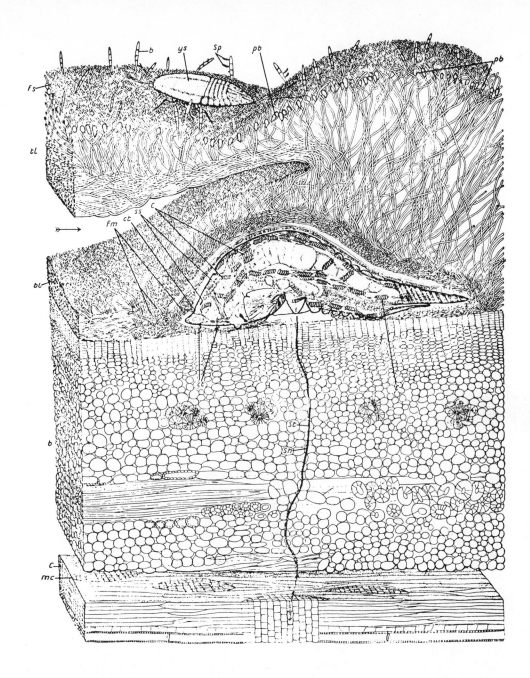

Fig. 13–13. Transverse section through portion of *Septobasidium* colony and scale insect with sucking apparatus inserted into medullary rays of wood. Key, beginning at top: *b,* basidium; *ys,* young insect; *sp,* basidiospores; *pb,* probasidium; *fs,* fruiting surface of fungus; *tl,* top layer of fungus; *bl,* bottom layer of fungus; *fm,* fungus mat; *ct,* thread connecting fungus mat with insect; *ss,* spindle-shaped hyphal threads in insect; *c,* coiled haustoria within insect; *st,* stylet of insect; *sh,* sheath secreted around stylet; *b,* bark of tree; *c,* cambium of tree; *mc,* medullary ray of tree. [From J. N. Couch, 1931, *Quart. J. Microscop. Sci.* II **74:**383–437.]

Ectosymbionts

SEPTOBASIDIUM

Septobasidium, a member of the Subclass Phragmobasidiomycetidae, is abundant in the tropics, semitropics, and southeastern United States. This fungus forms irregular, flattened, resupinate colonies that adhere closely to the bark or leaves of living trees. The colonies may be very inconspicuous (only a few millimeters in diameter) or may cover the entire lower side of the branches. The colonies may be dry and crustaceous or spongy and range in color from whitish through various shades of brown and black. Some species of *Septobasidium* are directly parasitic on the tree, causing a hypertrophy of the tissues. The majority of the species, however, share a symbiotic relationship with scale insects.

The scale insects occupy chambers within the middle layer of the fungus colony. These chambers are only slightly larger than the insect, and there are numerous tunnels connecting chambers. Over this labyrinth, there is a dense fungus mat, on which the hymenium is formed (Fig. 13–13). Some of the insects are parasitized by *Septobasidium,* and the fungus forms several coiled or knotted haustoria within their hemocoels; numerous hyphae extend from the insect into the tissues of the fungus colony. These parasitized individuals occupy a chamber, where they are attached to the plant by their sucking apparatus. The plant sap therefore directly nourishes the insects and indirectly nourishes the fungus. A parasitized insect is much smaller than uninfected individuals and usually cannot reproduce. Many insects within the colony are not parasitized, often outnumbering the parasitized individuals. The fungus colony overwinters in this manner.

Growth of the colony is renewed in the spring, and the colony grows only a few millimeters annually. Some species may form a new layer of mycelium over the old one. With the renewal of colony growth over the insect, females give birth to nymphs while the fungus probasidia germinate to form basidia and basidiospores. The basidiospores then bud in a yeastlike manner. The young nymphs crawl about and, if they venture on the surface of the colony, will pick up the fungus buds which adhere to their surface. These buds germinate and penetrate the integument of the scale insect, thereby infecting these individuals. The nymphs settle down under the same mother colony, crawl out and settle under other fungal colonies, or crawl out and settle down on the bark where there is no fungal growth. These latter individuals initiate new colonies.

Species of *Septobasidium* that are parasitic on the insects are unable to grow independently of the insects in nature, and the amount of fungus growth that occurs is directly dependent on the number of parasitized insects. In addition to depending on the insects for food, the fungus is dependent on this relationship for spore distribution.

Although the scale insects can grow in the absence of the fungus in nature, free-living individuals occur only rarely. Life is too precarious for the free-living insects, as they are subjected to extremes of heat and cold from which the fungus colony protects them. They are also much more likely to be eaten by birds and to be parasitized by their most deadly enemy, those wasps which lay eggs within the scale insect's body. Although some insects are parasitized and may be killed by the fungus, the majority of the insects benefit from this protective relationship.

AMBROSIA FUNGI

Some scolytid beetles are inhabitants of wood, forming tunnels in the sapwood of diseased trees or felled logs. These tunnels consist of an entrance passageway from the outside, and they widen into a number of cavelike chambers or elongated galleries where eggs are laid and larvae develop. The tunnels, chambers, and galleries are lined with a cushionlike growth of *ambrosia* fungi (Fig. 13-14), which usually is the sole source of food for the larvae and is also important in the diet of the adults, who feed on the wood in addition. This symbiotic relationship between the beetle and fungus is highly specific, with only certain species of the beetle or fungus occurring together. Neither the ambrosia fungi nor the beetles are found without the other in nature.

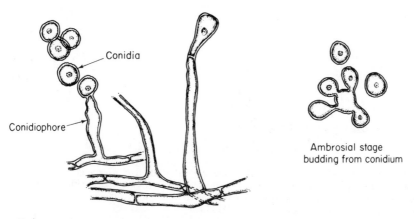

Conidia

Conidiophore

Ambrosial stage
budding from conidium

Fig. 13-14. The ambrosia fungus, *Ambrosiella canadensis.* × 900.
[From L. R. Batra, 1967, *Mycologia* **59:**976-1017.]

The ambrosia fungi were originally seen as a saltlike crust lining beetle tunnels and were thought to be dried, exuded sap (the "ambrosia"). Later fungi were found in the tunnels but were thought to be incidental invaders that fed on the ambrosia and acquired the name of ambrosia fungi. The ambrosia fungi are filamentous members of the Hemiascomycetes (especially *Endomycopsis* and *Ascoidea*) and members of the Deuteromycotina (especially *Monilia* and *Ambrosiella*). These fungi form hyphae with conidiophores, conidia, and yeastlike sprout cells that bud from the conidia. The yeastlike cells constitute the ambrosial stage on which the beetles feed.

The beetles carry spores of the ambrosia fungus with them during their overwinter hibernations and during their migrations. The spores are stored in specialized tubes or pouches (*mycetangia*) that occur in different locations of the body among the insect species (Fig. 13-15). The mycetangia occur in the one sex of the beetle that initiates the tunnel-building; this is usually the female who prepares the tunnels for egg-laying. The mycetangia produce secretions that protect the thin-walled spores from desiccation and provide nutrients needed for the germination of the spores. When the beetles bore new tunnels in the spring, they prepare a mixture of feces and wood fragments which they smear on the tunnel walls as a bed for the fungal growth. The beds are inoculated with ambrosia fungi through incidental contact with the

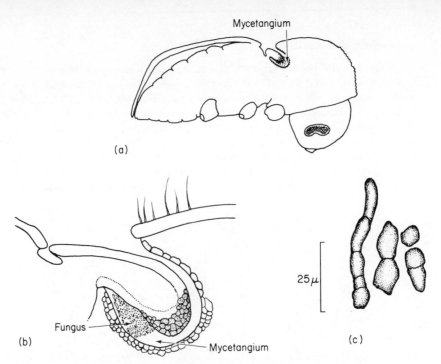

(a)

(b)

(c)

25μ

Mycetangium

Fungus

Mycetangium

Fig. 13-15. The mycetangium of *Anisandrus dispar:* (a) Position of mycetangium in insect. × 7.3; (b) Enlargement of mycetangium showing glandular cells associated with it, and the enclosed fungus. × 180; (c) Ambrosia fungus removed from the mycetangium. (From L. R. Batra, 1963, *Trans. Kansas Acad. Sci.* **66:**213-236.]

mycetangia. A palisade layer of ambrosia fungus then develops, and the yeast-like ambrosial stage predominates. The ambrosial stage is dominant as long as the larvae and beetles are present, but when the beetles vacate the tunnel, the mycelial form dominates (as it does in pure culture). The beetles tend to maintain a pure culture of the ambrosia fungus and are able to do so when the tunnels are fresh. The means by which the beetles are able to suppress growth of foreign fungi is unknown, but perhaps the larvae and beetles have secretions that promote growth of the ambrosia fungi, which in turn are antagonistic to other fungi. Alternatively, the constant enlargement of the tunnels, coupled with the discardation of the debris, may eliminate the foreign fungi. Often secondary ambrosia fungi may appear in the tunnel, and the beetles may feed on them (the Hemiascomycete *Dipodascus,* discussed in Chapter 4, is of this type). Unlike the primary ambrosia fungus, these secondary fungi are not transported in mycetangia, and the beetle is not dependent on their presence. When the beetles vacate the tunnels, contaminating fungi of several types develop profusely.

The relationship between the fungus and beetles is mutually beneficial and is obligatory for both partners. Larvae are dependent on the ambrosia fungus as a source of food and vitamins, and even the adults may starve in the absence of ambrosial growth. There is evidence that at least some of the ambrosia beetles may have an absolute requirement for the ergosterol produced by the fungi (Kok, 1979).

Advantages of a fungus diet are that the fungus provides a high nitrogen, low residue food that can be obtained with a minimum of tunneling as the nutrients are translocated for some distance through the mycelium. Those beetles which feed instead on wood (a low nitrogen, high residue food) must actively tunnel through the wood to obtain enough food. The fungus benefits because it is protected from desiccation during transport in the mycetangia, the number of cells available for inoculation of wood are increased by the mycetangial secretions, it is disseminated directly to and placed within a suitable substratum, injury of the wood by the beetle facilitates rapid penetration by the mycelium, and urea and uric acids in the beetle's feces serve as an important source of nitrogen. Attempts to establish the fungus in the host plant in the absence of the beetle have been unsuccessful.

FUNGUS-GROWING INSECTS

Fungi may be cultivated by termites and ants, which tend their fungal gardens with great diligence. These insects depend upon the fungi for their food (Fig. 13-16). The fungus-cultivating termites are especially prevalent in Africa and southern Asia, where they usually construct conspicuous clay mounds that are up to 6 meters high and cover their nests and fungal gardens (Batra and Batra, 1979). These fungus-cultivating termites share many similarities with the ants, which are discussed further below.

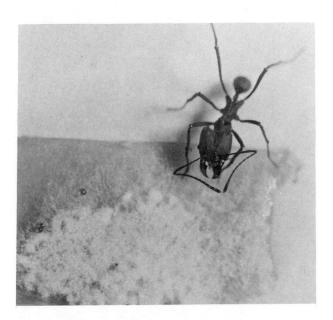

Fig. 13-16. An ant with its cultivated fungus. [Courtesy N. A. Weber, 1966, *Science* **153**:587–609.]

Fungus-Growing Ants. The fungus-cultivating ants belong to a New World tribe of ants, the Attini, which are widely distributed on the American continent but are especially prevalent in the rain forests of South America. The ants and fungi are dependent on each other for their maintenance, and neither is found alone in nature.

The fungi are cultivated in well-circumscribed gardens. The gardens consist of a light gray to brown flocculent, spongy mass of leaves overgrown with the fungus and may be honeycombed with several tunnels or cells. The ants grow their gardens either in protected places on the ground such as under rocks or overhanging logs or, more usually, in chambers within nests (Fig. 13-17). The chambers may be craterlike excavations at the surface of the ground, or they may be irregular or domelike cavities

(a)   **Fig. 13-17.**   Two fungus gardens cultivated by ants: (a) A section of a nest made by   (b)
Trachymyrmex wheeleri in Colombia. Compare the size of the nest with the forceps which are
113 millimeters long. (b) A section of a nest made by *Acromyrmex histrix* in Venezuela. There
are two gardens in this nest. Compare the size of the nest with the knife, which is 25 mm
long. [Courtesy of N. A. Weber. (a) 1966, *Monogr. 1, Prog. biol. del Suelo. Actas Prim. Coloq.
Latin. Biol. del Suelo.* (b) From W. M. Wheeler, 1937, *Mosaics and other anomalies among ants.*
Harvard University Press, Cambridge, Massachusetts.]

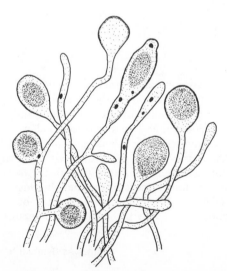

Fig. 13-18.   Fungi cultivated by
ants. The enlarged hyphal endings
may be as wide as 50 microns in diameter. [From N. A. Weber, 1966,
Science **153:**587-604.]

in the ground that are connected to the exterior by tunnels. A single fungus garden may vary from 20 millimeters to 50 centimeters in diameter. As many as 2,000 chambers may be found in a nest. The area of ground covered by the nest may be as large as 11 x 15 meters.

Fungi associated with the ant gardens belong in one instance to the yeasts (*Tyridomyces formicarum*), but otherwise they are mycelial fungi. While the identity of the fungus is unknown in most instances, basidiocarps of agarics in the genus *Lepiota* have been obtained by culturing the mycelium from several ant nests (Weber, 1979). The mycelia form clusters of inflated hyphae, which occur either in the ant nest or in pure culture in the absence of ants (Fig. 13–18). The protoplasm-rich inflations are the sole food of the larvae, although adult ants feed on both hyphae and hyphal inflations. The fungus is capable of actively degrading the cellulose in the substratum and of providing the ants with a nutritious diet. More than 50% of the dry weight of the fungus is available as soluble nutrients (including carbohydrates, amino acids, proteins and lipids) (Martin et al., 1969).

Some ants begin their fungus gardens by cutting leaves, while other genera scavenge for plant debris. The ants forage for plant material in the daytime and carry large cut pieces of leaves back to the nest. The ants maintain the cleanliness of the cut leaves by carrying the pieces high above the ground; also, a small ant may ride upon a section of a leaf carried by a larger ant, licking the leaf to cleanse it during its transport (Weber, 1972-b). When the ants reach the garden, they cut the leaf into small pieces (about 1 or 2 millimeters in diameter) and make these pieces pulpy by pinching them with their mandibles. They also lick the pieces repeatedly and then deposit a fecal droplet from the anus on it. Maceration of the leaves encourages the initial invasion of the tissue by the hyphae, and the application of fecal droplets facilitates this invasion because these droplets contain protein-degrading enzymes (the fungi themselves are deficient in these enzymes). The bits of plant material are packed into place and then the ant plants the garden with several tufts of mycelium carried in from an adjacent garden. Additional material may be placed in the nest, such as woody particles which may be dragged in, defecated on, and then inoculated with the fungus. Some ants also drag insect carcasses or insect excreta into the garden. The ants continue to deposit anal drops on the fungal garden even after it is established. These anal drops are an important source of nutrients for the fungus because they contain ammonia and a variety of amino acids. Continual deposition is required because the fungus is constantly utilizing the nutrients present within the anal drops (Martin and Martin, 1970). The ants also tend the garden by probing the fungus with their antennae to determine the condition of the fungus and by licking the hyphae. The ants can distinguish any invading fungi from those that they are cultivating and will remove hyphae of the unwanted species by pulling the hyphae out with their mandibles and placing the discarded hyphae on their refuse heap. Any unused substrata is carried out of the nest, and the ants may open or close the entrances of the tunnels to regulate the temperature. While the garden is maintained in good condition, the foreign bacteria and yeasts normally present exist at relatively low levels and are apparently simply suppressed in their growth by competition with the cultured fungus. Once the nest is abandoned, these bacteria, yeasts, and other fungi completely overrun the nest as the gardens deteriorate. The ants abandon their nest if it is disturbed or if they migrate. Whenever the ants leave their nest, they carry portions of their garden along with them to serve as inoculum for the gardens that they will build in their new nest.

References

Ahmadjian, V. 1964. Further studies on lichenized fungi. Bryologist **67**:87–98.

——. 1965. Lichens. Ann. Rev. Microbiol. **19**:1–20.

——. 1966-a. Lichens. *In:* Henry, S. M., Ed., Symbiosis. Academic Press, New York. **1**:35–97.

——. 1966-b. Cultural and physiological aspects of the lichen symbiosis. *In:* Jensen, W. A., and L. G. Kavaljian, Eds., Plant biology today—advances and challenges. Wadsworth Publishing Co. Inc., Belmont, Calif. pp. 148–163.

——. 1966-c. Artificial reestablishment of the lichen *Cladonia cristatella.* Science **151**:199–201.

——. 1967-a. The lichen symbiosis. Blaisdell Publishing Co., Waltham, Mass. 152 pp.

——. 1967-b. A guide to the algae occurring as lichen symbionts: isolation, culture, cultural physiology, and identification. Phycologia **6**:127–160.

——. 1973. Resynthesis of lichens. *In:* Ahmadjian, V., and M. E. Hale, Eds., The lichens. Academic Press, New York. pp. 565–579.

——, L. A. Russell, and K. C. Hildreth. 1980. Artificial establishment of lichens. I. Morphological interactions between the phycobionts of different lichens and the mycobionts *Cladonia cristatella* and *Lecanora chrysoleuca.* Mycologia **72**:73–89.

Arditti, J. 1979. Aspects of the physiology of orchids. Advan. Botan. Res. **7**:421–655.

Bailey, R. H. 1976. Ecological aspects of dispersal and establishment in lichens. *In:* Brown, D. H., D. L. Hawksworth, and R. H. Bailey, Eds., Lichenology: progress and problems. Academic Press, New York. pp. 215–247.

Baker, J. M. 1963. Ambrosia beetles and their fungi, with particular reference to *Platypus cylindrus* Fab. *In:* Symbiotic associations. Thirteenth Symp. Soc. Gen. Microbiol. **13**:232–265.

Barrows, J. B., and R. W. Roncadori. 1977. Endomycorrhizal synthesis by *Gigaspora margarita* in *Poinsettia.* Mycologia **69**:1173–1184.

Batra, L. R. 1967. Ambrosia fungi: a taxonomic revision, and nutritional studies of some species. Mycologia **59**:976–1017.

——, and S. W. T. Batra. 1979. Termite-fungus mutualism. *In:* Batra, L. R., Ed., Insect-fungus symbiosis—nutrition, mutualism, and commensalism. Allanheld, Osmun and Company, Montclair, N. J. pp. 117–163.

Batra, S. W. T., and L. R. Batra. 1967. The fungus gardens of insects. Sci. Am. **217**:112–120.

Bliss, L. C., and E. B. Hadley. 1964. Photosynthesis and respiration of alpine lichens. Am. J. Botany **51**:870–874.

Blum, O. B. 1973. Water relations. *In:* Ahmadjian, V., and M. E. Hale, Eds., The lichens. Academic Press, New York. pp. 381–400.

Bowen, G. D. 1973. Mineral nutrition of ectomycorrhizae. *In:* Marks, G. C., and T. T. Kozlowski, Eds., Ectomycorrhizae—their ecology and physiology. Academic Press, New York. pp. 151–205.

Brooks, M. A. 1963. Symbiosis and aposymbiosis in arthropods. *In:* Symbiotic associations. Thirteenth Symp. Soc. Gen. Microbiol. **13**:200–231.

Clowes, F. A. L. 1951. The structure of mycorrhizal roots of *Fagus sylvatica.* New Phytologist **50**:1–16.

Couch, J. N. 1935. *Septobasidium* in the United States. J. Elisha Mitchell Sci. Soc. **51**:1–77.

Downie, D. G. 1957. *Corticium solani*—an orchid endophyte. Nature **179**:160.

Francke-Grosmann, H. 1963. Some new aspects in forest entomology. Ann. Rev. Entomol. **8**:415–438.

——. 1967. Ectosymbiosis in wood-inhabiting insects. *In:* Henry, S. M., Ed., Symbiosis. Academic Press, New York. **2**:141–205.

Farrar, J. F. 1976. The lichen as an ecosystem: observation and experiment. *In:* Brown, D. H., D. L. Hawskworth, and R. H. Bailey, Eds., Lichenology: progress and problems. Academic Press, New York. pp. 385–406.

Garrett, S. D. 1950. Ecology of the root-inhabiting fungi. Biol. Rev. **25**:220–254.

Gerdemann, J. W. 1968. Vesicular-arbuscular mycorrhizae and plant growth. Ann. Rev. Phytopathol. **6**:397–418.

Gilbert, O. L. 1973. Lichens and air pollution. *In:* Ahmadjian, V., and M. E. Hale, Eds., The lichens. Academic Press, New York. pp. 443–472.

Graham, K. 1967. Fungal-insect mutualism in trees and timber. Ann. Rev. Entomol. **12**:105–126.

Hacskaylo, E. 1971. Metabolite exchange in ectomycorrhizae. *In:* Hacskaylo, E., Ed., Mycorrhizae. U. S. Department of Agriculture Forest Service Misc. Publ. No. 1189, pp. 175–182.

——. 1972. Mycorrhiza: the ultimate in reciprocal parasitism? Bioscience **22**:577–583.

——. 1973. Carbohydrate physiology of ectomycorrhizae. *In:* Marks, G. C., and T. T. Kozlowski, Eds., Ectomycorrhizae—their ecology and physiology. Academic Press, New York. pp. 207–230.

Hale, M. E. 1959. Studies on lichen growth rate and succession. Bull. Torrey Botan. Club **86**:126–129.

——. 1973. Growth. *In:* Ahmadjian, V., and M. E. Hale, Eds., The lichens. Academic Press, New York. pp. 473–492.

Harley, J. L. 1968. Mycorrhiza. *In:* Ainsworth, G. C., and A. S. Sussman, Eds., The fungi—an advanced treatise. Academic Press, New York. **3**:139–178.

——. 1969. Biology of the mycorrhizae, Second ed. Leonard Hill, London. 334 pp.

——, and C. C. McCready. 1950. The uptake of phosphate by excised mycorrhizal roots of beech. New Phytologist **49**:388–397.

——, ——, and J. K. Brierley. 1953. The uptake of phosphate by excised mycorrhizal roots of beech IV. The effect of oxygen concentration upon host and fungus. New Phytologist **52**:124–132.

——, and J. S. Waid. 1955. The effect of light upon the roots of beech and its surface population. Plant Soil **7**:96–112.

——, and J. M. Wilson. 1959. The absorption of potassium by beech mycorrhiza. New Phytologist **58**:281–298.

Harris, G. P., and K. A. Kershaw. 1971. Thallus growth and the distribution of stored metabolites in the phycobionts of the lichens *Parmelia sulcata* and *P. physodes.* Can. J. Botany **49**:1367–1372.

Hartzell, A. 1967. Insect ectosymbiosis. *In:* Henry, S. M., Ed., Symbiosis. Academic Press, New York. **2**:107–140.

Hatch, A. B. 1937. The physical basis of mycotrophy in *Pinus.* Black Rock Forest Bull. no. 6. 168 pp.

Henriksson, E. 1951. Nitrogen fixation by a bacteria-free, symbiotic *Nostoc* strain isolated from *Collema.* Physiol. Plant. **4**:542–545.

——. 1957. Studies in the physiology of the lichen *Collema* I. The production of extracellular nitrogenous substances by the algal partner under various conditions. Physiol. Plant. **10**:943–948.

——. 1961. Studies in the physiology of the lichen *Collema* IV. The occurrence of polysaccharides and some vitamins outside the cells of the phycobiont, *Nostoc* sp. Physiol. Plant. **14**:813–817.

Hill, D. J. 1976. The physiology of lichen symbiosis. *In:* Brown, D. H., D. L. Hawksworth, and R. H. Bailey, Eds., Lichenology: Progress and Problems. Academic Press, New York. pp. 457–496.

Huberman, M. A. 1940. Normal growth and development of southern pine seedlings in the nursery. Ecology **21**:323–334.

Jahns, H. M. 1973. Anatomy, morphology, and development. *In:* Ahmadjian, V., and M. E. Hale, Eds., The lichens. Academic Press, New York. pp. 3–58.

Jurzitza, G. 1979. The fungi symbiotic with anobiid beetles. *In:* Batra, L. R., Ed., Insect-fungus symbiosis —nutrition, mutualism, and commensalism. Allanheld, Osmun and Company, Montclair, N.J., pp. 65–76.

Koch, A. 1967. Insects and their endosymbionts. *In:* Henry, S. M., Ed., Symbiosis. Academic Press, New York. **2**:1–106.

Kok, L. T. 1979. Lipids of ambrosia fungi and the life of mutualistic beetles. *In:* Batra, L. R., Ed., Insect-fungus symbiosis—nutrition, mutualism, and commensalism. Allanheld, Osmun and Company, Montclair, N.J. pp. 33–52.

Marks, G. C., and R. C. Foster. 1973. Structure, morphogenesis, and ultrastructure of ectomycorrhizae. *In:* Marks, G. C., and T. T. Kozlowski, Eds., Ectomycorrhizae—their ecology and physiology. Academic Press, New York. pp. 1–41.

Martin, M. M., R. M. Carman, and J. G. MacConnell. 1969. Nutrients derived from the fungus cultured by the fungus-growing ant *Atta colombica tonsipes.* Ann. Entomol. Soc. Am. **62**:11–13.

——, and J. S. Martin. 1970. The biochemical basis for the symbiosis between the ant, *Atta colombica tonsipes,* and its food fungus. J. Insect Physiol. **16**:109–119.

Melin, E. 1953. Physiology of mycorrhizal relations in plants. Ann. Rev. Plant Physiol. **4**:325–346.

——. 1963. Some effects of forest tree roots on mycorrhizal basidiomycetes. *In:* Symbiotic Associations. Thirteenth Symp. Soc. Gen. Microbiol. **13**:125–145.

——, and S. Krupa. 1971. Studies on ectomycorrhizae of pine II. Growth inhibition of mycorrhizal fungi by volatile organic constituents of *Pinus silvestris* (Scots Pine) roots. Physiol. Plant. **25**:337–340.

——, and H. Nilsson. 1950. Transfer of radioactive phosphorus to pine seedlings by means of mycorrhizal hyphae. Physiol. Plant. **3**:88–92.

——, and ——. 1957. Transport of C^{14}-labelled photosynthate to the fungal associate of pine mycorrhiza. Svensk. Botan. Tidskr. **51**:166–186.

Meyer, F. H. 1973. Distribution of ectomycorrhizae in native and man-made forests. *In:* Marks, G. C., and T. T. Kozlowski, Eds., Ectomycorrhizae—their ecology and physiology. Academic Press, New York. pp. 79–105.

——, 1974. Physiology of mycorrhiza. Ann. Rev. Plant Physiol. **25**:567–586.

Mikola, P. 1973. Application of mycorrhizal symbiosis in forestry practice. *In:* Marks, G. C., and T. T. Kozlowski, Eds., Ectomycorrhizae—their ecology and physiology. Academic Press, New York. pp. 231–298.

Millbank, J. W. 1976. Aspects of nitrogen metabolism in lichens. *In:* Brown, D. H., D. L. Hawksworth, and R. H. Bailey, Eds., Lichenology: progress and problems. Academic Press, New York. pp. 385–406.

Mosse, B. 1963. Vesicular-arbuscular mycorrhiza: an extreme form of fungal adaptation. *In:* Symbiotic associations. Thirteenth Symp. Soc. Gen. Microbiol. **13**:146–170.

——. 1973. Advances in the study of vesicular-arbuscular mycorrhiza. Ann. Rev. Phytopathol. **11**:171–196.

Norris, D. M. 1979. The mutualistic fungi of Xyleborini beetles. *In:* Batra, L. R., Ed., Insect-fungus symbiosis—nutrition, mutualism, and commensalism. Allanheld, Osmun and Company, Montclair, N.J. pp. 53–63.

Pearson, L. C. 1970. Varying environmental conditions in order to grow intact lichens under laboratory conditions. Am. J. Botany **57**:659–664.

Perez-Llano, G. A. 1944. Lichens—their biological and economic significance. Botan. Rev. **10**:1–65.

——. 1948. Economic uses of lichens. Econ. Botany **2**:15–45.

Peyronel, B., B. Fassi, A. Fontana, and J. M. Trappe. 1969. Terminology of mycorrhizae. Mycologia **61**:410–411.

Richardson, D. H. S. 1973. Photosynthesis and carbohydrate movement. *In:* Ahmadjian, V., and M. E. Hale, Eds., The lichens. Academic Press, New York. pp. 249–288.

Scott, G. D. 1956. Further investigations of some lichens for fixation of nitrogen. New Phytologist **55**:111–116.

——, 1967. Studies of the lichen symbiosis: 3. The water relations of lichens on granite kopjes in central Africa. Lichenologist **3**:368–385.

Slankis, V. 1971. Formation of ectomycorrhizae of forest trees in relation to light, carbohydrates, and auxins. *In:* Hacskaylo, E., Ed., Mycorrhizae. U. S. Department of Agriculture Forest Service. Misc. Publ. No. 1189, pp. 151–167.

——. 1973. Hormonal relationships in mycorrhizal development. *In:* Marks, G. C., and T. T. Kozlowski, Eds., Ectomycorrhizae—their ecology and physiology. Academic Press, New York. pp. 231–298.

——. 1974. Soil factors influencing formation of mycorrhizae. Ann. Rev. Phytopathol. **12**:437–457.

Skye, E. 1979. Lichens as biological indicators of air pollution. Ann. Rev. Phytopathol. **17**:325–341.

Smith, A. H. 1971. Taxonomy of ectomycorrhizae-forming fungi. *In:* Hacaskaylo, E., Ed., Mycorrhizaes. U. S. Department of Agriculture, Forest Service, Misc. Publ. No. 1189, pp. 1–18.

Smith, D. C. 1962. The biology of lichen thalli. Biol. Rev. **37**:537–570.

——. 1963. Experimental studies of lichen physiology. *In:* Symbiotic associations. Thirteenth Symp. Soc. Gen. Microbiol. **13**:31–50.

——, and E. A. Drew. 1965. Studies in the physiology of lichens V. Translocation from the algal layer to the medulla in *Peltigera polydactyla*. New Phytologist **64**:195–200.

Tinker, P. B. H. 1975. Effects of vesicular-arbuscular mycorrhizas on higher plants. Symp. Soc. Exptl. Botany **29**:325–349.

Trappe, J. M. 1962. Fungus associates of ectotrophic mycorrhizae. Botan. Rev. **28**:538–606.

Voigt, G. K. 1971. Mycorrhizae and nutrient mobilization. *In:* Hacskaylo, E., Ed., Mycorrhizas. U. S. Department of Agriculture Forest Service Misc. Publ. No. 1189, pp. 122–131.

Weber, N. A. 1945. The biology of the fungus-growing ants. Part VIII. The Trinidad, B. W. I., species. Rev. Entomol. **16**:1–88.

——, 1972-a. Gardening ants—the attines. Mem. Am. Phil. Soc. **92**:1–146.

——, 1972-b. The fungus-culturing behavior of ants. Am. Zoologist **12**:577–587.

——, 1972—c. The attines: the fungus-culturing ants. Am. Sci. **60**:448–456.

——. 1979. Fungus-culturing by ants. *In:* Batra, L. R., Ed., Insect-fungus symbiosis—nutrition, mutualism, and commensalism. Allanheld, Osmun and Company, Montclair, N.J. pp. 77–116.

FOURTEEN

Fungi and Humans

Not only do fungi exploit other organisms (as saprophytes, parasites, or symbionts), but other organisms exploit fungi in numerous ways. The most usual of these is the use of fungi as food by various animals. It may be assumed that thousands of species consume fungal mycelium or the sporocarps of fungi. As we noted in Chapter 11, many soil animals consume fungi. Particularly avid mycophagists include the slugs and the red squirrel. Slugs will attack most fleshy Hymenomycete species, including *Amanita muscaria* and *A. phalloides,* which are poisonous to humans. The red squirrel collects large numbers of fungi in the late autumn and stores these fungi in tree trunks, in birds' nests, and on branches of trees whereupon the fungi dry out and serve as a supply of food for the winter (Buller, 1922). People are also avid mycophagists, but in addition they exploit the fungi in numerous ways which are not shared by other organisms. The remainder of the chapter deals with exploitation of fungi by humans.

Like other forms of life, fungi have long been objects of wonder and speculation. Some of the earliest recorded observations of the fungi are those of the Greeks and Romans, who were curious about the nature and origin of fungi. Pliny could not decide whether fungi were living or nonliving but nevertheless concluded that they were earthy concretions or imperfections having no vital connections with the earth (he was obviously unaware of the mycelium). A belief rather widely held by the Greeks (about 300 B.C.) was that truffles originate from thunder; that is, a generating fluid was contained in the thunder which, when mixed with heat, pierces the earth and forms the fungi (Buller, 1914). Even at the present, fungi are still objects of curiosity, although answers are being sought in a scientific manner by mycologists rather than by speculation as in the case of the ancients.

In a similar manner, fungi have been the subject for art, poetry, and literature since the time of the ancient Greeks and Romans. Traditionally, fungi have been treated with a great feeling of repulsion, probably because they were thought to be an "evil ferment of the earth" (as expressed by the physician and poet Nicander about 185 B.C.), and many species were known to be poisonous. This view is expressed in an excerpt of a translation of Nicander's poem "Alexipharmaca" (Buller, 1914):

> Let not the evil ferment of the earth which often causes swellings in the belly or strictures in the throat, distress a man; for when it has grown up under the viper's deep hollow track, it gives forth the poison and hard breathing of its mouth; an evil ferment is that; men generally call the ferment by the name of fungus . . . .

The interest of some cultural groups in fungi may have been based on religious beliefs. For example, mushrooms are illustrated in some ancient Mayan hieroglyphics, and a possible interpretation is that these are sacred mushrooms being held in the hand of an animal deity (Lowy, 1972). The Mayans also made stone carvings in the shape of mushrooms. These carvings are about 25 to 50 centimeters high and have some human features (many are pregnant females) or are effigies of animals such as toads (Fig. 14-1). Although these are relics of a culture that existed 1,000 to 4,000 years ago, they only came to the attention of anthropologists in the early part of this century and are being recovered from excavations in Guatemala. These carvings may be relics of a culture in which mushrooms were consumed during religious ceremonies, during which some may have been used as seats (Lowy, 1971).

Fig. 14.1 A mushroom stone featuring a human-animal effigy with claws on the base. The height is 29.5 centimeters, and the mushroom pileus is 16.5 centimeters. [Courtesy B. Lowy.]

Fungi have been utilized by humans for a variety of purposes since ancient times. Some puffballs were recovered from British archeological sites that were approximately 2,000 years old and are believed to have been used to stop the bleeding of wounds or as tinder (Watling and Seaward, 1976). Puffballs have long been utilized for these purposes, and as late as World War I were used to dress wounds when bandages were in short supply. Presently in Nigeria, fungi are extremely important to the Yoruba people, who have a culture rich in mycological folklore (Oso, 1975, 1976, 1977-a, 1977-b). These people may use various fungi as food; for medicinal purposes; as toothbrushes; in charms intended, for example, for use in making oneself disappear in the face of danger; or in a preparation that is rubbed on the gun, bow, or arrow along with incantations to make game drowsy and easy to kill.

FUNGI AS FOOD FOR HUMANS

One of the predominating interests that humans have had in fungi is their utilization as food. The Greeks and Romans were fond of truffles, mushrooms, and puffballs, all of which were such delicacies that they were the only foods that the wealthy insisted upon preparing themselves. The Roman Coelius Apicus gives us a recipe for the preparation of truffles (Buller, 1914):

> . . . slice, boil, sprinkle with salt, and transfix with a twig; partly roast, and place in a cooking-vessel with oil, liquor, sweet boiled wine, unmixed wine, pepper and honey; while boiling, beat up with fine flour, take out the twigs and serve.

Although methods of preparation have changed with time, the view that fungi are delicacies has not.

Fungi collected for food include truffles, morels, mushrooms, puffballs, species of *Clavaria,* and nonwoody polypores. The truffles are probably the most highly esteemed of all and in France are collected with the aid of trained dogs or pigs that locate the scent of the underground ascocarps. Truffles are also commonly sold in the markets in North Africa and the Near East.

Cultivation of Fungi

Fungi cultivated for food in the Orient include various species of mushrooms and the jelly fungus *Auricularia.* Truffles are cultivated in France. Spores of the truffles are introduced into the ground in which oak trees are growing, and large numbers of ascocarps may be harvested later. Only one species, *Agaricus brunnescens,* is commercially cultivated to any extent in the United States. Mushrooms are higher in protein than many vegetables and fruits, as about 3% to 4% of their fresh wet weight is protein. They are also a good source of the following nutrients for humans: fat, phosphorus, iron, thiamine, riboflavin, and niacin. *A. brunnescens* is also high in ascorbic acid (Chang, 1980).

The commercial mushroom, *A. brunnescens* has been cultivated in cellars, abandoned quarries and mines, and caves. It is most profitably cultivated in specially constructed windowless buildings in which shelf beds are arranged in tiers (Fig. 14-2). Such buildings are desirable because temperature, humidity, and ventilation can be easily controlled.

Fig. 14-2. Commercial mushroom farming. [Courtesy of Butler County Mushroom Farm, Inc.]

The substratum used to support growth of the mushrooms is compost, prepared from horse dung and straw. The horse dung is placed in large heaps (about 3 meters high) and is allowed to undergo a natural fermentation process during which bacteria and fungi decompose the sugars, starch, and hemicelluloses and leave behind the resistant lignins and cellulose. Also insoluble nitrogen compounds accumulate as a result of nitrogen assimilation by the microbes. At the end of this decomposition phase, the compost provides a medium favoring growth of *A. brunnescens* which is especially dependent on the high concentration of lignin and insoluble nitrogen compounds. Usually other fungi are not able to compete once *A. brunnescens* has become established.

After the compost is prepared, it is placed in tiers of flat beds and kept at a temperature of 55°C for a week or more to eliminate insects; then it is planted with the mushroom *spawn* or inoculum. Mushroom spawn is a pure culture of mycelium derived usually from germinated basidiospores grown under laboratory conditions in bottles on a medium of cereal grains, bran, or some other similar material. After the spawn is introduced into the compost, a shallow *casing* layer of soil is spread over the compost. This casing layer is necessary to obtain a good crop of basidiocarps, perhaps because it provides a nutrient-deficient layer that does not encourage vegetative growth but does encourage the reproductive stage (see Chapter 9).

Mycelium grows from the spawn and throughout the substratum. Rate of mycelial growth is affected by the moisture content of the substratum, pH, and temperature, all of which must be carefully controlled. Both the optimum moisture content and pH vary with the composition of the particular compost used, while the optimum temperature is 25°C, with the range supporting growth between 2°C and 33°C. The temperature range over which basidiocarp formation takes place is somewhat more limited, occurring between 7°C and 24°C. Commercial growers prefer the temperature range of 9°C to 13°C because the mushrooms are firmer and less likely to be diseased or attacked by insects when grown within this temperature range. Relative humidity is maintained between 70° and 80°. Adequate ventilation is required as a concentration greater than 1% of carbon dioxide in the air will arrest development of the mushrooms (especially of the pilei), and this amount of carbon dioxide could accumulate from respiration of fungal mycelium and bacteria in a mushroom house closed for more than 24 hours.

Bacteria and fungi may cause diseases of mushrooms and limit their successful cultivation. Fungus diseases include bubbles (caused by *Mycogone perniciosa*), brown spot (caused by *Verticillium malthousei*), mildew (caused by *Dactylium*), and damping-off (caused by *Fusarium*). These diseases may be combated by fumigating the mushroom house between crops and by disinfecting the equipment and surrounding area.

Poisonous Fungi

An unfortunate consequence of indulging oneself in the gastronomical delights of mushroom consumption is often poisoning. People have been aware of mushroom poisoning for centuries, with some of the earliest records from Greece and Rome. In the fifth century B.C. cases of accidental mushroom poisoning were recorded. Not all cases of mushroom poisoning were accidental, however. The Roman emperor Claudius Caesar was murdered by his wife Agrippina (A.D. 54), who poisoned his dish of edible mushrooms, probably by addition of the poisonous mushroom *Amanita phalloides* (Wasson and Wasson, 1957). Also the prefect of Nero's guard and a number of other officers were poisoned at a banquet by dishes to which a poisonous fungus had been added. In what were probably the first attempts at classifying the fungi, the Greek medicine writer Dioscorides in the first century A.D. divided the fungi into edible and poisonous varieties. This system of classification was followed later by the herbalists (Buller, 1914).

The line between poisonous and nonpoisonous fungi is often hazy. Some fungi may cause minor gastrointestinal upsets due to individual sensitivities or allergies (just as common foods such as chocolate, eggs, or strawberries might upset certain people), to overindulgence, or to consumption of spoiled mushrooms. Numerous fungus species must be considered "dangerous" as they are poisonous to many individuals, although others can eat them with no ill effects; perhaps the most notorious example of this is the beefsteak morel, *Helvella esculenta,* which is considered a delicacy by many but which has also caused death in others. Some fungi such as the inky caps [*Coprinus* spp. (Fig. 5–43)] or *Clitocybe clavipes* may make the victim violently ill if consumed with alcohol but have no effect if eaten alone. In the United States, the death rate from mushroom poisoning is comparatively low, with approximately 50 fatalities annually (Block et al., 1955), but 90% of these deaths are attributed to consumption of species of *Amanita* (Fig. 14-3), particularly the white *Amanita phalloides* (the "death angel") and its close relatives *A. verna* and *A. virosa.* These last fungi are universally poisonous, and as little as 1 cubic centimeter will cause death (Bessey, 1950).

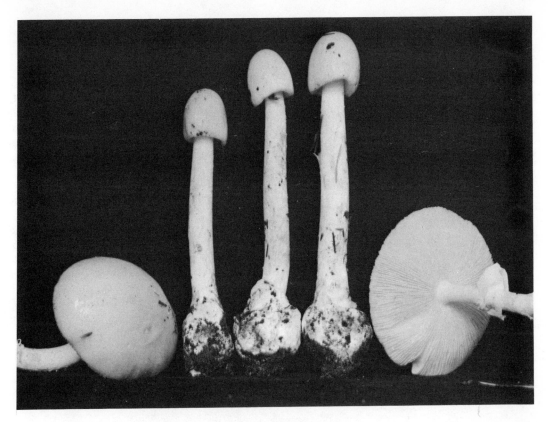

Fig. 14–3. *Amanita bisporigera,* one of the deadly white *Amanita* mushrooms. [Courtesy A. H. Smith.]

It has been commonly thought that a poisonous species will blacken a silver coin during cooking, that poisonous ones have a foul odor, or that poisonous fungi may be easily peeled while nonpoisonous ones cannot be. These tests are not reliable, and especially so when one considers that there is no clear distinction between "poisonous" and "safe" fungi because many species affect people differently. If one wants to collect fungi for consumption, the only safe way to proceed is to learn to identify species in the field, consume only those that are universally recognized as being safe, and eat at first only small quantities in the event that one is sensitive to that particular species.

CLINICAL ASPECTS OF FUNGUS POISONING

A large number of the fungi are poisonous and, upon analysis, the poisons are found to be biochemically diverse and sometimes unique for a fungal species (Benedict, 1972). Although a given fungal species may produce unique symptoms, the effects of some of these fungi are often generally similar. Partially to aid the physician who does not have either a specimen of the fungus or an accurate identification of it in cases of accidental poisonings, working classification schemes have been devised that are based on the similarities of symptoms resulting from poisonings. Although these schemes vary from author to author (e.g., Benedict, 1972; Lincoff and Mitchell, 1977), the three major groups discussed below are those generally recognized.

Fungi Affecting the Gastrointestinal Tract. Fungi such as *Russula, Lactarius, Boletus, Clavaria, Chlorophyllum molybdites,* and *Entoloma lividum* and most of those to which individual sensitivity occurs belong to this group. Typical symptoms are nausea, vomiting, and diarrhea, which may vary in severity but terminate spontaneously within a day or so. There are practically no fatalities from this type of poisoning although a person who is aged or debilitated by disease may succumb to it. In this group of fungi, the toxic principles are poorly known, and furthermore, it is not known whether partial decay of the basidiocarp or strain differences may affect the toxicity (Lincoff and Mitchell, 1977).

Fungi Affecting the Nervous System. Fungi belonging in this category are those that cause sensitization to alcohol (*Coprinus* spp., etc.), *Amanita muscaria,* and the hallucinogenic fungi.

The inky cap, *Coprinus atramentarius,* contains a toxic derivative of the amino acid *L*-glutamine. This toxin has chelating properties similar to those of disulfiram (Antibuse®); it binds molybdenum, blocks acetaldehyde dehydrogenase, and arrests ethanol metabolism at the acetaldehyde stage. The symptoms appear about $\frac{1}{2}$ to 1 hour after drinking alcohol if the mushroom has been eaten within the preceding 4 or 5 days or along with the alcohol. Symptoms include flushing of the face and neck, protrusion and throbbing of the neck veins, a sensation of swelling and paresthesia of the hands and feet, and often chest pain. Nausea, sweating, and vomiting then follow.

Fig. 14-4. *Amanita muscaria.* [Courtesy A. H. Smith.]

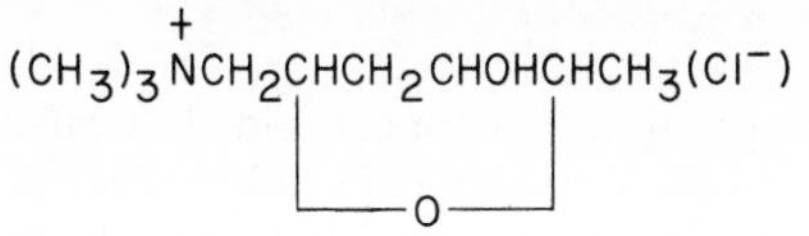

$(CH_3)_3\overset{+}{N}CH_2CHCH_2CHOHCHCH_3(Cl^-)$

Muscarine

$(CH_3)NCH_2CH_2CH_2CHOHCHOHCH_3(Cl^-)$

Muscaridine

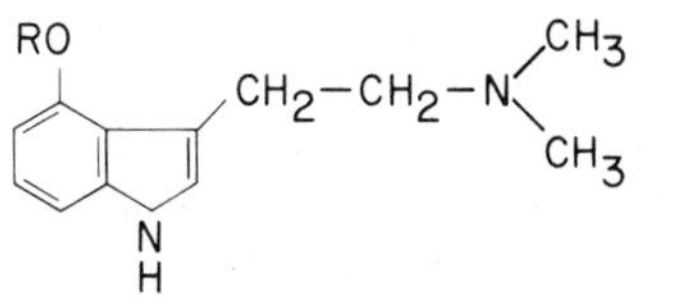

Psilocin, R=H
Psilocybin, R= PO_3H_2

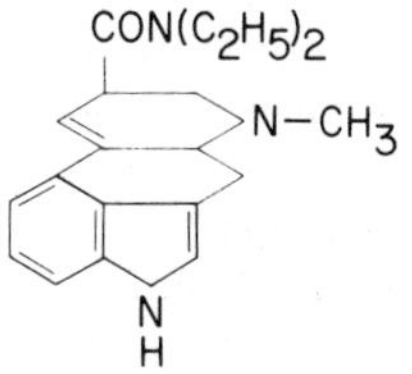

LSD

Fig. 14-5. Structural formulas of some fungal poisons.

Amatoxins

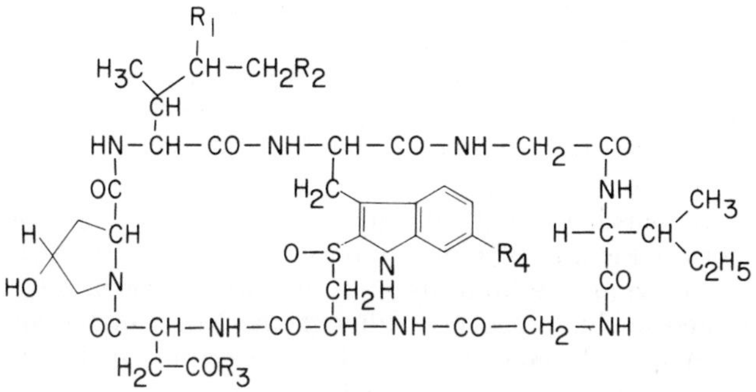

		R_1	R_2	R_3	R_4
a	α – Amanitin	OH	OH	NH_2	OH
b	β – Amanitin	OH	OH	OH	OH
c	γ – Amanitin	H	OH	NH_2	OH
d	Amanin	OH	OH	OH	H
e	Amanullin	H	H	NH_2	OH

Recovery usually occurs spontaneously and completely within 2 to 4 hours (Lincoff and Mitchell, 1977).

The red or yellow *Amanita muscaria* (Fig. 14-4) is responsible for a large number of mushroom poisonings. This species is popularly known as the fly-agaric because it was formerly used to kill flies (the pileus was brought indoors, and sprinkled with sugar to attract flies, which nibbled upon it). Toxic principles include the amines muscarine and muscaridine (Fig. 14-5). Muscarine may also occur in large quantities in other fungi, including various species of *Clitocybe* and *Inocybe*. After *A. muscaria* or another species containing muscarine is eaten, there is a latent period of 1 to 6 hours before symptoms begin. Symptoms include salivation, nausea, vomiting, abdominal pains, thirst, and mucous and bloody stools. Respiration is at first rapid, but later becomes slow. At first, the symptoms are similar to those of acute alcoholic intoxication, as the victim staggers, may lose consciousness and become delirious, and occasionally suffers from hallucinations, manic conditions, and stupor. Convulsions may occur in severe cases. In fatal cases death results from respiratory arrest. Death results in probably less than 1% of the cases of poisoning by *A. muscaria,* but the rate may be as high as 6% to 12% in the case of poisoning by *Clitocybe dealbata* or *Inocybe fastigiata.* Most of these deaths occur in children who have cardiac or pulmonary disease (Lincoff and Mitchell, 1977).

Ingestion of *A. muscaria* or other agarics containing muscarine is not ordinarily fatal because the symptoms begin to develop rapidly, and the earliest symptoms often include vomiting and diarrhea, which prevent further absorption of the poison through the gastrointestinal tract. Poisoning from these fungi is treated by prompt administration of emetics, gastric lavage, cathartic, and enemas, all of which clear the gastrointestinal tract. Antispasmodic agents or sedatives may be administered in addition.

Hallucinogenic Fungi. A number of fungi affect the nervous system in such a manner that the individual perceives nonexistent sights and sounds, or has hallucinations. *Amanita muscaria* is one of these hallucinogenic fungi and is eaten in orgies by some Siberian tribes in the Kamchatka peninsula. A similar usage of other hallucinogenic fungi in religious rites occurred among peoples over a wide geographic area (from northern Europe, eastward to Siberia, southward to Borneo and New Guinea, and westward to Peru).

The Mexican Indians have used hallucinogenic mushrooms, which they believe to be sacred, since before the time of Montezuma's coronation in 1502, the date of the first record. (The hieroglyphics and mushroom stone carvings mentioned on page 515 may have been related to this practice.) Although the Mexican Indians have been using the hallucinogenic mushrooms in sacred rites for centuries, they guarded their secret so well that only a handful of people (predominantly missionaries) knew of their practice until a few decades ago. Hallucinogenic mushrooms are species of *Psilocybe* (Fig. 14–6) and are consumed during religious rites when a serious problem must be solved through divine revelation. The active ingredients of the hallucinogenic mushrooms are the indole derivatives psilocybin and psilocin (Fig. 14–5). Although psilocybin occurs in greater quantities than psilocin, both exert similar effects. Other closely related toxins may occur in species of *Psilocybe* as well as some other agarics, including Japan's "laughing mushroom" (*Panaeolus papilionaceus*), which causes the euphoric victim to laugh.

Members of Occidental cultures are currently indulging in hallucinogenic experiences induced by LSD (*d*-lysergic acid diethylamide). One of the natural sources of LSD is *ergot,* which is treated more fully on page 541.

Hallucinogenic mushrooms, purified psilocybin, and LSD produce similar effects when consumed. They produce a euphoric, dreamlike state in which the body may feel heavy but there is a sensation of an extraordinary lightness or hovering. Perception of time and space are lost, and there are changes in the awareness of oneself. Visions of colorful patterns and objects may seem to fill all space and to dance past in an endless succession. All other senses (auditory, touch, taste, and smell) seem to be equally sharpened to produce a feeling of intense reality.

Fewer than 1% of the cases of poisoning by hallucinogenic fungi are fatal. Treatment includes emptying of the gastrointestinal tract as described for muscarine poisoning, administration of sedatives, and reassurance (Lincoff and Mitchell, 1977).

Fig. 14-6. A hallucinogenic mushroom, *Psilocybe strichper.* [Courtesy A. H. Smith.]

Fungi Causing Cell Destruction. Fungi having hemolytic poisons capable of destroying red blood cells include *Helvella esculenta,* species of *Galerina,* and *Amanita phalloides* and its close relatives *A. verna* and *A. virosa.*

The *Amanitas* are of particular concern here, as they are responsible for the majority of deaths (more than 95%) following ingestion of mushrooms; and the mortality rate after consumption of these fungi is over 50% and may be as high as 90%. *A. phalloides* contains several toxins that are divided into two families, the phallotoxins and the amatoxins, which contain at least five and six members, respectively. These toxins are cyclopeptides with a sulfur-containing bridge coupling the indole nucleus of a tryptophan and a hydroxylated amino acid. The phallotoxins and amatoxins differ in biochemical details of their structure and in the mode of their action. The phallotoxins act quickly and a large dose may cause death in a mouse within 1 hour. The amatoxins are much slower in their action and require more than 15 hours to exert their lethal effects, but they are about 10 to 20 times more toxic than the phallotoxins. The amatoxins are among the most lethal poisons known, and a dose of only 5 to 10 milligrams is lethal for an adult human. Because the phallotoxins may not be absorbed from the digestive tract and because of the extreme toxicity of the amatoxins, it is believed that the amatoxins are primarily responsible for the deaths from *Amanita* poisoning. One of the amatoxins, amanitin, interferes with DNA and RNA transcription by inhibiting RNA polymerase (Lincoff and Mitchell, 1977; Wieland and Wieland, 1972).

After ingestion of *A. phalloides,* there is a latent period of 10 to 12 hours during which the fungus is most likely to be thoroughly digested and much of its poison absorbed. During this time, the toxins reach the kidneys and liver and begin to do irreversible damage to these organs, but the victim is without outward symptoms. Symptoms begin with severe abdominal pains, nausea, vomiting, and diarrhea. The patient may become restless, suffer from delirium and hallucinations, and eventually collapse. Fragility of the blood cells increases. After 3 to 4 days jaundice may set in, as well as renal disturbances and toxic hepatitis. The liver becomes enlarged, dark red, and very brittle owing to an excessive accumulation of blood from hemorrhaging. Death results from destruction of the liver cells and occurs about 2 to 4 days after ingestion of the poisonous agarics. In addition to the damage to the kidneys and liver, damage sometimes occurs to the heart, adrenals, and skeletal muscles.

Treatment for *A. phalloides* poisoning consists of immediate evacuation of the gastrointestinal tract during the latent period, administration of cardiac stimulants in the event of circulatory failure, administration of fluids such as a glucose solution, and hemodialysis in the event of kidney failure. Intravenous administration of thioctic acid in glucose may be a lifesaving treatment (Lincoff and Mitchell, 1977).

Fungi and Food Processing

In addition to being directly consumed by humans, fungi are used in the processing of many foods. These include the worldwide production of bread, wine, beer, and cheese. The use of fungi in the processing of food has reached its greatest diversity in the Orient, especially Japan, where different kinds of foods are produced.

BREAD AND ALCOHOLIC BEVERAGES

The most widely consumed foods processed with the aid of fungi are bread and alcoholic beverages, which have been produced for centuries. The alcoholic beverages include beer, wine, whiskeys, and other fermented beverages, such as saké, made in Japan. The origins of bread and beverage making are lost in antiquity, but it is known that the Sumerians produced a beer consisting of moistened, fermented bread prior to 7,000 B.C. The Egyptians also made this "bread beer," and regarded it as a holy gift from Osiris, the god of the dead. They were also the first to make leavened bread. Both bread making and brewing were later refined by the Greeks and Romans, and numerous changes in the manner of making beer have been made since these early times. For example, hops have been used as an ingredient of beer only since the Middle Ages (Stewart, 1974). Both bread and alcoholic beverages have been held in great esteem: bread is often referred to as the "staff of life," and wines have traditionally had a role in sacramental rites.

The production of both leavened bread and alcoholic beverages utilizes yeasts, which are members of the Hemiascomycetes and often strains of *Saccharomyces cerevisiae.* As noted in Chapter 8, the yeasts ferment glucose to yield ethyl alcohol and carbon dioxide:

$$C_6H_{12}O_6 \longrightarrow 2C_2H_5OH + 2CO_2$$

The baker takes advantage of the carbon dioxide gas production to make his bread porous and of a light texture, while the brewer and wine maker rely on the production of ethyl alcohol to make the characteristic beverages. The yeast cells produced during the fermentation process may be harvested and dried for use as a nutritional supplement for humans or in livestock feeds.

Bread. The baker utilizes compressed yeast cakes or active dry yeast. Commercial production of bakers' yeast begins with the selection of a suitable strain of *Saccharomyces cerevisiae,* which should have an aerobic growth habit, a high yield when grown in a medium containing molasses, stability during storage, and a superior performance in dough. The selected strain is propagated in large closed, sterile tanks containing about 35,000 to 190,000 liters of diluted molasses, essential minerals and vitamins, and a source of amino nitrogen. A propagation period of about 12 hours is required, and during this time the tanks are maintained at 30°C, the pH is controlled, and additional minerals and sterilized molasses are added to support growth of additional cells. It is particularly important to pump sterile air into the tanks during this period to maintain strongly aerobic conditions which repress alcohol formation and encourage formation of cellular components. When the propagation period is completed, the medium containing the yeast cells is centrifuged and a slurry containing about 18% solids is obtained. This slurry may be further compressed to approximately 30% solids and this is used by the baker or processed with starch for public sales. Alternatively, the yeast slurry may be dried and packaged under oxygen-free conditions as active dry yeast (Peppler, 1979).

In preparation of bread dough, a combination of yeast, water, flour, salt, and possibly other ingredients is made. The mixture is then kneaded to cause the flour proteins (*gluten*) to form a network that will give the dough its elasticity and enable it to hold the carbon dioxide produced by the yeast. Fermentation begins, and the yeast at first uses the hexose sugars present in the flour and later utilizes some of the maltose released by hydrolysis of starch. A fermentation period of about 2 to 4 hours is needed at a temperature of 27°C. The carbon dioxide bubbles are trapped in the gluten network and produce a large number of pores that make the bread rise. As the dough rises, the yeast cells become separated from the nutrients and the dough rises much more slowly. It is then kneaded to place the yeast cells and the nutrients in close proximity so that the rising process may continue. Immediately before baking, the alcohol content of the bread is as high as 0.5%, about one-sixth that of ale. This alcohol is driven off during baking and helps give the bread a pleasant aroma (Duddington, 1961).

Wine. Wine is fermented fruit or vegetable juice and is most often made from grape juice. The characteristic and distinctive flavors of wine are partially determined by the variety of fruit used, the environmental conditions under which the fruits were grown, the parts of the fruit used, and the ability of the yeast to form flavored by-products (particularly esters).

Traditionally, the yeasts used to ferment the juices were those occurring naturally on the surface of the fruit, but recently the trend has been to utilize laboratory cultures of strains of *Saccharomyces cerevisiae* var. *ellipsoideus.* These strains are chosen for their ability to properly ferment the variety of grapes used and for their ability to impart flavors by formation of by-products which are characteristic for a particular wine. The wine maker may grow the yeasts in the winery by first inoculating about 1 to 5 liters of fruit pulp (*must*) from the laboratory culture slant and then serially transferring the inoculum to successively larger volumes of fresh must until a quantity suitable for bulk fermentation is obtained. The grape juice is treated with bisulfite to kill undesirable yeast and bacteria. The purity and final vigor of the yeast cultures produced in this way is often uncertain. Alternatively, quantities of the selected yeast may be produced on a large scale, like the bakers' yeast, and then either compressed or active dry wine yeast may be added to the tanks of must or strained juice. This latter method eliminates the laborious and often uncertain stepwise preparation of starter cultures and the need for the costly must required for the starter cultures.

Fermentation is carried out in vats, usually of wood or stone, that vary in capacity from a small keg to one that holds 190,000 liters or more. The vat is covered to maintain anaerobic conditions that encourage fermentation and also discourage growth of acetic acid bacteria which could convert the wine into vinegar. The course of fermentation (particularly the nature of by-products) and the concentration of alcohol produced are influenced by the temperature, sugar concentration, acidity, and tannin content of the must and the amount of sulfite added. Of these factors, temperature is especially important as the yield of alcohol is higher at the lower temperatures and it also encourages the formation of pleasant flavors. At higher temperatures the alcohol resistance of the yeast is decreased. A favorable temperature for fermentation is in the vicinity of 10°C. If the sugar content is over 30%, the fermentation is inefficient because alcohol accumulates and arrests the activity of the yeasts before all the sugar is converted to alcohol. A low acid content favors production of acetaldehyde, glycerol, and volatile and fixed acids, which have a detrimental effect on the flavor, color, and stability of the wine. In addition, a low acid content gives a lower yield of aromatic principles, which give the wine its bouquet.

After the wine is partially fermented, the *racking* operation begins and continues throughout the remainder of the fermentation period. In racking, wine is siphoned off from the sediment, which consists of dead yeast cells and other materials which settle at the bottom of the vat and is transferred to a fresh vat. This procedure is repeated a number of times until the desired clarity of the wine is achieved. Before the last racking, a final clarification is performed by adding some substance (such as isinglass, casein, or bentonite) that carries additional sediment to the bottom. Fermentation continues until the alcohol content is about 12%; above this point the alcohol kills the yeast cells so that no more fermentation takes place. The wine is then stored in completely filled and sealed tanks for aging. Additional blending, fortification with additional alcohol, heat treatments, refrigeration, or filtration may be required before the wine is bottled.

Beer. Beer is an alcoholic beverage prepared from fermented grains, usually barley. In the initial steps of beer production, barley is *malted* through a controlled germination. The barley grains are washed and soaked for 2 to 3 days to stimulate development of the embryo; the water is then drained off and the seeds germinated. The embryo is allowed to grow until the plumule attains a length equal to approximately three-fourths of the kernel length and is then dried to halt growth. During the brief period of germination, the embryo produces a number of digestive enzymes which begin to hydrolyze food reserves in the seed. From the brewer's standpoint, the most important of these are the carbohydrases which attack starches and sugars, and of these the most important are especially the amylases, which convert the starch into dextrins, maltose, and a little glucose. The maltose is the principal substratum for yeast fermentation.

The next major step in beer production is known as *mashing*. The dried malt is ground and mixed with hot water. Most of the enzymatic conversion of starch to maltose by amylases takes place during mashing, which requires about 2 hours. The aqueous extract is separated from insoluble materials and husks and is then boiled with hops, the female inflorescence of the hop plant containing essential oils and resins which impart characteristic flavors to the beer.

The aqueous extract is then cooled to a temperature between 8°C and 16°C (the temperature is partially determined by the type of beer being produced) and placed in fermentation vats. A selected strain of yeast is added. Traditionally *Saccharomyces carlsbergensis* is used to produce lager beer while *S. cerevisiae* is used to produce ales (in the United States, *S. carlsbergensis* may be used for both). During a fermentation period of about 5 to 14 days, the sugar is converted to alcohol. The yeasts settle to the bottom of the vat in a flocculent material in the production of the lager type of beer. In contrast, bubbles of carbon dioxide rise to the top of the vat and carry with them the yeast cells and dark flocculent materials from the liquid in the production of the ales. The resultant foamy scum must be periodically removed.

After the fermentation period is completed, the newly formed beer is allowed to rest for a few days to allow the yeast cells to settle out and is then drawn off into casks or large tanks for a maturation period of several weeks. The beer is clarified by addition of gelatinous materials in a manner similar to that in which wines are cleared and may also be filtered through diatomaceous earth. Antioxidants are added since beer changes flavor upon oxidation, and carbon dioxide is added either as the pure gas or by mixing in some freshly fermented beer. The beer is then bottled for sale.

Whiskey. Whiskeys are made from fermented grains (corn, wheat, barley malt, or rye malt), which are mixed in varying proportions according to the type of whiskey being produced, and fermented by yeasts. The processes involved are similar to those involved in beer production (Fig. 14–7); the major exception is that the fermented grain broth is distilled in order to concentrate the alcohol.

Fig. 14–7. Rectified cane spirit is passed through these nine charcoal filters at a very slow rate to remove any aromatic impurities so that final product, vodka, will be completely colorless, tasteless, and odorless. [Courtesy Gilbey-Santhagens (Pty.) Ltd.]

CHEESE

Cheese is a food product made from milk and it contains butterfat, lactose, the proteins albumin and casein, and the watery whey. Casein will coagulate to form a curd when it is acted upon by an acid or by the enzymes pepsin or rennin. Rennin was formerly obtained from calves' stomachs, but there has been a widespread shortage of calves' rennin because few calves are slaughtered and the consumption of cheese has been increasing. Fungi may produce extracellular enzymes, including rennin, which can coagulate milk. Rennins from fungi have become increasingly important in cheese manufacture (one fungus which is commercially important as a source of rennin is the plant pathogen *Endothia parasitica*) (Sternberg, 1976). The coagulated curd is the newly formed cheese that separates from a watery fluid, the whey (whey is sometimes used as a cheap substrate for the production of food yeasts). The cheese may be eaten in this form when freshly prepared, such as cottage cheese. Alternatively, the cheese may be further drained of moisture, pressed, or cooked, as in the production of Edam, Gouda, and Cheddar cheeses. Some cheeses may be processed with a characteristic bacterial or fungal flora, which will impart distinctive flavors during the ripening period.

Fungi are used to process cheeses of two principal types: (1) those of the Camembert type, which are soft in texture and are covered on the outside with a rind formed through the agency of the mold; and (2) those semihard cheeses of the Roquefort type which have streaks and pockets lined with a blue-green mold.

Camembert Cheese. Camembert cheese was first made in the community of Camembert in northwestern France. Milk is coagulated with rennin; the curds are placed in hoops and allowed to drain overnight and then turned over to drain evenly. After 2 or 3 days, each cheese is rolled in salt and sprinkled with spores of *Penicillium camemberti,* a white mold. The layer of salt extracts the whey from the curd and forms a shiny cover of brine which flows away. The surface of the cheese becomes hardened after this extraction and forms a rind. During these initial stages, the cheese consists of a sour curd surrounded by a salty rind, both of which inhibit the growth of bacteria and the fungus, *Geotrichum candidum.* The spores of *P. camemberti* germinate and mycelium penetrates the rind and eventually spores are produced on its surface. After *P. camemberti* has become well established, it secretes proteolytic enzymes into the curd which makes the curd become soft and buttery and develop a mild flavor. The acidity also disappears, allowing the establishment of *G. candidum* and bacteria during the later stages of ripening, thus accounting for the ammoniacal flavor. The ripening process requires about 4 weeks and is carried out in ripening rooms held at 10°C to 16°C with a humidity of 86% to 88% (Thom, 1944).

Roquefort Cheese. Roquefort cheese originated in the sheep-raising region surrounding Roquefort in southern France. This cheese is made from sheep's milk, while a similar one, blue cheese, is produced from cow's milk in the United States.

The milk is curdled with rennin at low temperatures, and the curd is broken into small masses and drained until it is rather dry and the pieces no longer merge completely when placed together. The curd is heaped loosely in cheese hoops so that cracks and channels remain. It is then inoculated with bread crumbs containing spores of *Penicillium roqueforti* which was grown on the bread. The curd is then maintained at a cool temperature 18°C to 20°C and at a high humidity (80% to 90%). During the next 3 to 10 days, the ripening curd is periodically scraped, washed, and salted. After the salting period, the cheese is perforated with additional holes that provide adequate aeration for the fungus to grow along the many interfaces within the cheese. A 2-month ripening period is required and then the cheese is ready for sale.

During its growth in the cheese, the fungus secretes protein and fat-digesting enzymes. The pungent flavors characteristic of Roquefort cheese result from the hydrolysis of fats and the liberation of high-molecular-weight fatty acids (capric, caproic, and caprylic).

ORIENTAL FOODS

Foods altered by bacterial or fungal growth have been part of the human diet since the beginning of recorded history and have been widely used in diverse cultures. The microbial growth often improves the texture, flavor, and nutritional value of the food while delaying spoilage. These fermented foods have often been valued in cultures where people do not have access to electricity or to commercially prepared foods because these foods add variety to the diets and have longer useful life than nonfermented foods. Such foods may be used as a beverage, a condiment, a vegetable, or a substitute for meat or bread. The use of these foods has been especially popular in Africa, Asia, and the Middle East. We shall consider the Oriental foods further.

In Japan and other Oriental countries, large quantities of diverse foods are prepared from soybeans, wheat, and rice through either the deliberate or accidental inoculation with members of the Mucorales, various yeasts, and *Aspergillus oryzae* (a member of the Deuteromycotina). These fungi may be used in combination with each other or in combination with bacteria to yield a specific food product. Foods of this type include tempeh, a solid food prepared with soybeans processed with species of *Rhizopus*; sufu, a Chinese cheese prepared from soybeans and *Actinomucor elegans* and *Mucor* spp.; miso, or soybean paste, prepared from rice and soybeans fermented by *Aspergillus oryzae* and *Saccharomyces rouxii;* and shoyu, or soya sauce, prepared from soybeans and wheat and processed with *Aspergillus oryzae* and other organisms. The production of tempeh and shoyu will be considered in more detail.

Tempeh. Tempeh is a food which is produced in great quantities in Indonesia, New Guinea, and Surinam. In tempeh preparation, soybeans are washed and soaked overnight in water. The following day the seed coats are removed and the dehulled soybeans are boiled for $\frac{1}{2}$ hour, drained, and cooled. When cool, the soybeans are inoculated with spores of a strain of *Rhizopus,* usually *R. oligosporus.* In Indonesia, this inoculation is accomplished by adding pieces of a previously made batch of tempeh or from the wrapper in which a former batch was held. Small amounts of inoculated soybeans (about the quantity that would fill a petri dish) are wrapped in leaves and tied with string. These soybeans are held at room temperature for about 24 hours and during this time the mycelium permeates the soybeans so that they are embedded in a whitish mycelium and the entire mass is held together. The newly prepared tempeh is consumed after this 24-hour period and is prepared fresh daily. The tempeh may be thinly sliced, dipped into salt water, and fried in vegetable oil; it may be roasted; or it may be cut into cubes and used as a meat substitute in soup.

Untreated soybeans do not soften well during cooking and are difficult to digest, but after treatment they are quite easy to digest and have a pleasant, newly acquired flavor. During the brief period of treatment, the fungus produces large quantities of proteolytic enzymes which break down the soybean proteins. The percentages of both soluble solids and soluble nitrogen increase during the period of fungal growth and thereby become more digestible upon consumption. There is also an increase in the riboflavin, niacin, and B_{12} content of the treated soybeans, although there is a decrease of thiamine (Hesseltine, 1965).

Shoyu. Shoyu, or soya sauce, is a brown liquid which is used for seasoning a large number of foods in China, Japan, the Philippines, Indonesia, Europe, and the United States; it is produced in all of these countries. Shoyu may be made through the acid hydrolysis of soybean protein with an inorganic acid (such as hydrochloric acid), by hydrolysis with fungus enzymes during a fermentation period, or through a combination of these methods. Although manufacture is easier using an inorganic acid hydrolysis, the product is inferior to that produced utilizing fungal fermentation.

Japan is the foremost producer of shoyu today, and 90% of the shoyu made there utilizes fungal processing of either defatted soybean meal or whole soybeans. Since 75% of all shoyu is made from defatted soybean meal, this mode of manufacture is discussed. Soybeans are washed, soaked for 15 hours at room temperature, and then autoclaved for 1 hour. Meanwhile wheat is cleaned, roasted, and crushed, or wheat bran is steamed and roasted. Wheat and soybeans are then mixed together in a ratio of 45% wheat to 55% soybean; the material is cooled and then inoculated with either *Aspergillus oryzae* or *A. soyae*. The molding grain mixture is placed in rooms with the temperature held at 25°C to 35°C for 3 to 4 days with at least two turnings of the grain. At the end of this period, the moldy grain is placed in large tanks and an equal amount of brine is added. The incubating mixture (mash) remains in the tanks for 3 to 4 months if it is warmed or up to 1 year if it is not warmed. The mash is occasionally stirred with compressed air. Bacteria produce lactic acid during the first part of the fermentation period. Later during the fermentation period, yeasts produce ethyl alcohol, carbon dioxide, and acetic acid. Finally there is a period of aging and relatively little fermentative activity. The microorganisms which carry out the last two fermentations are not deliberately introduced but carried along randomly in the procedures used. At the end of the aging period, the mash is compressed as a cake, and the liquid is extracted as shoyu, which is pasteurized to destroy any microorganisms and enzymes present (Hesseltine, 1965).

INDUSTRIAL PRODUCTION AND UTILIZATION OF FUNGUS METABOLITES

In the previous section, it was indicated that some fungus metabolites become involved in the processing of foods and change the flavor and composition of the food which served as a medium for the growth of the fungi. Many fungus metabolites are useful products alone and are separated from the substratum for further use in the pure form. Fungus products used commercially include organic acids, alcohol, antibiotics, pigments, vitamins, and enzymes. A more complete listing may be found in Table 14–1. Fungi are also used to convert steroids from one form to another, and as a direct source of alkaloids that they may contain.

The metabolites are biosynthesized along diverse pathways and, as we might expect, there is no single set of environmental or nutritional conditions common to all commercial syntheses. In addition, a single fungus is usually capable of producing several metabolites. Attainment of the metabolite of choice is accomplished by forcing the fungus to favor certain biosynthetic pathways over others. This may often be accomplished by adjusting the pH so that some enzymes act normally while others are inhibited, by adding antagonists that affect certain enzymes only, by adding nutrients or maintaining conditions that do not favor growth and thereby divert assimilated nutrients into metabolic waste products rather than into protoplasm, or by adding precursors that favor certain pathways.

Table 14–1: Commercially Important Fungus Metabolites

Metabolite	Produced By	Use
Organic acids		
Citric acid	*Aspergillus* spp. especially *A. niger*	Manufacture of citrates; flavoring ingredient; effervescent salts; in ink; in dyeing; medically; in silvering mirrors
Gluconic acid	*Aspergillus niger*	Medically—increases solubility of calcium when combined as a salt; manufacture of toothpaste
Itaconic acid	*Aspergillus terreus*	Manufacture of plastics
Alcohols		
Ethyl alcohol	Yeasts	Solvent; raw material in manufacture of ether, esters, acetic acid, synthetic rubber
Glycerol	*Saccharomyces cerevisiae*	Explosives
Fats	*Penicillium* spp. Yeasts	Soap; manufacture of food (useful in times of food shortage)
Antibiotics		
Griseofulvin	*Penicillium griseofulvum*	Oral and topical antibiotic
Penicillin	*Penicillium chrysogenum*	Oral and parenteral antibiotic
Drug		
1-Ephedrine	Yeasts	Drug for treatment of asthma; ingredient for nose drops and inhalants
Vitamins, pigments and growth factors		
B vitamins	Yeasts	Nutritional supplement and medical therapy
Riboflavin	*Ashbya gossypii* *Eremothecium ashbyii*	Nutritional supplement and medical therapy
Gibberellin	*Fusarium moniliforme*	Growth-promoting substance (vascular plants)
β-Carotene	*Blakeslea trispora*	Precursor for vitamin A; coloring agent for margarine, baked goods
Enzymes		
Amylase	*Aspergillus* spp.	Starch for hydrolysis for alcoholic fermentation; starch removal agent in food preparation; removal of starch sizing from fabrics; converting starch in sizing materials
Rennet	*Endothia parasitica* *Mucor* spp.	Milk coagulation in cheese manufacture
Invertase	*Saccharomyces cerevisiae*	Production of artificial honey and invert sugar
Pectic enzymes	*Penicillium* spp.	Pretreatment of fresh fruit juices to remove turbidity; removal of pectins before concentrating juice
Protease	*Asperillus* spp. *Penicillium roqueforti* *Agaricus campestris* *Cantharellus cibarius*	Food processing

Methods Involved in Industrial Mycology

PILOT STUDIES

A great deal of preliminary work is required to determine those conditions favoring the production of a desired metabolite before large-scale plant operations can take place. The first steps are selection of a suitable fungus strain and the determination of optimum aeration, temperature, pH, and nutritional requirements. Following this small-scale laboratory work, the next step is to expand to a large-scale laboratory simulation of the proposed plant method. The large-scale laboratory methods may or may not prove to be satisfactory as the optimum conditions established in the initial laboratory tests may not be those optimum in larger operations. Once the large-scale laboratory methods are successful, the techniques are next tested in a single pilot plant where it is determined whether they are satisfactory under industrial conditions; finally full-scale industrial production is carried out (Gray, 1959).

CULTURE VESSELS

In full-scale industrial production, cultures are usually grown in several shallow pans or in tanks with a capacity of thousands of liters.

Shallow pans are filled with a liquid medium, inoculated with a fungus, and set aside for the incubation period. The purpose of the shallow pan is to allow maximum exposure of the fungus mycelium to the air and is useful for highly aerobic organisms. A serious disadvantage of this method is that if a fungus is to benefit from exposure to the air, the pan must remain uncovered, thereby creating a great danger of contamination. The fungus grows as a dense mat on the surface of the medium. Those organisms profitably cultured in shallow trays are ones that are highly aerobic but also not easily overrun by contaminating organisms or that have metabolic by-products that are not ruined by contaminants.

Large tile, wooden, or metal tanks are widely used for industrial cultures. These tanks may be left open if the fungus utilized makes conditions unfavorable for contaminating organisms; for example, in alcohol production, the medium quickly becomes anaerobic and therefore unfavorable for growth of aerobic contaminants. If contamination is a potential hazard, the tanks may be entirely closed. Either aerobic or anaerobic conditions may be maintained in a closed tank. If aerobic conditions are required, air may be pumped through the tank or the medium may be agitated. If anaerobic conditions are necessary, the medium in a closed tank is neither agitated nor aerated. Under these conditions, the fungus develops as a submerged mycelium.

If it is feasible to do so, the submerged tank method is used as it offers several advantages over the tray method (and other less common methods). Thousands of liters of medium may be processed at once in a tank, which reduces the amount of expensive hand labor involved and greatly simplifies the entire operation. If required, aseptic conditions may be maintained, whereas it is impossible to have aseptic tray cultures. The rate of metabolism and yields of the product are greater as the mycelium is bathed in a circulating medium which makes the nutrients readily available to the fungus.

MEDIUM

In planning a medium for industrial processes, the prime considerations are cost and effectiveness in the desired operation. A good medium is inexpensive and consists of a cheap carbon source (such as molasses, which contains glucose) and a cheap nitrogen source (such as ammonia). Vitamins or purified amino acids may be added to meet the specific requirements of a given organism. If the organism does not grow well on a medium of the type above, owing to complex nutritional requirements, cheap biological materials may be added to the medium or used exclusively. These biological additives may be corn steep liquor, malt, fish solubles, and yeast autolysate. Other additives are used to control pH or to direct the course of biosynthesis.

Some products are associated with an actively growing and healthy organism. For these products, the medium is composed so that maximum growth of the cells will take place, and the product is harvested when growth ceases. If the product is not associated with growth, the medium is composed so that growth is limited by trace metals, unfavorable pH, or unfavorable nitrogen but not by carbon. In this latter case, an excess of nutrients may be required in the medium after growth ceases to provide carbon and energy for the synthesis of the product (Weinshank and Garver, 1967).

The medium should be sterilized only if it is likely that contaminant organisms will become established under the conditions imposed in the synthesis. Frequently the organism being cultured will make conditions unfavorable for contaminants, and sterilization is not required. Semiaseptic conditions may be required, and only certain additives that are believed to be a major source of contamination are sterilized. If aseptic conditions are required, the vessel and medium may be autoclaved together if a small vessel is used. In the case of large tanks, the tank is usually steam-sterilized while empty, and the medium is steam-sterilized while it is being pumped into the tank. Chemical sterilants (such as ozone, ethylene oxide, or formaldehyde may be used for materials that cannot be heat-sterilized.

Inoculum is prepared from stock cultures and basically involves culturing the organism in successively larger batches. Two or three buildup stages are cultured in the laboratory, followed by one to three stages in tanks of increasing size until enough cells are obtained to inoculate the final tank in which actual production will take place. The amount of inoculum may vary from 1% to 20% of the volume of the production tank. The object is to add enough healthy cells so that the growth phase and buildup of cells in the tank will occur rapidly, thereby reducing as much as possible the time required for production of the metabolite to take place.

A number of factors may need to be controlled during the large-scale tank culture. Aeration may be required and, if so, air that has been sterilized by passage through filters is piped into the tank or the medium is mechanically agitated. The actively metabolizing organisms release heat (as much as 2,000 calories/gram of dry yeast) and, in many processes, this rising temperature is controlled by cooling coils within the jacket of the tank or within the vessel. The pH is ordinarily controlled by constant monitoring and the addition of the required buffers.

At the completion of the incubation period, the cells are separated by filtration, and the product is separated by distillation or chemical processes.

We discuss in detail three industrial processes. Two of the most widely produced fungus metabolites are citric acid (an organic acid) and penicillin (an antibiotic). Both require highly aerobic conditions, and the latter requires strict asepsis. Ethyl alcohol serves as an example of a process that requires neither air nor aseptic conditions.

Citric Acid

The most important organic acid produced by fungi is citric acid (Fig. 14–8). It is used commercially as a flavoring ingredient in beverages and foods, especially in dry mixtures such as gelatins and soft drink powders or tablets, and as the principal acid in the preparation of soft drinks, desserts, jams, jellies, candies, wines, and frozen fruits. Citric acid is rapidly and almost completely metabolized in the human body and has wide pharmaceutical uses. These include its incorporation in effervescent products and as citrates in blood transfusion. Citric acid is also used in astringent lotions to adjust the pH, in hair rinses and hair-setting preparations, in electroplating, in leather tanning, and in reactivating old oil wells where the pores of the sand become clogged with iron.

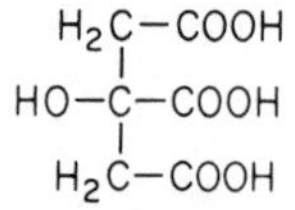

$$
\begin{array}{l}
H_2C{-}COOH \\
\quad | \\
HO{-}C{-}COOH \\
\quad | \\
H_2C{-}COOH
\end{array}
$$

Fig. 14-8. Citric acid.

Citric acid was originally produced from calcium citrate obtained from cull lemons. Methods for producing citric acid from fungus metabolism were introduced in the United States in 1923, and now practically all the citric acid is produced by fungi. In the United States and Europe in 1979, annual production of citric acid was more than 100 million kilograms (Lockwood, 1979).

CITRIC ACID PRODUCTION

Citric acid is produced commercially with selected strains of *Aspergillus niger,* which actively produce citric acid under certain conditions. The most important of these is utilization of a medium deficient in some essential metallic nutrient (especially manganese, iron, or zinc) and, generally, conditions which limit growth.

In citric acid production, *A. niger* may be cultured on the surface of shallow pans or as submerged mycelium in aerated vats (Fig. 14-9). Shallow-pan culture has been widely used but is being rapidly supplanted by more efficient vat cultures. The requirements and methodology for shallow pan and vat culture are generally similar, although in actual plant operation many minute details are different and govern the success of the operation. The carbohydrate source may be refined or crude sucrose, cane syrup, or beet molasses. Beet molasses is the least expensive carbohydrate source, but it has a high metallic ion content which may cause low yields of citric acid to be produced. Beet molasses has been widely used in the pan techniques, although pure sucrose or highly refined syrups give higher yields in submerged cultures in vats. Metallic cations are removed from the carbohydrate source by the use of cation-exchange processes and absorbents or (in the case of beet molasses) by treatment with ferricyanide. For submerged culture, it is particularly important to make the medium deficient in manganese and iron. The carbohydrate is diluted to a concentration of about 20% to 25%—this high sugar concentration inhibits formation of acids other than citric acid. Production of citric acid requires a low pH, and hydrochloric acid is added to the medium to reduce the medium to the range of pH 2 to 5 when the spores in the inoculum germinate. In the vat, the pH is kept below 3.5. Ammonium salts or urea are added to provide nitrogen. For vat cultures, copper or organic ions may be added as antagonists to iron and aid in the control of mycelial growth which increases citric acid production. The prepared medium is inoculated with spores from stock cultures. Highly aerobic conditions are required, and submerged cultures are aerated with sterile air and agitated. Also during the incubation period, temperature must be maintained in the range of 25° to 30°C. The incubation period is 7 to 10 days duration.

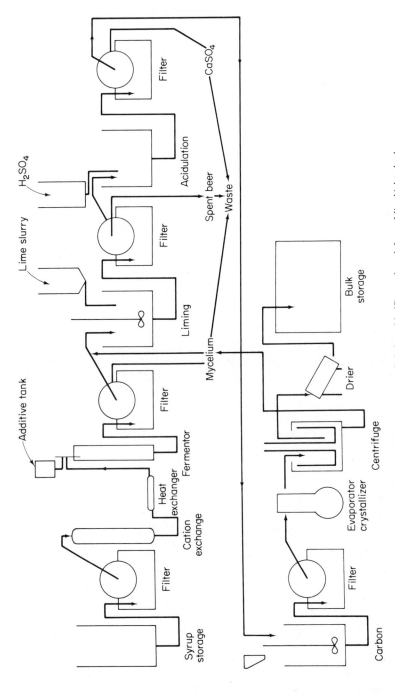

Fig. 14–9. Commercial production of citric acid. [Reproduced from *Microbial technology,* H. J. Peppler, ed., chapter by L. B. Lockwood and L. B. Schweiger, copyright 1967, by Reinhold Publishing Corporation, with permission.]

At the completion of the incubation period, the harvesting of the citric acid is started. Lime is added to the culture medium to precipitate any oxalic acid which formed, and the mycelium and calcium oxalate are filtered off. Additional filtration may be required to clarify the medium. A slurry of calcium hydroxide, $Ca(OH)_2$, is added to form calcium citrate, which precipitates out. The calcium citrate is filtered out and then treated with sulfuric acid which precipitates the calcium. The dilute citric acid solution is purified by treatment with carbon and is demineralized by successive passage through cation- and anion-exchange resins. This purified citric acid solution is then evaporated, leaving behind citric acid crystals, which are further purified by recrystallizations from water.

MECHANISM OF CITRIC ACID FORMATION

As indicated in Chapter 8, citric acid is one of the principal organic acids produced in the citric acid (CA) cycle. Although many pathways and interconversions have been suggested for citric acid biosynthesis and accumulation under the conditions outlined, recent evidence indicates that the probable origin of citric acid is directly from the CA cycle. When mycelium is actively growing, citric acid is produced as an intermediate in the CA cycle and is further diverted to growth-promoting biosyntheses or energy release. For industrial purposes, culture conditions that are inhibitory to growth of the fungus are maintained, and the CA cycle itself is inhibited or blocked by the enzymatic inhibition of the low pH or by specific enzyme inhibitors such as the copper ions. Biosyntheses leading to formation of citric acid are not inhibited, while those involving its conversion are inhibited, leading to the accumulation of citric acid as an "overflow" product.

During citric acid production, the activity of the condensing enzyme (operating in the condensation of acetyl CoA and oxaloacetic acid to citric acid) is increased in activity, while the activities of isocitric dehydrogenase and aconitase disappear. Aconitase controls biosynthesis of isocitric acid from citric acid, and in turn isocitric dehydrogenase mediates in the hydrogen removal which yields oxalosuccinic acid from isocitric acid. It is strongly suggested that the inactivity of these latter enzymes is responsible for the accumulation of citric acid.

Alcohol

Production of ethyl alcohol as an end product of anaerobic fermentation is the oldest known industrial process utilizing fungi (in the context of beer and wine production). As is to be expected, processes similar to those used in making alcoholic beverages can be used to obtain pure ethyl alcohol for utilization as a solvent or for other industrial purposes. As late as 1930, fermentation of molasses accounted for 75% of the industrial alcohol produced. Molasses is a comparatively expensive medium and at times it has been in short supply. For many years (into the late 1960s), alcohol could be produced less expensively by chemical means from petroleum, and the mycological production of industrial alcohol declined to a point at which less than 25% of the industrial alcohol was made through yeast fermentation. There has been a renewed interest in the production of alcohol from cheap and readily available substrates since the early 1970s. The rising cost and decreasing supply of petrochemicals has again made fermentation processes an attractive alternative, especially as the ethyl alcohol may be used as an automotive fuel. For production of industrial alcohol, any substrate that can be fermented to alcohol may be used, and some countries or regions are utilizing waste products or surplus crops locally to make alcohol. For example, alcohol is being produced from cane sugar in India and Cuba, while cereal starch has been seriously considered in Australia and South Africa (Bu'Lock, 1979).

Much of the industrial alcohol is produced by fermentation of industrial wastes. Industrial wastes that can be fermented include *sulfite liquor,* the aqueous material left from the manufacture of cellulose or pulp from wood. This sulfite liquor contains sugars and other organic materials (amounting to more than half the original weight of the wood) and most of the inorganic chemicals used in processing the pulp; its utilization is sound both economically and from the standpoint of conservation. If this sulfite liquor were simply discarded in streams, aquatic wildlife would be in peril since the sulfite liquor may deplete the oxygen supply. For these reasons, pulp and paper companies often build a full-sized alcoholic fermentation plant to dispose of their wastes.

Production of industrial alcohol from sulfite liquor may be carried out using the Ekström process, which is as follows. The hot sulfite liquor is collected, neutralized with limestone, and stirred with steam. The sediment is allowed to settle out and then the liquor is cooled to about 30°C. Yeast and nutrients (such as ammonium sulfate or phosphoric acid) are added and the liquor is fermented in tanks. At the end of the fermentation period the medium contains about 1% alcohol by volume. It is degassed, or freed from the carbon dioxide, and the alcohol concentrated by distillation until 95% alcohol is recovered. A yield of about 11 liters of 95% alcohol may be expected from 1,000 liters of waste sulfite liquor.

Penicillin

Production of antibiotics (antibacterial or antifungal compounds) is one of the largest and most important microbiological industries. In the United States in 1979, the value of the antibiotics produced from both bacteria and fungi was more than $1 billion, and more than 100,000 tons was produced worldwide annually at that time (Perlman, 1979). The role of antibiotics in ecology of fungi is discussed in Chapter 11.

Interest in utilization of antibiotics for therapy began in 1929 when Alexander Fleming found that a mold had contaminated his cultures of a pathogenic bacterium, *Staphylococcus aureus,* and had killed bacteria in its immediate vicinity leaving a clear ring of lysed bacteria. The mold was a *Penicillium* (later found to be *P. notatum*). Fleming named the unidentified lytic principle *penicillin* and realized the potential value of penicillin as a therapeutic agent. Little additional work was done as yields were low and the production of penicillin difficult (Fleming, 1929). It was not until World War II created large demands for effective therapeutic agents that the methodology for large-scale production and refinement of penicillin was developed.

Not only did penicillin engender interest in antibiotics and launch the antibiotic industry, but penicillin is still the most widely used antibiotic. It is the drug of choice when infection is caused by organisms susceptible to it. Penicillin is effective against Gram-positive bacteria and also against some of the larger viruses and rickettsia. Penicillin may be administered orally, intravenously, or intramuscularly. If penicillin is given orally, the dosage must be at least five times as large as that given by injection because the penicillin is partially destroyed in the digestive tract and is incompletely absorbed.

Penicillin is a generic term applied to an entire group of antibiotics which are closely related in structure. Most penicillins are produced by species of *Penicillium*

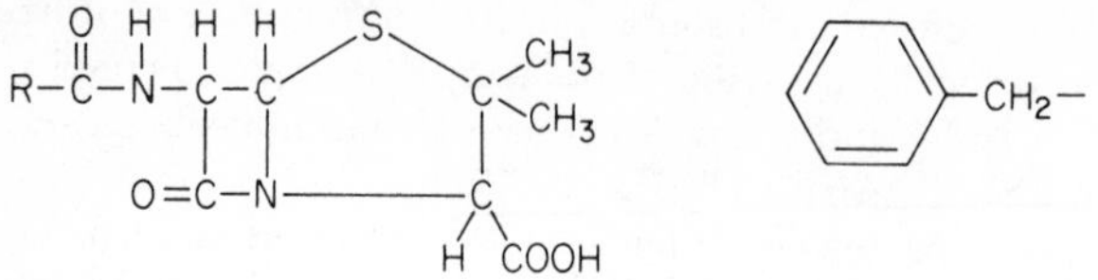

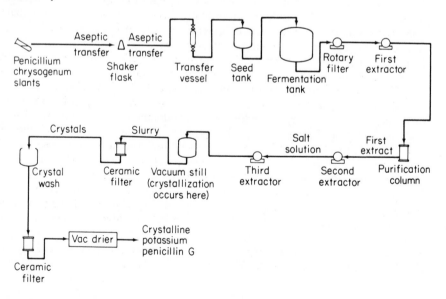

Fig. 14-10. The basic structure of penicillin. Penicillin G would have the radical shown as the R group.

while others are semisynthetic. The naturally occurring penicillins have the following basic structure and differ in their R groups. The basic structure and the R group of the most commonly used penicillin (penicillin G) are given in Fig. 14-10.

PENICILLIN PRODUCTION

The original *P. notatum* strain isolated by Fleming gave low yields of penicillin, as did many strains which were subsequently tested. Finally a strain of *P. chrysogenum* was isolated from a moldy cantaloupe by Raper and Alexander (1945) and was found to give superior yields of penicillin. This strain was subjected to irradiation, and a mutant that gave even higher yields of penicillin was obtained. With successive irradiation of high-yielding mutants and with the selection of mutants with superior ability to produce penicillin, the present commercial strains were developed. Strains of *Penicillium* may produce more than 180 times as much penicillin as the original isolate with which the breeding program was initiated.

Penicillin is produced in submerged vat cultures using a selected strain of *P. chrysogenum* (Fig. 14-11). Composition of the medium is of prime importance in obtaining a good yield. A medium giving an unusually good yield of penicillin is corn steep liquor, a cheap bulk material which is a byproduct of cornstarch manufacture. Corn steep liquor is a concentrate of water soluble materials (principally nitrogenous compounds) from the corn grain. These nitrogenous compounds are of great impor-

Fig. 14-11. Commercial production of penicillin. [Reproduced from *Microbial technology*, H. J. Peppler, ed., chapter by D. Perlman, 1967, Reinhold Publishing Corp., New York, with permission.]

tance in giving a good yield of penicillin. The principal carbohydrate in corn steep liquor is lactose, which is the most favorable carbon source for penicillin production as it is assimilated slowly, giving the fungus a steady carbon source over a period of time and does not lead to the accumulation of organic acids. Precursor molecules (phenethylamine and phenylanine) needed for penicillin production are present in the corn steep liquor.

About 30,000 liters of the medium are placed in a tank, sterilized, and inoculated with an aqueous suspension of *P. chrysogenum* conidia. During the incubation period, the medium is aerated with sterile air and agitated. The tank is equipped with devices that allow the continuous addition of glucose syrups, sodium hydroxide, and sulfuric acid to maintain the pH between 6.8 and 7.4, cooling coils to maintain the temperature between 23°C and 25°C, devices for the introduction of antifoam agents, and pumps for the addition of the acyl donor or precursor. The most commonly used precursor is phenylacetic acid which is required for penicillin G, but other precursors in the form of a salt, amide, or ester of the corresponding acid or amine may be added to yield penicillins bearing the desired acyl group.

At first, the mycelium grows actively and utilizes lactic acid and organic nitrogen compounds as sources of carbon. Ammonia is formed during the breakdown of these nitrogenous compounds. Upon exhaustion of the readily available nitrogenous compounds, the fungus then utilizes lactose as a carbon source and growth becomes substantially slower. Ammonia is now actively assimilated and the pH is lowered. Active penicillin production is associated with this phase of lactose and ammonia utilization. Addition of glucose increases the assimilation of ammonia, thereby increasing the yield of penicillin. Penicillin production ceases when the lactose is exhausted.

The duration of the incubation period is about 5 to 6 days. At the end of this time, the mycelium is filtered from the liquid medium which contains the penicillin. The filtrate, containing the penicillin, is mixed with a solvent (amyl or butyl acetate) and the resulting emulsion is centrifuged to extract the acetate solvent which now contains the penicillin. A phosphate buffer is added to the acetate, and the penicillin is extracted with the phosphate buffer by centrifugation. This last step may be repeated, using successively smaller quantities of liquid. Butanol is added to the aqueous mixture, and the potassium salt of penicillin is crystallized from the solution. Potassium penicillin may be further purified and used in this form or it may be converted to procaine penicillin (Perlman, 1979).

PENICILLIN BIOSYNTHESIS

Penicillin is produced along synthetic pathways not required for growth but superimposed upon those pathways required for maintenance. Key intermediates in the pathway are pyruvic acid, valine, and cysteine. As noted previously (Chapter 8), pyruvic acid is a compound required for the CA cycle to take place. Penicillin production is at its maximum when the CA cycle is not actively taking place or if a great deal of pyruvic acid is not being converted to acetyl CoA and thence to fatty acids. The accumulating pyruvic acid can then be available for synthesis of L-valine, the amino acid which joins with cysteine or a derivative of cysteine (presumably originating from the CA cycle) to form a dipeptide precursor of penicillin, L-cysteinyl-L-valine. This last intermediate is then combined with the acyl group from phenylacetic acid in the case of penicillin G or acyl groups from other specific precursors in the synthesis of the other penicillins. Many of the details concerning penicillin biosynthesis are not known.

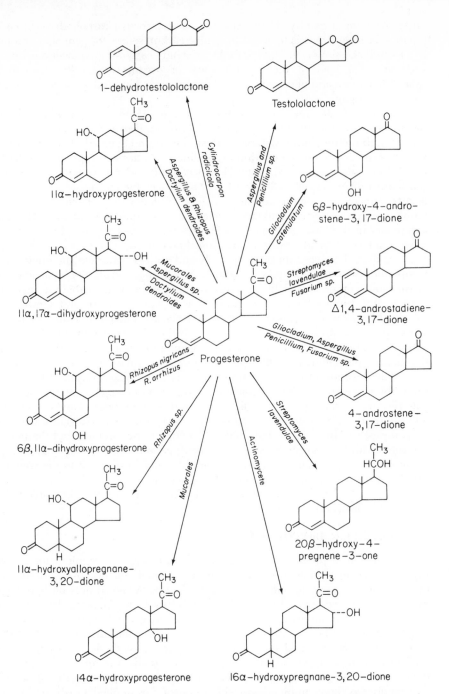

Fig. 14-12. Some steroids produced from progesterone by transformation by several fungi, a Streptomycete (*Streptomyces lavendulae*) and an Actinomycete. [From *Industrial microbiology* by S. C. Prescott and C. G. Dunn. Copyright 1959. Used with permission of Mc-Graw-Hill Book Company.]

Steroid Conversion

Steroids are a group of organic compounds which have the four-membered ring which appears in Fig. 14–12. Steroids are biologically active as hormones produced by the testes, ovaries, adrenal cortex, and placenta. The steroids differ in the nature of their side groups or side chains, and these differences in structure confer different biological properties on the steroids. Steroids are widely used medically as antiinflammatory agents, anesthetics, antifertility agents, or for the treatment of sterility.

Steroids may be obtained directly from a natural source or they may be synthesized. The steroid nucleus can be produced chemically, and many of the transformations or additions of side chains may also be accomplished chemically. Some changes are impossible to accomplish by chemical means or may be very expensive and time-consuming.

An example of a difficult conversion is the hydroxylation of carbon-11, necessary in the synthesis of cortisone and cortisol. It was discovered that *Rhizopus arrhizus* could accomplish the hydroxylation as a one-step process and produce 11-hydroxyprogesterone in low yields, while *R. stolonifer* could carry out the same conversion and give almost quantitative yields. After this pioneer discovery, it was found that many fungus and bacteria species can alter steroid compounds. Species of *Rhizopus* and *Aspergillus* are widely used for this purpose, but fungi capable of converting steroids are found in all the major fungal groups. Alterations that can be accomplished by fungi are hydrogenation, dehydrogenation, hydroxylation, epoxidation, side-chain cleavage, and the expansion of the five-membered steroid ring into a six-membered ring.

Microbiological transformation of steroids has been utilized greatly in the commercial production of steroid compounds, which are often synthesized using a combination of chemical and microbiological transformations. If a transformation can be readily accomplished chemically, it is done by the chemist. If such a reaction is chemically difficult, expensive, or otherwise impractical and if a microorganism can accomplish it readily and inexpensively, the transformation is done with the aid of a microorganism. In practice, chemical and microbiological transformations may be used stepwise and in any sequence to obtain the desired end product.

To carry out a microbiological transformation, the fungus or bacterium is grown in a flask or tank on a medium that will support good growth for a period of 17 to 48 hours with agitation and aeration. After adequate growth of the microorganism is attained, a measured quantity of the steroid precursor is dissolved in a solvent and added to the medium. The enzyme (or enzymes) produced by the microorganisms then effects the conversion of the steroid precursor. Optimum conditions of temperature, aeration, agitation, pH, and time differ with each conversion, and these factors must be carefully controlled. The usual period of transformation is 24 to 48 hours. The modified steroid is recovered from the medium by extraction with a solvent such as methylene chloride, ethylene chloride, or chloroform.

Ergot, Ergotism, and Ergot Alkaloids

Ergot is the sclerotium of the Pyrenomycete, *Claviceps purpurea,* which infects rye and, less commonly, other grains. Overwintered sclerotia give rise to a fleshy perithecia-bearing stroma. The discharged ascospores infect developing rye ovules. Normal development of the grain is suppressed, and a hardened, purple-black sclerotium develops in place of the grain (Fig. 14–13).

(a)

Fig. 14-13. *Claviceps purpurea,* which causes ergot of grains: (a) Fleshy ascocarps arising from scleotized rye grains; (b) An enlarged view of the apex of the ascocarp, showing perithecia embedded within the stroma. [Courtesy Plant Pathology Department, Cornell University.]

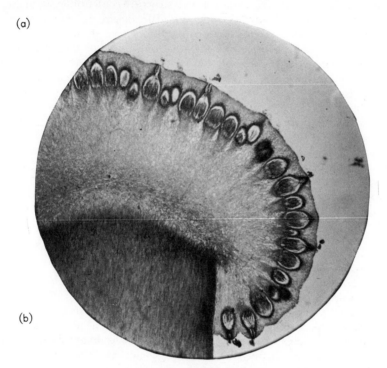

(b)

If consumed in large quantities, as might happen if the sclerotia are included in milled flour, a disease known as *ergotism* develops. Cattle in the field may also be poisoned by the sclerotia if they eat infected pasture grasses. The cattle stagger, develop lameness, gangrene, and possibly inflammation of the digestive tract (Brook and White, 1966).

Ergotism in humans is marked by vomiting, feelings of intense heat or cold, pain in the muscles of the calf, a yellow color in the face, lesions on the hands and feet, diarrhea, and an impairment of the mental functions. People with ergotism often become hysterical or hallucinate. Some have severe convulsions, while the limbs of others become gangrenous and drop off. Spontaneous abortion occurs in pregnant women. Severe cases of ergotism are fatal.

Epidemics of ergotism were common in Europe in the Middle Ages, when rye was consumed in large quantities. Ergotism changed the course of history in at least one event: Peter the Great in Russia was unable to gain control of some ports on the Black Sea in 1772 because his soldiers bought rye bread for themselves and fed rye to their horses. Many of the men developed convulsions, while the horses developed the "blind staggers" (Carefoot and Sprott, 1967). It is also possible that the women accused of witchcraft because of their bizarre behavior in the colonial Massachusetts village of Salem had ergotism (Caporael, 1976). As recently as 1951, an epidemic of ergotism occurred in Pont-Saint-Esprit, France, resulting in the death of four people and the hospitalization of over 150 who were afflicted with hysteria. This epidemic was traced to contaminated flour.

The active principles in ergot that are responsible for ergotism are the alkaloids ergometrine, ergometrinine, ergotamine, and ergotaminine, which occur in the sclerotium. These alkaloids stimulate smooth muscle and selectively block the sympathetic nervous system; they have found modern medical use in stimulating the uterus to contract to initiate childbirth, and also to hasten the return of the uterus to its normal size after childbirth, in the treatment of certain peripheral circulatory disorders, and in the treatment of migraine headaches.

A modern mycological problem has been one of supplying the demands of the pharmaceutical industry for a source of ergot alkaloids. Until about 1950, efforts to produce the alkaloids in culture by conventional laboratory techniques had failed, and the only source of the alkaloids were the sclerotia obtained from naturally infected rye, or rye which was inoculated for this purpose. Inoculation of rye under field conditions is a risky affair as it depends on suitable environmental conditions for development of the sclerotia and, at best, only one crop of sclerotia can be obtained in a season. The search has continued for more reliable and productive methods of obtaining the ergot alkaloids. A process for their laboratory production was patented in France in 1965, but a large-scale production method has yet to be developed.

References

Backus, M. P., and J. F. Stauffer. 1955. The production and selection of a family of strains in *Penicillium chrysogenum*. Mycologia **47**:429–463.

Batra, L. R., and P. D. Millner. 1974. Some Asian fermented foods and beverages and associated fungi. Mycologia **66**:942–950.

Benedict, R. G. 1972. Mushroom toxins other than Amanita. *In:* Kadis, S., A. Ciegler, and S. J. Ajl, Eds., Microbial toxins. Academic Press, New York. **8**:281–320.

Bessey, E. A. 1950. Morphology and taxonomy of fungi. Hafner Press, New York. 791 pp.

Block, S. S., R. L. Stephens, A. Bareto, and W. A. Murrill. 1955. Chemical identification of the *Amanita* toxin in mushrooms. Science **121**:505–506.

Brook, P. J., and E. P. White. 1966. Fungus toxins affecting mammals. Ann. Rev. Phytopathol. **4**:171–194.

Buck, R. W. 1961. Mushroom poisoning since 1924 in the United States. Mycologia **53**:537–538.

Bull, A. T., D. C. Ellwood, and C. Ratledge. 1979. The changing scene in microbial technology. *In:* Bull, A. T., D. C. Ellwood, and C. Ratledge, Eds., Microbial technology: current state, future prospects. Symp. Soc. Gen. Microbiol. **29**:1–28.

Buller, A. H. R. 1914. The fungus lore of the Greeks and Romans. Trans. Brit. Mycol. Soc. **5**:21–66.

———. 1922. Researches on fungi. Vol. 2. Longman, Inc. London. 492 pp.

Bu'Lock, J. D. 1979. Industrial alcohol. *In:* Bull, A. T., D. C. Ellwood, and C. Ratledge, Eds., Microbial technology: current state, future prospects. Symp. Soc. Gen. Microbiol. **29**:308–325.

Caporael, L. R. 1976. Ergotism: the satan loosed in Salem? Science **192**:21–26.

Carefoot, G. L., and E. R. Sprott. 1967. Famine on the wind. Rand McNally & Company, Skokie, Ill. 229 pp.

Chang, S. T. 1980. Mushrooms as human food. Bioscience **30**:399–401.

Cochran, K. W., and M. W. Cochran. 1978. *Clitocybe clavipes*: antabuse-like reaction to alcohol. Mycologia **70**:1124–1126.

Duddington, C. L. 1961. Micro-organisms as allies—the industrial use of fungi and bacteria. Macmillan, Inc., New York. 256 pp.

Dunn, C. G. 1958. The industrial utilization of yeasts. Econ. Botany **12**:145–163.

Fleming, A. 1929. On the antibacterial action of cultures of a *Penicillium,* with special reference to their use in the isolation of *B. influenzae.* Brit. J. Exptl. Pathol. **10:**226-236.

Foster, J. W. 1949. Chemical activities of fungi. Academic Press, New York. 648 pp.

Gray, W. D. 1959. The relation of fungi to human affairs. Holt, Rinehart, & Winston, New York. 510 pp.

Groves, J. W. 1964. Poisoning by morels when taken with alcohol. Mycologia **56:**779-780.

Hatch, R. T., and S. M. Finger. 1979. Mushroom fermentation. *In:* Peppler, H. J., and D. Perlman, Eds., Microbial technology, Second ed. Academic Press, New York. **2:**179-199.

Hesseltine, C. W. 1965. A millennium of fungi, food, and fermentation. Mycologia **57:**149-197.

Hockenhull, D. J. D. 1963. Antibiotics. *In:* Rainbow, C., and A. H. Rose, Eds., Biochemistry of industrial micro-organisms. Academic Press, New York. pp. 227-299.

Hoffer, A., and H. Osmond. 1967. The hallucinogens. Academic Press, New York. 626 pp.

Johnson, M. J. 1954. The citric acid fermentation. *In:* Underkofler, L. A., and R. J. Hickey, Eds., Industrial fermentations. Chemical Publishing Co., New York. **1:**420-445.

Joslyn, M. A., and M. W. Turbovsky. 1954. Commercial production of table and dessert wines. *In:* Underkofler, L. A., and R. J. Hickey, Eds., Industrial fermentations. Chemical Publishing Co., New York. **1:**196-251.

Lambert, E. B. 1932. Mushroom growing in the United States. U.S. Department of Agriculture Circ. 251, pp. 1-34.

——. 1933. Effect of excess carbon dioxide on growing mushrooms. J. Agr. Res. **47:**599-608.

——. 1938. Principles and problems of mushroom culture. Botan. Rev. **4:**397-426.

Lincoff, G., and D. H. Mitchell. 1977. Toxic and hallucinogenic mushroom poisoning. Van Nostrand Reinhold Company, New York. 267 pp.

Lockwood, L. B. 1979. Production of organic acids by fermentation. *In:* Peppler, H. J., and D. Perlman, Eds., Microbial technology, Second ed. Academic Press, New York. **1:**241-280.

Lowy, B. 1971. New records of mushroom stones from Guatemala. Mycologia **63:**983-993.

——. 1972. Mushroom symbolism in Maya codices. Mycologia **64:**816-821.

Maisch, W. F., M. Sobolov, and A. J. Petricola. 1979. Distilled beverages. *In:* Peppler, H. J., and D. Perlman, Eds., Microbial technology, Second ed. Academic Press, New York. **2:**79-94.

McCarthy, J. L. 1954. Alcoholic fermentation of sulfite waste liquor. *In:* Underkofler, L. A., and R. J. Hickey, Eds., Industrial fermentations. Chemical Publishing Co., New York. **1:**95-135.

Moskowitz, G. J. 1979. Inocula for blue-veined cheeses and blue cheese flavor. *In:* Peppler, H. J., and D. Perlman, Eds., Microbial technology, Second ed. Academic Press, New York. **2:**201-222.

Olson, N. F. 1979. Cheese. *In:* Peppler, H. J., and D. Perlman, Eds., Microbial technology, Second ed. Academic Press, New York. **2:**39-77.

Oso, B. A. 1975. Mushrooms and the Yoruba people of Nigeria. Mycologia **67:**311-319.

——. 1976. *Phallus aurantiacus* from Nigeria. Mycologia **68:**1076-1082.

——. 1977-a. *Pleurotus tuber-regium* from Nigeria. Mycologia **69:**271-279.

——. 1977-b. Mushrooms in Yoruba mythology and medicinal practices. Econ. Botany **31:**367-371.

Peppler, H. J. 1960. Yeast. *In:* Matz, S. A., Ed., Bakery technology and engineering. Avi, Westport, Conn. pp. 35-74.

——. 1967. Ethyl alcohol, lactic acid, acetone-butyl alcohol and other microbial products. *In:* Peppler, H. J., Ed., Microbial technology. Reinhold Publishing Co., New York. pp. 403-416.

——. 1979. Production of yeasts and yeast products. *In:* H. J. Peppler and D. Perlman, Eds., Microbial technology, Second ed. Academic Press, New York. **1:**157-185.

Perlman, D. 1979. Microbial production of antibiotics. *In:* H. J. Peppler, and D. Perlman, Eds., Microbial technology, Second ed. Academic Press, New York. **1:**241-280.

Peterson, D. H. 1963. Microbial transformations of steroids and their application to the preparation of hormones and derivatives. *In:* Rainbow, C., and A. H. Rose, Eds., Biochemistry of industrial micro-organisms. Academic Press, New York. pp. 537-606.

Phaff, H. J., and M. A. Amerine. 1979. Wine. *In:* Peppler, H. J., and D. Perlman, Eds., Microbial technology, Second ed. Academic Press, New York. **2:**131-153.

Prescott, S. C., and C. G. Dunn. 1959. Industrial microbiology. Third ed. McGraw-Hill Book Company, New York. 945 pp.

Raper, K. B., and D. F. Alexander. 1945. Penicillin: V. Mycological aspects of penicillin production. J. Elisha Mitchell Sci. Soc. **61:**74-113.

Rolf, R. T., and F. W. Rolfe. 1925. The romance of the fungus world. (1974 reprint.) Dover Publications, Inc., New York. 308 pp.

Sebek, O. K., and D. Perlman. 1979. Microbial transformation of steroids and sterols. *In:* Peppler, H. J., and D. Perlman, Eds., Microbial technology, Second ed. Academic Press, New York. **1:**483–496.

Sternberg, M. 1976. Microbial rennets. Advan. Appl. Microbiol. **20:**135–157.

Stewart, G. G. 1974. Some thoughts on the microbiological aspects of brewing and other industries utilizing yeast. Advan. Appl. Microbiol. **17:**233–264.

Sylvester, J. C., and R. D. Coghill. 1954. The penicillin fermentation. *In:* Underkofler, L. A., and R. J. Hickey, Eds., Industrial fermentations. Chemical Publishing Co., New York. **2:**219–263.

Tenney, R. I. 1954. The brewing industry. *In:* Underkofler, L. A., and R. J. Hickey, Eds., Industrial fermentations. Chemical Publishing Co., New York **1:**172–195.

Thom, C. 1944. Molds in the cheese industry. J. N. Y. Botan. Garden **45:**105–113.

Vining, L. C., and W. A. Taber, 1963. Alkaloids. *In:* Rainbow, C., and A. H. Rose, Eds., Biochemistry of industrial micro-organisms. Academic Press, New York. pp. 341–378.

von Oettingen, W. F. 1958. Poisoning—a guide to clinical diagnosis and treatment. W. B. Saunders Co., Philadelphia, Pa. 627 pp.

Wang, H. L., and C. W. Hesseltine. 1979. Mold-modified foods. *In:* Peppler, H. J., and D. Perlman, Eds., Microbial technology, Second ed. Academic Press, New York. **2:**95–129.

Watling, R., and M. R. D. Seaward. 1976. Some observations on puff-balls from British archeological sites. J. Archaeol. Sci. **3:**165–172.

Wasson, R. G. 1961. The hallucinogenic fungi of Mexico: an inquiry into the origins of the religious idea among primitive peoples. Botanical Museum Leaflets Harvard Univ. **19:**137–162.

——. 1962. The hallucinogenic mushrooms of Mexico and psilocybin: a bibliography. Botanical Museum Leaflets Harvard Univ. **20:**25–73.

Wasson, V. P., and R. G. Wasson. 1957. Mushrooms, Russia, and history. Pantheon Books, New York. Vols. 1 and 2. 433 pp.

Weinshank, D. J., and J. C. Garver. 1967. Theory and design of aerobic fermentations. *In:* Peppler, H. J., Ed., Microbial technology. Reinhold Publishing Co., New York. pp. 417–449.

Westermann, D. H., N. J. Huige. 1979. Beer brewing. *In:* Peppler, H. J., and D. Perlman, Eds., Microbial technology, Second ed. Academic Press, New York. **2:**1–37.

Wieland, T., and O. Wieland. 1972. The toxic peptides of *Amanita* species. *In:* Kadis, S., A. Ciegler, and S. J. Ajl, Eds., Microbial toxins. Academic Press, New York. **8:**249–280.

Young, F. M., and B. J. B. Wood. 1974. Microbiology and biochemistry of soy sauce fermentation. Advan. Appl. Microbiol. **17:**157–194.

Glossary

Acervulus, pl.* acervuli (dim. of L.* *acervus* = heap). A subcuticular or subepidermal mass of closely clustered conidiophores and conidia which are not covered by fungal tissue.

Aeciospore (Gr.* *aikia* = injury, *sporos* = seed). A binucleate spore borne in an aecium.

Aecium, pl. aecia (Gr. *aikia* = injury). The first sorus that is formed after plasmogamy and bears binucleate aeciospores, stage I (in the Uredinales).

Aerobe (Gr. *aēr* = air, *bios* = life). An organism which requires free oxygen for respiration.

Aerobic. With the qualities of an aerobe.

Agaric. A gill-bearing mushroom, or member of the Agaricaceae.

Alternation of generations. Succession of gamete- and spore-bearing thalli in the life cycle of an organism.

Anaerobic. Not requiring free, molecular oxygen for respiration.

Anamorph (Gr. *ana* = anew, *morphe* = form). The nonsexual (usually conidial) stage in the life cycle of a fungus (cf. telomorph).

Annulus, pl. annuli (L., ring). A membranous skirt surrounding the stipe of a Hymenomycete.

Antagonism. An ecological association between organisms in which one or more of the participants is harmed or has its activities limited.

Antheridium, pl. antheridia (diminutive of Gr. *anthos* = flower). A male gametangium.

Antibiosis. (Gr. *anti* = against, *bios* = life). Production of antibiotic(s) that affects other organisms adversely.

Antibiotic. A compound of high molecular weight produced by a microorganism that has specific toxic effects on other microorganisms sensitive to it.

Apomictic (Gr. *apo* = lack, *mixis* = mingling). Developing by apomixis.

Apomixis. Development of a product that is normally the result of sexual reproduction without plasmogamy, karyogamy, or meiosis.

* Abbreviations: pl. = plural; L. = Latin; Gr. = Greek.

547

Apophysis (Gr. *apo* = lack, *physis* = growth). A swelling of the vegetative cell beneath the sporangium in some chytrids.

Apothecium, pl. **apothecia** (Gr. *apo* = lack, *theke* = sheath). An ascocarp in which the hymenium is not covered by fungal tissue at maturity.

Appressorium, pl. **appressoria** (L. *apprimere* = to press against). An enlargement on a hypha or germ tube which attaches itself to the host before penetration takes place.

Arbuscule. A swollen, vesiclelike ending of the fungal hypha formed by some mycorrhizal fungi and of unknown function.

Archicarp (Gr. *archi* = first, *karpos* = fruit). The female organ(s) and supporting cells (in the Ascomycotina).

Ascocarp (Gr., *askos* = sac, *karpos* = fruit). The ascospore-bearing, multicellular sporocarp formed by a member of the Ascomycotina.

Ascogenous hyphae (Gr. *askos* = sac, *genesis* = being produced). Those hyphae, often arising from the archicarp, which give rise to asci.

Ascogonium, pl. **ascogonia** (Gr. *askos* = sac, *gone* = offspring). The cell or cells of the archicarp which function as the female gametangium and which usually receive the male nuclei and then give rise to the ascogenous hyphae.

Ascospore (Gr. *askos* = sac, *sporos* = seed). A spore borne in an ascus.

Ascostroma, pl. **ascostromata** (Gr. *askos* = sac, *stroma* = cover). A stroma bearing asci.

Ascus, pl. **asci** (Gr. *askos* = sac). A cell which is the site of meiosis and in which endogenous spores of sexual origin are formed.

Assimilation. The uptake of nutrients and their incorporation into the organism.

Autoecious (Gr. *autos* = self, *oikos* = house). Capable of completing a life cycle on one host (of Uredinales) (*cf.* heteroecious).

Autotrophic (Gr. *autos* = self, *trophe* = nourishment). Independent of an external source of organic food (*cf.* heterotrophic).

Azygospore (Gr. *azygos* = unmatched, *sporos* = seed). A body morphologically similar to a zygospore but which is of nonsexual origin (in the Zygomycetes).

Basidiocarp (Gr. *basidion* = small base, *karpos* = fruit). A sporocarp produced by a member of the Basidiomycotina and which bears basidiospores.

Basidiole (dim. of Gr. *basidion* = small base). A structure in a hymenium that is morphologically similar to a basidium without sterigmata and may be an immature basidium or a permanently sterile structure in the hymenium of a member of the Basidiomycotina (*cf.* cystidium).

Basidiospore (Gr. *basidion* = small base, *sporos* = seed). An exogenous sexual spore borne on a basidium.

Basidium, pl. **basidia** (Gr. *basidion* = small base). A cell in which karyogamy and meiosis take place and which bears exogenous spores of sexual origin.

Binding hyphae. Thick-walled, highly branched, aseptate hyphae in a basidiocarp that bind other hyphal types together (cf. generative, lactiferous, and skeletal hyphae).

Biotrophic parasite (Gr. *bios* = life, *trophe* = nourishment). A fungus parasitic on other fungi which does not kill its host but feeds on living cells.

Bipolar sexuality (L. *bi* = two, *polus* = pole). A type of sexual compatibility controlled by multiple-allelic genes at a single locus (*cf.* tetrapolar sexuality).

Bitunicate (L. *bi* = two, *tunica* = covering). Having a double ascus wall (*cf.* unitunicate).

Blastic development (Gr. *blastos* = bud, embryonic cell). Formation of a conidium by enlargement and maturation of a conidial initial that is distinct from the vegetative hyphae or conidiophore and enlarges before septation separates the conidium from the parent cell. In **enteroblastic** development (Gr., *enteron* = that within), some of the wall layers of the new conidium are not continuous with the wall layers of the parent cell. In **holoblastic** development (Gr. *holos* = whole or entire), all wall layers of the parent cell are continuous with those of the conidium (*cf.* thallic development).

Bolete (Gr. *bolos* = a throw). A member of the Agaricales which possesses tubes and not gills.

Bromatium, pl. **bromatia** (Gr. *bromatos* = food). An enlarged hyphal ending, induced and used as food by fungi-cultivating ants.

Budding. Production of a new cell from a small outgrowth or protrusion from the mother cells (of yeasts or spores) (*cf.* fission).

Capillitium, pl. **capillitia** (L. *capillus* = hair). A mass of sterile fibers interspersed among spores within a sporocarp (in the Gasteromycetes).

Carpogenic cell (Gr. *karpos* = fruit, *genesis* = being produced). A cell that gives rise to the ascogonium and supporting cells in the Laboulbeniales.

Catahymenium (Gr. *kata* = under or lower, complete; *hymen* = membrane). A hymenium type initiated as a palisade of cystidia through which basidia eventually penetrate; non-thickening (*cf.* euhymenium).

Centrum (Gr. *kentron* = center). The contents of an ascocarp locule, including the ascogenous system and sterile cells.

Chemotropism. See **tropism.**

Chlamydospore (Gr. *chlamydos* = cloak, *sporos* = seed). A thick-walled, nonsexual spore which represents a transformed hyphal cell.

Clamp (connection). A recurving outgrowth of a cell that, at cell division, acts as a bridge to allow passage of one of the products of nuclear division into the penultimate cell, thereby ensuring maintenance of the dikaryotic condition (of members of the Basidiomycotina).

Cleistohymenial (Gr. *kleistos* = closed, *hymen* = membrane). Of apothecia in which the hymenium is covered by excipular tissue during part of the development (*cf.* gymnohymenial).

Cleistothecium, pl. **cleistothecia** (Gr. *kleistos* = closed, *theke* = case). An ascocarp with the asci surrounded by fungal tissue and without regularly formed openings.

Colonization. The invasion of a substratum or habitat by a fungus.

Colony. A discrete mass of cells or hyphae developing from a single cell, hyphal fragment, or spore.

Columella, pl. **columellae** (L. *columen* = column). A sterile column within a spore-bearing structure, often an extension of a supporting stalk.

Compatible. Describes fungal sexual partners that are able to produce a fertile cross with each other.

Competition. A more or less active demand on the part of two organisms for some commodity (space, food, etc.) that is inadequate to provide for all organisms present.

Competitive saprophytic ability. Relative capability of successfully competing with other saprophytes.

Complementation. Compensation by a nucleus for deficiencies of other nuclear type(s) in a heterokaryon.

Conidiophore (Gr. *konis* = dust, *phoros* = bearing). A hypha, often specialized in structure, that bears one or more conidia.

Conidium, pl. **conidia** (Gr. *konis* = dust). A thin-walled, nonsexual spore that is borne both terminally and exogenously on a conidiophore and is deciduous at maturity.

Constitutive dormancy. See **dormancy.**

Coprophilous (Gr. *kopros* = dung, *philein* = to love). Living on dung.

Cortina (dim. of L. *cortis* = curtain). A veil extending from the margin of the pileus in some Hymenomycetes.

Crozier. A recurved hook at the tip of an ascogenous hypha which will become the ascus.

Crustose thallus (L. *crusta* = crust). A flat, crustlike thallus that adheres closely to the substratum (of lichens) (*cf.* foliose thallus, fruticose thallus).

Culture (L. *colere* = to cultivate). (1) To grow an organism or (2) the resulting growth. Usually on artificial medium for experimental or industrial use.

Cystidium, pl. **cystidia** (Gr. *kyst* = sac). A sterile cell occurring among the basidia and often projecting beyond the hymenium and differing morphologically from the basidium (*cf.* basidiole).

Decay. See **decomposition.**

Decomposition. Decay; separation of a material into its component parts, for example, breakdown of organic matter by microorganisms.

Dermatophyte (Gr. *dermatos* = skin, *phyton* = plant). A parasitic fungus which attacks and causes a disease of the skin or its appendages.

Dikaryon (L. *di* = two, Gr. *karyon* = nucleus). A cell or hypha in which haploid, genetically different nuclei of two types are closely associated in pairs following plasmogamy (*cf.* heterokaryon, monokaryon).

Dikaryotic. The condition of being a dikaryon.

Dimorphic (L. *di* = two, Gr. *morphe* = form). Producing two morphologically different forms, not necessarily required to complete the life cycle (of animal parasites which can exist as single-celled and mycelial forms).

Dimorphism. Existence of two morphologically different forms in a dimorphic organism.

Dioecious (L. *di* = two, Gr. *oikos* = house). Having male and female reproductive structures on separate thalli.

Diplanetic (L. *di* = two, Gr. *planos* = wandering). Having a succession of two morphologically different zoospore stages separated by a resting stage (in the Oomycetes) (*cf.* monoplanetic).

Diplanetism. The succession of two zoospore stages in a diplanetic organism.

Diploid (Gr. *diploos* = double). Having the *2N* number of chromosomes within a single nucleus.

Diploidization. Karyogamy of haploid nuclei to form a diploid nucleus, especially in a heterokaryon.

Disease. A malfunctioning of an organism, caused by external agents, which results in abnormalities.

Dissimilation. Metabolic decomposition within an organism.

Dormancy. A period of rest or minimal metabolic activity, especially in spores. Constitutive dormancy is imposed by the organism, while exogenous dormancy is imposed by external conditions.

Ectomycorrhiza (Gr. *ecto* = on the outside, *mykes* = fungus, *rhiza* = root). A mycorrhizal type in which a sheath of fungal hyphae is formed on the outside of the root (*cf.* endomycorrhizae).

Endobiotic (Gr. *endon* = within, *bios* = life). Growing entirely within a host (*cf.*epibiotic, interbiotic).

Endogenous (Gr. *endon* = within, *genesis* = being produced). Produced or borne within a structure.

Endomycorrhiza (Gr. *endon* = within, *mykes* = fungus, *rhiza* = root). A mycorrhizal type in which the fungal hyphae penetrate the root and a sheath is lacking (*cf.* ectomycorrhiza).

Endospore (Gr. *endon* = within, *sporos* = seed). The innermost layer of a spore wall (*cf.* epispore, perispore).

Enteroblastic development (see **blastic development**).

Enterothallic development (see **thallic development**).

Epibiotic (Gr. *epi* = upon, *bios* = life). Growing on the outside of a host (*cf.* endobiotic, interbiotic).

Epispore (Gr. *epi* = upon, *sporos* = seed). The outermost layer of the spore wall, which is often ornamented (*cf.* endospore, perispore).

Epithecium (Gr. *epi* = upon, *theke* = sheath or cover). A layer of tissue that covers the asci in some Discomycetes.

Eucarpic (Gr. *eu* = normal, *karpos* = fruit). A condition in which only part of an assimilative thallus is converted into reproductive structure(s).

Euhymenium (Gr. *eu* = normal, *hymen* = membrane). A hymenium type initiated as a palisade of basidia; sometimes thickening (*cf.* catahymenium).

Excipulum, pl. **excipula** (L., *vessel*). In the Discomycetes, the tissue beneath the subhymenium and forming the outermost sterile margin of the apothecium.

Exogenous (L. *ex* = out of, Gr. *genesis* = being produced). Borne or produced on the outside of a structure.

Exogenous dormancy. See **dormancy.**

Exploitation. An ecological interrelationship in which one organism inflicts harm on another for its own benefit (parasitism or predation).

Facultative parasite (L. *facultas* = ability). An organism that normally lives as a saprophyte but under certain conditions can live as a parasite (*cf.* obligate parasite).

Fermentation (L. *fermentum* = yeast). Anaerobic respiration; anaerobic dissimilation of an organic compound.

Fertilization. Sexual union of both cytoplasm (plasmogamy) and nuclei (karyogamy).

Fertilization tube. An extension of the antheridium which penetrates the wall of the oogonium to reach the oosphere and through which the male protoplast passes (in the Oomycetes).

Fission. Cell division by a cleavage (splitting) of the cell into two parts (*cf.* budding).

Flagellum, pl. **flagella** (L. *flagellum* = whip). A hairlike structure that propels a swimming cell.

Flexuous hyphae (L. *flexuosus* = bent). Those hyphae in a pycnium that undergo plasmogamy with pycniospores (in the Uredinales).

Foliose thallus (L. *folium* = leaf). A flat, leaflike

thallus that is spread over the substratum but is attached at localized points only (of lichens) (*cf.* crustose thallus, fruticose thallus).

Fruticose thallus (L. *fruticosus* = shrub). An upright or pendulous thallus (of lichens) (*cf.* crustose thallus, foliose thallus).

Fungicide. A chemical or physical agent that kills fungi (*cf.* fungistasis).

Fungistasis. Inhibition of fungal growth that is not lethal (*cf.* fungicide).

Fungus, pl.**fungi** (L., *fungus* or *mushroom*). A heterotrophic organism whose usually walled, threadlike cells absorb nutrients in the form of a watery solution.

Funiculus, pl.**funiculi** (L. *funis* = rope). A hyphal cord that attaches a peridiole to the peridium (in the Nidulariales).

Gametangium, pl. **gametangia** (Gr. *gamete* = germ, *angeion* = case). A differentiated cell that produces discrete gametes or whose undifferentiated protoplast functions in the stead of discrete gametes.

Gamete (Gr. *gamete* = germ). A differentiated reproductive cell that is capable of undergoing plasmogamy with another similar or dissimilar reproductive cell.

Generative cell. In the Trichomycetes, a cell which bears a sporangiole.

Generative hyphae. Nonspecialized, thin-walled hyphae in a basidiocarp (*cf.* lactiferous, binding, and skeletal hyphae).

Geotropism. See **tropism.**

Germination (L. *germen* = sprout). The emergence of a protoplast, usually vegetative, from a spore.

Gill. A hymenium-covered, platelike appendage that hangs from the underneath surface of the basidiocarp of some Hymenomycetes.

Gleba (L., *clod*). Spore-bearing tissue that is enclosed by a peridium throughout development (in the Gasteromycetes).

Gymnocarpic development (Gr. *gymnos* = naked, *karpos* = fruit). Development of basidiocarps with the hymenium exposed throughout development (*cf.* hemiangiocarpic, pseudoangiocarpic).

Gymnohymenial development (Gr. *gymnos* = naked, *hymen* = membrane). Development of apothecia in which the hymenium is exposed throughout development (*cf.* cleistohymenial).

Haploid (Gr. *haploos* = single). Having a single set (*N*), or the reduced number, of chromosomes in a single nucleus.

Haploidization. In a heterokaryon, the transformation of a diploid nucleus to a haploid nucleus through a series of mitotic divisions in which aneuploid nuclei are formed.

Hapteron (Gr. *haptein* = fasten upon). An organ of attachment in some lichens or in the Nidulariales. In the latter, the hapteron is a mass of highly adhesive hyphae at the base of the funiculus.

Haustorium, pl. **haustoria** (L. *haustus* = drawing of water). A specialized hyphal extension, formed by some parasitic fungi, which penetrates the host cell and absorbs nutrients.

Hemiangiocarpic development (Gr. *hemi* = partly, *angeion* = case, *karpos* = fruit.) Development of basidiocarps in which the hymenium is formed endogenously during the early stages of development, but which becomes exposed before maturity (*cf.* gymnocarpic, pseudoangiocarpic).

Heteroecious (Gr. *heteros* = different from, *oikos* = house). Requiring more than one host species to complete a life cycle (of Uredinales) (*cf.* autoecious).

Heterogamy (Gr. *heteros* = different from, *gamete* = germ). Plasmogamy between gametes which are morphologically different (*cf.* isogamy).

Heterokaryon (Gr. *heteros* = different from, *karyon* = nucleus). A cell which contains genetically different nuclei or a thallus made up of cells (*cf.* dikaryon, homokaryon).

Heterokaryosis. See **Heterokaryotic.**

Heterokaryotic. The condition of being a heterokaryon.

Heterothallic (Gr. *heteros* = different from, *thallos* = shoot). The condition of being self-sterile, requiring a partner for sexual reproduction. (*cf.* homothallic).

Heterothallism. Self-sterility (*cf.* homothallism).

Heterotrophic (Gr. *heteros* = different from, *trophe* = nourishment). Deriving nourishment from organic matter formed by other organisms, living as a saprophyte or parasite (*cf.* autotrophic).

Hilar appendix. The hilum and associated structures at the base of the basidiospore.

Holobasidium (Gr. *holos* = entire, *basidion* = small base). A single-celled basidium (*cf.* phragmobasidium.

Holoblastic development. See **Blastic development.**

Holocarpic (Gr. *holos* = entire, *karpos* = fruit). A condition in which the entire assimilative thallus converts into a reproductive structure (*cf.* eucarpic).

Homokaryon (Gr. *homos* = alike, *karyon* = nucleus). A cell which contains nuclei of one genetic constitution only or a thallus made up of such cells (*cf.* heterokaryon).

Homokaryotic. The condition of being a homokaryon.

Homothallic (Gr. *homos* = alike, *thallos* = shoot). The condition of being self-fertile, able to reproduce sexually without a partner (*cf.* heterothallic).

Homothallism. Self-fertility (*cf.* heterothallism).

Host. A living organism which provides nourishment, and possibly protection, to a fungus which lives in or on it.

Humus (L., earth). The dark organic portion of soil consisting largely of partially decayed plant material.

Hymenium, pl. **hymenia** (Gr. *hymen* = membrane). A palisadelike layer of asci or basidia, including any sterile cells such as basidioles, paraphyses, or cystidia.

Hymenophore (Gr. *hymen* = membrane, *phoros* = carrying). The trama immediately beneath the hymenium and which bears the hymenium in a basidiocarp.

Hypertrophy (Gr. *hyper* = over, *trophe* = nourishment). Abnormal overdevelopment of a tissue, cell, or components of a cell.

Hypha, pl. **hyphae** (Gr. *hyphē* = a web). Filamentous part of a fungus, usually septate and consisting of several cells in linear succession.

Hyphal body. A segment of a hypha which becomes separated and reproduces by budding or fission (in the Entomophthorales).

Hypogeous fungi (Gr. *hypos* = below, *ge* = earth). Fungi that form sporocarps within the soil.

Inoculate (L. *inoculare* = to engraft). To introduce microorganisms, or a substance containing microorganisms, within or upon an organism or substratum.

Inoculum. That portion used in inoculation.

Inoperculate (L. *in* = not, *operculum* = lid). Of asci or sporangia that open by means of a pore lacking an operculum (*cf.* operculate).

Interbiotic (L. *Inter* = between, Gr. *bios* = life). With part of the thallus within a host and part external to the host (of Chytridiales) (*cf.* endobiotic, epibiotic).

Isidium (Gr. *is* = tissue, *idios* = one's own). A protuberance of the lichen cortex that may be cylindrical, clavate, or other shapes.

Isogamy (Gr. *isos* = equal, *gamete* = germ). Plasmogamy between gametes that are morphologically similar (*cf.* heterogamy).

Karyogamy (Gr. *karyon* = nucleus, *gamos* = marriage). Fusion of two nuclei.

Kinetosome (Gr. *kinein* = to move, *soma* = body). A basal body lying within the cell that gives rise to the fibers within the flagellum.

Lactiferous hyphae (L. *lactis* = milk). Hyphae filled with a milky fluid (*cf.* binding, generative, and skeletal hyphae).

Lacunar development (L. *lacuna* = small pit or gap). Development of Gasteromycetes in which the cavities are formed by tearing or stretching of the glebal tissue.

Lamella, pl. **lamellae** (dim. of L. *lamina* = a thin plate). See **gill.**

Lichen (Gr. *leichen* = tree-moss). A thallus consisting of an alga and fungus intermixed and living in a symbiotic relationship.

Life cycle. The succession of stages following a particular phase of development or spore form, which culminates with the production of that same phase or spore form.

Loculate (dim. of L. *locus* = place). Containing locules.

Locule. A cavity in a stoma which contains asci and is not bounded by a wall originating from the archicarp or other sexual structures.

Lomasome (Gr. *loma* = border, *soma* = body). A body of various forms which occurs between the plasmalemma and cell wall.

Long-cycled. Having an alternation of generations and all five spore stages in the life cycle (of the Uredinales) (*cf.* short-cycled).

Lysis (Gr. *lyein* = to break up). Breakdown of a substance or cells by an enzyme or other agent.

Metabasidium (Gr. *meta* = change, *basidion* = small base). The cell in which meiosis occurs in members of the Basidiomycotina (*cf.* probasidium).

Mitotic crossing over. Exchange of chromosome segments during haploidization.

Mold. A downy fungal growth on a substratum, usually consisting of a Hyphomycete mycelium.

Monoecious (Gr. *monos* = single, *oikos* = house). Having male and female reproductive organs on a single thallus (*cf.* dioecious).

Monokaryon (Gr. *monos* = single, *karyon* = nucleus). A cell or hypha in which there are one or more haploid nuclei, all of a single genetic type (*cf.* dikaryon, heterokaryon).

Monokaryotic. With the qualities of a monokaryon.

Monoplanetic (Gr. *monos* = single, *planos* = wandering). Having zoospores of one type only (of Oomycetes) (*cf.* diplanetic).

Monoplanetism. Formation of one type of zoospores only (of Oomycetes (*cf.* diplanetism).

Mushroom. A fleshy Hymenomycete with gills, popularly used to refer to an edible variety.

Mycelium, pl. **mycelia** (Gr. *mykes* = fungus). A mass of hyphae, often used to denote all hyphae comprising a thallus.

Mycetangium, pl. **mycetangia** (Gr. *mykes* = fungus, *angeion* = vessel). In some beetles, a specialized pouch where their fungal symbionts can be harbored.

Mycology (Gr. *mykes* = fungus, *logos* = discourse). The science dealing with fungi.

Mycorrhiza, pl. **mycorrhizae** (Gr. *mykes* = fungus, *rhiza* = root). A symbiotic relationship between a fungus and the root of a higher plant.

Necrosis (Gr. *nekros* = dead body). Death of cells or tissues, especially as a result of disease.

Necrotrophic parasite (Gr. *nekros* = dead body, *trophe* = nourishment). A fungus parasitic on other fungi which kills its host and feeds on the dead cells (*cf.* biotrophic parasite).

Nonsexual reproduction. Production of progeny without the formation of spores of sexual origin.

Nuclear cap. A body, containing a large quantity of RNA, at the side of the nucleus in a zoospore or gamete.

Obligate parasite. An organism that is incapable of living as a saprophyte and must live as a parasite (*cf.* facultative parasite).

Oidium, pl. **oidia** (Gr. *oidion* = small egg). A thin-walled cell released by the breakdown of a hypha into its component cells. An oidium may function as a nonsexual spore or as a fertilizing agent.

Oogamy (Gr. *oon* = egg). A type of heterogamy in which plasmogamy takes place between a large nonmotile egg and a small motile male gamete or cytoplasm from an antheridium.

Oogonium, pl. **oogonia** (Gr. *oon* = egg, *gone* = offspring). A female gametangium which contains one or more discrete gametes.

Oosphere (Gr. *oon* = egg, *sphaira* = sphere). A large naked sphere of protoplasm which functions as the egg or female gamete in the oogonium.

Oospore (Gr. *oon* = egg, *sporos* = seed). A thick-walled sexual spore which develops from the oosphere after plasmogamy or sometimes by parthenogenesis.

Operculate (L. *operculum* = lid). An ascus or sporangium bearing an operculum (*cf.* inoperculate).

Operculum. A flap or lidlike covering over the opening of an ascus or sporangium.

Ostiole (L. *ostium* = small opening). A tubular passage terminating in a pore which extends through the neck of a perithecium; an opening or pore allowing the release of spores from any sporocarp.

Papilla, pl. **papillae.** A nipplelike extension, for example, on a zoosporangium.

Paraphysis, pl. **paraphyses** (Gr. *para* = beside, *physis* = a growth). Sterile, elongated cells that may occur in the hymenium, intermixed with asci; they grow upward and have free apices.

Parasexual cycle (Gr. *para* = besides). A sequence involving heterokaryon formation, diploidization, and haploidization, often resulting in the formation of recombinant nuclei. Unlike the sexual cycle, the parasexual cycle may occur at any point or continuously throughout the life cycle.

Parasite (L. *parasitus* = one who eats at the table of another). An organism that derives its nourishment from another living organism, usually (but not always) causing harm to the host organism.

Parthenogenesis (Gr. *parthenos* = virgin, *genesis* = being produced). Reproduction involving the development of female gametes without fertilization taking place.

Partial veil. A veil in some Hymenomycetes extending from the stipe to the margin of the pileus.

Pathogen (Gr. *pathos* = disease, *genesis* = being produced). A parasite which causes disease in its host.

Pathogenicity. Ability to produce a disease; capacity to be a pathogen.

Peridiole (Gr. *peridion* = small pouch). In the Nidulariales, a segregant of the gleba, containing basidiospores and surrounded by a distinct wall, which functions as a unit of dispersal.

Peridium, pl. **peridia** (Gr. *peridion* = small pouch). The outermost membranous covering of a sporocarp.

Periphysis, pl. **periphyses** (Gr. *peri* = around, *physis* = a growth). Short, hairlike filaments that occur in the ostiole of some Pyrenomycetes.

Periplasm (Gr. *peri* = around, *plasma* = something formed). In some Oomycetes, the protoplasm which surrounds the oosphere within the oogonium.

Perispore (Gr. *peri* = around, *sporos* = seed). A membrane surrounding the spore walls which is not permanently attached (*cf.* exospore, endospore).

Perithecium, pl. **perithecia** (Gr. *peri* = around, *theke* = case). A closed ascocarp whose wall is developed from the archicarp or other sexual structures, which has a regular means of opening, such as by a pore or slit.

Phialide (Gr. *phiale* = drinking vessel). A terminal cell of a conidiophore within or upon which blastogenously produced conidia are formed without a change in length of the conidiophore and in a basipetal fashion.

Phototropism. See **tropism.**

Phragmobasidium (Gr. *phragmos* = partition, fence; *basidion* = small base). A basidium that is divided into more than one cell by transverse or longitudinal septa (*cf.* holobasidium).

Pileus, pl. **pilei** (L. *pileus* = cap). The expanded caplike portion of some basidiocarps or ascocarps that supports the hymenium.

Plasmodium (Gr. *plasma* = something formed). A naked, multinucleate thallus that moves in an amoeboid fashion (in the Plasmodiophoromycetes).

Plasmogamy (Gr. *plasma* = something formed, *gamos* = marriage). The fusion of cytoplasm of two protoplasts, often sexual in nature and followed by karyogamy.

Polypore (Gr. *polys* = many, *poros* = pore). A common name for a member of the Aphylophorales that has pores; a bracket fungus.

Pore (Gr. *poros* = pore). A small opening; in the Hymenomycetes, the mouth of a tube.

Predator. An organism that captures other organisms which are then killed and used as food.

Probasidium (Gr. *pro* = before, *basidion* = a small base). The cell in which karyogamy occurs in the basidiomycetes. (*cf.* metabasidium).

Progamete (Gr. *pro* = before, *gamete* = germ). In the Zygomycetes, a swollen hyphal tip that will differentiate into the gametangium and the suspensor.

Protoperithecium, pl. **protoperithecia** (Gr. *protos* = first). Early haploid stage that will develop into a perithecium after plasmogamy.

Pseudoangiocarpic development (Gr. *pseudes* = false, *angeion* = case, *karpos* = fruit). Development of basidiocarps in which the immature hymenium is exposed early in development but becomes covered by the incurving pileus margin and partial veil (*cf.* gymnocarpic, hemiangiocarpic).

Pseudoparaphysis, pl. **pseudoparaphyses** (Gr. *pseudes* = false, *physis* = growth). A sterile thread that grows downward in the cavity of some ascocarps, usually becoming attached at the bottom.

Pseudoparenchyma (Gr. *pseudes* = false). A fungal tissue consisting of more or less isodiametric cells and resembling the parenchyma of higher plants.

Pseudoseptum (Gr. *pseudes* = false, L. *septum* = hedge). A sievelike septum with numerous pores.

Pseudothecium (Gr. *pseudes* = false, *theke* = sheath or cover). An ascostroma resembling a flask-shaped perithecium.

Pycnidium, pl. **pycnidia** (dim. of Gr. *pyknos* = dense). A closed sporocarp which contains a cavity bearing conidia.

Pycniospore (Gr. *pyknos* = dense). A spore (spermatium) borne in a pycnium in the Uredinales.

Pycnium, pl. **pycnia** (Gr. *pyknos* = dense). In the Uredinales, stage 0, consisting of the male fertilizing elements (pycniospores) and female elements, the flexuous hyphae.

Receptacle. A stalk bearing a fertile region with

asci or basidia (in the Laboulbeniales and Phallales).

Resting spore. A thick-walled spore which germinates after a resting (dormant) period, often winter.

Rhizoid (Gr. *rhiza* = root). A rootlike anchoring hypha; in the Chytridiales, a filamentous portion of the thallus which extends into the substratum.

Rhizomorph (Gr. *rhiza* = root, *morphe* = form). A compact, macroscopic strand of fungus tissue which resembles a root.

Rhizoplast (Gr. *rhiza* = root, *plastos* = formed). A strand which extends from the nucleus to the blepharoplast in motile cells.

Saprophyte (Gr. *sapros* = decayed, *phyton* = plant). A fungus, bacterium, or plant that feeds upon dead organic matter.

Sclerotium, pl. **sclerotia** (Gr. *skleros* = hard). A hardened, resistant or resting body, consisting of a mass of fungal tissue, either with or without host tissue and which may give rise to a mycelium, a sporocarp, or a stroma.

Septal pore cap. A domelike configuration consisting of endoplasmic reticulum, which occurs directly over the pore in a septum (in the basidiomycetes).

Septum, pl. **septa** (L., fence). A cross wall in a hypha or spore.

Sexual reproduction. Production of progeny after karyogamy and meiosis take place at a regular point in the life cycle. Plasmogamy may or may not occur and may or may not be accomplished by specialized organs.

Short-cycled. Having fewer than five spore stages in the life cycle (of the Uredinales) (*cf.* long-cycled).

Skeletal hyphae. Thick-walled, branched or unbranched, straight or slightly flexed hyphae within a basidiocarp (*cf.* binding, generative, and lactiferous hyphae).

Soredium, pl. **soredia** (Gr. *sorus* = heap). A discrete group of algal and fungal cells formed by a lichen which can be disseminated and initiate a new lichen thallus.

Sorus, pl. **sori** (Gr., heap). A cluster or mass of spores, as in the Plasmodiophorales, Uredinales, or Ustilaginales.

Spermatium, pl. **spermatia** (Gr. *sperma* = seed or germ). A nonmotile, uninucleate male gamete which usually becomes detached before plasmogamy.

Sporangiole (Gr. *spora* = spore, *angeion* = vessel, L. *olus* = diminutive suffix). A small sporangium containing only one or a few sporangiospores and without a columella.

Sporangiophore (Gr. *spora* = spore, *phoros* = carrying). A modified hypha that supports the sporangium.

Sporangiospore (Gr. *sporos* = seed). A nonsexual spore produced within a sporangium.

Sporangium, pl. **sporangia** (Gr. *sporos* = seed, *angeion* = vessel). A sac which bears endogenous nonsexual spores (sporangiospores).

Spore (Gr. *sporos* = seed). A discrete sexual or nonsexual reproductive unit, usually enclosed by a rigid wall, which is capable of being disseminated. Unlike seeds of the higher plant, the spore does not contain an embryo.

Sporocarp (Gr. *spora* = spore, *karpos* = fruit). A general term for a multicellular body or organ that contains or bears spores, e.g., an ascocarp or basidiocarp.

Sporodochium, pl. **sporodochia** (Gr. *sporos* = seed, *doche* = receptacle). A cluster of conidiophores arising from a stroma or mass of hyphae.

Sporophore (Gr. *spora* = spore, *phoros* = carrying). A spore-producing or spore-bearing structure such as a conidiophore, ascocarp, or basidiocarp.

Sterigma, pl. **sterigmata** (Gr., support). A spiculelike structure on a basidium which supports the basidiospore.

Sterile. (1) Tissues or structures which are not producing spores; (2) uncontaminated by microorganisms.

Stipe (L. *stipes* = stem or stalk). The stalklike portion of some large ascocarps or basidiocarps; any spore-bearing stalk.

Stolon. A horizontal hypha that becomes anchored in the substratum at points of contact (in the Mucorales).

Stroma, pl. **stromata** (Gr., cover). A compact mass of vegetative tissue, sometimes intermixed with host tissue, which often bears sporocarps either within or upon its surface.

Subhymenium, pl. **subhymenia** (L. *sub* = under, Gr. *hymen* = membrane). A differentiated tissue immediately beneath the hymenium which gives rise to the asci or basidia and sterile elements in the hymenium.

Substratum. A medium which serves as a food source and/or support for fungi.

Succession. Progressive changes in the fungal population on a substratum.

Suspensor. A specialized hyphal tip which supports a gametangium and eventually a zygospore (in the Mucorales).

Symbiont (Gr. *sym* = with, *bios* = life). A partner in a symbiotic relationship.

Symbiosis. The relationship in which dissimilar organisms live together. This relationship may be harmful, beneficial, or neutral to one or both of the partners.

Symptom. Detectable abnormalities in either physiology or morphology produced during the course of a disease.

Synnema, pl. **synnemata** (Gr. *syn* = together, *nema* = thread). A fascicle of conidiophores, usually upright.

Teliospore (Gr. *telos* = completion, fulfillment, *sporos* = seed). A thick-walled resting spore which is the site of karyogamy and produces the basidium (in the Uredinales and Ustilaginales).

Telium, pl. **telia** (Gr. *telos* = completion, fulfillment). The final sorus (stage III) produced in the life cycle of the Uredinales and resulting in teliospores.

Telomorph (Gr. *telos* = completion, fulfillment, *morphe* = form). The sexual stage in the life cycle of a fungus (*cf.* anamorph).

Tetrapolar sexuality (Gr. *tetra* = four, L. *polus* = pole). A type of sexual compatibility controlled by multiple-allelic genes at two loci borne on separate chromosomes (*cf.* bipolar sexuality).

Thallic development (Gr. *thallos* = shoot). Formation of conidia by septation of a vegetative hypha or conidiophore into discrete conidial initials, which may then enlarge. In **enterothallic** development (Gr. *enteron* = that within), additional wall layers are laid down within walls of the parent cell and are not incorporated into the spore. In **holothallic** development (Gr. *holos* = whole or entire), all wall layers of the parent cell are incorporated into the new spore (*cf.* blastic development).

Thallus, pl. **thalli** (Gr. *thallos* = young shoot). Any simple vegetative plant body that lacks roots, stems, and leaves.

Tinsel flagellum. A flagellum which bears small filaments which extend from the main axis of the flagellum (*cf.* whiplash flagellum).

Tissue. An aggregate of hyphae or of shortened, more or less rounded, hyphal cells that are attached to each other along all sides.

Trama (L. *trama* = woof). The sterile tissue of a basidiocarp.

Trichogyne (Gr. *trichos* = hair, *gyne* = woman). That portion of the archicarp that initially receives the male protoplasm and is often long and filamentous.

Trichophoric cell (Gr. *trichos* = hair, *phoros* = carrying). The cell that bears the trichogyne in the Laboulbeniales.

Tropism (Gr. *trope* = an attraction for). A bending or growth in response to a one-sided stimulus, either toward (positive) or away from (negative), frequently used as a suffix (for example: chemotropism, response to a chemical agent; geotropism, response to gravity; phototropism, response to light).

Unitunicate (L. *unus* = one, *tunica* = covering). Having a single layered ascus wall (*cf.* bitunicate).

Universal veil. A veil enveloping the entire basidiocarp during early developmental stages (in some Hymenomycetes).

Urediniospore (L. *urere* = to burn, Gr. *sporos* = seed). A binucleate repeating spore borne in the uredinium in the Uredinales.

Uredinium, pl. **uredinia** (L. *urere* = to burn). The sorus that bears urediniospores in the Uredinales (stage II).

Vegetative. A cell or structure which is not producing reproductive structures, usually in the assimilative state.

Vesicle (L. *vesica* = bladder). A bladderlike sac or an evanescent bubble within which zoospores mature; any bubblelike cell or bubblelike membranous structure within the cell.

Volva (L. *volvere* = to roll). A cuplike structure at the base of a basidiocarp (in the basidiomycetes).

Whiplash flagellum. A smooth flagellum (*cf.* tinsel flagellum).

Woronin body. A rounded organelle occurring near septa in at least some Ascomycetes or deuteromycetes.

Yeast. A unicellular member of the Endomycetales; sometimes used for a nonmotile unicellular stage, for example, of the dimorphic animal parasites.

Zoosporangium, pl. **zoosporangia** (Gr. *zoon* = animal). A sporangium within which zoospores are produced (in the lower fungi).

Zoospore (Gr. *zoon* = animal, *sporos* = seed). A nonsexual, motile spore which bears one or two flagella.

Zygospore (Gr. *zygon* = yoke, *sporos* = seed). A thick-walled sexual resting spore resulting from the fusion of gametangia (in the Zygomycetes).

Zygosporophore (Gr. *zygon* = yoke, *phoros* = bearing). A specialized hyphal branch that first bears a gametangium and then the zygospore in the Zygomycetes.

Zygote (Gr. *zygon* = yoke). A diploid cell resulting from the union of two gametes.

Author Index

559

Emmons, C. W., 134, 135, 184, 459–461, 464, 465, 480
Engler, A., 114, 123, 170, 237, 390
Entner, N., 337
Esser, K., 370
Evans, L. V., 147

Fargues, J., 482
Faro, S., 353, 370
Farr, M. L., 176
Farrar, J. F., 491, 510
Fassi, B., 512
Fayod, V., 242
Federici, B. A., 480
Fennel, D. I., 125, 132, 133, 184, 186
Fenton, E., 436
Fergus, C. L., 184, 304–306, 370
Ferron, P., 457, 458, 480
Feuell, A. J., 437
Fielding, A. H., 22
Fields, W. G., 20
Fincham, J. R. S., 331, 337
Findlay, W. P. K., 300, 305
Finger, S. M., 544
Finn, R. K., 322, 337
Fischer, E., 237
Fischer, G. W., 254, 480
Fisher, W. S., 480, 482
Fitzpatrick, H. M., 85, 114, 140, 184, 254, 480
Flaig, W., 414, 437
Fleer, A. B., 437
Flegler, S. L., 20, 112, 254
Fleming, A., 537, 544
Fletcher, J., 20, 93, 94, 110
Flor, H. H., 468, 469, 480
Flores-Carreon, A., 323, 338
Foister, C. E., 480
Fontaine, C. T., 481
Fontana, A., 512
Ford, E. J., 419, 439
Forgacs, J., 433, 437
Forsyth, F. R., 403
Foster, J. W., 544
Foster, R. C., 511
Francke-Grosmann, H., 510
Frederick, S. E., 10, 20
Fries, L., 300, 301, 306
Fries, N., 305, 306, 346, 370, 401, 403
Fuller, M. S., 20, 64, 67, 80, 82, 112
Funk, A., 184
Furtado, J. S., 212
Fuson, G. B., 21

Gabriel, B. P., 456, 480
Galbraith, J. C., 373
Gallegly, M. E., 360, 371
Galleymore, H. B., 364, 370
Gancedo, C., 338
Garber, E. D., 18, 20, 373
Garner, G. E., 385, 405, 486
Garnjorbst, L., 332, 338, 370
Garratt, G. A., 429, 437
Garrett, M. K., 398, 404, 438
Garrett, R. H., 288, 306
Garrett, S. D., 419, 422, 423, 437, 510

Garver, J. C., 533, 545
Gascoigne, J. A., 431, 437
Gascoigne, M. M., 431, 437
Gäumann, E. A., 238, 242, 251
Gay, J. L., 63, 80
Georgopoulos, S. G., 348, 353, 369
Gerard, J., 242
Gerdemann, J. W., 95–98, 110–112, 510
Gessner, R. V., 306
Gibbons, I. R., 24
Gielink, A. J., 81
Gieseking, J. E., 437
Giesey, R. M., 7, 20
Gilbert, O. L., 510
Gilbertson, R. L., 208, 255
Gilkey, H. M., 184
Gilliam, M. S., 215
Gilman, J. C., 271
Ginther, O. J., 183
Goddard, D. R., 477
Godfrey, R. M., 96, 111
Goldblatt, L. A., 434, 436–438
Goldstein, A., 364, 370
Golesorkhi, R., 482
Golumbic, C., 437
Gooday, G. W., 337, 354, 370
Gooday, M. A., 89, 111
Goodwin, T. W., 337, 354, 373
Goos, R. D., 146, 184, 271
Gorman, C., 125, 183
Gotelli, D., 65
Gottlieb, D., 311, 313, 337, 338, 403
Graff, P. W., 139, 184
Graham, K., 510
Granados, R. R., 398, 404
Grant, C. L., 290, 306
Gray, T. R. G., 419, 422, 437
Gray, W. D., 303, 307, 415, 437, 532, 544
Greathouse, G. A., 437, 438
Green, D. M., 351, 369
Green, G. J., 254
Greenhalgh, G. N., 147
Greenwood, A. D., 12, 13, 20, 80
Gregory, P. H., 375, 391, 392, 395, 403
Griffin, D. H., 81
Griffin, D. M., 420, 436, 437
Griffith, N. T., 480
Griffiths, D. A., 120, 187, 271
Griffiths, H. B., 146, 184
Grimstone, A. V., 24
Gross, S. R., 332, 238
Grove, S. N., 10, 20
Groves, J. W., 544
Gruber, F. J., 10, 20
Gruen, H. E., 277
Guilliermond, A., 130, 184
Gull, K., 20, 405
Guthrie, E. J., 395, 403

Haard, R. T., 385, 404
Hacskaylo, E., 511, 512
Hadland-Hartmann, V. E., 26–28, 79
Hadley, E. B., 510
Hafiz, A., 366, 370

Hagen, K. S., 458, 480, 481
Hale, M. E., 486, 510–512
Halisky, P. M., 254, 346, 370
Hall, R., 20
Halver, J. E., 434, 437
Halvorson, H. D., 17, 398–399, 401, 405
Hamamoto, S. T., 75, 81
Hamilton, E., 367, 371
Hammill, T. M., 6, 20, 266, 267, 271
Hanlin, R. T., 146, 151, 152, 184, 266, 267, 272
Hansen, H. N., 348, 370, 373
Hansford, C. G., 139, 184
Harley, J. L., 437, 493, 496, 497, 500, 511
Harnish, W. N., 370
Harper, R. A., 138
Harrar, J. G., 467, 482
Harris, B. J., 60, 82
Harris, G. C. M., 369, 373
Harris, G. P., 491, 492, 511
Harris, J. L., 306
Harris, J. W., 21, 22
Harrison, J. L., 306
Harrison, J. S., 128, 183, 186, 307, 338
Harrison, K. A., 254
Harvey, A. E., 21
Hartmann, G. C., 272
Hartmann, V. E., 28
Hartzell, A., 511
Harvey, R., 405
Hasija, S. K., 306, 359, 370
Haskins, R. H., 63, 80
Hastie, A. C., 360, 371
Hatch, A. B., 496, 511
Hatch, R. T., 544
Hatch, W. R., 80
Hawker, L. E., 84, 89, 111, 167, 169, 184, 239, 254, 359–361, 367–369, 371, 404
Hawksworth, D. L., 510–512
Heald, F. D., 395, 404
Heath, I. B., 9–13, 15, 20, 21, 24, 31, 58, 69, 80, 81
Heck, F. L., 332, 337
Held, A. A., 80, 306
Hemmes, D. E., 21, 61, 75, 81
Henderson, J. A., 461, 463, 465, 479
Henderson, L. S., 438
Henderson, M. E. K., 437
Hendey, N. I., 427, 437
Hendrix, F. F., Jr., 372
Hendrix, J. W., 361, 371
Henriksson, E., 511
Henry, S. M., 510, 511
Hervey, A., 303, 307
Hess, W. M., 21, 405
Hesseltine, C. W., 88, 89, 95, 107, 111, 112, 529, 530, 544, 545
Hickey, R. J., 544, 545
Higgins, B. B., 179, 180, 184
Higgins, N. P., 336, 338
Hildreth, K. C., 510
Hill, D. J., 511
Hill, E. P., 306
Hill, T. W., 10, 21, 174, 184
Hinson, W. H., 437
Hiratsuka, Y., 198, 254

Reynolds, D. R., 179, 186
Ribeiro, O. K., 61, 81
Richardson, D. H. S., 512
Rietz, E., 414, 437
Rippon, J. W., 459, 460, 482
Roane, M. K., 36, 82
Robb, J., 21
Robbins, W. J., 293, 303, 307
Roberts, D. W., 480
Robertson, N. F., 307
Robinow, C. F., 21, 126, 128, 186
Robinson, H. M., 479, 480, 483
Robinson, P. M., 398, 404, 438
Robinson, W., 367, 372
Rockcastle, V. N., 153, 164, 220, 222, 223, 235, 246, 247, 368, 488
Rockett, T. R., 405
Rogers, J. D., 272
Rogerson, C. T., 152, 186
Rolfe, F. W., 216, 248, 255, 545
Rolfe, R. T., 216, 248, 255, 545
Roncardori, R. W., 510
Roper, J. A., 372
Rose, A. H., 128, 183, 186, 307, 316, 334, 336, 338, 544, 545
Roseman, S., 337
Rosinski, M. A., 146, 167, 186
Ross, I. S., 290, 307
Rothstein, A., 307
Rovira, A. D., 438
Ruiz-Herrera, J., 9, 20
Russell, L. A., 510
Ryan, F. J., 276, 307
Ryan, M. H., 187

Sachs, J., 251
Safeulla, K. M., 61
Saikawa, M., 86, 112
St. George, A. A., 438
Salmon, E. S., 137
Saltarelli, C. G., 303, 307
Salvin, S. B., 358
Samson, R. A., 89, 112, 265, 271
Samuels, G. J., 146, 152, 186
Samuelson, D. A., 156, 186
San Antonio, J. P., 346, 372
Sánchez, A., 163, 186
Sanders, F. E., 111
Sanderson, K., 439
Sansome, E., 60, 82
Sappin-Trouffy, P., 199, 200
Sarachek, A., 336, 338
Sarkanen, K. V., 438
Sassen, M. M., 112
Savile, D. B. O., 251, 252, 255
Scherr, H., 482
Schibeci, A., 338
Schipper, M. A. A., 89, 112
Schisler, L. C., 334, 338
Schleser, R. A., 480, 482
Schneider, A., 487
Schoknecht, J. D., 158, 186, 271
Schuerch, C., 438
Schweiger, L. B., 535
Scott, G. D., 512
Scott, W. W., 66, 82

Seaburg, F., 21
Seaward, M. R. D., 516, 545
Sebek, O. K., 545
Seshadri, R., 351, 369, 371
Seymour, R. L., 67
Shaffer, R. L., 255
Shannon, M. C., 22
Shanor, L., 171, 183, 186
Shapiro, R. E., 435, 438
Shapovalov, M., 299
Shaw, C. G., 61, 163, 187
Shaw, M., 21
Shaw, P. D., 337
Shay, J. R., 468, 470, 481
Sherwood, W. A., 372
Shigo, A. L., 449, 482
Shoemaker, R. A., 147, 179, 184, 187
Shrum, R. D., 482
Sietsma, J. H., 10, 22
Silver, J. C., 372
Silverberg, B. A., 161, 187
Simon, F. G., 401, 403
Singer, R., 255
Siu, R. G. H., 438, 439
Skucas, G. P., 46, 47, 82
Skye, E., 512
Slankis, V., 495, 497, 512
Slifkin, M. E., 482
Small, E., 463, 482
Smalley, H. M., 326, 337
Smith, A. H., 192, 207, 209, 225, 227, 229, 231, 237, 255, 256, 512, 519, 520, 523
Smith, D. C., 488, 512
Smith, G., 137
Smith, J. E., 337, 338, 373, 405
Smith, M. T., 187
Snell, E. E., 307
Snell, W. H., 254, 300, 307
Snyder, T. E., 438
Snyder, W. C., 348, 370, 373, 436–438
Sobolov, M., 544
Sols, A., 338
Spaeth, J., 401, 403
Sparrow, F. K., 23, 32, 33, 36, 37, 52, 63, 64, 69, 71, 77, 79, 83, 110, 111, 183–187, 253–256, 272
Speare, A. T., 454
Sprott, E. R., 542, 543
Srb, A. M., 331, 338
Stahmann, M. A., 482
Stakman, E. C., 467, 482
Staley, J. M., 163, 187
Stalpers, J. A., 89, 112
Stambaugh, W. J., 370
Staples, R. C., 398, 404
States, J. S., 363, 373
Stauffer, J. F., 543
Stechert, G. M., 439
Stedman, O. J., 404
Steinberg, R. A., 290, 307
Steinhaus, E. A., 456, 457, 480–482
Stephens, R. L., 518, 543
Sternberg, M., 528, 545
Stevens, F. L., 139, 187
Stewart, G. G., 524, 545
Stiers, D. L., 10, 22, 120, 146, 187

Stocks, D. L., 21
Stotzky, G., 305, 306
Strandberg, J., 58, 83
Studhalter, R. A., 395, 404
Subramanian, C. V., 183, 264, 272
Sundeen, J., 351, 369
Suomalainen, H., 307, 311, 323, 338
Sussman, A. S., 17, 19, 21, 78, 80–83, 110, 111, 183–186, 253–256, 272, 306, 307, 337, 338, 370–373, 398, 399, 400–405, 479, 481, 483
Sutton, B. C., 270, 272
Swart, H. H., 259, 261, 269
Sylvester, J. C., 545
Syrop, M., 125, 163, 183, 187

Tabak, H. H., 304, 305, 307
Taber, W. A., 306, 373, 545
Takahashi, H., 288, 307
Talbot, P. H. B., 84, 93, 111, 193, 211, 215, 216, 255, 256, 395, 405
Talboys, P. W., 476, 482
Tan, K. K., 363, 373
Tanada, Y., 482
Tanaka, K., 272
Tatum, E. L., 276, 307, 332, 338
Tavares, I. I., 171, 187
Taylor, B. L., 320, 338
Taylor, E. E., 371
Teixeira, A. R., 256
Temmink, J. H. M., 30
Tenney, R., 545
Thaxter, R., 46, 53, 170–173, 187
Thielke, C., 22
Thiers, H. D., 256
Thind, K. S., 154
Thirumalachar, M. J., 61, 82
Thom, C., 262, 528, 545
Thomas, B., 84, 111
Thomas, D. M., 354, 373
Thornton, R. H., 436
Thurston, E. L., 21
Thyr, B. D., 163, 187
Timberlake, W. E., 352, 353, 373
Timnick, M. B., 357, 373
Tinker, P. B. H., 111, 501, 512
Tinline, R. D., 373
Tomiyama, K., 482
Tommerup, I. C., 58, 82
Towers, G. H. N., 338
Townsend, B. B., 256
Trappe, J. M., 95–97, 111, 167, 187, 256, 512
Travland, L. B., 82
Trinci, A. P. J., 20, 267, 272
Trione, E. J., 373
Truesdell, L. C., 25, 47, 82
Tubaki, K., 264, 272
Tulasne, C., 238
Turbovsky, M. W., 544
Turian, G., 13, 22, 321, 337, 373
Turner, W. B., 338
Tuveson, R. W., 373
Tyrrell, D., 22
Tyson, K., 120, 187

Subject Index

Emden-Meyerhof pathway, 311–314
Emericella, 135
Emericellopsis microspora, 133
Emmonsiella capsulata, 133
Endocarpon pusillum, 490, 491
Endochytrium operculatum, 36, 37
Endocochlus, 100
Endogonaceae, Family, 95–98
Endogone, 96, 98, 108, 110, 397, 499
Endogone incrassata, 97
Endogone flammicorona, 95
Endogone lactiflua, 95
Endomycetales, Order:
  in classification, 125
  growth, 275
  morphology, 126–129
  vitamin requirements, 295
Endophyllum sempervivi, 203
Endoplasmic reticulum,
  characteristics, 9–11 (*see also* Ultrastructure)
Endothia, 142
Endothia parasitica, 143, 154, 395, 473, 528, 530
Enteroblastic conidium
  development, 265, 266
Enterothallic conidium
  development, 264, 266, 267
Entner-Doudoroff
  pathway, 313–315
Entoloma lividum, 520
Entomophthora, 98, 110, 451–453, 458
Entomophthora coronata, 98, 384, 385, 459
Entomophthora fumosa, 457
Entomophthora grylli, 98, 385
Entomophthora muscae, 377, 380, 451, 452
Entomophthora sphaerosperma, 453
Entomophthorales, Order:
  in classification, 86
  morphology, 98, 99
Entophylictis-type development, 35, 36
Entophlyctis confervae f. *marina,* 35
Enzymes:
  adaptive, 282
  digestive, 281, 282
  element requirement, 290
  industrial production, 530, 531
  pectic, 471, 476
  proteolytic, 471
  temperature effects, 298, 300
Eocronartium muscicola, 466
Epichloë, 142
Epicoccum andropogonis, 269
Epidermophyton, 464
Epithecium, 167
Eremascus albus, 303
Eremothecium ashbyii, 531
Ergot and ergotism, 541–543
  alkaloids, 543
  fungus, 154, 542
  medical uses, 543
  source LSD, 522
  symptoms, 542
Erysiphales, Order:
  in classification, 125

morphology, 137–139
Erysiphe graminis, 136, 357, 477
Euallomyces, 48
Euglena, 42, 44
Eurotiales, Order:
  in classification, 125
  in ectomycorrhizae, 494
  morphology, 132–136
Eurotium, 133, 135
Eurychasmales, in evolution, 77
Eutypa, 142
Excipulum, types, 154, 155
Exidia nucleata, 213
Exit tube, 31
Exobasidiales, Order:
  in classification, 196
  morphology, 214, 215

False truffles, 237 (*see also* Hymenogastrales)
Fertilization tube, 61
Filobasidiella neoformans, 203
Fission, 127
Fomes, 218, 224
Food, fungi as:
  for animals, 514
  for humans, 514–524
Food processing, 524–530
  beer, 524, 526–527
  bread, 524, 525
  cheese, 528, 529
    camembert, 528
    Roquefort, 528, 529
  Oriental foods, 529, 530
    Shoyu, 529, 530
    Tempeh, 529
  whiskey, 527
  wine, 525, 526
Fossilized fungi, 32, 466, 467
Fungistasis in soil, 419
Funiculus, characteristics, 248, 249
Fusarium, 258, 315, 333, 359, 360, 451, 473, 475, 477, 518, 540
Fusarium coeruleum, 299, 473
Fusarium eumartii, 299, 305, 472
Fusarium lycopersici, 476
Fusarium moniliforme, 475, 530
Fusarium oxysporum, 299, 305
Fusarium radicola, 299
Fusarium solani phaseoli, 401
Fusarium sporotrichoides, 433
Fusarium trichothecoides, 299
Fusicoccum, 424

Galerina, 523
Gametangia:
  in Ascomycotina, 114–117
  characteristics, 16
  in Chytridiomycetes, 39
  in Oomycetes, 58–63, 66, 67
  in Zygomycotina, 87–90
Gamma particles, 25, 47
Ganoderma applanatum, 388
Gasteromycetes, Class:
  in classification, 196
  in ectomycorrhizae, 494
  in evolution, 252–253

spore dispersal by animals, 397
spore release:
  active, 389–391
  passive, 375–376
Gautieria, 240
Gautieriales, Order:
  in classification, 196
  morphology, 240
Geastrum, 247
Gelasinospora, 142, 148, 150
Gelasinospora adjuncta, 148
Gelasinospora calospora, 143, 148
Gemmae, 69
Gene-for-gene mechanism, 468–469
Generative cell, 104, 105
Genetic control of reproduction, 345–350
Genistellospora homothallica, 103–106
Geoglossum, 166
Geolegnia, 69
Geotrichum, 261
Geotrichum candidum, 432, 528
Geotropism, 387, 388
Gibberella, 142
Gibberella fujikuroi, 475
Gilbertella persicaria, 88, 91, 92, 449, 451
Gills, 217, 226, 231
Gleba:
  characteristics, 236
  development, 236, 237, 239
Gliocladium catenulatum, 540
Gloeophyllum saepiarium, 363
Glomerella, 142, 148
Glomerella cingulata, 349
Glotzia, 106, 110
Glycerol production, 316, 317
Glycogen:
  as reserve, 324
  in spore discharge, 389
Glyoxylic bypass, 319, 320, 323, 335, 358
Glyoxysomes, 63
Gnomonia, 142, 151
Golgi apparatus, characteristics, 10, 11 (*see also* Ultrastructure)
Gonapodya, 52
Gonatobotrys, 258
Gonatobotrys simplex, 447, 448
Griseofulvin:
  as antifungal antibiotic, 466
  formula, 467
  industrial production, 530
  as metabolite, 333
Growth, 275–307
  in culture, 280
  curves:
    cardinal points, 297
    effects of physical factors, 297, 299, 302
  factors, 293
  lag phase in nature, 369
  measurement, 278
  media, 280, 281
  in multicellular fungi, 275–278
  nutritional requirements, 281–295

Insects (*cont.*)
  termite gardens, 507
Irish potato famine, 73, 467
Iron requirement, 290
Isolation of fungi from nature:
  aquatic, 23
  predacious, 440
  soil, 416–418

Jack-o'-lantern fungus, 235

Karyogamy (*see also*
    Reproduction, sexual):
  characteristics, 16
  genetic control, 349
Kickxella, 108
Kickxellaceae, Family:
  in evolution, 102, 110
  septation, 102

Laboulbeniales, Order (*see*
    Laboulbeniomycetes)
Laboulbeniomycetes, Class:
  in classification, 125
  in evolution, 182
  morphology, 170–174
Lactarius, 225, 520
Lactarius deliciosus, 494
Lactic acid production, 316, 317
Lacunar development,
    Gasteromycetes, 236
Lagenidiales, Order:
  in classification, 32
  morphology, 69–71
Lagenidium, 69
Lagenidium rabenhorstii, 70
Lambertella pruni, 292
Lasiobolus monascus, 166
Lecanora esculenta, 485
Lecanorales, Order, 490
Lentinus tuber-regium, 364
Leotia, 166
Leotia, stipitata, 165
Lepiota, 225, 509
Lepiotaceae, Family, 225
Leptomitaceae, 64–66
Leptomitus lacteus, 289
Leptosphaeria, 177
Leptosphaeria avenaria, 179
Leptosphaeria coniothyrium, 473
Leptosphaeria sengalensis, 177
Leptosphaerulina briosiana, 360
Lichens, 485–492
  characteristics of lichens, 485
  characteristics of symbionts,
      489–490
  distribution, 485
  growth, 489
  interrelationships of symbionts,
      491–492
  isolation and culturing of
      symbionts, 489
  nutrition, 490
  reproduction, 488

synthesis, 490–491
thallus morphology, 485–488
Life cycles (*see also* Reproduction,
    sexual):
  Ascomycotina (general), 124
  Basidiomycotina (general), 189
  Dipodascus aggregatus, 129, 130
  general characteristics, 16, 257
  Mucor mucedo, 87
  Oomycetes, 62
  Plasmodiophora brassicea, 57, 58
  Polyphagus euglenea, 42–45
  Puccina graminis, 202
  Pythium debaryanum, 74
  Saccharomyces cerevisiae, 129,
      130
  Saccharomycodes ludwigii, 129,
      130
  Synchytrium brownii, 41–43
  Taphrina deformans, 131, 132
  Tilletia eleusines, 205
Light:
  effects on ascospore discharge,
      381–382
  effects on growth, 303–304
  effects on lichens, 488, 489, 491,
      492
  effects on metabolism, 364
  effects on mycorrhizae, 496–497
  effects on reproduction:
    mechanism, 363
    photoinduction, 364–366
    photoinhibition, 364–366
  spore activation by, 400
Lignin:
  decomposition, 413, 425
  in humus formation, 414
Lipid:
  in cells, 17, 25, 27, 47, 58, 60, 63,
      335, 336
  metabolism, 333–336
  in spore germination, 403
Loculoascomycetes, Class, 178–181
    (*see also* Dothideales)
  in classification, 125
  morphology, 174–181
  in evolution, 182
Lomasomes, 11, 12
Long-cycled Uredinales, 201
Lophodermella morbida, 163
Lophodermella sulcigena, 163
Lophodermium, 161, 424
Lophodermium pinastri, 161
LSD, 522
Lycomarasmin, 333
Lycoperdales, Order:
  in classification, 196
  morphology, 245–248
Lycoperdon, 246, 248
Lysosomes, characteristics, 10

Macrophoma, 270
Magnesium requirements, 289, 290
Malus, 468
Manganese requirement, 290, 291
Marasmius, 225
Marasmius rotula, 233

Massospora, 98, 99
Massospora cicadina, 453, 454
Mastigomycota, Division:
  characteristics, 23
  in classification, 19
  evolution, 77–79
  heterokaryosis, absence, 344
  heterothallism, 345
  homothallism, 348
  lactic acid production by, 316
  septation, 5
Medeolariales, Order, 125
Medullary excipulum, 155
Meiosis (*see also* Reproduction,
    sexual):
  characteristics, 14–15
  genetic control, 349
  in sexual cycles, 16
Melampsora lini, 468, 469
Melanconiales, Form-Order, 270
Melanconium fuligineum, 357
Melanogastrales, Order:
  in classification, 196
  morphology, 240
Melanospora destrunes, 359
Melanosporaceae, Family, 142
Meliola, 139
Meliola circinans, 139
Meliola ptaeroxli, 140
Meliolales, Order:
  in classification, 125
  morphology, 139–140
Membraneous (myelin-like)
    configurations, 13
Memnoniella echinata, 263
Merosporangia, 93, 94
Merulius lacrymans, 428
Metabasidium, characteristics, 193,
    194, 213
Metabolites (*see also* Vitamins):
  carbon:
    acetate pathway, 326, 328
    aliphatics, 325–326
    aromatics, 326–328
    pigments, 326
    shikimic acid pathway, 326
    storage, 333
  nitrogen, 333
Metarhizium anisopliae, 458
Methionine-requiring mutants, 331
Microbodies, characteristics, 10
    (*see also* Ultrastructure)
Microsporum, 132, 431, 465
Microthyriaceae, Family, 175
Microthyrium, 175
Mineral uptake (*see also* Elements):
  by hyphae, 282, 283
  by mycorrhizae, 496, 501
Mitochondria, characteristics, 11,
    12 (*see also* Ultrastructure)
Mitotic crossing-over, 342, 343
Moisture, affecting:
  ascospore discharge, 384
  growth, 302–303
  lichens, 488, 489–492
  mushroom cultivation, 518
  reproduction, 361
  soil fungi, 420
  spore activation, 399
  spore dormancy, 401